IFAE 2005

To learn more about the AIP Conference Proceedings, including the
Conference Proceedings Series, please visit the webpage
http://proceedings.aip.org/proceedings

IFAE 2005

XVII Incontri di Fisica delle Alte Energie
17th Italian Meeting on High Energy Physics

Catania, Italy   30 March – 2 April 2005

EDITORS
Alessia Tricomi
Sebastiano Albergo
Massimiliano Chiorboli
*University of Catania and INFN
Catania, Italy*

SPONSORING ORGANIZATIONS
Istituto Nazionale di Fisica Nucleare (INFN)
Università di Catania

Melville, New York, 2005
AIP CONFERENCE PROCEEDINGS ■ VOLUME 794

Editors:

Alessia Tricomi
Sebastiano Albergo
Massimiliano Chiorboli

Dipartimento di Fisica e Astronomia
Via S. Sofia 64
I-95123 Catania
Italy

E-mail: alessia.tricomi@ct.infn.it
sebastiano.albergo@ct.infn.it
massimiliano.chiorboli@ct.infn.it

Cover: A pictorial view of Catania and the Etna volcano.
Etna Photo: 2002 Eruption

Authorization to photocopy items for internal or personal use, beyond the free copying permitted under the 1978 U.S. Copyright Law (see statement below), is granted by the American Institute of Physics for users registered with the Copyright Clearance Center (CCC) Transactional Reporting Service, provided that the base fee of $22.50 per copy is paid directly to CCC, 222 Rosewood Drive, Danvers, MA 01923, USA. For those organizations that have been granted a photocopy license by CCC, a separate system of payment has been arranged. The fee code for users of the Transactional Reporting Services is: ISBN/0-7354-0285-X/05/$22.50.

© 2005 American Institute of Physics

Permission is granted to quote from the AIP Conference Proceedings with the customary acknowledgment of the source. Republication of an article or portions thereof (e.g., extensive excerpts, figures, tables, etc.) in original form or in translation, as well as other types of reuse (e.g., in course packs) require formal permission from AIP and may be subject to fees. As a courtesy, the author of the original proceedings article should be informed of any request for republication/reuse. Permission may be obtained online using Rightslink. Locate the article online at http://proceedings.aip.org, then simply click on the Rightslink icon/"Permission for Reuse" link found in the article abstract. You may also address requests to: AIP Office of Rights and Permissions, Suite 1NO1, 2 Huntington Quadrangle, Melville, NY 11747-4502, USA; Fax: 516-576-2450; Tel.: 516-576-2268; E-mail: rights@aip.org.

L.C. Catalog Card No. 2005933689
ISBN 0-7354-0285-X
ISSN 0094-243X

Printed in the United States of America

CONTENTS

Preface ... ix
Committees ... xi
Funding Agencies and Sponsors ... xiii
Photo Gallery ... xv

PLENARY SESSIONS

Status and Perspectives in Italian HEP with Accelerators ... 3
    F. Ferroni
Which Future for High-Energy Physics? ... 8
    M. L. Mangano
Possible Scenarios for the LHC Luminosity Upgrade ... 15
    W. Scandale
First Months of Data Taking at LHC ... 20
    F. Parodi
Results from the RUN II at the Tevatron ... 24
    F. Bedeschi
B Physics at the B-Factories ... 30
    R. Faccini
B Physics at Hadron Colliders: Present and Future ... 36
    M. Calvi
Status of Electroweak Precision Measurements and Perspectives at LHC ... 42
    R. Tenchini
The Pierre Auger Observatory: Present Status and Future Prospects ... 48
    S. Petrera (*On behalf of the Pierre Auger Collaboration*)

PARELLEL SESSION I: STANDARD MODEL

The Measurement of α_{QED} in e^+e^-: An Alternative Approach ... 55
    L. Trentadue
Multiple Photon Corrections to W and Z Boson Production at Hadron Colliders ... 58
    C. M. Carloni Calame, G. Montagna, O. Nicrosini, and M. Treccani
Electroweak Physics at the Tevatron ... 62
    S. Leone (*On behalf of the CDF and DØ Collaborations*)
Top Quark Measurements at Tevatron Run II ... 66
    P. Azzi
QCD Physics at the Tevatron Collider ... 70
    A. Messina
The Muon g-2 in the Standard Model ... 74
    M. Passera
Bottom-Quark Fragmentation in Higgs Decay ... 78
    G. Corcella
Electroweak Physics at LHC ... 81
    A. Giammanco
Top Physics at ATLAS ... 85
    M. Barisonzi
Measurement of $\Delta\Gamma_S$ in $B_S^0 \to J/\psi\phi$ with CMS ... 89
    N. Magini
A Preliminary Study of Z+b Production at LHC ... 93
    M. Verducci, A. Tonazzo, and S. Diglio
VV-Fusion Process to Investigate EWSB; Interplay LC-LHC ... 97
    G. Cerminara
Pentaquark Searches in e-p Interactions at HERA ... 101
    A. Polini

PARALLEL SESSION II: SUSY AND BEYOND THE STANDARD MODEL

Particle Dark Matter: Searching for New Physics without Accelerators 107
    N. Fornengo

Search for the SM and MSSM Higgs Bosons at CDF 111
    G. Cortiana

Searches for New Physics at CDF 115
    M. P. Giordani

Higgs Boson Searches at LHC 119
    D. Giordano

Search for New Physics at ATLAS 123
    G. Comune

Sudakov Expansions and Top Quark Physics at LHC 127
    M. Beccaria

Reconstruction of the Sparticle Masses at the LHC 131
    T. Lari

Searches for New Physics in CMS 135
    M. Galanti

The Higgs Boson as a Gauge Field in Extra Dimensions 139
    M. Serone

Theories with Extra Dimensions at Finite Temperature 143
    A. Gruzza

Search for Extra Dimensions at the LHC 146
    A. Ghezzi

Split Supersymmetry 150
    G. F. Guidice

PARALLEL SESSION III: HEAVY FLAVOUR PHYSICS

Recent Theoretical Progress in Kaon Physics 157
    F. Mescia

Recent KLOE Results on Kaon Physics 161
    M. Martini (*On behalf of the KLOE Collaboration*)

Kaon Physics at CERN: Recent Results by NA48/2 Experiment 165
    G. Lamanna

Charmonium Spectroscopy and Production: New Results 169
    E. Robutti

Semi-Inclusive Charmless B Decays 173
    G. Ricciardi

$b \to s\gamma$ Decays at BABAR 177
    F. Bucci

Measurements of $|V_{cb}|$ and $|V_{ub}|$ at BaBar 181
    M. Rotondo

γ Measurement at the B-Factories 185
    L. Li Gioi

Status and Future Perspectives on the Measurement of the Unitarity Triangle Angle α 189
    A. Carbone

Unitarity Triangle Analysis beyond the Standard Model 193
    M. Pierini

Search for B_S^0 Flavour Oscillations with the CDF Experiment 197
    S. De Cecco

Branching Fractions and CP Asymmetries in Hadronic Charmless Decays of B Mesons at CDF 201
    M. Casarsa

PARALLEL SESSION IV: NEUTRINOS AND COSMIC RAYS

Neutrino Mass Phenomenology 207
    A. Marrone

Long-Baseline Neutrino Oscillation Experiments in Europe 211
    F. Terranova
Solar Neutrinos: Present and Future 215
    A. Ianni
Neutrinos from Supernovas and Supernova Remnants 219
    M. L. Costantini and F. Vissani
Seesaw and Leptogenesis 224
    P. Di Bari
Neutrinoless Double Beta Decay and ν-Mass Determination 228
    M. Pedretti
Neutrinos and Cosmology: An Update 232
    O. Pisanti and P. D. Serpico
Anti-GZK Effect in UHECR Spectrum 236
    R. Aloisio and V. S. Berezinsky
Ultrahigh Energy Cosmic Rays Detection 240
    C. Aramo
Gamma-Ray Astronomy with Cherenkov Telescopes 244
    A. Stamerra
Experimental High Energy Neutrino Astrophysics 248
    C. Distefano
Cosmogenic Neutrinos and New Physics Signal at Neutrino Telescopes 252
    J. I. Illana, M. Masip, and D. Meloni

PARALLEL SESSION V: DETECTORS AND NEW TECHNOLOGIES

Detector R&D for the International Linear Collider 259
    E. Garutti
Modern Detectors for Astroparticle Physics 263
    O. Adriani
Neutrino Detectors Review 267
    N. D'Ambrosio
Data Analysis Techniques at LHC 271
    T. Boccali
The CDF Analysis Farm 275
    M. Casarsa, S.-C. Hsu, E. Lipeles, M. Neubauer, S. Sarkar, I. Sfiligoi, and F. Würthwein
The Silicon Drift Detector of the ALICE Experiment 279
    G. Batigne (*On behalf of the ALICE Collaboration*)
The Tracker of the CMS Experiment 283
    E. Migliore (*On behalf of the CMS Collaboration*)
The Monitored Drift Tube Chambers of Atlas 287
    S. Ventura
The CMS Electromagnetic Calorimeter 291
    R. Paramatti
The ATLAS Liquid Argon Electromagnetic Calorimeter 295
    L. Carminati
The LHCb Muon System 299
    W. Baldini (*On behalf of the LHCb Muon Group*)
The CERN RD50 Collaboration: Development of Radiation-Hard Semiconductor Detectors for Super-LHC 302
    A. Macchiolo (*On behalf of the RD50 Collaboration*)
The Virgo Detector 307
    M. Punturo (*On behalf of the Virgo Collaboration*)
Triple-GEM Detectors for the Innermost Region of the LHCb Muon Apparatus 311
    M. Poli Lener (*On behalf of the LHCb-GEM Group*)
Liquid Xenon Calorimetry: Status and Perspectives 315
    D. Nicolò

List of Participants 319
Author Index 323

PREFACE

The "XVII Incontri di Fisica delle Alte Energie - IFAE 2005" was held in Catania from March 30 - April 2, 2005.

The Conference was organized by the University of Catania and by the Istituto Nazionale di Fisica Nucleare, INFN.

The IFAE 2005 Conference is part of a series of meetings begun in 1989, just at the start up of the LEP experiments, which aimed to bring together the Italian physics community to discuss current problems of elementary particle physics.

The 2005 edition was the fourth meeting after the conclusion of LEP. The area of interest of IFAE has been broadened with respect to the LEP series, however it mantains the aim to be an annual meeting between Italian specialists to discuss the present status and the future perspectives of high energy physics.

The characteristic feature of IFAE is to take the broadest possible view of the discipline by inviting distinguished speakers, both experimentalists from many different collaborations and theoreticians. One of the main goal of this Meeting is indeed to promote effective collaboration between experimentalists and theorists working in this field. The active participation of young physicists is strongly encouraged. This kind of formula naturally results in deep discussions both on the talks and among the speakers, boosted by the unformal style given by the use of Italian language. The Proceedings have been written in English preserving however the conference style. Many new ideas and opinions are reported, making this volume an interesting tool to understand the present debates in the HEP community.

IFAE 2005 was divided into three half day plenary sessions and in two half day parallel sessions, in which discussions could touch in deeper detail the topics presented in the plenary sessions. Five parallel sessions have been organised:

1. Standard Model
2. Supersymmetry and Beyond the Standard Model
3. Heavy Flavour Physics
4. Neutrinos and Cosmic Rays
5. Detectors and New Technologies.

In total there were 12 plenary talks and 72 parallel session talks. About 150 participants were registered for IFAE 2005. The scientific programme was complemented by a "Round Table" devoted to discuss the status of universities and research agencies in Italy as well as the new draft bill recently proposed on this subject by the Italian Government. The round table session was introduced by Prof. Ferdinando Latteri, Rector of the Catania University, and was held in the "Aula Magna of Siculorum Gymnasium". The Conference was concluded by a talk given by Prof. Renato Cristofolini who introduced IFAE participants to a review of studies on the geological evolution of the Etna volcano. A guided excursion to the volcano was also organized for the conference participants.

Besides the interesting scientific programme, the conference participants could also enjoy the architectural and historical heritage of Catania. The social dinner was hosted in Palazzo Biscari, which is considered the most beautiful palace in Catania, a precious relic of the Sicilian rococo period.

We would like to thank again the members of the Scientific Committee and the conveners of the parallel sessions who helped us in the preparation of the rich scientific programme. We would like to express our gratitude to all the speakers for their very interesting talks and to the chairpersons for the efficient organization of each session. The Organizers would like to thank all the local staff for all their efforts to host IFAE 2005 in our Physics and Astronomy Department. In particular, we would like to thank Prof. Francesco Porto, Director of the Department, for his kind hospitality, and Prof. Francesco Catara, Director of the local INFN Section, for all the support given to the Conference. We also like to thank the Rector of the Catania University and the Dean of our Faculty of Science. Special mention deserve those people who helped us in the everyday organization: in particular, IFAE 2005 secretaries, Anna Linda Magrì and Cettina Lombardo, members of our local INFN Administrative Staff, and also Mario Galanti and Agata Cosentino. We thank Ruggero Moncada Paternò Castello, prince of Biscari, who voluntiered to be the guide for our participants during their visit to Palazzo Biscari.

Thanks, of course, are also due to the agencies, both scientific institutions and administrative bodies which contributed either with their financial support or with their work, or all of the above, to make IFAE 2005 possible. All of them are listed and acknowledged on a specific page of the present proceedings.

Alessia Tricomi, Sebastiano Albergo and Massimiliano Chiorboli
(The Editors)

IFAE 2005 – XVII Incontri di Fisica delle Alte Energie

Scientific Committee

Guido Altarelli (CERN)

Alessandro Ballestrero (University of Torino)

Matteo Cacciari (LPTHE)

Mario Calvetti (University of Firenze and LNF)

Tiziano Camporesi (CERN)

Giorgio Chiarelli (INFN Pisa)

Paolo Ciafaloni (INFN Lecce)

Fabrizio Fabbri (INFN Bologna)

Fernando Ferroni (University of Roma "La Sapienza")

Chiara Mariotti (INFN Torino)

Antonio Masiero (University of Padova)

Antonino Pullia (University of Milano Bicocca)

Alessandra Romero (University of Torino)

Luca Trentadue (University of Parma)

Local Organising Committee

Sebastiano Albergo

Massimiliano Chiorboli

Alessia Tricomi

IFAE 2005 – XVII Incontri di Fisica delle Alte Energie

Funding Agencies and Sponsors

Istituto Nazionale di Fisica Nucleare (INFN)
Università di Catania
Presidenza della Regione Siciliana
Azienda Provinciale del Turismo di Catania

Plenary Sessions

Chairpersons:

G. Chiarelli
INFN Pisa, Italy

F. Ferroni
University of Roma "La Sapienza" and INFN, Italy

L. Foà
SNS and INFN Pisa, Italy

E. Migneco
University of Catania and INFN LNS, Italy

R. Potenza
University of Catania and INFN, Italy

Status And Perspectives In Italian HEP With Accelerators

Fernando Ferroni

Universita` di Roma 'La Sapienza' & INFN Roma

Abstract. In this paper I give an overview on the status of elementary particle physics research performed with accelerators in INFN. Recent successes, potential problems and a look to the future are discussed.

Keywords: Elementary Particles.
PACS: 29.

INTRODUCTION

The talk given at this workshop covers several aspects of the High Energy Physics research performed with particle accelerators by the Italian community supported by Istituto Nazionale di Fisica Nucleare (INFN).

Here you can find some statistics describing the sharing of people and resources among experiments, a brief review of the scientific successes harvested in the last year and a discussion on what is to come. Clear about the near future and rather fuzzy on what will come at middle and long term.

STATISTICS OF CSN I

Experiments performed by Italian groups at accelerator complexes are historically scrutinized by a committee called CSN I and possibly given a budget by INFN after the committee recommendation. The Committee can express recommendations in the frame of a yearly budget of the order of 30Meuro. The physicist belonging to groups in this field are in exess of 900 corresponding to more than 700 Full Time Equivalent with a ratio FTE/physicist of around 78%.

The pro-capite budget results therefore in about 50kEuro/FTE. The budget is shared amongst several experiments. The largest part (2/3) of it is absorbed by the preparation of the LHC experimentation. The same proportion is found if one looks at the experiment purposes, 2/3 of the budget goes to the energy frontier, much of the rest to flavor physics and some to spin physics and leptonic probes.

The magic number also describes the relative size of the experimentation carried out at CERN lab versus the others (FNAL and SLAC in USA, DESY in German, PSI in Switzerland and Laboratori Nazionali di Frascati at home).

One third of the budget goes into construction, another third in supporting travels and the rest in consumables, maintenance, publications, transports.

The biggest participation is in the CMS experiment at LHC with ~ 200 FTE, the committee however in special cases supports even individual participations. Half of the groups coming from a section or lab counts however for more than 7 FTEs.

One element of worry for the future is the fact that between 35 and 40% of the FTEs are either Ph.D. students or PostDocs. Given the restrictive rules existing in the country for giving permanent jobs this situation has to be considered with extreme attention since it might give rise to a weakness in the future forbidding for example to ripe the LHC physics fruits in proportion to the investment made in construction.

2004 HIGHLIGHTS

There are quite a few experiment running in different facilities around the world that see an INFN participation. COMPASS at CERN for spin and hadronic physics, CDF2 at the FNAL Tevatron at the energy frontier, BaBar at SLAC for B-physics, NA48 at CERN and KLOE at Dafne for K-physics.

All of them collects data successfully and produces important results. I have chosen here to report on very few of them that are real highlights. KLOE has given a crucial contribution to resolve a puzzling issue concerning the unitarity of CKM matrix.

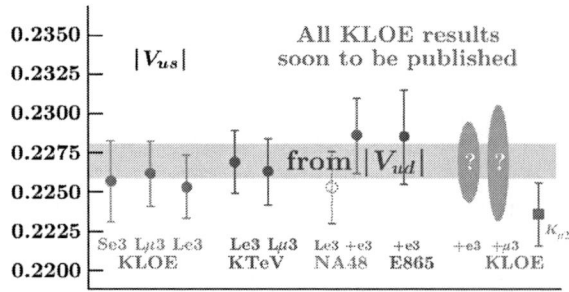

FIGURE 1. Several determination of Vus. The band shows the unitary constrain.

As seen in the above picture (Fig. 1) the most recent measurement and amongst them many of KLOE show an agreement between the measured value of Vus [1] and its value predicted by the unitarity of the CKM matrix.

This cut short a dispute originated by previous measurements quoted in the PDG.

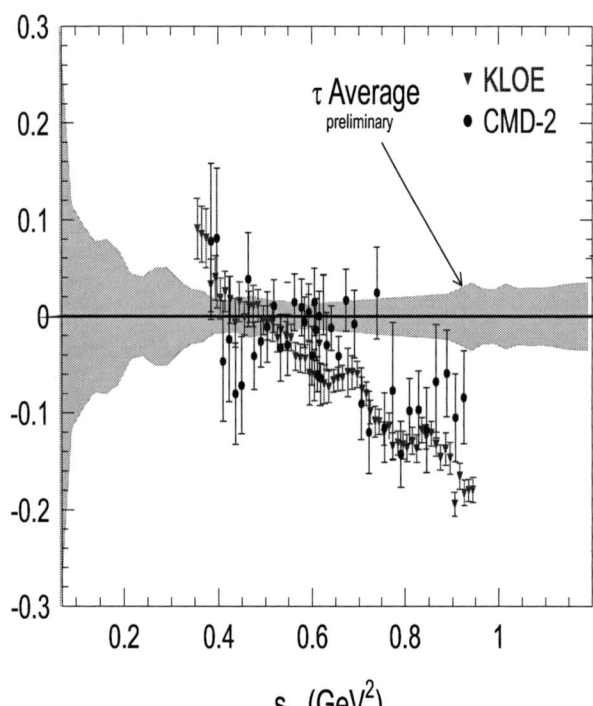

FIGURE 2. e+e- results are shown in the region below 1GeV² wrt to t-data

Another important result obtained by KLOE is the measurement [2] of the e+e- cross section at center of mass energies below 1 GeV (Fig. 2). This measurement is of paramount importance for the calculation of the correction to the muon (g-2) measurement. The results of KLOE when used in the calculation confirm that there is a disagreement, although of only 2.7 standard deviations) between the measured value of (g-2) and the QED prediction.

The second experiment that is producing important results at an impressive rate is BaBar. In the course of year 2004 they announced the observation [1] of the direct CP violation in B-decays. As shown in Figure 3 below:

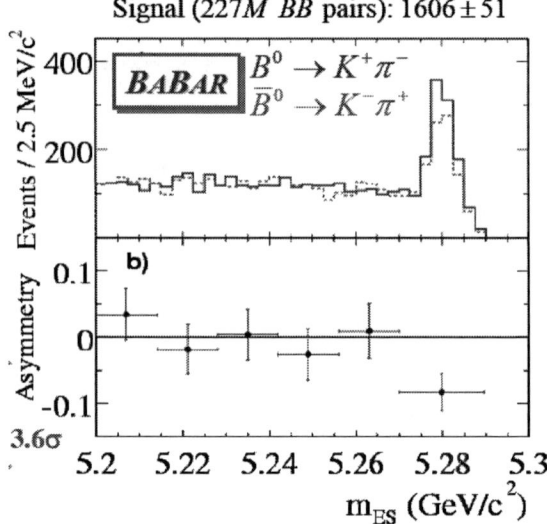

FIGURE 3. Difference (above) and asymmetry (below) in the decay of B⁰ and its antiparticle in Kπ.

The decay rate of B⁰ and its antiparticle in Kπ indeed differs as expected in case of CP violation. It has to be noticed that this phenomenon was already observed in K-physics (ϵ'/ϵ) but it took there quite a long time (40 years of intense efforts) while in B meson decay it took only a couple of years.

The B-factories are integrating luminosities at an even faster pace than the more optimistic expectation so that the measurement of all the angles of the unitary triangle is at hand.

For example BaBar by making use of the combined results of the B decay into ππ, ρπ and ρρ has reached a precision of 10^0 in the value of α (Fig, 4).

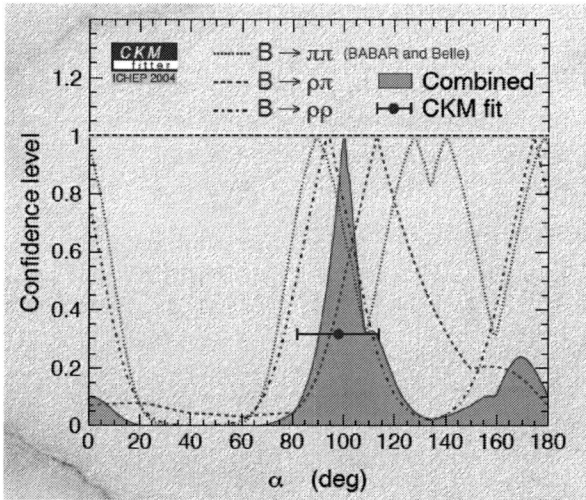

FIGURE 4. The confidence level as a function of the value of the angle.

It is clearly seen that the peak of probability at around 100^0 is in remarkable agreement with the prediction (CKM fitter [3]) of the fit to the unitary triangle.

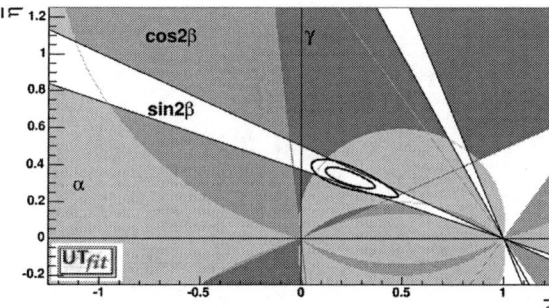

FIGURE 5. Constrained on the apex of the unitary triangle as obtained by fitting all together the results of BaBar and Belle on the angles α, β and γ.

It can be seen by the picture shown above (Fig. 5) how by making use only of the measurement of the angles of the triangle (namely the CP violating quantities: α, $\sin 2\beta$, $\cos 2\beta$, γ) the apex of the triangle is pretty well determined (Utfit [4]) . A couple of year ago only measurements of the sides were available.

The most ambitious goal of today high energy physics is of course in looking for and possibly finding evidence of New Physics beyond the Standard Model.
The B-factories make use of the measurement of CP violation in the decay b-> sssbar (the most important of them being B-> ΦK). These decay amplitude are mostly governed by a penguin diagram whose weak phase would eventually produce a measurement very close to the one precisely obtained in B-> J/Ψ K ($\sin 2\beta$).

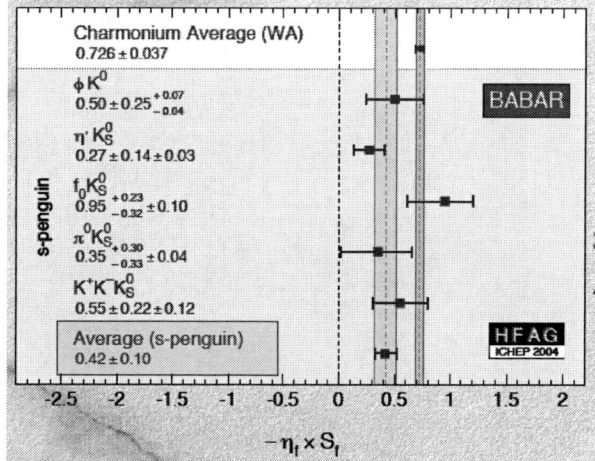

FIGURE 6. $\sin 2\beta$ as determined in various b->s penguin dominated channels wrt to the value obtained in charmonium

Figure 6 shows the values obtained in several channels somewhat similar each other compared with the expectation [6]. There is a hint of a deviation. However none of the channel is significant per-se and their average is a risky if not illegal operation due to the fact that each of them gets different correction to the basic decay diagram. However with more statistics an intriguing picture might emerge with more clarity.

THE NEAR FUTURE: THE LHC

LHC era is coming. Sometime in 2007 the two most energetic proton beams ever put in operation by a human intelligence will start to collide in the former LEP tunnel at CERN. Out of the billions of events that the LHC experiment will record we expect:
- the completion of the Standard Model picture with the discovery of the Higgs boson
- the observation of the superparticles of the the New Physics that necessarily exists beyond the Standard Model in order to explain the existence of one of the ingredients of our universe: the Dark Matter
- whatever it might come from sailing in the unknown lands of the energy frontier

The machine has a solid schedule and the experiments have started to be installed (Fig. 7).

FIGURE 7. Details of the installation of ATLAS (top) and CMS(bottom)

Soon the commissioning will start.

It is crucial to remember that a fast *mise en route* of the detector might allow for early discoveries given the expected performances of the machine in term of luminosity. Indeed once the detectors will be understood and calibrated and a set of standard measurements (top, W, Z...) performed the background to the potential New Physics discoveries will have been understood so that the road will open for discoveries.

As an example the production cross section for SUSY particles with a mass of ~ 1.5 TeV is such that a single month of running of the LHC with a luminosity of 10^{33} cm^{-2}s^{-1} (considered low luminosity and possibly achievable within the first year of running of the accelerator) will be enough to allow to claim an observation for a well understood detector. It is also true that extending the search to a substantially higher mass (say 2.5 TeV) would cost more than a factor 100 in luminosity, definitely calling for a luminosity upgrade of the machine.

A COMPLEMENTARY FUTURE

When hopefully LHC will unveil the SuperSymmetric nature of our microscopic world a problem will arise. Which, amongst the hundred of models, is the kind of SUSY we will be observing ?

Can we measure the equivalent SuperCKM matrix ? The answer is that we could but going to different machines. Machines at high intensity, where very rare events can be studied and that even in the unlucky case that SUSY lies beyond the reach of LHC could give indication of New Physics scale.

The SUSY scale might be at reach of the International Linear Collider in which case the road to follow would be crystal clear. However it might be beyond it that would make attractive to study with precision the effect induced at low energy. There are of course cases very difficult to threat such a SUSY of Minimal Flavor Violation type.

A research program to be carried out at a high intensity protosynchrotron (of the kind being built now at Tokai in Japan) might allow to complete the K-physics through the measurement of the apex of the unitary triangle in the super rare K-> πνν decays (Fig. 8),

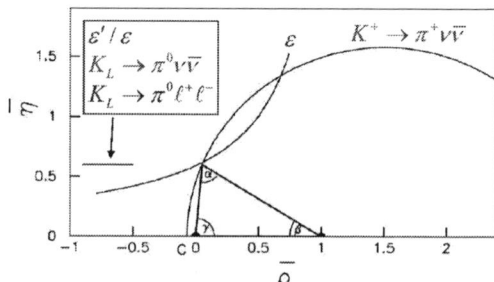

FIGURE 8. Unitary triangle as it could be potentially determined by K-physics only

support a new generation of flavor violation experiments (μ->e, μ->τ conversions), improve the precision on the EDM experiment beside being functional to ν-physics.

Of course the best profut of this machine could be obtained if at the same time the accelerator is part of the upgrade program for the LHC that in almost all of the scenario needs either a PS++ or a Superconducting SPS.

In the picture below (Fig. 9) it is shown where a High Intensity Proton Synchrotron might be situated in terms of power.

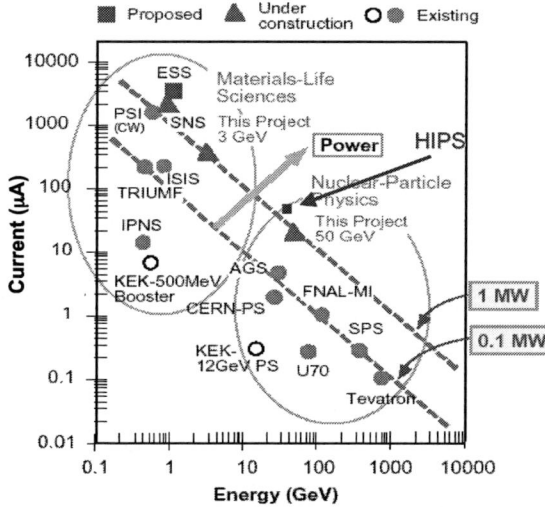

FIGURE 9. Diagram current vs energy of existing (square) and in construction (triangles) accelerators

Other projects that should be pursued with determination are:

-the study of the transversity structure function and in general the spin physics that could be hosted at future GSI improved complex of machines

- a super B factory that would exhaust all the potentialities of B-physics only partially uncovered by BaBar abd Belle

-the upgrade of the accelerator running in Laboratori Nazionali di Frascati that would allow to complete the Ks physics, study the K-interferometry and improve the determination of R at low energy.

CONCLUSION

The conclusion of this talk is that the sector of HEP with accelerators is pretty healthy in our country with beautiful results already secured and others expected till almost the end of the decade from the currently running experiments.
The future of our field shall proceed on the double rail of the energy and intensity frontier.
We are waiting with excitement for the start-up of LHC that might open a window on the uncharted lands of the energy frontier as well as we see a future for the flavor physics that might certainly profit from the construction of an accelerator at the intensity frontier.

REFERENCES

1. G. Lanfranchi et al., hep-ex/0505089
2. A. Aloisio et al., *Nucl. Phys. Proc. Suppl.* **144**, 231 (2005)
3. http://ckmfitter.in2p3.fr/
4. http://utfit.roma1.infn.it/
5. http://www.slac.stanford.edu/xorg/hfag/
6. B. Aubert et al., *Phys. Rev. Letters* **93**, 131801 (2004).

Which Future for High-Energy Physics?

Michelangelo L. Mangano

CERN, PH-TH, 1211 Geneva 23, Switzerland

Abstract. I will review here some of the outstanding issues facing high-energy physics, and the prospects to tackle them with the current and future generation of experiments.

Keywords: LHC, Supersymmetry, neutrinos, flavour
PACS: 12.60.-i, 12.10.-g

INTRODUCTION

More and more frequently we find ourselves engaged in discussions not about physics itself, but about its future. The reason is (at least) twofold. On one side, and in spite of the unequivocal indication from neutrino masses that there is physics beyond the Standard Model (SM), we start becoming impatient at the long wait for concrete manifestations of this new physics. On the other, the community is getting worried at the series of cancellations, terminations, downscopings, delays, etc, which we have witnessed in the major HEP laboratories over the past few years. There is a strong fear that the support from the funding agencies, which made it possible for our field to come this far, is coming to an end. As we all wait for the LHC to at least address the first issue and ensure flashy headlines, most of us are already thinking ahead and planning for the next major projects (ILC, CLIC, neutrino factories, etc). But while there appears to be consensus (at least among those who share it) on the desirability of the ILC, a more comprehensive roadmap for the field, which properly weighs the different possible directions, their role in contributing to the progress of the field, and the mechanism by which their prioritization and implementation will be carried out once the data from the LHC are available, is still lacking. In this short report I will present some of the open issues and stress the need for diversification in our field, pointing out how experiments of scale reduced compared to the multibillion dollar ensterprises provide an essential component of a healthy HEP programme, both scientifically and sociologically.

To start a discussion on the future of HEP, it is useful to focus on its present. In a nutshell, the past 50 years led us to a theory, our SM, with the following features:

- the underlying mathematical structure is given by a renormalizable quantum field theory;
- the main dynamical principle is provided by a local gauge symmetry, with a group structure $SU(3) \times SU(2) \times U(1)$ dictated by experimental evidence;
- in addition to the particles dictated by the gauge structure (the gauge bosons), "matter" fields exist grouped into three families of quarks and leptons, plus a sector required to break the electroweak gauge symmetry.

The above scheme leads to a reliable perturbation theory, which is extremely well tested experimentally. The strong-coupling regime of the $SU(3)$ sector can furthermore be studied theoretically using the techniques of lattice field theory, once again providing excellent agreement with the data for observables which can be calculated in practice (like the mass spectra).

The future of HEP should be driven by the key questions left unanswered by the above picture. Some of these questions are of formal nature: why a gauge theory? Are particles really pointlike or rather strings or more complex extended objects? Other questions are more phenomenological:

- Why 3 families of quarks and leptons?
- What is the origin of the particle mass hierarchies, and in particular why neutrinos are so much lighter than the others?
- Why is there a matter-antimatter asymmetry in the Universe?
- What is the origin of dark matter (DM)? Of dark energy?
- Is there a grand unification of forces?
- Why is gravity so weak?
- Why do we live in 3+1 dimensions?

All of the above questions can be classified under a smaller number of headings: what is the origin of EW symmetry breaking (EWSB)? What is the origin of flavour? How do quantum mechanics and gravity merge into a consistent and calculable theory capable of describing the Universe vacuum state? Most likely a better understanding of the origin of EWSB will also shed light on the issue of flavour and its ramifications (neutrino masses, CP violation). Pragmatically, it is reasonable to expect that on the short term, namely within the reach of the LHC, we should be able to make significant advances on at least the problem of EWSB, and that these advances will point to a more clear direction for the future of HEP.

Before I move to a more detailed discussion, I'd like to close this introduction with some comments on a branch of HEP which is not covered by the above set of questions, namely the study of strongly interacting matter. This includes the continued exploration of the proton structure, hadronic spectroscopy, and the field of relativistic heavy ion collisions. The understanding of the proton appears as a "fundamental" problem, since the proton is the key component of what the visible Universe is made of. I personally see the understanding of the proton mostly as a tool which allows us to use protons as probes of the small scales where the new physics is hiding. The accurate knowledge of the quark and gluon distributions is critical to calculate production cross sections at the LHC and to connect the observed event rates with the coupling parameters of the new particles. Even diffractive studies can provide useful tools for fundamental searches, such as the production of Higgs bosons in diffractive proton-proton collisions. To the extent that what we know today is still insufficient to achieve the required precision at the LHC and possible future hadron colliders, the study of the proton structure can therefore be considered a priority for HEP. On the other hand, and in spite of the great excitement caused by the recent observations of new exotic mesonic and baryonic states, I cannot find a solid reason to promote hadron spectroscopy to the level of a "priority" for HEP. It should be acknowledged nevertheless that there is an extremely lively community of both experimentalists and theorists fully devoted to this research, and that there are potential spinoffs in other areas of physics (for example the possibility that diquarks provide an important degree of freedom in the equation of state of neutron stars). It could also turn out that, should a strongly interacting force be responsible for the EWSB, unexplored territories in the non-perturbative regime of QCD could help shed light on the dynamics of these new forces.

Relativistic heavy ion collisions are a new entry in HEP. They will open a new window on QCD at extreme densities and temperature. This is rather unknown territory, with room for interesting dynamical surprises. I am confident that the LHC will expose very interesting features of the deconfined phase of QCD, adding to what RHIC has been unveiling so far. Nevertheless I am disturbed by the fact that no future is being layed out for this field. The LHC heavy ion programme will terminate at the time of moving to the very high-luminosity upgrade. No other machine in this energy range is being planned, or can reasonably be expected to be built. What will happen to the field afterwards? Perhaps after the big jump in energy this community will decide that the remaining open issues (say the precise identification of the QCD critical point) are better addressed at low energy, and the field will move to non-HEP laboratories and out of our horizon.

EWSB AFTER LEP AND TEVATRON

The Tevatron and LEP's heritage is a strong confirmation of the SM, and at the same time an apparent paradox[1], illustrated in the following paragraphs. Electroweak precision tests and the value of the top mass are consistent with a rather light Higgs mass: $m_H = 98^{+52}_{-36}$ GeV; EW radiative corrections in the SM, integrated up to a scale Λ, shift the bare value of m_H by:

$$\begin{aligned}
\delta m_H^2 &= \frac{6 G_F \Lambda^2}{\sqrt{2}\pi^2} (m_t^2 - \frac{1}{2}m_W^2 - \frac{1}{4}m_Z^2 - \frac{1}{4}m_H^2) \\
&\sim (115 \text{ GeV})^2 \left(\frac{\Lambda}{400 \text{ GeV}}\right)^2.
\end{aligned} \quad (1)$$

The integration in principle can extend up to very large values of Λ, where new particles may appear, changing eq. (1). As Λ gets significantly larger than 400 GeV, however, the presence of a counterterm (CT) should be assumed, to ensure that the overall value of m_H is consistent with its bounds. This CT can be interpreted as a low-energy manifestation of the physical mechanisms which, at some scale $\overline{\Lambda}$, modify eq. (1). Ensuring that the residual of the cancellation between eq. (1) and the CT is in the 100 GeV range, however, forces a fine tuning which becomes more and more unbelievable as $\overline{\Lambda}$ grows. Assuming that no new physics appears before the GUT scale of 10^{16} GeV would lead to a level of fine tuning of 10^{-28}! By and large theorists believe that this is unlikely enough to call for the existence of new physics at scales in the range of 1–few TeV, so as to maintain the fine tuning level to within $O(10^{-3})$. This belief however clashes (and this is where the paradox arises) with the staggering agreement between EW data and the SM. The inclusion of generic new physics, parameterized in terms of low-energy effective couplings between the SM particles, and the analysis of the effects induced on EW observables, set lower limits to the scale $\overline{\Lambda}$ in the range of 5-10 TeV[2], at the extreme limit of the fine-tuning window. The solution to the paradox could only be obtained with new physics which cancels the large radiative contributions to m_H and, at the same time, manages to leave all other EW parameters and observables unaffected. Supersymmetry (SUSY) provides one such example! The cancellation of large loop effects between SM particles and their SUSY partners modifies eq. (1) and leads to an upper limit on m_H, given in a simplified approximation here:

$$m_H^2 \lesssim m_Z^2 + \frac{3G_F m_t^4}{\sqrt{2}\pi^2} \log\left(\frac{m_{\tilde{t}}^2}{m_t^2}\right) \quad (2)$$

where $m_{\tilde{t}}^2$ is the average squared mass of the two stop states. At the same time, the structure of the theory is such that indeed generic choices of the SUSY parameters, consistent with current experimental limits on new particles, lead to negligible effects in the EW observables. In the minimal realization of SUSY (MSSM), when eq. (2) is improved with 2-loop and non-logarithmic corrections, the experimental limit on m_H pushes however the scale of SUSY in the TeV domain. Once again this is at the edge of being acceptable as a "natural" solution to the fine-tuning problem, and for many theorists the room left for SUSY is becoming too tight. As a result, new scenarios for EW symmetry breaking, particularly some where the upper limits on the Higgs mass are looser compared to the MSSM, have been proposed (as reviewed in[1]). Some of these ideas lead to rather artificial structures, where the problem of the Higgs naturalness is shifted to slightly higher scales, via the introduction of a new sector of particles around the TeV. Few of them however offer the appeal of Supersymmetry, with its clear predictions (calculability), and connections with the other outstanding problems of the Standard Model (Dark Matter, Flavour, CP violation). Little Higgs models, for example, predict the existence of heavy top-like quarks, and of new heavy gauge bosons (W' and Z'). Extra-dimensional, or Higgsless models, predict strong WW scattering at about 1 TeV, and Kaluza-Klein like excitations starting from the TeV scale.

While these alternative scenarios could take longer to be identified experimentally at the LHC, the SUSY framework provides a strong and appealing physics case for a possible early discovery, and therefore should be given maximum priority in the planning for the first LHC data analyses. SUSY is in fact expected to manifest itself with abundant and striking signals, such as the production of multijets with large missing transverse energy ($\not{E}_T$), multileptons (possibly same-charge), or prompt photons with large $\not{E}_T$. Because of rates, background levels, and nature of the observables, searches for SUSY are expected to be less demanding from the experimental point of view than the quest for the Higgs in the $m_H < 140$ GeV range. In addition, SUSY provides a natural candidate for dark matter, namely the lightest neutralino χ_1^0, the neutral SUSY partner of the photon and Z. Proving the direct link between dark matter and SUSY would be, perhaps even more than the Higgs discovery, the flagship achievement of the LHC! Last but not least, an early detection of SUSY could immediately provide clear directions to the field of experimental high-energy physics, and allow a robust planning for future facilities.

The large value of m_H shows that room is getting very tight now for SUSY, at least in its *minimal* realizations. This makes the case for an early observation of SUSY at the LHC quite compelling.

SUSY and its parameters

Many people argue that SUSY is ugly because of its many arbitrary parameters. It is important to point out that, when SUSY is realised as a symmetry, it has actually one less parameter than the SM: this because all new particles are degenerate with the SM ones, and the Higgs self-coupling, which in teh SM is a free parameter, in SUSY is related to the gauge couplings. All complexity and parameter proliferation of SUSY are therefore a consequence of SUSY breaking (SSB). A minimal SUSY extension of the SM, with arbitrary pattern of spontaneous SUSY breaking, has over 100 extra parameters (scalar and gauge-fermion masses, mixings among SUSY partners of quarks and leptons). This is not much worse than an arbitrary extension to leptons and hadrons of Fermi's theory of weak interactions, before Feynman, Gell-Mann and Cabibbo, or even before LEP/SLC firmly established the parameters of the SM. One could have needed parameters to describe non V-A couplings (S, P, T, V+A), as well as non-universal couplings to individual hadronic and leptonic currents. In the context of the SM, one could likewise assume more complex Higgs structures or, more in general, different realisations of EWSB. It is therefore reasonable to assume that the parameter proliferation in SUSY is the consequence of our current ignorance of the specific dynamics leading to SUSY breaking, and that once this is clarified, the spectrum of SUSY particles and their interactions will follow from some deeper considerations. To get clues on the mechanism of SUSY breaking should therefore be a crucial goal of HEP planning for the future. As a corollary, one would like to prove the SUSY origin of DM.

Constraints on SUSY breaking

No SUSY state has been observed as yet. This automatically sets constraints on the spectrum of SUSY particles, which must have masses typically larger than 100 GeV. Nevertheless they cannot be arbitrarily large, to prevent the artificial fine tuning which justified SUSY in first place, setting upper limits for at least some of the new states in the TeV region. As in the case of the SM, flavour physics, and in particular flavour-changing neutral currents both in the hadronic and leptonic sectors, place the strongest constraints on the most general inter-generational mixing. Generic SUSY breaking

(SSB) leads in fact to unacceptable FCNC. For example:

$$\mu \not\to e\gamma \Rightarrow \sin 2\theta_{\tilde{e}\tilde{\mu}} \frac{\Delta m^2_{\tilde{e}\tilde{\mu}}}{m^2_{\tilde{e}}} < 0.01 \quad (3)$$

and

$$\varepsilon_K \sim \left(\frac{100\text{ TeV}}{m_{\tilde{q}}}\right)^2 \text{Im}\left(\frac{\Delta m^2_{\tilde{d}_L \tilde{s}_L}}{m^2_{\tilde{d}}} \frac{\Delta m^2_{\tilde{d}_R \tilde{s}_R}}{m^2_{\tilde{d}}}\right) < 2 \cdot 10^{-3} \quad (4)$$

where Δm^2_{ij} are mass splittings between scalar leptons or quarks, and θ_{ij} are mixign angles. Some examples of SSB scenarios provide an automatic suppression of FCNC. For example, in the so called gauge-mediated SSB the SUSY breaking arises in a strongly coupled sector, and is transferred to the low energy sector only via gauge interactions at an intermediate scale $m_{SSB} \sim 1 - 100$ TeV. Since different flavours have the same gauge couplings, SSB is flavour independent and there are no extra contributions to FCNC processes. Typical additional predictions of GMSB include relations among SSB parameters determined by gauge couplings:

$$\frac{m(\tilde{q})}{m(\tilde{\ell})} \sim \frac{\alpha_s}{\alpha_w} \quad (5)$$

$$\frac{m(\tilde{g})}{m(\tilde{\chi})} \sim \frac{\alpha_s}{\alpha_w} \quad (6)$$

$$m(\tilde{q}) \sim m(\tilde{g}), \quad m(\tilde{\ell}) \sim m(\tilde{\chi}) \quad (7)$$

$$m(\tilde{\chi}^\pm_1) \sim m(\chi^0_2). \quad (8)$$

Furthermore the gravitino is the lighest SUSY particle (LSP), and the production of any SUSY particle leads to one of the following characteristic decays:

$$\chi^0 \to \tilde{G}\gamma \quad \text{or} \quad \tilde{\ell} \to \tilde{G}\ell \quad (9)$$

depending on which one is the next-to-lightest SUSY state.

Alternative SSB scenarios include the best explored one, namely minimal supergravity. Here SUSY is broken at an intermediate scale, $M_{SSB} \sim \sqrt{m_W m_{\text{Plank}}} \sim 10^{11}$ GeV, and one assumes universal scalar and fermion SSB masses at the Planck scale:

$$m_{\tilde{f}} = m_0 \quad \forall \text{ flavours } f \quad (10)$$

$$m_H = m_0 \quad (11)$$

$$m_{\tilde{V}} = m_{1/2} \quad \forall V = g, \gamma, W, Z \quad (12)$$

At the EW scale, these imply the following mass relations:

$$m(\tilde{g})/m(\tilde{\chi}) \sim \alpha_s/\alpha_W \quad (13)$$

$$m(\tilde{B}) = (5g'^2/3g^2)m(\tilde{W}) \sim 0.5 m(\tilde{W}) \quad (14)$$

In particular, this last relation is different than in the case of GMSB.

A new scenario ha recently been advocated, the so called Split SUSY[3]. This scenario gives up SUSY as a solution of the hierarchy problem, assuming that some yet unknown dynamics takes care of the fine-tuning (as must be the case for the cosmological problem). SUSY is then broken at the unification scale, all scalar SUSY partners are very heavy, well beyond the reach of any conceivable accelerator, and only the gauge fermions – protected by chiral symmetry – retain a mass in the range below the TeV. Nevertheless the main features of SUSY which give it a phenomenogical appeal (the possible connection with DM and the hint of grand unification) survive, and are promoted to being the only mandatory ingredients of model building. The main features of the spectrum are summarised here:

$$m^2(\tilde{f}) \sim \tilde{m}^2 \quad \tilde{m} \sim 10^{4-10} \text{GeV} \quad (15)$$

$$m(\tilde{V}) \sim \frac{\tilde{m}^2}{M}, \text{ with } M \text{ such that} \quad m(\tilde{\chi}^0) \sim 10^{2-3} \text{GeV} \quad (16)$$

The very heavy scalars remove all FCNC problems, gauge coupling unification is improved, and there is by construction a good DM candidate. Furthermore, there is room for a new source of CP violation in the complex higgsino/gaugino couplings and manifestations are possible in electric dipole moments. In these scenarios, the gluino is typically very long lived, providing spectacular (when not elusive!) signatures.

The role of flavour physics

As mentioned above, flavour phenomena provide a crucial testbench for most BSM theories[4]. This has been true also for the development of the SM, which was tailored to accomodate $K^0 - \bar{K}^0$ mixing, FCNC and the prediction of charm, CP violation. In the SM the key element is the GIM mechanism:

$$A(s\bar{d} \to Z/\gamma) \propto (m_t/m_W)^2 \lambda^5 \quad (17)$$

where λ is the Cabibbo angle. In absence of GIM cancellations (as would happen in generic extensions of the SM, like Supersymmetry with arbitrary squark mass matrices), the Cabibbo suppression is absent, and the mass factors reflect the masses of the new particles circulating in the loops. In the case of Supersymmetry, the mass factor is proportional to the mass splittings between squarks of the different generations. Essential probes of the flavour structure of theories BSM are given, for example, by the four very rare decays of kaons: $K^+ \to \pi^+ \nu \bar{\nu}$, $K^0_L \to \pi^0 \nu \bar{\nu}$ and $K^0_L \to \pi^0 \ell^+ \ell^-$, with $\ell = e, \mu$, as well as by the rare decays of B mesons: $B_{d,s} \to \ell^+ \ell^-$. Accuracies at the level of 10% in the determination of these BR would provide essential inputs in the study of the flavour structure of any new physics model.

The discovery of Supersymmetry or other new phenomena at the LHC will dramatically increase the motivation for searches of new phenomena in flavour physics. While there is no guarantee that any deviation from the SM will be found, the existence of physics BSM will demand and fully justify these studies: we'll be measuring the properties of something which we know exists, as opposed to frustratingly looking for "we don't know what" as we are unfortunately doing today!

The B physics studies at the LHC and at the B-factories will find therefore a natural complement in a revived and strenghtened K physics programme, as well as in new studies of the charm sector and searches for Lepton Flavour Violation (LFV) phenomena. It is inconceivable that a complete clarification of the nature of the new phenomena to be observed at the LHC can be obtained without further exploration of their implications in flavour physics.

Dark Matter

The evidence for the existence of dark matter is very strong today, with the new recent findings from studies of structure formation and CMB fluctuations. It is important to realize that, whatever its origin, the existence and properties of DM must be encoded somewhere in the Lagrangian of HEP. So it is "our" problem to find out what it is, and not the astrophysicists' problem. From our point of view, the main ingredients of DM are: a stable weakly interacting particle, with mass vs annihilation rates such as to decouple during the Big Bang from the other states at the appropriate time and with the appropriate density. It so happens that the required numerics works out to match the expected behaviour of particles with mass $O(100 \text{ GeV})$ and weak coupling:

$$\sigma \sim \frac{\alpha_W^2}{M_W^2}. \quad (18)$$

It is unavoidable to speculate that the origin of DM is directly linked to the phenomena responsible for EWSB. It is therefore not surprising that most alternative approaches to EWSB (little Higgs, extra-dimensions, Higgsless theories) provide a possible DM candidate. The mass vs coupling relations are inherited by the link to EWSB and the stability is associated to discrete symmetries (like SUSY's R parity). In the case of extra-dimensions, for example, DM could originate from the first photon or neutrino Kaluza-Klein mode. Once again: it is a crucial goal of the future in HEP to firmly establish the true nature of DM, and the hunt might optimistically end already during the first running of the LHC!

NEUTRINOS

The physics case for studies of the neutrino sector is clear and strong: the connection with GUT-scale physics, via the see-saw mechanism, is as strong as, if not stronger than, the unification of the gauge couplings; the failure of the SM to accomodate the baryon asymmetry of the Universe makes leptogenesis (lepton-driven B asymmetry of the Universe) a very exciting possibility; the connection with GUT, and the fact that only within SUSY one can so far quantitativily accomodate the gauge coupling unification, strenghtens the case for SUSY itself, and allows to explore the joint implications of SUSY and neutrino masses and mixings (e.g. a calculable framework to predict and interpret possible LFV transitions).

Neutrino physics is based on some rather straightforward ideas, both theoretically and experimentally. Theoretically, the interpretation of the data is simple, one is measuring the entries of a 3x3 matrix and some mass parameters. There is one absolute mass scale (still unknown), 2 mass differeces (known up to a discrete ambiguity), 3 mixing angles (two of which are known with reasonable accuracy) and at least one CP violating phase δ (unknown) plus two more phases if neutrinos are of Majorana type. Clear criteria drive the experimental design. One aims to measure oscillation probabilities, by examining appearance or disappearance rates P:

$$P(\nu_i \to \nu_j) = S \times \sin(\Delta m^2 E/L) \quad (19)$$

where the size of the signal S is driven by the source power, the detector volume and distance L from the source; the energy spectrum is driven by the neutrino source; and the purity of the initial state ν_i is again determined by the nature of the source. The optimization of these different ingredients, vis a vis the technological difficulties and the financial constraints, is where opinions may differ. In the current planning, covering time scales up to about 2010, we have K2K (Japan), Opera and Icarus (GSNL) and Minos (USA). These are optimized to confirm the evidence of atmospheric neutrino oscillations via ν_μ disappearance or ν_τ appearance. They have limited potential in measuring oscillation parameters and not optimized for ν_e appearance (θ_{13} determination). Between 2009–2015: T2K, Nova, Double Chooz. They are optimized for measurements of θ_{13} (reach up to a factor of 20 beyond the Chooz sensitivity) via ν_e appearance or disappearance and aim at a measurement of the atmospheric oscillation parameters to 1%. The reach for the CP violation phase δ is however limited. Beyond 2015, the menu of options offers Superbeams (driven by multi-MW proton sources) and/or Betabeams (pure electron (anti)neutrino beams from decay of radioactive nuclei), with sensitivity for θ_{13} in the few$\times 10^{-3}$ range, and CP violation discovery potential for $\theta_{13} > 1°$. Beyond that, a neutrino factory provides the ultimate playground for neutrino physics.

It is crucial for the HEP community to accept that neutrino physics will play a major role in our future. The size of the detectors being planned, and the power of the sources, provide immense technological challenges, which will require very large funding and human resources. The planning of new facilities should be far-sighted, and should anticipate the escalation of needs (namely tha path from standard beams to superbeams – possibly β beams – and finally a neutrino factory). The prospect of using a new detector or accelerator as a step towards a second more powerful facility should the firmly established before proceeding. These considerations tend to favour the idea of the establishment of a large international neutrino laboratory, in which to focus all neutrino-related activities. This is similar to what has happened with the LHC for hadron collisions, and to what is being planned for the ILC. This will allow to focus and consolidate the attention of the community, to optimize the resources and to identify the best location for detectors. The presence of a variety of components in the accelerator complex for a high-power neutrino laboratory opens the possibility of synergy with other HEP programmes[5], for example kaon or muon physics, providing a very healthy diversification, a physics programme on-site, and an optimal utilization of the resources in the course of major accelerator upgrades which could keep the main neutrino beams idle. The size of future neutrino experiments (Super-beams and then the neutrino factory) place this activity in direct competition for resources with the exploration of the high energy frontier (ILC, SuperLHC. CLIC), and it is difficult to imagine that it can be achieved without a coherent and well focused international effort.

A DREAM SCENARIO FOR HEP

I shall outline here a dream scenario, in which an early detection of new physics signals at the LHC drives a chain-reaction of observations and discoveries, and motivates measurements to be performed with a multitude of experiments both at the high and low-energy end of the spectrum. In this scenario, all the topics reviewed before acquire a primary role in exploring in full the parameters and features of the new physics.

Let me therefore assume that the LHC will give indications of a gluino in the TeV mass range (for example in the multijet plus $\not{E}_T$ final state). Measurements of the kinematics of these events will provide a determination of the mass of gluinos, and will possibly lead to a separation between gluinos and squarks. Several measurements will then follow: the identification of charginos and neutralinos, with a determination of their masses.

The determination of the parameters of the neutralino could establish the neutralino as the particle responsible for dark matter. The relation between the gluino mass and the mass of the charginos and neutralinos will provide important information to select the structure of the SUSY breaking mechanism. Additional information will come from other properties of the final states: for example, if hard photons are detected, and the missing energy is associated to very light particles, then scenarios such as gauge-mediated SUSY breaking are favoured. In this case the neutralino is unstable (and therefore could not be held responsible for dark matter), and decays to a photon and a light gravitino (which would therefore become the first detected particle in the quantum spectrum of gravity!). Further studies should focus on the identification of individual squarks, and in particular on the identification and mass determination of the stop and sbottom, together with the mass splitting between the two separate state of each squark. The slepton spectrum can be explored either through the direct production of sleptons (via e.g. DY-like processes), or through their emission during the chain decays of charginos and neutralinos. Their masses can be measured, and the relation between their masses and those of squarks will provide additional information on the structure of SUSY breaking. The detection of the Higgs boson, and of the additional Higgs particles present on SUSY models, will allow to verify the mass relations involving the Higgs, the stop and $\tan\beta$. This study will be complemented by the study of the ttH and bbH couplings, e.g. via the detection of ttH and bbH associated production. To the extent that the masses are low enough, the ILC will provide a unique facility to carry out a good fraction of the above measurements with excellent precision.

The LHC will complement the studies of high-mass phenomena with the observation of rare decays of B mesons, for example $B_{d,s} \to \mu^+\mu^-$. A more complete exploration of the flavour structure of SUSY will require however a new generation of rare kaon decay experiments, which will detect and measure the four golden modes listed before: $K^+ \to \pi^+\nu\bar{\nu}$, $K_L^0 \to \pi^0\nu\bar{\nu}$ and $K_L^0 \to \pi^0\ell^+\ell^-$. In addition, it will be mandatory to search for possible flavour-changing effects in the lepton sector, such as $\mu \to e\gamma$, $\mu \to 3e$, $\tau \to \mu\gamma$, $\tau \to 3\ell$, $\mu \to \tau$ transitions. The presence of SUSY could provide new sources of CP violation, for example in the Higgs sector, or in the gluino or chargino/neutralino couplings. These sources of CP violation could be exposed already at the LHC, but would likely appear as electric dipole moments, either in the neutron (if present in the gluino sector) or in the leptons (if present in the weak gauge fermion sector).

In parallel, the study of the neutrino sector will need to continue, to complete the determination of the remaining parameters of the neutrino mass matrix and mixing. Among other things, this will be required to complete the exploration of the sources of CP violation in particle physics, and to understand whether this is consistent with a quantitative picture of the baryon asymmetry of the Universe.

CONCLUSIONS

It is hard to imagine a future for our field should the LHC not detect new phenomena. In particular, it will be extremely hard in this case to identify the right direction and the best machine to succeed the LHC. The default option of pushing for yet higher energies (whether via a hadronic of leptonic collider) may not be viable, due to the immense costs and, even worse, the very long time scale necessary for their construction: it will be very difficult to motivate the young researchers to invest in our field if the hopes for exciting discoveries are postponed to their senior years. An e^+e^- linear collider is an obvious complement to the LHC: if it has access to the new particles seen at the LHC it will allow measurements of precision unattainable at the LHC. If it doesn't, it can still explore their properties via very precise measurements of their virtual effects. But it is utopistic in my opinion to expect that it will be supported by the funding agencies in absence of a clear indication of new phenomena nearby. While neutrino experiments will without doubt make clear progress and hopefully address and solve the long-standing problem of the baryon asymmetry of the Universe, any fruitful speculation about the future of HEP should start, in my view, from the assumption that the LHC will indeed observe new physics. In my ideal scenario outlined above the future of the field will be driven not just by the giant facilities, but also by smaller and less expensive experiments, which will explore crucial issues such as the flavour and CP properties of the new phenomena. It is essential, therefore, that we start planning for these facilities, and keep alive the R&D which will make their timely and effective deployment possible once the first data from the LHC become available. Should things turn out like what dreamt above, or in other similar scenarios, our field will witness an unprecedented boost, and the quest for the understanding of the ultimate laws of Nature will be guaranteed a bright and long-term future.

REFERENCES

1. R. Barbieri, arXiv:hep-ph/0410223.
2. R. Barbieri and A. Strumia, Phys. Lett. B **462** (1999) 144; R. Barbieri, A. Pomarol, R. Rattazzi and A. Strumia, Nucl. Phys. B **703** (2004) 127.
3. N. Arkani-Hamed and S. Dimopoulos, JHEP **0506** (2005) 073; G. F. Giudice and A. Romanino, Nucl. Phys. B **699** (2004) 65; see also G.Giudice, in these proceedings.
4. A. J. Buras, arXiv:hep-ph/0505175.
5. S. Geer, arXiv:hep-ph/0507223.

Possible Scenarios for the LHC Luminosity Upgrade

W. Scandale

CERN
1211 Geneva 23, Switzerland

Abstract. In this paper we consider possible scenarios to upgrade the LHC luminosity beyond 10^{34} cm^{-2}s^{-1}. By pushing the accelerator parameters to the ultimate performance we can increase to 1.7 10^{11} the bunch population and eventually reach a peak luminosity of 2.3×10^{34} cm^{-2}s^{-1}. To go beyond, a considerable improvement of the LHC parameters, such as ß*, beam intensity, bunch length, number of circulating bunches is required. Finally, the upgrade of the injector complex and of the injection energy is another important ingredient to upgrade both peak and integrated luminosity up to an order of magnitude above the nominal value.

Keywords:. LHC, accelerators
PACS: 29.20.-c

LHC PERFORMANCE LIMITATIONS

To reach the nominal LHC performance several steps will be needed. The crucial ones should allow running LHC at about one tenth of the nominal luminosity already during the first months of operation. In Table 1 we show two possible intermediate scenarios towards nominal luminosity. One of them assumes 75 ns bunch spacing and ß* = 1 m, in view of exploring and mastering multibunch operation, cleaning of the beam pipe to mitigate electron-cloud effect, ß-squeezing and collisions with finite crossing angle. The other scenario with 25 ns bunch spacing and less than half of bunch population is intended to investigate and control operation with nominal values of ß*, crossing angle and bunch spacing at reduced current. The path towards nominal performance may last up to four years, also in consideration of the staged installation of collimators and dilution kickers in the beam disposal system and of the progressive cleaning of the vacuum pipe. Refs. [1-2] describe in details the phenomena inducing performance limitations in LHC and the possible way to nominal performance and beyond.

MOTIVATION FOR A LUMINOSITY UPGRADE

There are twofold motivations for launching a vigorous LHC luminosity upgrade programme before the end of the first decade of operation, as originally suggested by J. Strait (FNAL) in 2003.

The expected run-time halving the statistical errors is a rather steep function of time. We can evaluate it, assuming that the first-year luminosity is a tenth of the nominal one, and that its increase is almost linear over four year, up to the nominal value of 10^{34} cm^{-2} s^{-1}. Four years after the LHC start-up when the nominal luminosity is reached, less than two years will be required to multiply by four the set of collected data. Another three years later, however, more that five years will be requested for a fourfold increase of the data. In these conditions, the stability of the experimental apparatus and the consistency of the data set may become serious issues.

On the other hand, collision debris induce radiation damages in the inner triplet eventually reducing the expected lifetime to below an upper bound, depending on the integrated dose. In the LHC insertions, the damage threshold typically corresponds to 700 nb^{-1} integrated luminosity. This limit is eventually reached in about seven years. In this eventuality, installing new quadrupoles with higher gradient and larger aperture will also ensure a tighter beam focusing and hence a substantial increase of the LHC peak luminosity.

TABLE 1. main steps to nominal performance in LHC.

Parameter	Units	75 ns	25 ns	Nominal
No. of bunches	n_b	936	2808	2808
Proton per bunch	N_b [10^{11}]	0.9	0.4	1.15
Normalis. emittance	ε_n [μm]	3.75	3.75	3.75
rms bunch length	σ_s [cm]	7.55	7.55	7.55
rms energy spread	σ_E [10^{-4}]	1.13	1.13	1.13
IBS growth time	τ_x^{IBS} [h]	135	304	106
Beta at IP	β^* [m]	1.0	0.55	0.55
Full crossing angle	θ_c [μrad]	250	285	285
Luminosity lifetime	τ_L [h]	22	26	15
Peak luminosity	L [10^{34}cm^{-2}s^{-1}]	0.12	0.12	1.0
Events per crossing		7.1	2.3	19.2
$\int$ over 200 runs L dt	L_{int} [fb^{-1}]	9.3	9.5	66.2

BEYOND THE NOMINAL LUMINOSITY

Table 2 shows parameters and expected performance of the nominal scenario and of the so-called ultimate luminosity scenario, with a bunch population of 1.7×10^{11} protons and a peak luminosity of 2.3×10^{34}cm^{-2}s^{-1}. It also contains three possible upgrade scenarios: (1) with 12.5 ns bunch spacing, (2) with a large Piwinski angle and 75 ns bunch spacing, and (3) the so-called superbunch scheme, with a single very long and dense bunch per beam. In all of them the peak luminosity is expected to become about 10 times larger than nominal, provided the ß* is reduced by a factor 2 to 0.25 m and the crossing angle increased by an appropriate amount. In addition, the circulating current will increase by at least a factor 1.7, the crossing angle by at least a factor 1.5 and the number of events per crossing by at least a factor of 2.3.

In the superbunch scenario the latter quantity will have the prohibitive value of 5×10^5, incompatible with the state of the art of today's particle detector technology.

TABLE 2. Possible scenarios for luminosity upgrade

Parameter	Symbol	Nominal luminosity	Ultimate luminosity	Shorter bunch	Longer bunch	Super bunch
No of bunches	n_b	2808	2808	5616	936	1
Proton per bunch	N_b [10^{11}]	1.15	1.7	1.7	6.0	5600
Bunch spacing	Δt_{sep} [ns]	25	25	12.5	75	89×10^3
Average current	I [A]	0.58	0.86	1.72	1.0	1.0
Normalized emittance	ε_n [μm]	3.75	3.75	3.75	3.75	3.75
Longitudinal profile		Gaussian	Gaussian	Gauss.	flat	flat
rms bunch length	σ_s [cm]	7.55	7.55	3.78	14.4	6×10^3
ß* at IP1&IP5	β^* [m]	0.55	0.50	0.25	0.25	0.25
Full crossing angle	θ_c [μrad]	285	315	445	430	1 510^3
Piwinski parameter	$\theta_c \sigma_s/(2\sigma^*)$	0.64	0.75	0.75	2.8	2.7×10^3
Luminosity	L [10^{34}cm^{-2} s^{-1}]	1.0	2.3	9.2	8.9	9.0
Events per crossing		19	44	88	510	5×10^5
l_{rms} of luminous region	σ_{lum} [mm]	44.9	42.8	21.8	36.2	16.7

UPGRADE OF THE LHC INJECTION ENERGY

Motivations to upgrade the LHC injection energy are based on the following argumentations.

Higher injection energy should guarantee from one side a larger circulating current and hence a larger peak luminosity, from another side a shorter turnaround time, i.e. the average time elapsed between two consecutive runs, with a consequent increase of the integrated luminosity. Indeed, the expected luminosity gain should be larger if the collimation and the protection systems have no constraints in handling larger currents and if reducing the dynamic effects in the superconducting magnets can substantially shorten the turnaround time. The operational experience of LHC will clarify these issues in due time.

In addition, increasing the injection energy is a necessary step to increase the LHC beam energy beyond the ultimate value presently achievable of 7.7 TeV per beam, without changing the energy swing and hence the span of magnetic field during the ramp of the main magnets.

Reduction Of The Turnaround Time

In a superconducting ring, such as LHC, the checks and the adjustments requested between two fills will strongly depend on machine reproducibility and should be somehow reduced when increasing the energy of the injection plateau and of the snap-back. Indeed, the range of variation of the cable magnetization should decrease substantially with the injection energy, with a consequent reduction of all kind of dynamic effects in superconducting cables.

By doubling the LHC injection energy we should have a more stable and reproducible magnetic cycle, and hence a shorter turnaround time. Presently we expect a turnaround time of 10 hours in LHC. By injecting at 1 TeV, we should reduce the turnaround time most likely by a factor of two, from 10 to 5 hours.

In Table 3 we consider two situations, one with nominal luminosity and another with ten times higher peak luminosity, and we show how a twofold shortening of the turnaround time may affect the LHC performance.

We based our calculation of the gain on the hypothesis that the luminosity decay is exponential and hence the optimal run time Trun will be computed using equation (1):

$$1 + \frac{T_{run} + T_{turnaround}}{\tau_L} = e^{\frac{T_{run}}{\tau_L}} \quad (1)$$

where $T_{turnaround}$ is the turnaround time and τ_L is the e-folding decay constant of the instantaneous luminosity.

In these circumstances, the integrated luminosity per run is:

$$\int_0^{T_{run}} L\,dt \approx \frac{L_0 \tau_L}{T_{run} + T_{turnaround} + \tau_L} \quad (2)$$

Increase Of The Beam Current

Injecting in LHC more intense proton beams with constant brightness, within the same physical aperture will increase the peak luminosity. Indeed, at the beam-beam limit regime, the peak luminosity is proportional to the normalized emittance $\varepsilon_n = \gamma\varepsilon$, as shown in the approximate formula (3):

$$L \approx \gamma \Delta Q_{bb}^2 \frac{\pi \varepsilon_n f_{rep}}{r_p^2 \beta^*} \sqrt{1 + \left(\frac{\theta_c \sigma_s}{2\sigma^*}\right)^2} \quad (3)$$

Where ΔQ_{bb} is the linear tune shift induced by beam-beam interactions, frep is the collision repetition rate, r_p is the classical radius of the proton, σ^* is the rms transverse beam size, and σ_s, θ_c, β^* have the same meaning as in Table 2.

By increasing the LHC injection energy, it is possible to inject beams with larger normalized emittance within the same physical aperture. The beam will be more intense and of a larger size at the collision energy, thereby imposing some compensation scheme for the far beam-beam interaction effects [3-4] or a larger crossing angle satisfying the equation (4):

$$\frac{d_{sep}}{\sigma} \approx \theta_c \sqrt{\frac{\varepsilon_n}{\gamma \beta^*}} \quad (4)$$

TABLE 3. integrated luminosity versus turnaround time.

L_0 [cm^{-2}s^{-1}]	t_L [h]	$T_{turnaround}$ [h]	T_{run} [h]	200 runs L dt [fb-1]	gain
10^{34}	15	10	14.6	66	51.0
10^{34}	15	5	10.8	85	51.3
10^{35}	6.1	10	8.5	434	56.6
10^{35}	6.1	5	6.5	608	59.2

Where d_{sep} is the full size beam-beam separation induced by the crossing angle, whilst the other symbols have been defined above.

At 1 TeV injection energy, the peak luminosity should double and, without compensation, the crossing angle should increase by a factor $\sqrt{2}$.

Increase Of The Beam Energy

The LHC collision energy is expected to slowly ramp up to the ultimate value of 7.7 TeV corresponding to a peak field of 9 T in the main dipoles. Any farther energy increase will require replacing the main magnets probably introducing Nb_3Tn superconductors to reach 15 T in the main dipoles and 320 Tm^{-1} in the main quadrupoles.

Should this approach become of interest for the high-energy physics community, it would be mandatory to increase the injection energy in view of reducing the energy swing during the acceleration ramp. Main dipoles of 15 T will at least require doubling the LHC injection energy and installing twice more powerful transfer lines.

A SCENARIO TO UPGRADE OF THE LHC INJECTOR CHAIN

In this section we present initial considerations to upgrade to 1 TeV the energy of the LHC injector chain.

Basic Assumptions

We assume that the low-energy part of the injector chain, including the PS, will be fully refurbished to provide bunches of the appropriate intensity, emittance and brilliance, with the required inter-bunch separation. We also assume a PS extraction energy of 25 GeV.

The PS should be able to provide a bunch population of 2×10^{11} within 3.5 µm emittance and of $2\sqrt{2}\times10^{11}$ within 7 µm, with a bunch separation of 12.5 ns (or 10 ns, if the impact on RF system should be minimised).

Energy swing

To accelerate particles from 25 to 1000 GeV energy we need two rings. Our preferred solution will be to put both of them in the SPS tunnel, one on top of the other. To evenly spread the energy swing, the first ring should reach 150 GeV and the second 1 TeV. Since the SPS tunnel is about 6 km long, the first ring will use normal conducting and the second superconducting magnets. With the present dipole occupancy-factor we will need SC dipoles with $B_{oper.} \approx 4.5$ T operating field and $B_{peak} \approx 5.0$ T peak field for the higer-energy ring, assuming 10 % operational margin.

Superconducting magnet aperture

To compute the magnet aperture we should make assumptions on lattice and beam parameters of the SPS at 1 TeV. We expect the optical layout to be very close to the present one, with $\beta_{max} \approx 100$ m, $D \approx 4$ m and a peak value for the closed orbit of about 5 mm. We also assume a normalized emittance of $\varepsilon_n \approx 7$ µm and a relative bucket height of $\delta_{bucket} \approx 10^{-3}$.

The rms transverse beam size will be of $\sigma \approx 2.2$ mm at injection and ≈ 0.8 mm at top energy. The dispersive beam size will be at most of $D\delta_{bucket} \approx 12$ mm, the betatron beam size (at $6\times\sigma$) ≈ 13 mm at injection and ≈ 5 mm at top energy. If a slow extraction will be required, the resonant separatrix size should be ≈ 20 mm long. Finally a radial clearance of at least 6 mm should be kept from the beam pipe.

Adding in quadrature the betatron and the dispersive beam size and linearly the closed orbit, the separatrix size, and the clearance one will need a radial aperture of at least 29 mm at injection and 44 mm at top energy.

Our conclusion is that an inner coil diameter of 70 to 100 mm should be adequate for the SC magnets.

Ramp rate and cycle duration

The ramp rate and cycle duration of the 1 TeV SPS will be determined by the envisaged use of the accelerator complex, beyond the pure LHC filling. We expect that a high duty cycle and a fast ramping rate will be required to maximise the availability of test beams and the potential physics reach with extracted beams.

A cycle 10 s is compatible with ascending and descending ramps of 3 s and with rates of 1 to 1.5 Ts^{-1}. In these conditions, one can envisage an injection plateau of 3 s and a flat top of 1 s for a slow extraction.

The main issue will be the cryogenic load induced by the ramp transient and by beam loss. Ideally the two contributions should be as small as possible and eventually of the same order of magnitude.

With a well thought collimation system, we hope reducing to about 20 kW the beam loss in the superconducting material. This corresponds to about 10^{12} protons of 1 TeV escaping from the collimation system in a 10 s cycle. The escaping halo will be lost in the whole accelerator, depositing about 5 W per meter of superconducting magnet.

Correspondingly, the thermal loss induced by the magnetic cycle should be of the same order of magnitude, which is by far not an easy goal. With the present SC technology at the requested ramp rate the thermal loss are an order of magnitude larger and an aggressive R&D program will be necessary to master the problem [5].

OPEN ISSUES

The scenarios of LHC upgrade presented here are still superficial sketches needing several clarifications and improvements. A non-exhaustive list of the open issues is given below and may need an important effort to be fully addressed.

- Installation staging in the SPS tunnel: it is required to propose scenarios to minimise the SPS shutdown during the upgrade of the injection chain.
- Lattice design: a realistic lattice design of the SPS and of its injector should be investigated, also considering the partial use the present SPS ring.
- Slow extraction design: we need a realistic simulation of the resonant extraction process in the 1 TeV SPS, also to compute the expected extraction loss and to refine the estimate of the magnet aperture.
- Injection optics: this issue is relevant both for the 1 TeV SPS and for the LHC itself since the space available in the extraction/injection regions is limited.
- Optimal extraction/injection channel: extraction kickers and septa will operate on more energetic particles within serious space occupancy constraints.
- Optimal design for the SC magnets: nominal parameters should be proposed for the SC pulsed magnets and a road map for the requested R & D presented.
- Cryogenic system: solution should be investigated for the installation of cryogenics in the SPS tunnel.
- RF systems: the optimal choice of the RF parameter is not yet available. We expect serious constraints, which may require a full iteration both on lattice design, on the choice of the injector energy and of the ramp-rate of the magnetic cycle.

ACKNOWEDGMENTS

We gratefully acknowledge F. Ruggiero, F. Zimmermann and D. Tommasini from CERN for discussions and suggestions.

REFERENCES

1. F. Ruggiero and F. Zimmermann, "Possible scenario for an LHC upgrade", CARE-Conf-05-002-HHH, Proceedings of the HHH 2004 Workshop, Geneva, November 2004, http://care-hhh.web.cern.ch/care-hhh/HHH-2004/Proceedings/proceedings_hhh2004.htm.
2. F. Ruggiero, "Parameters for First Physics and for 10^{33}", LHC Performance Workshop – Chamonix XII, 2004, http://ab-div.web.cern.ch/ab-div/Conferences/Chamonix/chamx2003/PAPERS/8_2_RF.pdf.
3. J.P. Koutchouk, "Correction of the Long-Range Beam-Beam Effect in LHC using Electro-Magnetic Lenses", IEEE Particle Accelerator Conference (PAC 2001), Chicago, Illinois, 18-22 Jun 2001.
4. F. Zimmermann, "Beam-beam Compensation Schemes", 1st CARE-HHH-APD Workshop on Beam Dynamics in Future Hadron Colliders and Rapidly Cycling High-Intensity Synchrotrons, Geneva, Switzerland, 8-11 Nov 2004.
5. J.E.Kaugerts et al., "Design of a 6T, 1T/S fast ramping Synchrotron Magnet for GSI's Planned SIS 300 Accelerator", IEEE Trans on Applied Superconductivity, Vol 15, No2, June 2005.

First Months of Data Taking at LHC

Fabrizio Parodi

Università di Genova, Dipartimento di Fisica, Via Dodecanneso 33, 16132 Genova

Abstract.
The ATLAS and CMS detector will start taking data at the LHC collider (proton-proton collider working at a center of mass energy of 14 TeV) in summer 2007. In this article I will review the commissioning of the two detectors before the starting of LHC and the analysis of the first pp collisions data ($10\ pb^{-1}$) devoted, mainly, to calibration purposes. I will also briefly review the first physics measurements aiming at the understanding of the detectors performance.

Keywords: ATLAS, CMS, Calibration, Alignment, LHC

INTRODUCTION

This article will review the calibration and alignment strategy of the two LHC (Large Hadron Collider) general purpose experiments, ATLAS [2] (A Toroidal LHC ApparatuS) and CMS [1] (Compact Muon Solenoid), before the beginning of pp collisions and with the first pp collisions data ($10\,\text{pb}^{-1}$). Detector specific procedures will be described for tracker detectors and calorimeters.
Early physics results will be discussed in view of their impact on the detector understanding.

LHC STARTUP SCENARIO

The LHC proton-proton (pp) collider will start operating in spring 2007 with single beam; the first collisions are expected in summer 2007.
The most probable scenario for the LHC startup is:

- 43+43 bunches, luminosity $\mathscr{L} = 6 \times 10^{31} cm^{-2} s^{-1}$: tuning machine parameters
- pilot run: 963+963 bunches, bunch crossing 75 ns, luminosity $\mathscr{L} > 5 \times 10^{32} cm^{-2} s^{-1}$
- 2-3 months shutdown
- physics run ($\sim$7 months): 2808+2808 bunches, bunch crossing 25 ns, luminosity $\mathscr{L} = 2 \times 10^{33} cm^{-2} s^{-1}$

The data collected before the shutdown (corresponding to $\sim 10 - 100\ pb^{-1}$) will be used for calibration purpose while the physics run should accumulate about $10\ fb^{-1}$.

STATUS OF THE EXPERIMENTS AT THE STARTUP

The experiments will not be complete at the startup:

- CMS: RPC muon chambers will cover only $|\eta| < 1.6$ (instead of 2.1), the pixel detector and the end-cap electromagnetic calorimeter (ECAL) will be installed only during the first shutdown;
- ATLAS: the Transition Radiation Tracker (TRT) will cover only $|\eta| < 2.0$ (instead of 2.4).

Both detectors will have reduced trigger bandwidth at the startup due to deferrals on the High Level Trigger (HLT) processors.

PRE-COLLISIONS PHASE

The first opportunity to understand the detector before the commissioning with real collisions is represented by the cosmics run and the events collected with one single beam circulating in the collider.

The run on cosmics will allow initial alignment and timing of the detector with particles, debugging of the sub-systems including mapping of dead channels.

When only one beam is circulating two type of interactions are possible: beam-halo, beam-gas. The first process produces muon particles with low transverse momentum with respect to beam direction (p_T) and will be mainly used to align and calibrate the end-caps. The second process due to the non perfect vacuum in the pipe consists in the interaction of one proton of the beam (7 TeV) with a nucleon at rest: this kind of events is similar to the collisions events but has a much softer spectrum.

General strategy to calibrate and align in the pre-collisions phase

The strategy to achieve a good precision in calibration and alignment begins with strict quality control on the tolerances of the detector construction and proceed through the following steps:

- redundant hardware calibration and alignment systems;
- extensive test beam characterization of the prototypes and final modules;
- "in situ" detector calibration: cosmics runs (end 2006-2007), single beam runs during LHC commissioning

The procedure is valid for all sub-detectors, tracking detectors, hadronic and electromagnetic calorimeters, muon chambers; as an example I will focus on inner silicon trackers and electromagnetic (EM) calorimeters.

Initial alignment of the Inner Silicon Trackers

After installation the expected alignment for the Inner Silicon Trackers is dominated by the construction tolerances and it is very similar for the two experiments:

- ATLAS: the module will be positioned on supports with a precision of $17-100$ μm, while the support will be positioned in the nominal position within $20-200$ μm. Overall the Inner Detector will be centered on the beam axis with a tolerance of ± 3 mm and with a rotation angle with respect to the solenoid axis that should smaller than 1 mrad.
- CMS: construction tolerances within $1-2$ mm, optical alignment with respect to fixed points will guarantee an initial precision of 100 μm

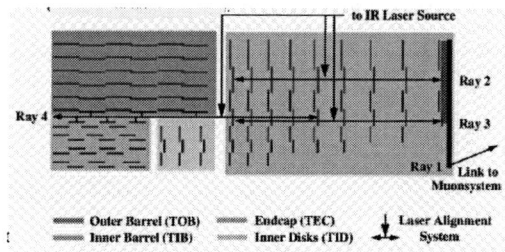

FIGURE 1. CMS Inner Tracker

With the foreseen misalignment tracks can be reconstructed with an efficiency of 40-60%.

Pre-collisions alignment of the ATLAS ID

As an example of the alignment procedure before pp collisions the ATLAS strategy for Inner Detector is presented:

- cosmics (about 1 tracks/sec in silicon detector (Pixel+SCT) and TRT): enough statistics for debugging readout and dead modules; the relative position of the Inner Detector with respect to the calorimeter and the muon spectrometer will be checked and first alignment studies performed;
- beam-gas: about 25 Hz of reconstructed tracks with $p_T > 1$ GeV and $z < 20$ cm; the statistics accumulated in 2 months will correspond to 10^7 tracks similar to the produced in collisions events and should be enough to achieve alignment precision exceeding the initial survey ($10-100$ μm).

Calibration of the ATLAS EM calorimeter

The ATLAS EM calorimeter is a Pb-liquid argon calorimeter with accordion shape, the main requirement, driven by $H \to \gamma\gamma$ search, is to have uniformity in the response within 0.7% over $|\eta| < 2.5$. The steps need to achieve this results are[3]

- Tight control mechanical tolerances: 1% more lead in cell leads to a response drop of 0.7%. The plate thickness has to be controlled to 0.5% (~ 1 μm): a measurement of the thickness (on 1536 absorber plates) has shown a spread of 9 μm over an average value of 2.2 mm well matching the requirement.
- Test beam uniformity studies: 4 barrel modules (out of 32) and 3 end-cap modules (out of 16) have been tested. The uniformity over "units" of size $\Delta\eta \times \Delta\phi = 0.2 \times 0.4$ turned out to be ~ 0.5% (Figure 2)

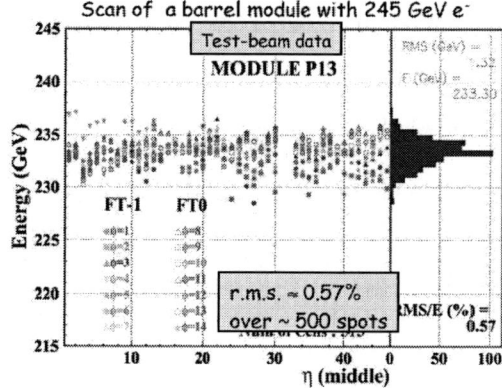

FIGURE 2. ATLAS EM calorimeter: energy response uniformity response as a function of η.

- Calibration check with cosmic muons: expected $\sim 10^6$ events in ~ 3 months data taking. From test-beam results this statistics will allow to check calorimeter response up 0.5%.

The uniformity is mainly related to the constant term in the energy resolution $c_{tot} = c_L + c_{LR}$:

- c_L: local term (checked using test-beam results);
- c_{LR}: long-range response non uniformities from unit to unit. It depends on module-to-module variations and different upstream material.

before the pp collisions $c_L \sim 1.3\%$ $c_{LR} \sim 1.5\%$ are expected giving c_{tot} about 2%. Without further corrrection with real data the impact on the $H \to \gamma\gamma$ significance at $m_H \sim 115$ GeV is degraded by 25% (50% more integrated luminosity for discovery).

TRIGGERING THE FIRST COLLISIONS

The initial trigger menu has the same signatures (with relaxed cuts) as the initial luminosity menu, each item accounts for about 10 Hz:

- μ: $p_T > 10 - 20$ GeV (good efficiency on W and Z, the rate is saturated by b-jets), $p_T > 10 - 20$ GeV (need to select J/ψ, it has to be prescaled);
- $e\gamma$: $E_T > 10$ GeV (W,Z, calibrations)
- Jet: $E_T > 150$ GeV (QCD studies, calibrations)
- Missing energy: jet $E_T > 40 - 50$ GeV + E_T missing $> 40 - 50$ GeV
- Minimum bias (prescaled)

Assuming a luminosity of $\mathscr{L} = 10^{31} cm^{-2} s^{-1}$ and an output rate of the High Level Triggers of 100 Hz the pilot run should collect 10 pb^{-1}.

CALIBRATION AND ALIGNMENT WITH FIRST DATA

Inner Silicon Tracker alignment (ATLAS)

The alignment can be performed using all the tracks or only the overlaps of tracks in the same module; the statistics needed to align pixel at $1 - 2\mu m$ and SCT at $2 - 3\mu m$ can be collected in few days.

The most important effect will come from systematics: to bring them under control the information coming from monitoring have to be integrated in the alignment strategy. Thermal instability (highly probable at the beginning) will be relevant below 100 μm.

EM Calibration (CMS)

The channel $Z \to e^+e^-$ will be used to intercalibrate different regions of the calorimeter and to set absolute energy scale. The long range term in the energy resolution will be reduced to design value of 0.5% in the first month in both experiments.

The ϕ symmetry of deposited energy in minimum bias event will be used to intercalibrate crystal of the CMS EM calorimeter within rings of constant η. The precision will only be limited by the inhomogeneity of the material. This method[4] can provide precision at percent level in a few hours (Figure 3).

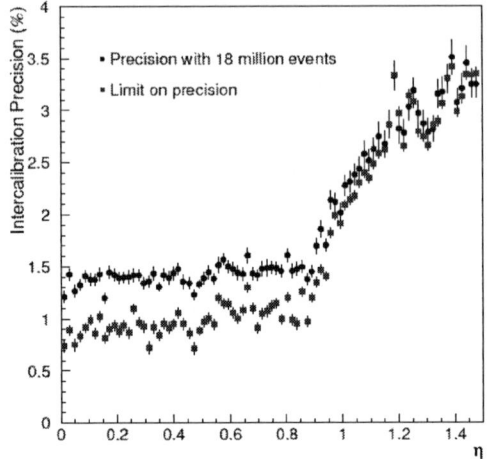

FIGURE 3. CMS EM calorimeter: intercalibration precision as function of η

EARLY PHYSICS

The first collected data are potentially rich in interesting physics events (Table 1) but it is important to remember that most of those will be probably collected with uncalibrated and misaligned detectors.

TABLE 1. Expected physics events in the first 10 pb^{-1}

Process	Events/s	Events in 10 pb^{-1}	Statistics from previous exp.
$W \to e\nu$	0.07	7×10^4	10^4 LEP 10^6 Tevatron
$Z \to ee$	0.011	10^4	10^6 LEP
$tt \to \mu X$	0.008	8×10^3	10^6 Tevatron
$H(130)$		100	

It is hence important to identify the road to move from calibration to physics.

From calibrations to physics

The physics processes will be initially used to calibrate the detectors then to understand Standard Model (SM) at the TeV scale and finally to characterize the backgrounds to New Physics).

- Calibrate and understand detectors and trigger using known processes
  - $Z \to ee$, $\mu\mu$: calibration and alignment of tracking detectors, calorimeters and muon detectors.
  - $tt \to b\ell\nu bjj$: hadronic scale and b-tagging performance
- Understand SM physics process at $\sqrt{s} = 14$ TeV and compare them with Monte Carlo (MC)
  - cross section measurements (W, Z, tt, QCD) at 10-20% precision
  - top mass measurement at 5-7 GeV
- Prepare the road to discovery:
  - measure backgrounds to New Physics: tt and W/Z + jets
  - look at specific "control samples" for the individual channels ($ttjj$ with $j \neq b$ "calibrates" the $ttbb$ irreducible background to $ttH \to ttbb$.

Initial measurement of top mass

An example of the above strategy is the measurement of the top mass in the standard channel $tt \to bWbW \to b\ell\nu bjj$ [3]. The selection is very simple and uses the following cuts

- isolated leptons $p_T > 20$ GeV
- 4 jets $p_T > 40$ GeV
- no kinematic fit
- no b-tagging required (pessimistic hypothesis, assumes that the trackers not yet understood)

The invariant mass plot of 3 jets with highest p_T is shown in Figure 4. It can be seen that the top signal will be visible in few days with simple selections and without b-tagging; the cross section wil be known at about 20% precision and the top mass at 7 GeV.

This measurement will be very useful to get feedback on the detector performance: if the top mass will not compatible with the (more precise) world average the jet energy scale have to be checked.

This sample will be also the ideal candidate to commission the b-tagging selection.

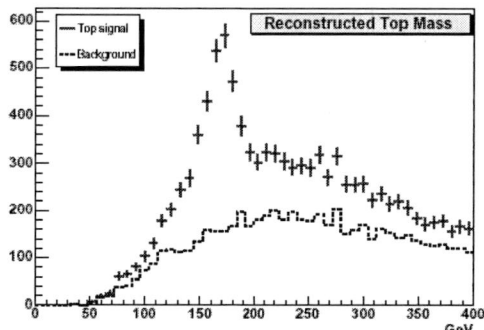

FIGURE 4. Initial measurement of the top mass in ATLAS with 150 pb^{-1}: invariant mass plot of 3 jets with highest p_T.

CONCLUSIONS

The first months of data taking at LHC, as well as the pre-collisions data, will be crucial to calibrate and align the detectors, to tune the trigger and test the data acquisition chain. Moving from calibration to physics the measurement of the known processes of Standard Model physics will be the first step to understand the backgrounds and to prepare the road to discovery.

REFERENCES

1. CMS Collaboration, Technical Proposal, CERN/LHCC/94-38.
2. ATLAS Collaboration, Technical Proposal, CERN/LHCC/94-43.
3. F. Gianotti and M. Mangano, hep-ph/0504221
4. D. Fuytan and C. Seez, CMS NOTE-2002/31.

Results from the Run II at the Tevatron

Franco Bedeschi

INFN - Sezione di Pisa, Largo B. Pontecorvo 3, I-56100 Pisa, Italy

Abstract. We present several new measurements obtained by the CDF and D0 experiments from the data samples collected until August 2004 at the Tevatron. These results include: jet and B meson cross-sections, limits on B_s mixing and BR($B_s \to \mu\mu$), improved top mass measurement and searches for new physics. Extrapolations of the future physics reach at the Tevatron are given for the most relevant cases.

Keywords: Tevatron, CDF, D0, Jets, B Physics, B Mixing, Rare Decays, W Boson, Z boson, Supersymmetry, Higgs, Top Quark
PACS: 13.85.-t

INTRODUCTION

We discuss results obtained from the study of proton-antiproton collisions at 1.96 GeV center of mass energy produced by the Tevatron collider at Fermilab. Two experiments, CDF and D0, operate in each of the two collision points.

The Tevatron luminosity has been steadily increasing since the start of operation in its upgraded configuration in 2001 (Run II). Maximum initial store luminosity of 1.3×10^{32} cm^{-2}sec^{-1} has been reached recently. This is close to the asymptotic expectation of 3×10^{32}. At the time of this writing the total delivered luminosity per experiment was about 1 fb^{-1} and $\sim$80% has actually been recorded to tape.

The CDF and D0 experiments have been quite significantly upgraded since Run I, in particular with regards to their tracking, trigger and DAQ systems. Currently available results are typically based on data collected until August 2004 which amount to 300 - 400 pb^{-1}.

PHYSICS RESULTS

The study $p\bar{p}$ collisions gives access to a wide range of physics topics. In the following sections we present highlights of the most relevant results currently obtained.

QCD

Hadronic jets are produced at the Tevatron with cross-section at the mbarn level and can therefore be studied with great detail. In particular we compare typical jet distributions with theory predictions. Establishing the reliability of NLO perturbative QCD in the description of jet production is an important test of the theory, which we can extend up to jet energies $\sim$600 GeV. Significant

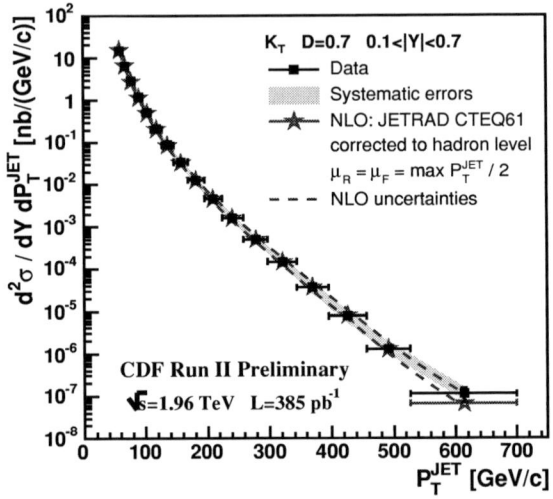

FIGURE 1. Inclusive jet production compared with QCD prediction.

discrepancies could signal the presence of new physics. In addition Tevatron data are sensitive to the gluon pdf at high x and Q^2 in a region only partially covered by the measurements at HERA.

In fig. 1 we show a comparison of the jet E_T distribution, as obtained by CDF, with the theory predictions indicating good agreement within the quoted errors.

Many other aspects of jet productions have been studied so far including angular correlations, production of heavy flavored jets and jet production in association with other particles such as W and Z bosons. All in all the agreement with QCD is quite consistent and these studies have been very useful in tuning the available Monte Carlo's and to understand their range of validity.

B physics

The production of b quarks at the Tevatron is only about a factor 10 less likely than ordinary jets. This results in a large variety of b physics being accessible to the experiments provided they are able to trigger on the events of interest. Major improvements to the b triggers have been included in the Run II detectors. In particular CDF is operating a secondary vertex trigger (SVT) that selects events containing tracks with large impact parameter relative to the primary interaction vertex. This allows the collection of large samples of fully hadronic B meson decays previously not accessible at hadron machines.

The areas of b physics covered at the Tevatron include: study of the production mechanisms, spectroscopy, measurements of lifetimes of all b-hadron species, search for rare decays, measurements of mixing in the B_d and B_s sectors and of quantities related to CP violation. We can show here just a few highlights from this large body of results.

with the analysis of Run 1 data [3]. This analysis is being extended now to b-jets of higher transverse momentum and preliminary results indicate the same level of overall consistency between experiments and theory calculations.

The search for rare decays of B mesons into leptonic final states is particularly favorable at the Tevatron due to the large b cross-section and the good signal to noise provided by the leptonic signatures of the events of interest. In the special case of the decays $B_d(s) \to \mu^+\mu^-$ the theoretical calculation for the branching ratio is in the $10^{-10}(10^{-9})$ range [4]. Indeed within the Standard Model these decays can occur only via higher order loop diagrams involving heavy quarks and vector bosons. These branching ratios could be substantially higher if we allow some of the virtual exchanges to be made in a non-standard way such as with supersymmetric particles. A measurement or a limit on these branching ratios can therefore seriously constrain the parameter space of supersymmetry. Both the CDF and D0 Collaboration have recently set interesting limits on these branching ratios in the 10^{-7} range [5] and it is projected that a limit on the $B_s \to \mu^+\mu^-$ at the 10^{-8} level should be feasible after a few more years of data taking.

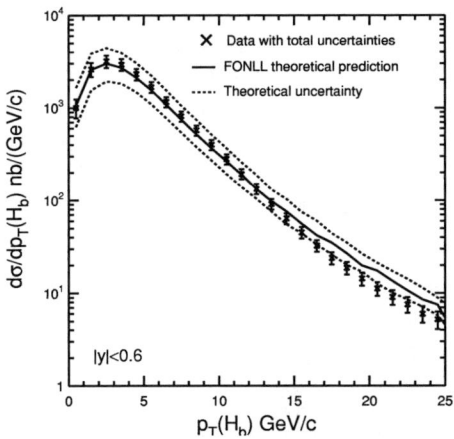

FIGURE 2. Differential cross-section of b-hadron production as a function of b-hadron p_t. The solid line is the FONLL calculation outlined in [2], the dashed line is the theoretical uncertainty.

Starting from a sample of a few million J/ψ's the CDF collaboration has measured [1] the differential b-hadron (see fig. 2) and b quark production cross-sections for transverse momentum, p_t from 0 to ~25 GeV/c. This is a high statistics measurement that for the first time is extended up to 0 transverse momentum. This measurement is in excellent agreement with recently updated QCD calculations [2] which include: resummation of leading logs, up-to-date parton pdf's and different approach in the extraction of the b fragmentation function from LEP data. This resolves a longstanding discrepancy between theory predictions and experimental results that started

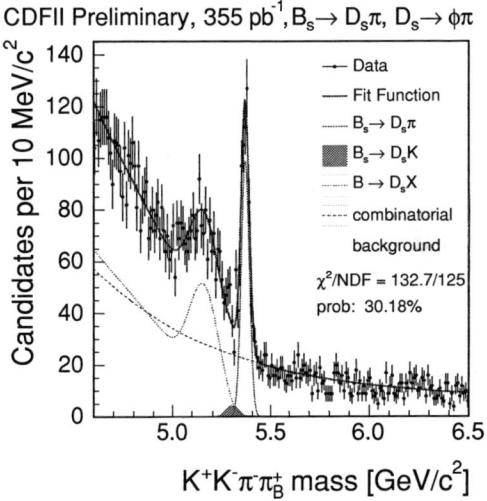

FIGURE 3. $B_s^0 \to D_s\pi$ ($D_s \to \varphi\pi$) signal observed by CDF.

B_d and B_s mesons can turn into their anti-particle with a frequency which is related to the CKM matrix elements V_{td} and V_{ts} respectively. A measurement of both mixing frequencies, Δm_d and Δm_s, would yield a measurement of the ratio of these two elements with about 5% theory uncertainty, thus providing a strong constraint in global fits of the Unitary Triangle [6]. Very precise measurements of Δm_d have been available for some time, and are currently dominated by the results of the B-factories [7]. Recent CDF and D0 measurement [8] are consistent with those results.

Δm_s is constrained to be larger than 14.5 ps^{-1} at the 95% C. L. from previous work by the LEP experiments, SLD and CDF. Both CDF [9] and D0 [10] have recently obtained new limits on Δm_s which significantly improve the Run 1 results but do not have yet much effect on the world average. In fig. 4 we show the result of the amplitude scan performed by the CDF experiment using a combination of semileptonic and fully reconstructed hadronic B_s decays. An example of the hadronic signals used for this analysis is shown in fig. 3.

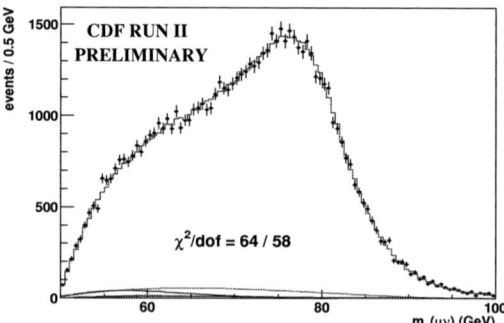

FIGURE 5. The M_T distribution in W boson decays to muons. The points represent the data, the histogram the simulation with the backgrounds added. The region between 60 and 90 GeV is used to fit the W mass.

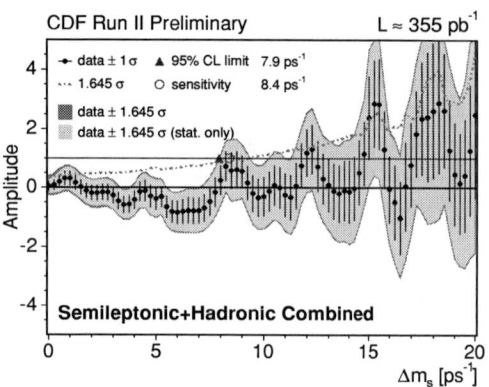

FIGURE 4. Amplitude scan combining hadronic and semileptonic B_s^0 signals by CDF.

Beyond just adding statistics, major improvements to the current analysis can be achieved by including more trigger paths and B_s decay modes, an additional reduction of the uncertainty on the measured B_s decay time and by adding same side kaon tagging, which could more than double the total flavor tagging performance. We expect these results to reach the level of the current world average in the course of the next year and to have a positive measurement before the end of Run II, if Δm_s is in the range of current Standard Model expectations.

Electroweak physics

With hundreds of thousands W bosons produced at the Tevatron, CDF and D0 are in a unique position to study in detail the weak boson production and properties. Precise measurements of W and Z production cross-sections and asymmetries [11] as well as low statistics studies of di-boson production [12] have been done and will continue with increasing precision as more data becomes available. So far these results are in excellent agreement with the predictions of the Standard Model.

Much work is ongoing to improve the current precision on the measurement of the W boson mass [13]. A fit of the W transverse mass distribution has been done using 200 pb^{-1} of data by CDF. The fit result for the muon channel is shown in fig. 5. A result however has not been released yet as we are still working on a further reduction of the systematic uncertainty. The total error currently evaluated after combining the electron and muon channel is 76 MeV. We expect to reduce this error by about 50% by the end of Run 2.

The top quark discovery has been firmly reconfirmed in Run 2 by both experiments. Exhaustive measurements of the top pair production cross-section using several different final states and signal selection methods (see ref. [15] for a brief review) have been performed yielding results consistent with recent theoretical predictions [14].

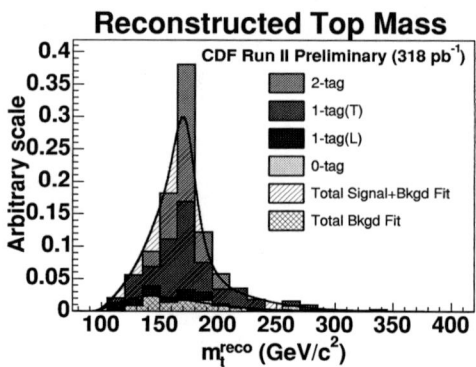

FIGURE 6. Reconstructed top mass distributions for several classes of events used in the mass fit by the CDF collaboration.

These top signals are being used also to measure the top quark mass. This is an important parameter that is needed for many calculations, moreover in association with the W mass it can provide model dependent constraints to the Higgs mass. CDF has recently updated its top mass measurement using the final state where one top quark decays to three hadronic jets and the other to a jet, a lepton and a neutrino. For the first time in this measurement the W mass constraint has been used to reduce the jet energy scale uncertainty, which has also been independently and significantly improved recently to a level of ∼3% as shown in fig. 7.

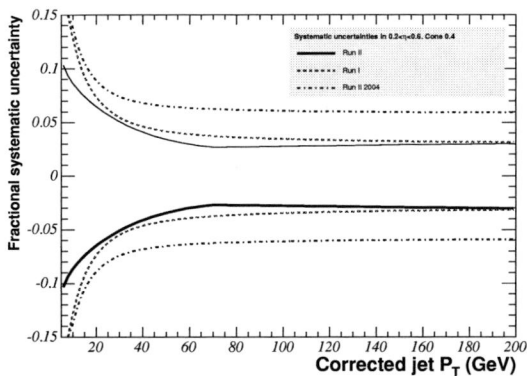

FIGURE 7. Fractional energy scale uncertainty as a function of jet p_t at CDF. The continuous curve is the most recent result. For comparison the curve for Run 1 (dashed) and that used for previous CDF Run 2 results (dot-dashed) are shown.

This measurement returns a top quark mass of $173.5^{+4.1}_{-4.0}$ (total) GeV/c^2. This is the best single measurement in the world and it is more accurate than the previous world average from Run 1. The CDF projection for the total expected top quark mass resolution as a function of the integrated luminosity is shown in fig. 8.

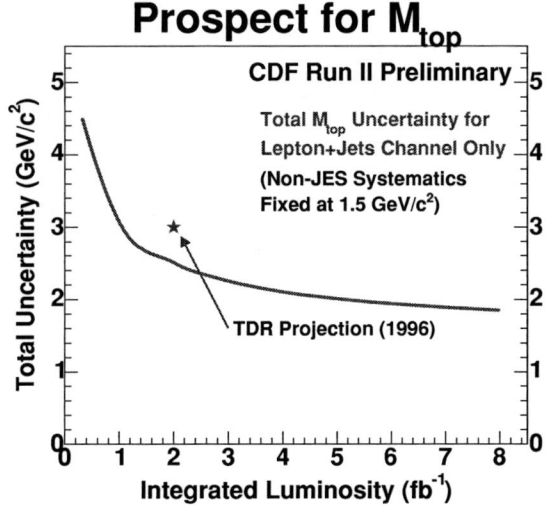

FIGURE 8. Total top mass error expected from the CDF experiment as a function of the total integrated luminosity.

The global electroweak fits have been recently updated [16] to include this measurement and yield for the Standard Model Higgs: $M_{Higgs} = 98^{+52}_{-36}\ GeV/c^2$ ($M_{Higgs} < 208\ GeV/c^2$ at 95% CL).

The search continues for single top production. While no experiment has yet observed a signal, the experimental limits on the cross-section are getting tighter. The current best results [17] are from the D0 experiment that using a neural net to separate signal and background sets 95% CL limits of 6.4 and 5.4 pb for the s-channel and t-channel production modes respectively.

New particle searches

Hints for new physics are being searched for at the Tevatron in a number of possible final states. So far no evidence for discrepancies from Standard Model behavior have been observed, however these searches are improving in many cases the previously available experimental limit on masses of new particles and cross-sections of many non-standard processes. We expect these searches to become more sensitive as more data is accumulated.

Some noteworthy examples of this ongoing work are the searches for supersymmetry in the three lepton plus missing E_T final state and the multiple hadronic jet plus missing E_T final state. The former being a very background free signature of chargino and neutralino production, the latter of squark and gluino production. Current best results are from the D0 experiment [18]. Based on the observation of three tri-lepton events with an expected background of $2.93 \pm 0.54 (stat) \pm 0.57 (syst)$ events, cross-section limits on chargino-neutralino production are set for various supersymmetric models. As shown in fig. 9.

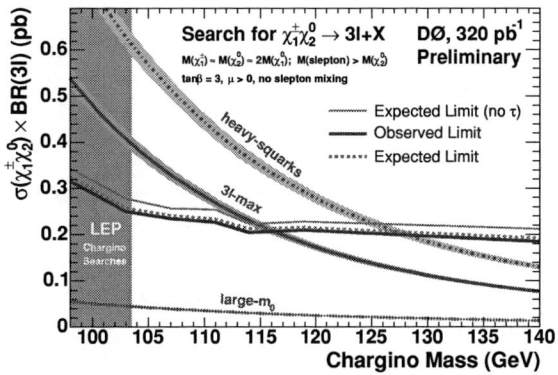

FIGURE 9. Limit on $\sigma \times BR(3l)$ as a function of the chargino mass in comparison with the expectations from several SUSY scenarios.

In fig. 10 we show results of a D0 search [19] for squarks and gluinos interpreted in the context of an mSUGRA model. Their observations are consistent with the Standard Model and further constrain the theory relative to previous results.

In spite of the limited available statistics both experimental collaborations are strongly pursuing any evidence for Standard Model Higgs boson production. For masses below 130 GeV/c^2 the highest sensitivity comes from Higgs produced in association with a vector boson, with the Higgs decaying to a $b\bar{b}$ pair, while at larger masses

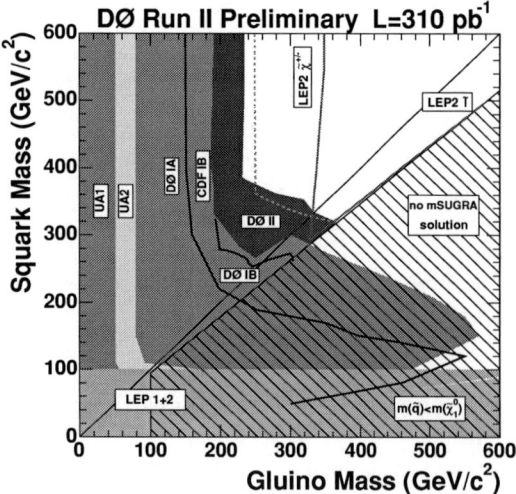

FIGURE 10. In the squark and gluino mass plane, excluded regions are at the 95% confidence level in the mSUGRA framework for $\tan\beta = 3$, $A_0 = 0$, $\mu < 0$.

direct Higgs production followed by its decay into pairs of vector bosons is preferable. Searches are carried on using many final states corresponding to various leptonic decay modes of the vector bosons involved. A summary of the current situation including both CDF and D0 results is shown in fig. 11.

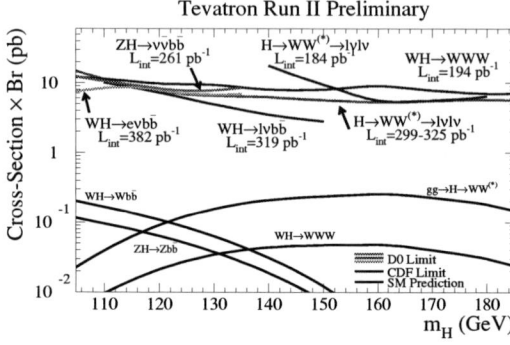

FIGURE 11. Summary of current limits on Higgs boson production cross-sections compared to Standard Model expectations.

CONCLUSIONS

The Tevatron has solved its initial problems and is currently delivering luminosity at the expected rates. The instantaneous luminosity is continuously growing thanks to the aggressive ongoing accelerator upgrade program. Given the current performance we can expect the Tevatron to deliver in the order of 8 fb^{-1} by the end of Run II.

The CDF and D0 detectors are working extremely well and are delivering physics results of high quality on a wide range of physics topics.

By the end of Run II we expect that Tevatron experiments will provide:

- detailed tests of QCD;
- top quark mass measurement with resolution ~ 2 GeV/c^2 per experiment;
- W mass measurement with resolution ~ 30 MeV/c^2 per experiment;
- B_s mixing observation and many more results in the B sector;
- strong limits on new physics in many different final states;
- the development of a sophisticated technology for Higgs searches in many final state configurations.

With some luck and a consistent Tevatron performance we may observe a low mass Higgs boson or indications of new physics before the LHC turn on.

REFERENCES

1. D. Acosta et al., The CDF Collaboration, *Phys. Rev.* **D71**, 032001 (2005).
2. M. Cacciari, S. Frixione, M. L. Mangano, P. Nason and G. Ridolfi, *JHEP* 0407:033, 2004.
3. T. Affolder et al., The CDF Collaboration, *Phys. Rev.* **D65**, 052006 (2002).
4. G. Buchalla and A. J. Buras, *Nucl.Phys.* **B400**, 225 (1993); G. Buchalla, A. J. Buras and M. E. Lautenbacher, *Rev. Mod. Phys.* **68**, 1125 (1996).
5. V. M. Abazov et al., The D0 Collaboration, *Phys. Rev. Lett.* **94**, 071802 (2005); D. Acosta et al., The CDF Collaboration, *Phys. Rev. Lett.* **93**, 032001 (2004); The CDF Collaboration, http://www-cdf.fnal.gov/physics/new/bottom/050407.blessed-bsmumu/bsmumupub_v01.pdf.
6. M. Bona et al., The UTfit Collaboration, arXiv:hep-ph/0501199; J. Charles et al., The CKMfitter Group Collaboration, arXiv:hep-ph/0406184.
7. K. Abe et al., The BELLE Collaboration, arXiv:hep-ex/0408111; B. Aubert et al., The BABAR Collaboration, arXiv:hep-ex/0408039.
8. The CDF Collaboration, February 2005, http://www-cdf.fnal.gov/physics/new/-bottom/050224.blessed-mix-bld/bmix-ld-pub.ps, http://www-cdf.fnal.gov/physics/new/bottom/050224.blessed-hadr-ost/mix-excl-pub.ps;

The D0 Collaboration, August 2004,
http://www-d0.fnal.gov/Run2Physics/WWW/results/prelim/B/B15/B15.pdf

9. F. Bedeschi, arXiv:hep-ex/0505015.
10. S. Burdin, arXiv:hep-ex/0505043.
11. D. Acosta *et al.*, The CDF Collaboration, Submitted to PRD-RC, arXiv:hep-ex/0501023;
 D. Acosta *et al.*, The CDF Collaboration, *Phys. Rev. Lett.* **94**, 091803 (2005);
 D. Acosta *et al.*, The CDF Collaboration, *Phys. Rev. D***71**, 052002 (2005).
12. D. Acosta *et al.*, The CDF Collaboration, FERMILAB-PUB-05-009-E, Submitted to PRL;
 D. Acosta *et al.*, The CDF Collaboration, Submitted to PRD-RC, arXiv:hep-ex/050121;
 D. Acosta *et al.*, The CDF Collaboration, *Phys. Rev. Lett.* **94**, 041803 (2005).
13. C. Hays, arXiv:hep-ex/0505064
14. M. Cacciari *et al.*, *JHEP* **404**, 68 (2004).
15. J. Nielsen, arXiv:hep-ex/0505051.
16. M. Grunewald, member of LEPEWWG/TEVEWWG. Private communication.
17. V. M. Abazov *et al.*, The D0 Collaboration, arXiv:hep-ex/0505063.
18. V. M. Abazov *et al.*, The D0 Collaboration, arXiv:hep-ex/0504032;
19. The D0 Collaboration,
http://www-d0.fnal.gov/Run2Physics/WWW/results/prelim/NP/N31/N31.pdf.

B Physics at the B-Factories

Riccardo Faccini

Università "La Sapienza" and INFN Roma
Dipartimento di Fisica, P.le Aldo Moro 2, 00185 Roma
E-mail: rfaccini@slac.stanford.edu

Abstract. After six years of data taking the B-Factories are now providing measurements with an accuracy which is beyond expectations. All the angles and sides of the unitarity triangle are measured to unprecedented accuracy and with several different techniques. This redundancy of measurements, in agreement with the Standard Model, allows to probe models for new physics. This paper summarizes the current results with particular emphasis on novel techniques.

Keywords: B-Factories, Unitarity triangle
PACS: 13.25.Hw, 14.40.Nd

INTRODUCTION

One consequence of the flavor non conservation of the weak decays is that the quarks coupled to the charged currents are flavor eigenstates while the of mass eigenstates are linear combinations of them [1]. Kobayashi and Maskawa [2] extended the Cabibbo mechanism to the 3 quark families introducing an unitary matrix (V) which can be expressed with a minimum of four independent quantities (3 rotation angles and an irreducible phase). The unitarity condition $V^\dagger V = I$ implies that the off diagonal elements of this matrix product are set equal to zero, and in particular that $V_{ub}^* V_{ud} + V_{cb}^* V_{cd} + V_{tb}^* V_{td} = 0$. This condition can be represented in a complex plane ($\rho, i\eta$) as a triangle (the "unitarity triangle", see Fig. 1)

The two existing asymmetric B-factories started their operations in 1999, with the goal of establishing CP violation in *B* decays and measuring sides and angles of the unitarity triangle. Four years after the first observation of CP violation [3] the measurement of $sin2\beta$ has become a precision measurement, new techniques have been found to measure the other two angles of the unitarity triangle, the sides of the triangle are being accurately measured, and hints of new physics are being searched in the consistency of the redundant determinations of the unitarity triangle parameters.

This paper will concentrate on the aspects that have seen the most innovative developments in the last years.

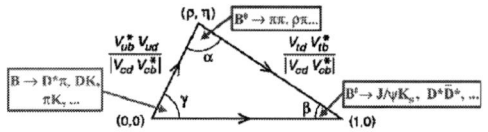

FIGURE 1. The unitarity triangle in the ($\rho, i\eta$) plane.

DATA SAMPLE

The two asymmetric B-Factories, PEPII and KEKB are now running with luminosities well above their design, their peak luminosities being 9.2×10^{33} and 14.0×10^{33} $cm^{-2} s^{-1}$, respectively. PEPII has been inactive from August 2004 to April 2005 due to a safety accident but it is now restarting. This paper, if not otherwise specified, will contain results on 244 (286) fb^{-1} of data collected by *BABAR*(Belle) detector at PEPII (KEKB).

Both detectors [4, 5] are typical hermetic apparata for e^+e^- colliders, made of an inner Vertex Detector,that contributes to vertexing and tracking, a Tracking system that measures angles and momenta of most charged particles, a Čerenkov detector that gives more than 3.6σ separation for particles with $p < 3.5 GeV/c$, an Electromagnetic Calorimeter, a Solenoidal Magnet and a Muon/Hadron system that allows muon and neutral hadron identification.

MEASUREMENTS OF *SIN2β*

The angles of the unitarity triangle can be measured directly in the interference between amplitudes which have different phases but comparable amplitudes. One possible way to achieve this is to consider the time dependent asymmetry in the decay of neutral *B* mesons in to a state which is accessible both by the B^0 and the $\bar{B}^0$ mesons.

In this case if one indicates as $N(t)$ ($\bar{N}(t)$) the number of B^0 ($\bar{B}^0$) mesons decaying into the final state f as a function of the time difference between the decays of the two mesons in the event (t), considering the B^0-$\bar{B}^0$

mixing of frequency Δm, one finds

$$A_{CP}(t) = \frac{N(t) - \bar{N}(t)}{N(t) + \bar{N}(t)} = -C_f \cos(\Delta mt) + S_f \sin(\Delta mt) \quad (1)$$

where $S_f = 2\frac{Im\lambda_f}{1+|\lambda_f|^2}$ and $C_f = \frac{1-|\lambda_f|^2}{1+|\lambda_f|^2}$. Indicating the amplitude for the B^0 ($\bar{B}^0$) meson to decay into f as A_f ($\bar{A}_f$), the observable that carries all the CP violation information is $\lambda_f = \frac{q}{p}\frac{\bar{A}_f}{A_f}$. Depending on the final state f one can probe different angles of the unitarity triangle because the phase of λ_f will be different. Equation 1 shows also that the close λ_f is to one, i.e. the most similar A_f is to $\bar{A}_f$, the highest is the sensitivity to the angles.

Several final states f have been explored so far and the experimental technique is common to all such analyses (see Ref. [6] for details). The final state with the smallest theoretical uncertainties and experimental challenges is the "golden mode" $f = X_{cc}K_S$, where X_{cc} indicates any charmonium state. In this case, in the Standard Model, $\lambda = -e^{-2i\beta}$ and therefore $S_f = \sin 2\beta$, $|\lambda| = 1$. Figure 2 shows the measured time distribution separately in B^0 and $\bar{B}^0$ decays and the corresponding A_{CP}. The sinusoidal behavior of the asymmetry is clearly visible.

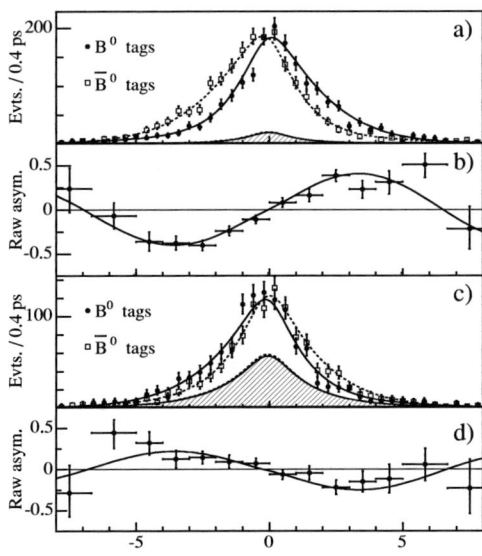

FIGURE 2. Time distribution of $(c\bar{c})K_S$, CP odd (top) and CP even (bottom), events and the raw asymmetry.

BABAR finds [7] $\sin 2\beta = 0.722 \pm 0.040_{Stat} \pm 0.023_{Syst}$ while Belle, based on 152 million $B\bar{B}$ pairs [8] $\sin 2\beta = 0.728 \pm 0.056_{Stat} \pm 0.023_{Syst}$. The World Averaged value

$$[\sin 2\beta]_{WA} = 0.726 \pm 0.037_{(Stat+Syst)}$$

is essentially determined by the B-Factories. This result is in good agreement with the fit to the indirect measurements (Fig. 3) showing that the Standard Model is able to describe CP violation at the few percent level.

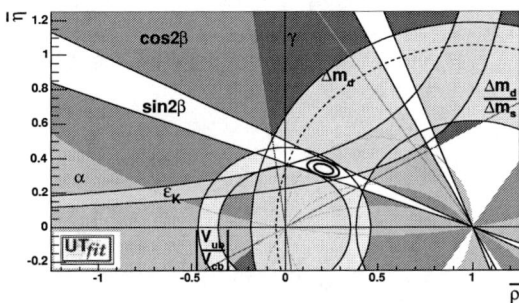

FIGURE 3. Fit to the unitarity triangle apex in the $(\rho, i\eta)$ plane [9]

Modes dominated by $b \to s$ penguins

Time dependent CP violation measurements can also be sensitive to the presence of new physics beyond the Standard Model. In particular those channels that proceed mostly through "penguin" diagrams ($B^0 \to \phi K^0, \eta^{(\prime)}, (KK)_{CP}K^0, \omega K^0, \pi^0 K^0...$, see Fig. 4) are such that in the Standard Model the phase of λ_f is the same as for the "golden mode" (a part from the sign of the CP eigenvalue), but there might be contributions from new physics that spoil this simple relationship. Measuring S_f in all this modes and checking their consistency is sensitive to circulation of new bosonic states in the loops.

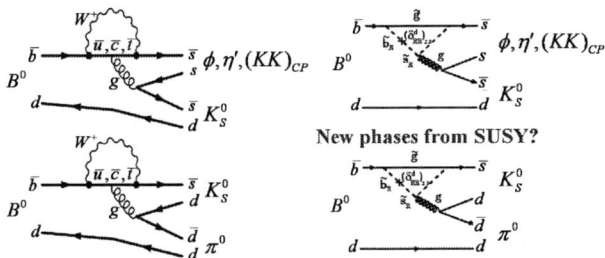

FIGURE 4. Loop contribution to $(b \to (s\bar{s}s))$ and $(u\bar{u}s))$ decays, on the left contribution in SM and on the right hand side a possible SUSY contribution

There are nevertheless both theoretical and experimental difficulties: theoretically some of these modes (e.g. $B \to \omega K^0, \pi^0 K^0,...$) can have significant $b \to u\bar{u}d$ tree contributions. The two amplitudes (A_f and $\bar{A}_f$) do no more have the same phase and therefore even in the Standard Model $|S_f| \neq \sin 2\beta$. This theoretical problem has already been addressed by several authors [10] and the Standard Model predictions have now errors below 10%.

Experimentally, the decays are much rarer and therefore the level of backgrounds is much higher. Fig. 5 shows the quality of the signal for representative modes involving the reconstruction of a K_S and a K_L.

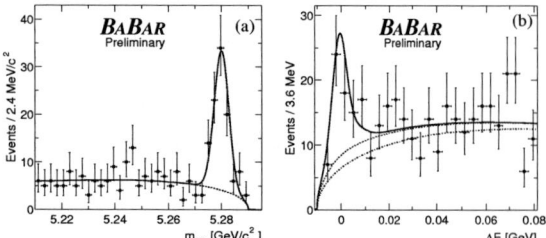

FIGURE 5. Left: B meson invariant mass distribution for ϕK_S candidates. Right: distribution of the difference between measured and expected B meson energy for ϕK_L candidates.

Figure 6 shows a compilation of all the measurements S_f, corrected for the sign of the CP eigenvalue in order to have similar expectation values in the Standard Model. Several observations are interesting:

- although it does not make sense to compare the average among all $b \to s$ penguin modes with the charmonium because there are different contributions, there is a general indication that the $b \to s$ penguin modes have a value of S_f which is lower that the Standard Model prediction. The significance of this statement depends on the hypotheses on the errors on the Standard Model predictions for each mode, but the estimates range around 3σ.

- some of the modes have no prompt tracks (e.g. $B^0 \to K_S \pi^0$). The production vertex reconstruction is possible anyhow utilizing the precision of the silicon microvertex detectors (in particular BABAR's) with a technique pioneered by BABAR in 2003 [11].

- some of the three body modes have a definite prediction for S_f. Although the amplitudes for three body decays are the superposition of amplitudes with different phases, it has been shown recently [12] that in the case in which three neutral pseudoscalars, two of which identical, are involved, the final state is a CP eigenstate. This has allowed to broaden the range of modes of interest.

MEASUREMENTS OF α

The angle α is measured with a time dependent analysis of several charmless B decays. In fact in the case of $B^0 \to \pi^+\pi^-$, considering only the $b \to (u\bar{u}d)$ tree contributions $\lambda_{\pi\pi} = e^{2i\alpha}$ (Fig. 7). Unfortunately, in this case there is more than one amplitude contributing to the decay, since also the so called "penguin" diagrams are giving

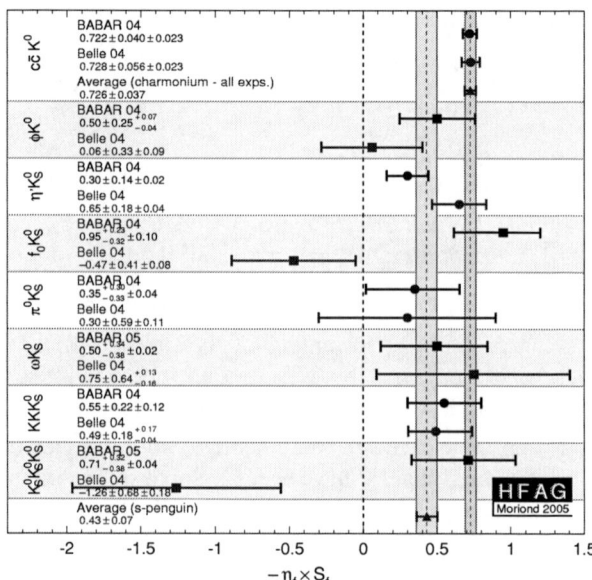

FIGURE 6. Summary of $\sin 2\beta$ measurements

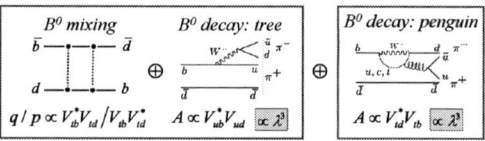

FIGURE 7. Contributions to the $b \to (u\bar{u}d)$ decay: mixing, "tree" and "penguin"

contributions. The time-dependent analysis is therefore not measuring α but an effective angle α_{eff}: $S_{\pi\pi} = \sqrt{1 - C_{\pi\pi}^2} \sin 2\alpha_{eff}$ and $C_{\pi\pi} \propto \sin\delta$. Table 1 summarizes the available results [13] The agreement between the BABAR and Belle results is improving with time.

To extract the strong phase which causes the difference between α and α_{eff}, Gronau and London [14] suggested a method based on full Isospin Analysis. This method requires the measurements of the branching fractions and direct CP asymmetries for $B^+ \to \pi^0 \pi^+$ and $B^0 \to \pi^0 \pi^0$. From a full Isospin Analysis on the inputs shown in Tab. 1 one obtains $\alpha - \alpha_{eff} \leq 35°$ at $68\% C.L.$. The largeness of the $B \to \pi^0 \pi^0$ branching fraction causes a very large uncertainty on α that makes this mode, which was considered golden in the past, quite insensitive to the unitarity triangle angle.

On the contrary, the $\rho^+\rho^-$ is now unexpectedly the best mode because it has been found to be is almost 100 % longitudinally polarized, to have a large branching fraction, and above all small penguin pollution ($B \to \rho^0\rho^0$ has not yet been seen). After ascertaining that the quasi-two-body approximation is accurate for the current

TABLE 1. Branching ratio and CP asymmetry measurements for the $B \to \pi\pi$ and $B \to \rho\rho$ isospin analysis.

	$h = \pi$(BABAR)	$h = \pi$(Belle)	$h = \rho$(BABAR)	$h = \rho$(Belle)
$\mathcal{B}(h^+h^-)(10^{-6})$	$4.8 \pm 0.6 \pm 0.2$	$4.4 \pm 0.6 \pm 0.3$	$30 \pm 4 \pm 5$	-
$\mathcal{B}(h^+h^0)(10^{-6})$	$5.8 \pm 0.6 \pm 0.4$	$5.0 \pm 1.2 \pm 0.5$	$9.4 \pm 1.3 \pm 1.2$	$8.0^{+2.3}_{-2.0} \pm 0.7$
$\mathcal{B}(h^0h^0)(10^{-6})$	$1.17 \pm 0.32 \pm 0.10$	$2.3^{+0.4+0.2}_{-0.5-0.3}$	<1.1@90%C.L.	-
$S(h^+h^-)$	$-0.30 \pm 0.17 \pm 0.03$	$-0.67 \pm 0.16 \pm 0.06$	$-0.33 \pm 0.24^{+0.08}_{-0.14}$	-
$C(h^+h^-)$	$-0.09 \pm 0.15 \pm 0.04$	$-0.56 \pm 0.12 \pm 0.06$	$-0.03 \pm 0.18 \pm 0.09$	-
$C(h^+h^0)$	$-0.01 +/- 0.10 \pm 0.02$	$-0.02 \pm 0.10 \pm 0.01$	$-0.19 \pm 0.23 \pm 0.03$	$0.00 \pm 0.22 \pm 0.03$
$C(h^0h^0)$	$0.12 \pm 0.56 \pm 0.06$	$0.44 \pm 0.53 \pm 0.17$	-	-

statistics, the same strategy as in $\pi\pi$ has been applied to the existing measurements [15] (summarized in Tab. 1). It returns $\alpha = (100 \pm 13)^o$ as shown also in Fig. 8.

There is a third approach to the measurement of α, based on the time-dependent Dalitz plot analysis of $B \to \pi^+\pi^-\pi^0$ decays [16]. It exploits the interference between the different amplitudes that contribute to this decay. The impact of this measurement on α is also shown in Fig. 8.

The combination of these three techniques allows an accurate measurement of α:

$$\alpha = (101^{+16}_{-9})^o \qquad (2)$$

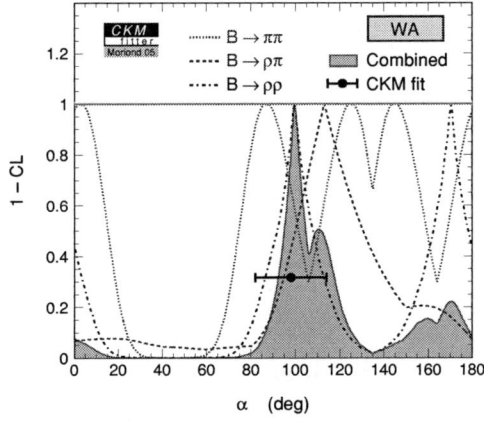

FIGURE 8. Confidence level as a function of the value of α from the $\pi\pi$, $\rho\rho$, and $\rho\pi$ analyses

MEASUREMENTS OF γ

The angle γ, see figure (1), is the phase between the $b \to c$ ($\propto V_{cb}$) and $b \to u$ ($\propto V_{ub}$) amplitudes. The cleanest way to measure γ is to study the $B^+ \to D^{(*)}K^+$ decays. Two different amplitudes generate this decay depending on whether the D meson is a D^0 or a $\bar{D}^0$, the former

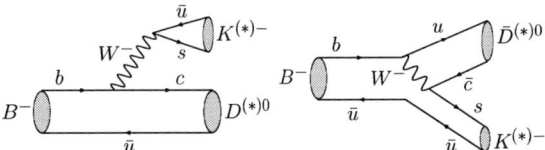

FIGURE 9. $B^- \to D^0 K^-$ and $B^- \to \bar{D}^0 K^-$ processes with D^0 and $\bar{D}^0$ decaying into the same final state interfere giving an asymmetry that contains γ

being suppressed by V_{ub} (see Fig. 9). The relative phase between the two amplitude is the sum of a strong phase, which is the same for B^+ and B^- decays, and the weak phase γ, which depends on the B meson charge.

If one chooses D decay modes (f) allowed to both D^0 and $\bar{D}^0$ mesons, the decay branching fraction and the asymmetry between the B^+ and the B^- decay rates are sensitive to γ. One difficulty in this approach is that the B decay amplitude proportional to V_{ub} is an order of magnitude smaller than the other one ($r_b = A(V_{ub})/A(V_{cb})$ would be, neglecting hadronic effects, around 0.2). This implies the interference between two amplitudes is in general small since the absolute value of their ratio is

$$R = r_b \frac{A(D^0 \to f)}{A(\bar{D}^0 \to f)}. \qquad (3)$$

Moreover, to measure the relative phases it is critical to know the small r_b parameter accurately. Both this aspects are experimentally challenging but these measurements have the advantage of being theoretically very clean since only tree diagrams are involved.

Three are the types of final state f that have been utilized so far, with different sensitivities and problematics:

- *CP eigenstates.* In the so-called "Gronau-London-Whyler" (GLW) [17, 18] method f is a CP eigenstate. Since the amplitudes of the D^0 and $\bar{D}^0$ decay into f are equal, the interference (R in Eq. 3) is small and there is almost no sensitivity to r_b.

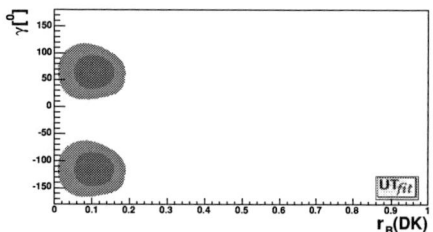 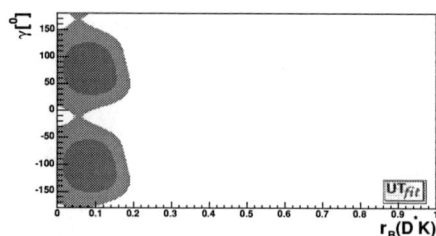

FIGURE 10. *BABAR* and Belle combined results on r_b and γ in $B \to DK$ (left) and $B \to D^*K$ (right).

- *Cabibbo allowed D^0 decay.* In the so-called Atwood-Dunietz-Soni (ADS) [19] method f is chosen in order to maximize the interference since $A(D^0 \to f) >> A(\bar{D}^0 \to f)$ (e.g $f = K^-\pi^+$). The ADS method is very sensitive to r_b but cannot be used by itself to measure γ because of the unknown strong phase in the D decay.
- *3 body decay.* In the so-called "Dalitz" [20] method, whose most powerful application is the case $f = K_S^0\pi^+\pi^-$. In the Dalitz plane $(m_{K_S^0\pi^+}, m_{K_S^0\pi^-})$ there will be regions where f is effectively a CP eigenstate (e.g. $f = K_S^0\rho^0$) and regions where it will be Cabibbo suppressed (e.g. $f = K^{*-}\pi^+$). This method is sensitive to γ and r_b at the same time.

The combined results [21] are summarized in Fig. 10 and correspond to the measurements

$$\gamma = (64 \pm 18)^o$$
$$r_b(DK) = 0.10 \pm 0.04$$
$$r_b(D^8K) = 0.09 \pm 0.04$$

The low r_b values imply that much more statistics is needed for a precision measurement of γ

MEASUREMENTS OF V_{UB}

The accurate measurement of the angles of the unitarity triangle needs to be complemented by the measurement of V_{ub}/V_{cb} because on one side this is an orthogonal limit in the $\rho - \eta$ plane and on the other side it is the measurement that is least affected by new physics and therefore can be used as SM reference.

Since V_{cb} is now known with a precision of about 2% [22], the critical point is the measurement of V_{ub}. This is achieved by measuring the branching fractions of $b \to u\ell\nu$ decays and the measurements can be grouped in two sets, exclusive and inclusive ones.

Exclusive Measurements

In the exclusive measurements the full decay chains $B \to X_u\ell\nu$ where $X_u = \pi,\rho,\omega,\eta'$ have been reconstructed by estimating the four-momentum of the neutrino as the difference of the center-of-mass four-momentum and the total detected one. To fight the copious background from $b \to c\ell\nu$ events and also feed-down from other $b \to \ell\nu$ decays two different approaches are followed: either a tight selection on the lepton energy is applied or the semileptonic decay is reconstructed on the side of fully reconstructed B decays.

Among all possible X_u final states only the pion modes can currently be used to measure V_{ub} since the corresponding form factors are the only one lattice calculations can estimate. Figure 11 summarizes the most recent measurements of $\mathscr{B}B \to \pi\ell\nu$ [23]. Translating the average into a measurement of $|V_{ub}|$ yields

$$|V_{ub}| = (3.75 \pm 0.27^{+0.64}_{-0.42}) \times 10^{-3} \quad (4)$$

where the first error is experimental and the second one is due to the form factors normalization uncertainty.

Inclusive Measurements

In the inclusive measurements only the lepton is identified and the neutrino four momentum is reconstructed as difference between the total and the reconstructed four-momentum. The biggest experimental difficulty in this kind of analyses is the reduction of the $b \to c\ell\nu$ background which is almost two order of magnitudes larger. More or less severe selection criteria need then to be applied which select only a small fraction of the $b \to u\ell\nu$ decays. Several techniques have been developed which have different background levels and different signal efficiencies. Figure 11 summarizes the measured values of V_{ub} [24]. The dominant error is due to the theoretical extrapolation of the branching fraction from the selected phase space to the full one. The biggest uncertainty comes from the ignorance of the b quark fermi motion inside the B meson, which is described by a non-

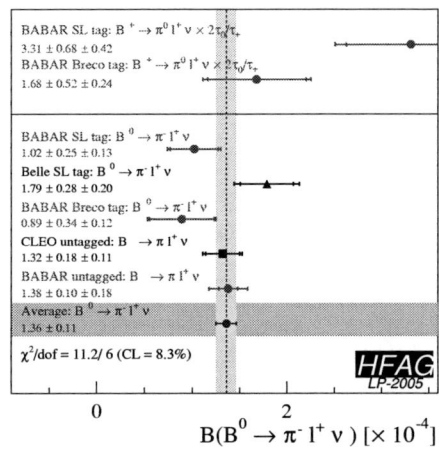 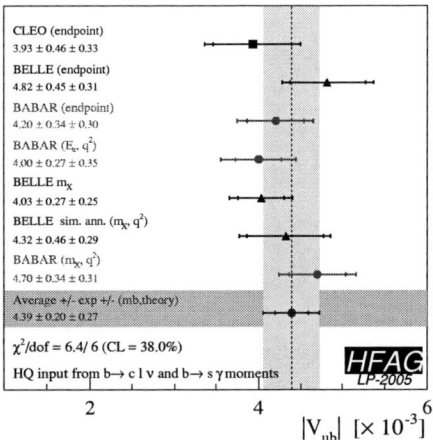

FIGURE 11. Combination of exclusive (left) and inclusive (right) semileptonic measurements.

perturbative function, called Shape Function in literature. There is no theoretical prediction for such a function, but only theoretical relationships between its moments and observables in $b \to c\ell\nu$ and $b \to s\gamma$ decays [25].

The current measurement

$$|V_{ub}| = (4.39 \pm 0.20 \pm 0.27) \times 10^{-3} \qquad (5)$$

has an 8% error, due mostly to the $40\,\text{MeV}/c^2$ uncertainty in the b quark mass (m_b). The discrepancy with the exclusive measurements is no more significant.

SUMMARY

Six years of running of the B-Factories together with novel analysis ideas brought to unexpectedly good accuracies in the measurements of the sides and angles of the unitarity triangle: a 2% accuracy on $|V_{cb}|$ and an 8% accuracy on $|V_{ub}|$, a 10° error on α and an 18° error on γ. The measurement of $sin2\beta$ is now a 4% precision measurent and several modes dominated by $b \to s$ penguins are being considered in order to probe new physics scenarios.

REFERENCES

1. N. Cabibbo, *Phys. ReV. Lett.* **10**, 531 (1963)
2. M. Kobayashi and T. Maskawa, *Prog. Theor. Phys.* **49**, 652 (1973)
3. B. Aubert *et al.* Phys. Rev. Lett. **87**, 091801 (2001); K. Abe *et al.* Phys.Rev.Lett. 87 091802 (2001)
4. B. Aubert *et al.*,Nucl. Instrum. Meth. A **479**, 1 (2002)
5. Nucl. Instrum. Meth. A **479**, 117 (2002)
6. T.E. Browder and R. Faccini Ann.Rev.Nucl.Part.Sci.53:353-386,2003
7. B. Aubert *et al.* Phys.Rev.Lett.94:161803 (2005)
8. K. Abe *et al.* Phys.Rev.D71:072003(2005)
9. M. Ciuchini *et al.* JHEP 0107 (2001) 013, J. Charles *et al.* Eur.Phys.J.C41:1-131,2005
10. e.g. M. Beneke, hep-ph/0505075; M. Gronau and J. L. Rosner Phys.Rev. D71 (2005) 074019, M. Ciuchini, *et. al.* AIP Conf.Proc.698:422-426,2004 and references therein
11. B. Aubert *et al.* Phys.Rev.Lett. **93** (2004) 131805
12. T. Gershon and M. Hazumi Phys.Lett. B596 (2004) 163
13. B. Aubert *et al.* hep-ex/0501071; K. Abe *et al.* hep-ex/0502035; B. Aubert *et al.* Phys.Rev.Lett.94:181802,2005; K. Abe *et al.* Phys.Rev.Lett.94:181803,2005;
14. M. Gronau and D. London,Phys. Rev. Lett. **65**, 3381 (1990)
15. B. Aubert *et al.* Phys.Rev.Lett.94:131801,2005;B. Aubert *et al.* Phys.Rev.Lett.93:231801,2004; B. Aubert *et al.* hep-ex/0408032 ;A. Gordon *et al.* Phys.Lett.B542:183-192,2002; B. Aubert *et al.* 0503049.
16. B. Aubert *et al.* hep-ex/0408099
17. M. Gronau and D. London.,Phys. Lett. B **253**, 483 (1991)
18. M. Gronau and D. Wyler,Phys. Lett. B **265**, 172 (1991)
19. D. Atwood, I. Dunietz and A. Soni,Phys. Rev. Lett. **78**, 3257 (1997)
20. A. Giri, Y. Grossman, A. Soffer and J. Zupan,Phys. Rev. D **68**, 054018 (2003)
21. M. Saigo *et al.*, Phys. Rev. Lett. 94 091601 (2005); B. Aubert *et al.* hep-ex/0504047; B. Aubert *et al.* hep-ex/0408082; A.Poluektov*et al.*, Phys. Rev. D70 072003 (2004); B. Aubert *et al.* hep-ex/0504039
22. B. Aubert *et al.* Phys. Rev. Lett. **93**, 011803 (2004)
23. B. Aubert *et al.* hep-ex/0507003; B. Aubert *et al.* hep-ex/0506065; B. Aubert *et al.* hep-ex/0506064; B. Aubert *et al.* hep-ex/0408068;K. Abe *et al.* hep-ex/0408145; S. B. Athar *et al.* Phys. Rev. D 68, 072003 (2003)
24. A. Bornheim *et al.* Phys.Rev.Lett.88:231803,2002; A. Limosani *et al.* hep-ex/0504046; B. Aubert *et al.* hep-ex/0408075; I. Bizjak *et al.* hep-ex/0505088; H. Kakuno *et al.* Phys.Rev.Lett.92:101801,2004;B. Aubert *et al.* hep-ex/0507017.
25. P. Gambino and N. Uraltsev Eur.Phys.J.C34:181-189,2004; S.W.Bosh, B.O. Lange, M. Neubert, and G. Paz Nucl.Phys.B699:335-386,2004; B. Benson, I. Bigi, and N. Uraltsev Nucl.Phys.B710:371-401,2005;

B Physics at Hadron Colliders: Present and Future

Marta Calvi

University of Milano-Bicocca and INFN Milano
Piazza della Scienza 3, I-20126 Milano, Italy.

Abstract. An extensive program of B physics and CP violation measurements can be performed at Hadron Colliders. Results from the experiments CDF and DØ at the Tevatron and prospects for future measurements from experiments at the LHC are presented here.

Keywords: B-Physics, Hadron Colliders, CP violation
PACS: 12.15.Hh

INTRODUCTION

In the last few years, outstanding results from B-factories have provided a successful test of the CKM paradigm of flavour structure and CP violation. However New Physics could be hidden in B decays, specially those involving box and penguin diagrams. On the other hand, if New Physics will be found in direct searches, B physics measurements will have to sort out the corresponding flavour structure.

The B physics program at hadron colliders is rich. It will include precise measurement of $B_s^0 \bar{B}_s^0$ mixing (mass difference, width difference and phase), precise measurements of the angle γ of the Unitarity Triangle and several measurements of other CP phases in different channels for over-constraining the Unitarity Triangles, search for New Physics effects appearing in rare exclusive and inclusive B decays, studies of b-baryons, B_c physics and also studies of b production.

B physics at hadron colliders has the great advantage of a high $b\bar{b}$ cross section ($\sigma_{bb} \sim 100 - 500 \mu$b at 2-14 TeV), several orders of magnitude higher than at the $\Upsilon(4S)$, and of the production of all species of b-hadrons. The challenge in the analysis is related to the presence of the underlying event, to the high particle multiplicity and to the high rate of background events (the inelastic cross section is $\sim 50 - 80$ mb). These features demand to the experiments an excellent trigger capability, with good efficiency also on fully hadronic decay modes of b-hadrons, excellent tracking and vertexing performance, allowing for high mass resolution and proper time resolution, and excellent particle identification to separate exclusive decays.

Results obtained at the Tevatron Collider Run II by the CDF and DØ experiments, have demonstrated that precision B physics is possible at hadron machines. This program will be further developed and completed in the coming years by experiments at LHC, ATLAS and CMS and in particular LHCb.

PRESENT AND FUTURE EXPERIMENTS

CDF and DØ Experiments

The Tevatron accelerator complex has undergone an extensive upgrade for the Run II phase. An integrated luminosity of about 600 pb^{-1} per experiment has been collected up to now and projection for the future are to accumulate from 4 to 8 fb^{-1} for the year 2009. A description of the major upgrades undergone by CDF and DØ Detectors can be found elsewhere[1]. We just mention here for CDF the use of a Silicon Vertex detector to trigger on displaced tracks which allows to select purely hadronic events.

LHC Experiments

First pp collisions are foreseen in the Large Hadron Collider at Cern in summer 2007. Protons will collide at a center-of-mass energy of 14 TeV with a bunch crossing rate of 40MHz. Two experiments, ATLAS and CMS, are omni-purpose and optimized for discovery physics. Their B physics program is mainly pursued in the first years of running, when the LHC luminosity is expected to be $1-2\times 10^{33}$cm^{-2}s^{-1}. In the subsequent years at high luminosity (10^{34}cm^{-2}s^{-1}), when several pp collisions per bunch crossing will pile up, only search for very rare B decays with clear signatures will be performed. Reaches in B Physics will depend on the chosen trigger strategy and allocated bandwidth. B events will be mainly triggered by high p_T muons or di-muon triggers. CMS also exploits on-line tracking for the selection of exclusive B events at High Level Trigger [2], while ATLAS foresees a flexible trigger strategy with the progressive addition of other triggers [3] to the muon one.

LHCb is the LHC experiment dedicated to B physics. It will locally tune the luminosity, by de-focusing the beams, to 2×10^{32} cm^{-2}s^{-1}, in order to limit the pile up of pp interactions. In a nominal year of 10^7s, corresponding to an integrated luminosity of 2 fb^{-1}, 10^{12} $b\bar{b}$ events are expected. LHCb is a single-arm forward detector in the polar region 10-300 mrad, with good acceptance for b events due to the forward peaked production of b-hadrons at LHC. A description of the detector and its performances can be found in [4]. The LHCb trigger is operating in three stages. The Level-0 reduces the rate to 1 MHz requiring the presence of leptons or photons or hadrons with high p_T while the Level-1 selects on high impact parameter, high p_T tracks. The High Level Trigger is a software trigger using the full information on the event. Its output contains 200 Hz of exclusive B candidates and about 1.8 KHz of inclusive channels to be used also for calibration purposes and systematic studies.

$B_s^0 \bar{B}_s^0$ MIXING

The mixing of neutral B mesons has been clearly established and precisely measured in the B_d sector. However the mixing of B_s mesons has not yet been observed. The SM interprets B mixing with second order box diagrams. Contributions from new particles can affect both the amplitude and the phase of mixing.

Measurements of Δm_s

The best limit available on the $B_s^0 \bar{B}_s^0$ oscillation frequency, obtained from direct searches at LEP and SLC experiments, is $\Delta m_s > 14.5$ ps^{-1} at 95 %CL. On the other hand, from CKM fits a value of $\Delta m_s = 20.5 \pm 3.2$ ps^{-1} has been obtained [5]. As a consequence, the non-observation of oscillations up to $\Delta m_s = 30$ ps^{-1} would mean an evidence of the presence of new physics at 3σ.

The Tevatron is at present the only available source of B_s mesons and CDF and DØ have the chance to find a mixing signal in the coming years.

With the present statistics of 355 pb^{-1} of data, CDF has the best sensitivity in the semileptonic channel $B_s^0 \to D_s^- \ell^+ \nu X$, since the large event yield compensates the limited proper time resolution caused by the missing momentum of the neutrino [6]. The D_s is reconstructed in several channels: $D_s^- \to \phi \pi^-$, $D_s^- \to K^{0*} K^-$ and $D_s^- \to \pi^- \pi^+ \pi^-$ with a total yield of about 7700 events. The b flavour at production is determined from opposite side tag algorithms using electrons, muons and jet charge, with a total tagging power $\varepsilon D^2 = (1.43 \pm 0.09)\%$. The

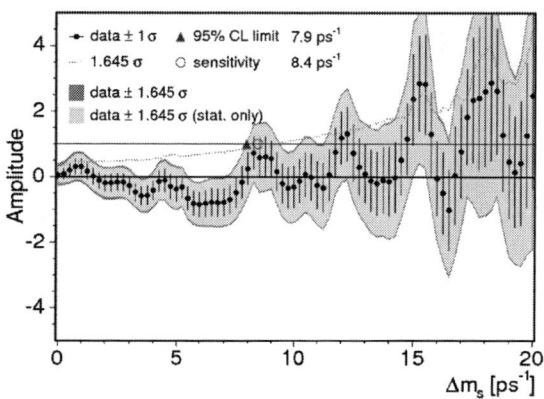

FIGURE 1. Amplitude scan for B_s oscillation combining semileptonic and hadronic channels, performed by CDF with 355 pb^{-1} of data.

limit of $\Delta m_s > 7.7$ ps^{-1}, at 95% C.L. has been obtained. In the same data sample, about 900 events are fully reconstructed in the hadronic channel $B_s^0 \to D_s^- \pi^+$. The total tagging power in this case is $\varepsilon D^2 = (1.12 \pm 0.18)\%$. The present statistics is not sufficient to provide a relevant limit with the hadronic channel alone, but in the combination with the semileptonic one the limit is improved to $\Delta m_s > 7.9$ ps^{-1}, at 95% C.L. The amplitude scan is shown in Figure 1. For the future, CDF expects to increase the flavour tagging power, thanks to the addition of same side kaon tags and to improve the proper time resolution. With the same data set a sensitivity up to Δm_s=15-16 ps^{-1} might be reached. Further improvement will come with more integrated luminosity. While the present result is limited by statistics and dominated by the semileptonic modes, at large Δm_s the proper time resolution will be the limiting effect and the hadronic modes will have the larger weight.

The result of the DØ experiment is currently based on semileptonic decays and gives $\Delta m_s > 5.0$ ps^{-1} at 95% C.L. with 460 pb^{-1} of data. Several improvements are expected from the use of additional decay channels, increased tagging power and increased proper time resolution and from future upgrades in the trigger.

But the definitive answer on $B_s^0 \bar{B}_s^0$ mixing may come from LHC. The hadronic channel, $B_s^0 \to D_s^- \pi^+$, giving best proper time resolution will be used. Results of LHCb full simulation indicate a proper time resolution of $\sigma_\tau \simeq$ 40 fs and an annual yield of 80.000 events with a signal over background ratio of about 3. The effective efficiency for flavour tagging is estimated to be about 7% in this channel. It is the result of the combination of several algorithms, using opposite side leptons, kaons and jet charge, and same side kaons. The good performances on kaon tagging, thanks to the particle identification available from RICH detectors, have large weight. In

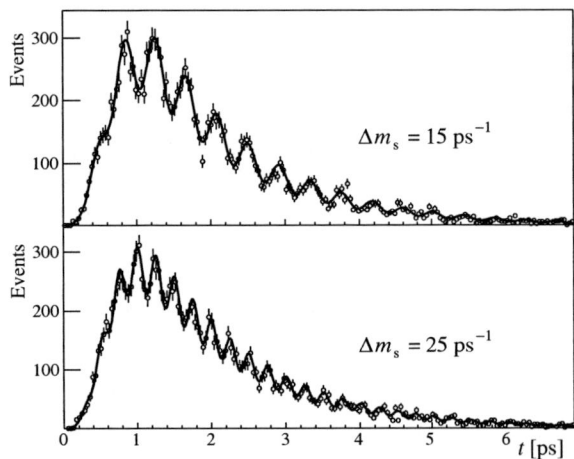

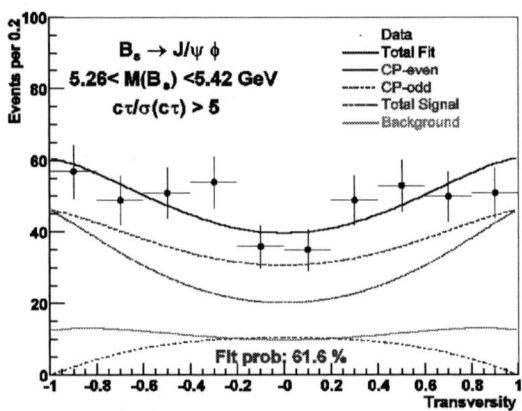

FIGURE 2. Expected proper time distribution of simulated $B_s \to D_s\pi$ candidates that have been flavour tagged as having not oscillated, for two different values of Δm_s. Points corresponds to one year of data-taking in LHCb.

FIGURE 3. Distribution of the transversity angle for $B_s \to J/\psi\phi$ events as measured by DØ. Also shown are the prediction for pure CP-even and pure CP-odd states in the signal.

Figure 2 the expected proper time distribution of tagged events is shown for two values of Δm_s. In one year of data-taking a 5σ observation of oscillation is expected if $\Delta m_s < 68$ ps^{-1}. Once observed, the precision to measure Δm_s is ~ 0.01 ps^{-1}.

ATLAS will also make a 5σ observation of oscillations if $\Delta m_s < 22$ ps^{-1}, in 10 fb^{-1}. Most recent expectation of CMS is lower, due to restriction to the trigger bandwidth allocated to this channel.

$\Delta\Gamma_s/\Gamma_s$ and ϕ_s measurements

Hints of New Physics could also be found in the measurement of the decay width difference between the two mass eigenstates of the B_s mesons $\Delta\Gamma_s = \Gamma(B_L) - \Gamma(B_H)$. In the Standard Model $\Delta\Gamma_s$ is expected to be of the order of 10%.

CDF and DØ have used untagged $B_s^0 \to J/\psi\phi$, ($J/\psi \to \mu^+\mu^-$) events to measure $\Delta\Gamma_s/\Gamma_s$ and $\bar{\Gamma}_s = [\Gamma(B_L) + \Gamma(B_H)]/2$. In a decay to two vector mesons three distinct amplitudes contribute, two CP-even and one CP-odd. The relative contribution of CP states can be disentangled with a combined fit to the distributions of mass, proper time and transversity angle. The distribution of the transversity angle, which is defined as the angle between the positive muon and the ϕ decay plane, in the J/ψ rest frame, is shown in Figure 3.

Result for $\Delta\Gamma_s/\Gamma_s$ versus $1/\bar{\Gamma}_s$ obtained by DØ are shown in Figure 4. They are compared with the previous result obtained by CDF and with the theoretical prediction. The relation with the B_s lifetime to flavour specific decay channels can be added as a constraint:

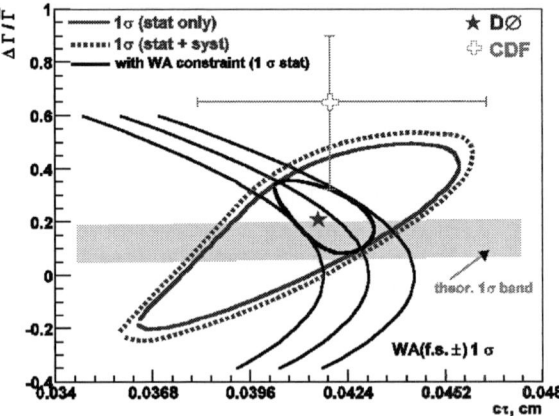

FIGURE 4. DØ result for $\Delta\Gamma_s/\Gamma_s$ versus $1/\bar{\Gamma}_s$. Also shown the previous result from CDF, the theory band and the improved constraint obtained including the world average value of B_s lifetime to flavour specific decays.

$1/\tau_{fs} = \bar{\Gamma}_s[1 - \frac{1}{2}(\Delta\Gamma_s/\bar{\Gamma}_s)^2]$. Taking the world average $\tau_{fs} = 1.43 \pm 0.05$ ps, $\Delta\Gamma_s/\Gamma_s = 0.23^{+0.16}_{-0.17}$ is obtained.

The phase ϕ_s of $B_s^0\bar{B}_s^0$ mixing is expected to be very small in the Standard Model $\phi_s = -2\chi = -2\lambda^2\eta \simeq -0.04$ resulting in a high sensitivity to possible New Physics contributions in $b \to s$ transitions. LHC experiments will have enough sensitivity for this measurement. They plan to use flavour tagged $B_s \to J/\psi\phi$ events and a simultaneous fit to the proper time, the transversity angle and the Δm_s distributions. In one year of data-taking LHCb expects to collect 100.000 $J/\psi(\mu\mu)\phi$ decays to obtain a precision on $sin(\phi_s)$ of about 0.06 and precision on $\Delta\Gamma_s/\Gamma_s$ of about 0.018 (for $\Delta m_s=20$ ps^{-1}). The sensitivity will be increased by adding $B_s^0 \to J/\psi\eta$ events, which are pure CP eigenstates. About 7000 events per

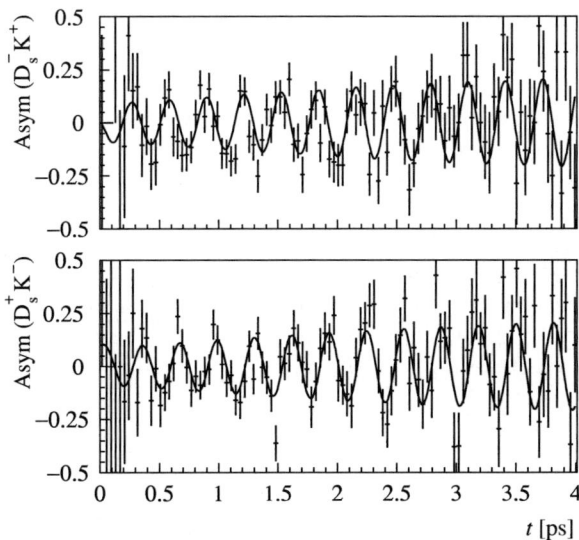

FIGURE 5. Time-dependent $B_s^0 \bar{B}_s^0$ asymmetry of simulated $D_s^- K^+$ (top) and $D_s^+ K^-$ (bottom) candidates, for $\Delta m_s = 20$ ps^{-1}. Points correspond to 5 years of data-taking.

year are expected in this channel.

CMS and ATLAS, with 30 fb^{-1}, expect a sensitivity on $\sin(\phi_s)$ of 0.03-0.04 and a sensitivity on $\Delta\Gamma_s/\Gamma_s$ of 0.015-0.012, respectively.

MEASUREMENTS OF γ

γ measurements from $B_s \to D_S K$ decays

$B_s^0 \to D_s^\pm K^\mp$ and $\bar{B}_s^0 \to D_s^\mp K^\pm$ decays can proceed through tree diagrams which can interfere via mixing. From the measurement of the time-dependent decay asymmetries the phase $\gamma + \phi_s$ can be extracted, together with a strong phase. If ϕ_s has been determined otherwise (eg. as shown in the previous section), the γ angle can be extracted, with little theoretical uncertainty, and insensitive to New Physics.

Strong particle identification capabilities are required to separate $B_s \to D_s K$ decays from $B_s \to D_s \pi$ background, having ~12 times larger branching fraction. The performance of the LHCb RICH detectors will be fully adequate, the residual contamination from $B_s \to D_s \pi$ events becoming about 10% after selection cuts. Monte Carlo studies have shown that 5400 $D_s^\mp K^\pm$ events will be collected in one year of data-taking, with S/B ratio, estimated from $b\bar{b}$ events, larger than 1. Hereafter only limits on S/B are quoted, due to the limited statistics with which the value of B has been determined. The $D_s^\mp K^\pm$ asymmetries are shown in Figure 5. A sensitivity of $\sigma_\gamma = 14$ degrees can be obtained if $\Delta m_s = 20$ ps^{-1}.

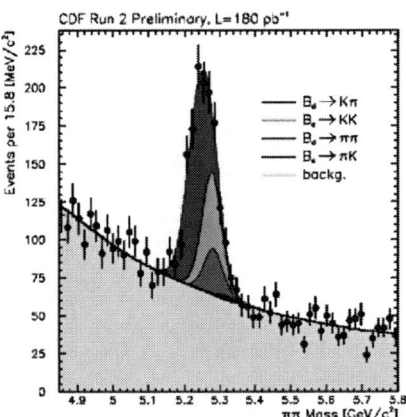

FIGURE 6. Invariant mass distribution of $B_{d,s}^0 \to h^+ h'^-$ candidates with π mass assumed for both tracks.

γ measurements from $B^0 \to D^0 K^{*0}$ decays

A theoretically clean determination of the angle γ can be performed using $B^0 \to D^0 K^{*0}$ decays. The method, described in [8], is based on the measurement of six time-integrated decay rates for $B_d^0 \to D^0 K^{*0}, \bar{D}^0 K^{*0}, D_{CP} K^{*0}$ and their CP conjugates; the decays are self-tagged through $K^{*0} \to K^+ \pi^-$, while the CP auto-states D_{CP} can be reconstructed in $K^+ K^-$ and $\pi^+ \pi^-$ modes. This method is similar to the analysis of $B^\pm \to D^0 K^\pm$ decays, already performed at the B-Factories, but has the advantage of using two colour-suppressed diagrams with a larger ratio of amplitudes expected
$r = |A(B^0 \to D^0 K^{*0})|/|A(B^0 \to \bar{D}^0 K^{*0})| \sim 0.4$.

In one year of data taking LHCb expects to collect a total of about 4.000 signal events leading to a sensitivity on γ of about 8 degrees.

γ measurements from $B \to h^+ h^-$ decays

Several strategies have been proposed [9] to extract informations on the angle γ from two body charmless decays of B mesons, some of them make use of assumptions on dynamics or on U-spin flavour symmetry.

A sample of $B_{d,s}^0 \to h^+ h'^-$ events has been collected by the CDF experiment, the invariant mass spectrum is shown in Figure 6 [7]. Different channels contribute to the mass peak: $B_d^0 \to \pi^+ \pi^-, K^+ \pi^-$; $B_s^0 \to K^+ K^-, \pi^+ K^-$ and a statistical separation has been performed combining particle identification from specific ionization and kinematic variables. In a sample of 180 pb^{-1} CDF extracts 134 ± 28 $B_d^0 \to \pi^+ \pi^-$, 232 ± 29 $B_s^0 \to K^+ K^-$ and 509 ± 28 $B_d^0 \to K^+ \pi^-$ decays. These events are used to measure the relative branching ratios:

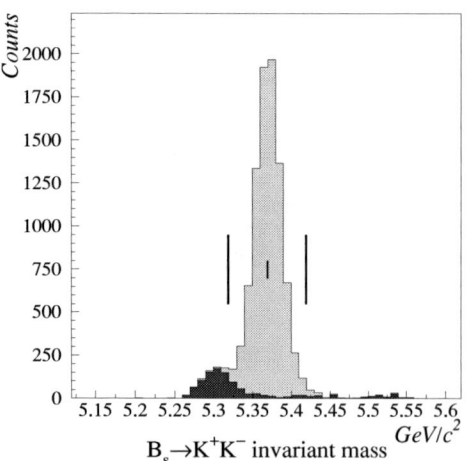

FIGURE 7. Invariant mass distribution of selected $B_s^0 \to K^+K^-$ candidates in LHCb. The light shaded histogram shows the signal and the dark one represents the background from $B^0 \to \pi^+\pi^-, K^+\pi^-$; $B_s^0 \to \pi^+K^-$; $\Lambda_b \to pK^-, p\pi^-$.

$[f_s \times BR(B_s^0 \to K^+K^-)]/[f_d \times BR(B_d^0 \to K^+\pi^-)] = 0.50 \pm 0.08(stat) \pm 0.07(syst)$

and $BR(B_d^0 \to \pi^+\pi^-)/BR(B_s^0 \to K^+K^-) = 0.24 \pm 0.06(stat) \pm 0.05(syst)$ and the direct CP asymmetry $A_{CP}(B_d^0 \to K^+\pi^-) = -0.04 \pm 0.08(stat) \pm 0.01(syst)$. The first result is unique to CDF while the other two are compatible with similar results from BaBar and with competitive uncertainties. The next step will be the measurement of time dependent asymmetries in flavour tagged samples.

At LHCb RICH detectors allow to separate the K/π channels with high efficiency and purity, as shown in Figure 7. LHCb in one year of data taking expects to collect 26.000 $B_d^0 \to \pi^+\pi^-$, 37.000 $B_s^0 \to K^+K^-$ and 135.000 $B_d^0 \to K^+\pi^-$ decays, with mass resolution $\sigma(M_B) \sim 17$ MeV and a proper time resolution of $\sigma_\tau \sim 30$ fs. The two time dependent CP asymmetries

$A_{CP}(B_d \to \pi^+\pi^-) = A_{CP}^{dir,\pi\pi} cos(\Delta m_d t) + A_{CP}^{mix,\pi\pi} sin(\Delta m_d t)$

$A_{CP}(B_s \to K^+K^-) = A_{CP}^{dir,KK} cos(\Delta m_s t) + A_{CP}^{mix,KK} sin(\Delta m_s t)$

will be used to fit the four CP asymmetries, which will be extracted with a precision of about 6%. Following the method suggested in [10], U-spin symmetry can be exploited to constrain the penguin to tree ratios in the two decays to be the same. Assuming the knowledge of the mixing phases ϕ_d and ϕ_s from previous measurements, the γ angle can be extracted with a precision $\sigma_\gamma \sim 5$ degrees in one year of data-taking. Additional measurements can be used to test the uncertainty related to the U-spin assumptions.

α MEASUREMENTS FROM $B^0 \to \rho\pi$ DECAYS

As it was suggested in [11], a time dependent Dalitz plot analysis of the three body decay $B^0 \to \rho\pi \to \pi^+\pi^-\pi^0$ allows a clean extraction of the angle $\alpha = \pi - \beta - \gamma$.

LHCb expects to reconstruct 14000 decays per year (with S/B> 1.3) in this channel. In a simulation study, an 11-parameter fit has been used to get an independent measurement of tree and penguin parameters, taking into account resonant and non resonant background sources. A sensitivity $\sigma_\alpha <10$ degrees is expected in one year of data-taking.

RARE DECAYS

$$B_d^0 \to K^{*0}\mu^+\mu^-$$

$B_d^0 \to K^{*0}\mu^+\mu^-$ is a rare decay with branching fraction of the order of 10^{-6} which has a clear experimental signature. The forward-backward asymmetry is defined as

$$A_{FB}(\hat{s}) = \left(\int_0^1 dcos\theta - \int_{-1}^0 dcos\theta \right) \frac{d\Gamma^2}{d\hat{s}dcos\theta}$$

where θ is the angle between the μ^+ and the K^{*0} in the di-muon rest frame, and $\hat{s}=(m_{\mu^+\mu^-}/m_B)^2$. The forward-backward asymmetry is a sensitive probe of New Physics. In the Standard Model the value of $\hat{s}$ for which $A_{FB}(\hat{s})$ is zero can be calculated with a 5% precision. Models with non-standard values of Wilson coefficients C_7, C_9, C_{10} predict $A_{FB}(\hat{s})$ of opposite sign or without zero point.

LHCb will select 4400 decays per year (with S/B> 0.4), this allows a determination of branching fractions and CP asymmetries with a precision of few percent. Using a toy Monte Carlo to determine the sensitivity in the forward-backward asymmetry measurement, including background subtraction, an uncertainty of 0.06 on the location of $\hat{s}_0$ is found, in 1 year of data-taking.

ATLAS will also collect about 2000 events of $B_d^0 \to K^{*0}\mu^+\mu^-$, with S/B> 7 in 30 fb^{-1}.

$$B_s \to \mu^+\mu^-$$

$B_s \to \mu^+\mu^-$ is a rare decay involving flavour changing neutral currents whose branching ratio is estimated to be $BR(B_s \to \mu^+\mu^-) = (3.5 \pm 0.1) \times 10^{-9}$ in the Standard Model [12]. In various supersymmetric extensions of the Standard model it can be enhanced by one to three orders of magnitude, being $BR \sim (tan\beta)^6$, for large $tan\beta$.

The best upper limit on the branching ratio at present come from experiments at Tevatron: $BR(B_s \to \mu^+\mu^-) < 7.5 \times 10^{-7}$ at 95% CL from CDF with 171 pb^{-1} of data, and $BR(B_s \to \mu^+\mu^-) < 3.7 \times 10^{-7}$ at 95% CL from DØ with 300 pb^{-1} of data.

In the SM context, LHCb expects to select 17 events per year, with a resolution on the B_s mass of 18 MeV/c^2. The background determination is still incomplete and require additional Monte Carlo statistics. No events were selected in the 10^7 $b\bar{b}$ event sample used so far.

CMS has studied a selection at the High Level Trigger, giving $B_s \to \mu^+\mu^-$ candidates with a rate smaller than 1.7 Hz, and a resolution on the B_s mass of 74 MeV/c^2. In 10 fb^{-1} 47 signal events are selected. With a refined selection at the offline level 7 signal events are expected to be retained, with less than 1 background event.

For this search ATLAS and CMS will also exploit the high luminosity runs. In 100 fb^{-1} (1 year at 10^{34}) 92 signal events are expected (with 660 of background) and 26 signal events (with ~ 6 of background) respectively. The different levels of background can be attributed to different vertex reconstruction and selection, however an update on these estimations is expected.

In conclusion there are good prospects of significant measurement in this channel, even for the SM value of the branching ratios.

PHYSICS OF B_C

The first observation of B_c meson was made by CDF in 1998 in the semileptonic channel $B_c^+ \to J\psi l^+ \nu$. Recently they made the first observation in the hadronic channel $B_c^+ \to J\psi \pi^+$. The mass distribution of fully reconstructed events is shown in Figure 8, giving $M(B_c) = 6287.0 \pm 4.8(stat) \pm 1.1(syst)$ MeV/c^2. DØ has performed a new measurement of lifetime with inclusive semileptonic decays $B_c^+ \to J/\psi \mu^+ X$ obtaining $\tau(B_c) = 0.448^{+0.123}_{-0.096}(stat) \pm 0.121(syst)$ps.

At LHC 10^9 (5×10^{10}) B_c will be produced per year at 2×10^{32} (10^{34}) cm^{-2}s^{-1} luminosity, allowing for precision measurements of mass and lifetime. As an example, LHCb expects to reconstruct 14.000 $B_c^+ \to J\psi(\mu\mu)\pi^+$ events per year with S/B> 1.3. B_c mesons, which are unique for being a two heavy quark state, will allow several studies to be performed, related to the interplay of strong and weak interactions, Heavy Quark Expansion, Non-Relativistic QCD, Factorization models etc. The possibility to perform γ measurements with $B_c^+ \to D_s^+ D^0$ using triangle relations has also been proposed by Fleischer and Wyler [13].

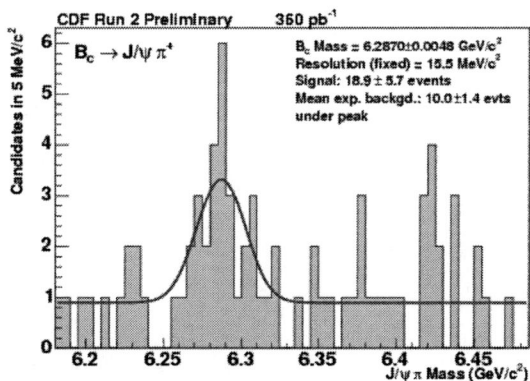

FIGURE 8. Invariant mass distribution of $B_c^+ \to J\psi\pi^+$ events as measured by CDF with 360 pb^{-1} of data.

CONCLUSION

An extensive program on B-Physics is being developed by CDF and DØ experiments at Tevatron and a significant step further is expected in the future by LHC experiments. The access to all b-hadron species and the high statistics will be the main advantage over B-Factories. LHCb combines excellent vertexing and particle identification capabilities with a flexible trigger, dedicated to B-Physics. ATLAS and CMS will also contribute significantly, in particular for decay modes with muons or with very small decay rates.

REFERENCES

1. R. Blair *et al*, CDF Collaboration; FERMILAB-PUB-96/360-E(1996); S. Abachi *et al*, DØ Collaboration; FERMILAB-PUB-96/357-E(1996)
2. V. Ciulli, Eur. Phys. J. *C* 34, s01, s379-s384 (2004);
3. M. Smizanska, Eur. Phys. J. *C* 34, s01, s385-s392 (2004);
4. R. Antunes *et al*, LHCb Collaboration; CERN/LHCC 2003-030 (2003)
5. M. Pierini, these Proceedings.
6. S. De Cecco, these Proceedings.
7. M. Casarsa, these Proceedings.
8. M. Gronau and D. Wyler, Phys. Lett. **B 265**, 172 (1991); I. Dunietz, Phys.Lett. **B 270**, 75 (1991).
9. See for example M. Battaglia, A.J. Buras, P. Gambino and A. Stocchi, eds. CERN-2003-002 and references therein.
10. R. Fleischer, Phys. Lett. **B 459**, 306 (1999); R. Fleischer and J. Matias, Phys. Rev. **D 66**, 054009 (2002).
11. A. Snyder, H. Quinn, Phys. Rev.**D 48** 2139 (1993).
12. A.Ali, Nucl.Phys.Proc.Suppl. 59 (1997) 86-100.
13. R. Fleischer and D. Wyler, Phys. Rev.**D 62** 057503 (2000).

Status of Electroweak Precision Measurements and Perspectives at LHC

Roberto Tenchini

Istituto Nazionale di Fisica Nucleare, Sezione di Pisa
Largo Bruno Pontecorvo 3, I-56100 Pisa, Italy

Abstract. The status of precision measurements in the electroweak sector is briefly reviewed. Emphasis is given on final LEP results using data collected at the Z and beyond the WW production threshold. A common analysis of electroweak observables sets at 260 GeV/c^2 the upper limit on the mass of the Standard Model Higgs boson (95% CL). The current bounds on trilinear WWZ and WWγ anomalous couplings at LEP are discussed and compared with perspectives at future accelerators. The rôle of LHC as a top factory is briefly discussed.

Keywords: Electroweak, Precision Measurements, LEP, LHC.
PACS: 14.70.-e, 14.65.Ha, 14.80.Bn.

INTRODUCTION

The end of last century has seen a rapid increase of our knowledge of gauge bosons. A decade after its discovery [1,2] the mass of the Z boson was measured, thanks to LEP, with a relative precision of 10^5, becoming one of the best known physical constants. Such a precision has opened the possibility of testing the electroweak theory at one-loop level. As an example the W mass (m_W), that is currently known with a precision of 39 MeV/c^2 [3] can be put in relation with the Z mass, (m_Z), the running fine-structure constant ($\alpha(m_Z)$), and the Fermi constant (G_F), through

$$m_W^2 (1 - \frac{m_W^2}{m_Z^2}) = \frac{\pi \alpha(m_Z)}{\sqrt{2} G_F} \frac{1}{1 + \Delta r}$$

where the Δr term modifies the tree-level relation and effectively measures the radiative corrections. The current measurements give a 12σ evidence for the need of Δr. Within the Standard Model the electroweak radiative corrections can be used to indirectly measure the less known parameters of the theory, in particular the mass of the Higgs boson (m_H).

In these proceedings I will briefly discuss some important eletroweak observables very sensitive to the Higgs mass (the asymmetries at the Z), I will cover recent results on W physics from LEP and Tevatron and show the present constraints on the Standard Model Higgs mass. The perspectives at future accelerators, in particular at LHC, will be discussed.

ASYMMETRIES AT THE Z POLE

Parity violation at production and decay yields asymmetries in various distributions at the Z pole. These asymmetries are sensitive to the ratio between the vector and the axial lepton couplings, in other words to the effective electroweak mixing angle

$$\sin^2 \vartheta_{eff}^l = \frac{1}{4}(1 - \frac{g_{Vl}}{g_{Al}}).$$

The precision on the measurement of the couplings has increased by two orders of magnitudes since the times of deep inelastic scattering, thanks to the accurate measurements made at LEP and at SLC. The most recent (and final) results are shown in Fig. 1, the combined value has un uncertainty of about half permil. This plot reveals a disagreement of about 3σ between the two most precise determinations, coming from the left-right asymmetry measured at SLD (A$_l$(SLD)) [4] and from the forward-backward asymmetry of the $Z \to b\bar{b}$ process (A$_{fb}^{0,b}$) measured at LEP [5]. This discrepancy is a long standing issue. The left-right asymmetry is a direct probe of the Zee vertex, with systematic uncertainties essentially limited to the uncertainty on the polarization of the electron beam. As far as the b couplings are concerned there is no independent evidence of any anomaly, that could explain the disagreement. The b couplings can be determined by combining the measurement of the b polarized forward-backward asymmetry at SLD, giving the ratio of vector to axial couplings, with the precision measurement of R_b, that is related to the sum of the squares of the couplings. Both measurements agree with the Standard Model expected values [6]. From the experimental side the b asymmetry is measured by four experiments with two almost uncorrelated techniques, based on lepton tagging and on vertex-inclusive tagging. The two set of results are in agreements and have very different systematic effects. All measurements and the average itself are dominated by statistics. The most important common error (from QCD corrections) is a factor five lower than the statistical error and over the years an impressive number of experimental cross checks have been done [5] leaving little room for unexpected experimental effects.

It is likely that this discrepancy will stand for a long time. It will presumably be solved at future linear colliders if polarization of both beams is available and a very high statistics run at the Z takes place.

At LHC the forward-backward asymmetry of lepton pairs from the $q\bar{q} \to Z \to \ell^+\ell^-$ Drell-Yan process can provide interesting additional information. The signature will be clear (a lepton pair with invariant mass consistent with the Z mass) and the rate huge (about 1.5 million Z decays to electron pairs in ten days of low luminosity). The crucial experimental point is the determination of the initial quark direction. Since at LHC anti-quarks can be produced only by the sea the direction of the Z boost can be taken as a measurement of the initial quark direction. A careful study of the quark and anti-quark PDF at LHC is required in order to assess the limitation of this assumption and compute the related systematic uncertainties.

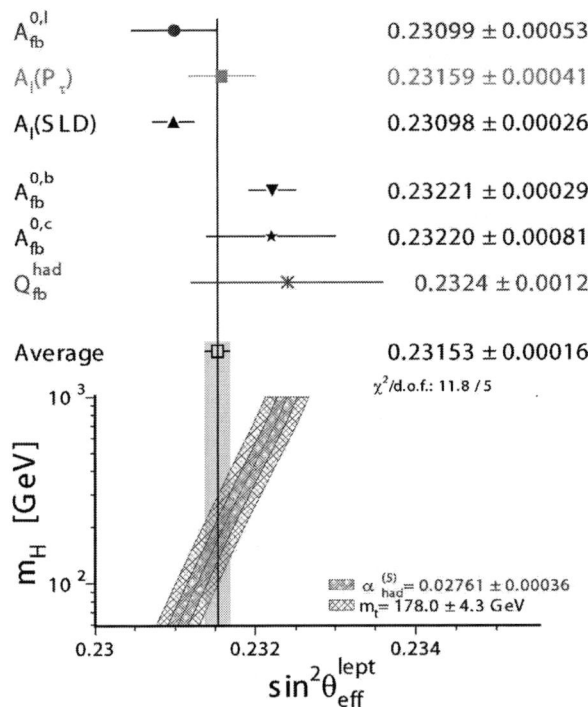

FIGURE 1. Measurements of the electroweak mixing angle, $\sin^2 \vartheta_{eff}^l$, at the Z pole and their combination. The dependence of the Higgs boson mass on $\sin^2 \vartheta_{eff}^l$ is indicated. (Courtesy of the LEP Electroweak Working Group).

W PHYSICS

At hadron colliders the Drell-Yan process, $q\bar{q}' \to W \to \ell^+\nu$, is the main W production mechanism. This process represents at the same time the past and the future of W physics, since it allowed the W discovery and it is currently providing large samples of W's at the Tevatron Run II. We should have soon news concerning the W mass from the Tevatron. The preliminary transverse mass from high p_t leptons from CDF and D0, shown at this conference, is clearly very promising for the mass measurement. Run II results for the W and Z cross sections are available [7].

The $e^+e^- \to W^+W^-$ process is the source of most W's at LEP. Three diagrams, the s-channel γ and Z diagrams and the t-channel neutrino exchange diagram contribute to WW production at Born level. One has to bear in mind that experimentally the final state is a 4-fermion process due to the subsequent decay into 4-fermions.

Event selections are needed to disentangle WW events from the competing 2-fermion processes and to classify the events in three different channels (hadronic, leptonic and semileptonics). Decays to electrons, muons and taus can further be tagged.

The typical signatures of the three topologies allow to separate WW events among each others and from background with good efficiencies and excellent purities. The cross sections can be readily computed from the events selected, from the total luminosity and from the efficiency matrix where cross contamination among channels is taken into account. The residual background and the four fermion interference must be considered. The four fermion interference arises from different processes with the same final state, as may happen, for instance, for WW and single-W production. The correction is small, of the order of 1 percent relative, and the most important non WW 4-fermion processes can be directly measured. The new combination of LEP results on the total $e^+e^- \rightarrow W^+W^-$ cross section (Fig. 2) reaches a 9 permil precision, challenging theoretical predictions: computations without an adequate treatment of $O(\alpha)$ radiative corrections are definitively discarded.

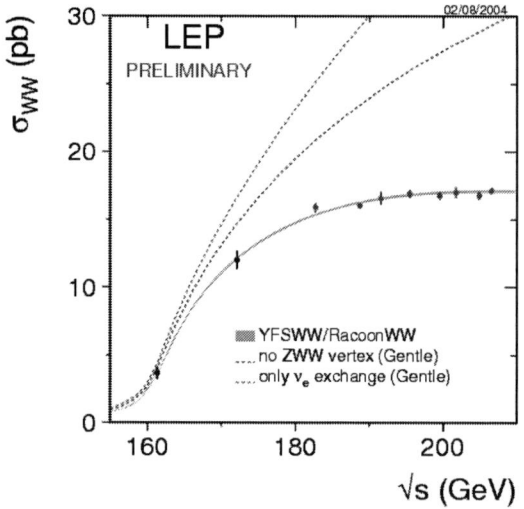

FIGURE 2. The $e^+e^- \rightarrow W^+W^-$ cross section as a function of the centre-of-mass energy. Predictions from recent calculations [8] are shown. (Courtesy of the LEP Electroweak Working Group).

At LEP the W mass is measured from the direct reconstruction of the invariant mass in the semileptonic and hadronic channels. Kinematic fits imposing energy momentum conservation are a key tool to improve the invariant mass resolution. For this a precise determination of the beam energy is required. The final LEP energy files are now available, opening the road for the final W mass measurements at LEP. The remaining hot issue is the fight against an important systematic error depressing the weight of the four-quark channel in the LEP combination: the uncertainty due to final state hadronic interactions, that is by far the largest error in this channel.

Interconnection phenomena in the final states (colour reconnection and Bose-Einstein correlations) are a non perturbative effect and data must be used to get a robust limit on the related uncertainties.

The present LEP result uses constraints from a particle flow analysis to set limits on colour reconnection (CR) effects. Colour Reconnection models predict modified particle flow in the intra-W regions. By measuring the particle flow ratio between the inter- and intra-W regions limits can be set on extreme scenarios. This method sets an uncertainty of 90 MeV/c^2, used in the present LEP combination.

A more sensitive method will be used for the forthcoming final result. A general feature of all models is the correlation between the mass shift predicted by a model and the difference between the mass measured in a standard analysis and the mass measured using only particles in the jet core. To a larger mass shift caused by CR a larger mass difference corresponds. Since the mass variation can be measured in data, direct limits on the mass shift can be set. The CR systematic uncertainty is expected to go down to about 50 MeV/c^2 in the final LEP result.

The current combined W mass result from LEP and Tevatron is [9]
$$m_W = 80.425 \pm 0.034 \text{ GeV}/c^2.$$
This combination can be compared with the Standard Model prediction computed using electroweak radiative corrections, measured by Z-pole observables as the asymmetries described in the previous section. The indirect determination gives
$$m_W = 80.373 \pm 0.033 \text{ GeV}/c^2,$$
in agreement with the direct measurement, showing consistency of the electroweak theory at one-loop level.

An improvement of more than a factor two on the W mass uncertainty (15 MeV) is expected at LHC from the measurement of the transverse mass of Drell-Yan leptons. This precision will be achieved if some key sources of systematic uncertainties, such as the knowledge of the lepton energy scale, are kept under control.

CONSTRAINTS ON THE STANDARD MODEL HIGGS MASS

The W mass and Z-pole asymmetries depend quadratically on the top mass through electroweak radiative corrections. The W mass, for instance, is modified by the already mentioned Δr term, whose top mass dependence is $\Delta r \approx G_F m_t^2$. The top mass can be indirectly evaluated by the measured observables yielding $m_t = 178.9^{+12}$ GeV/c^2 [9], in agreement with the direct measurement, given by the combination of the CDF/D0 Run I results, $m_t = 178 \pm 4.3$ GeV/c^2.

The dependence of the electroweak observables on the Higgs mass shows, instead, a logarithmic ansatz. Taking again as an example the W mass

$$\Delta r \approx G_F m_W^2 \log(\frac{m_H^2}{m_W^2})$$

The constraints on the Higgs mass, as computed using various electroweak observables, are combined together to give [9]

$$m_H = 124^{+207}_{-69} \text{ GeV/c}^2.$$

The direct determination of the top mass greatly helps in constraining the mass of the Higgs, since the large quadratic term is fixed by the measurement.
If this is done one gets

$$m_H = 114^{+69}_{-45} \text{ GeV/c}^2$$

corresponding to $m_H < 260$ GeV/c^2 at 95% CL. The exact value of this limit depends on many assumptions, such as the data used to extrapolate the fine-structure constant to m_Z, or the central value of the top mass; in the past two years it has oscillated in the range $220 \div 280$ GeV/c^2. Nevertheless a significant indication for a relatively light Higgs boson mass exists. The Higgs constraints from the W and top masses and from Z-pole observables are shown in Fig.3. Direct and indirect measurements are consistent, showing a success of the electroweak theory.

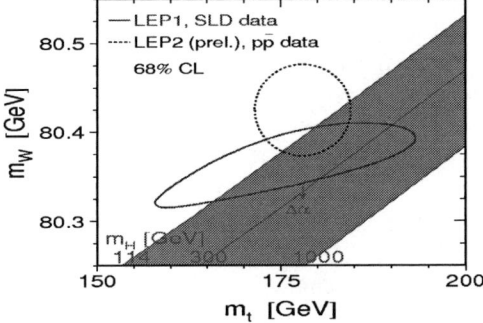

FIGURE 3. The direct (LEP2, Tevatron) and indirect (LEP1, SLC) determinations of the mass of the W boson (m_W) and of the top quark (m_t). The 68% CL allowed regions are indicated. The band represents the Standard Model expectations for different values of the Higgs boson mass (m_H) (Courtesy of the LEP Electroweak Working Group).

TRIPLE GAUGE COUPLINGS

The presence of the tree-level triple-gauge-boson WWγ and WWZ couplings is related to the non-abelian nature of the Standard Model. The behaviour of the $e^+e^- \to W^+W^-$ total cross section (Fig.2) is a clear evidence of their existence. Triple gauge interactions are also seen in other processes, such as single W production in e^+e^- interactions or associated Wγ production at hadron colliders, both testing the WWγ vertex.

Triple gauge couplings do not only affect productionrates, but also angular distributions of emitted weak bosons and of their decay products. Precision tests of triple gauge couplings, and of $SU(2)_L \times U(1)_Y$ predictions, require detailed measurements of differential cross sections. Angular distributions analyse different weak boson helicity states: since the longitudinal component of the helicity is directly linked to the symmetry breaking mechanism and to the mass generation, these are fundamental tests of the theory.

The most general form for an effective charged TGC lagrangian consistent with Lorentz invariance involves 14 complex couplings, 7 for the WWγ vertex and 7 for the WWZ vertex [10]. Some couplings are C- or P-violating while in the Standard Model C- and P-conservation is predicted in triple gauge couplings.

All couplings have been experimentally tested with the $e^+e^- \to W^+W^-$ sample collected at LEP2. A fit of the total cross section and the angular distributions has been performed to models where only one coupling at the time is allowed to vary, while all the others were set to zero [11]. In most analyses C- and P-conservation, as well as electromagnetic gauge invariance, is assumed and the 14 couplings are reduced to five

$$g_1^z, \kappa_\gamma, \kappa_Z, \lambda_\gamma, \lambda_Z.$$

Within the Standard Model the first 3 couplings are equal to unity at tree level, while the λ's are equal to zero. Standard Model loop contributions to the triple gauge couplings are of the order of 10^{-3}. The couplings can be related to physical properties of gauge bosons, for instance the W magnetic dipole moment and the W electric quadrupole moment may be written in term of κ_γ and λ_γ. The additional requirement of $SU(2)_L \times U(1)_Y$ local gauge invariance, often used in LEP analyses, introduces a relation between γ and Z couplings leaving, in practice, only three to be measured. These three couplings have been measured at LEP [12] by means of three-parameter fits to $e^+e^- \to W^+W^-$ and $e^+e^- \to W^+e\upsilon$ data.

The three couplings are consistent with the Standard Model expectation, in particular the g_1^Z and κ_γ

measurements confirm the presence of triple gauge interactions with the expected strength, with a precision of about 2%.

At LHC triple gauge couplings will be studied during the high-luminosity phase (10^{34} cm^{-2}s^{-1}) by selecting gauge-boson pairs. The selection of WZ or Wγ pairs produced through the processes $q\bar{q}' \to W \to WZ$ or $q\bar{q}' \to W \to W\gamma$ allows independent measurements of (κ_γ, λ_γ) and (κ_Z, λ_Z). Anomalous contributions are expected to show up as an excess of high p_t gauge-boson pairs. Very interesting precision can be reached for the λ couplings, at the permil level (300 fb^{-1}), testing the expected Standard Model and SUSY loop contributions. At future e$^+$e$^-$ linear colliders all couplings will be tested with a precision well beyond 10^{-3}, by analyzing at higher energy and luminosity the processes already studied at LEP.

TOP PHYSICS AT LHC

The cross section for $t\bar{t}$ production in pp collisions at 14 TeV is about 800 pb, two order of magnitudes larger than at the Tevatron. About one $t\bar{t}$ pair per second will be produced at LHC during the low luminosity phase (10^{33} cm^{-2}s^{-1}). Semileptonic and dileptonic $t\bar{t}$ events (i.e. events where one or two W's decay leptonically) can be triggered by LHC experiments with high efficiency. Fully hadronic $t\bar{t}$ events will be selected, too, but with lower efficiency, because of the higher thresholds needed to fight the huge multi-jet background. The importance of a precise measurement of the top mass in constraining the predicted Higgs-boson mass, and hence in testing the models once the Higgs boson is discovered, has already been stressed. The top mass will be measured by direct reconstruction of the top decay products, taking advantage of kinematics constraints, such as momentum conservation in the transverse plane or the known value of the W mass, to improve resolution. Alternative methods, based on the selection of semi-exclusive $t \to Wb$ decays, will provide additional information. An uncertainty of about 1 GeV/c^2 on the top mass seems reachable at LHC, provided that the jet energy scale is controlled at an adequate level.

Beyond the measurement of top mass the large LHC top sample can be used to perform detailed studies of electroweak interactions. An example is the measurement of the fraction of longitudinally polarized W bosons in top decays. As longitudinal W bosons are closely related to the mechanism of electroweak symmetry breaking this is an important test of the theory. Since the $t \to Wb$ decay occurs before hadronization takes place this fraction (about 70%) can accurately be predicted by the electroweak calculations.

Another electroweak process taking place with relatively high cross section at LHC is single top production. Single top can be produced through b exchange in t-channel W-gluon fusion ($Wg \to tb$) with a cross section of about 250 pb. The top quark gets almost completely polarized, allowing precision measurements of the helicity both at production and decay. The s-channel ($q\bar{q}' \to W \to tb$) production has a lower cross section (about 10 pb), but is potentially very interesting for a precise determination of the CKM matrix element $|V_{tb}|$.

CONCLUSIONS

Final results on the electroweak mixing angle from asimmetries at the Z-pole are now available. The combination of these measurements yields a precision of 1.6 10^{-4} (absolute) on $\sin^2 \vartheta^l_{eff}$. The combined LEP-Tevatron measurement of the W mass has currently an uncertainty of 34 MeV/c^2 [9]. The forthcoming final LEP results and the new measurements at Tevatron RUN II will soon decrease this error. The combination of several electroweak observables gives indications for a light Higgs boson: m_H < 260 GeV/c^2 at 95% CL. Trilinear couplings have been measured with a precision of few percent at LEP and found to agree with Standard Model expectations.

At LHC improved measurements of the W and top mass will constraint the mass of the Higgs boson to an uncertainty around 30%, allowing a stringent test of the theory once the Higgs is discovered. LHC will be a top factory allowing detailed tests of the electroweak theory through the properties of the top quark.

ACKNOWLEDGMENTS

I am grateful to my colleagues of the LEP Electroweak Working Group. A special thank to the organizers of this conference.

REFERENCES

1. UA1 Collaboration, *Phys. Lett.* **B122**, 103 (1983).
2. UA2 Collaboration, *Phys. Lett.* **B122**, 476 (1983).
3. The Particle Data Group, *Phys. Rev.* **D66**, 010001 (2002).
4. SLD Collaboration, *Phys. Rev. Lett.* **5945**, 84 (2000).
5. ALEPH Collaboration, *Eur. Phys. J.* **C24**, 177 (2002), DELPHI Collaboration, *Eur. Phys. J.* **C34** (2004) 109,

L3 Collaboration, *Phys. Lett.* **B448** 152 (1999),
OPAL Collaboration, *Phys. Lett.* **B577** 18 (2003),
ALEPH Collaboration, *Eur. Phys. J.* **C22** 201 (2001),
DELPHI Collaboration, *Eur. Phys. J.* **C40** 1 (2005),
L3 Collaboration, *Phys. Lett.* **B439** 225 (1998),
OPAL Collaboration, *Phys. Lett.* **B546** 29 (2002).

6. SLD Collaboration, *Phys. Rev. Lett.* **88** 151801 (2002),
   SLD Collaboration, *Phys. Rev. Lett.* **90** 141804 (2003),
   ALEPH Collaboration, *Phys. Lett.* **B401** 150 (1997),
   ALEPH Collaboration, *Phys. Lett.* **B401** 163 (1997),
   DELPHI Collaboration, *Eur. Phys. J.* **C10** 415 (1999),
   L3 Collaboration, *Eur. Phys. J.* **C13** 47 (2000),
   OPAL Collaboration, *Eur. Phys. J.* **C8** 217 (1999).
7. CDF Collaboration, FERMILAB-PUB-05-009-E, submitted to Phys. Rev. Lett.
8. S. Jadach et al., *Phys. Rev.* **D61** 113010 (2000),
   A. Denner et al., *Phys. Lett.* **B475** 127 (2000).
9. The LEP Electroweak Working Group (2004), http://cern.ch/LEPEWWG/
10. K. Hagiwara, R.D.Peccei, D. Zeppenfeld and K. Hikasa, *Nucl. Phys.* **B282** 253 (1987).
11. ALEPH Collaboration, *Eur. Phys. J.* **C 21** 423 (2001).
12. ALEPH Collaboration, *Phys. Lett.* **B614** 7 (2005),
    L3 Collaboration, *Phys. Lett.* **B586** 151 (2004),
    DELPHI Collaboration, *Phys. Lett.* **B502** 9 (2001),
    OPAL Collaboration, *Eur. Phys. J.* **C33** 463 (2004).

The Pierre Auger Observatory: Present Status and Future Prospects

Sergio Petrera
for the Pierre Auger Collaboration

INFN and Dipartimento di Fisica, Università di L'Aquila, via Vetoio, 67010, Italy

Abstract. The Pierre Auger Observatory is in advanced stage of construction in the southern site of Malargüe, Argentina. This progress report mainly focuses on hybrid events, a remarkable subset of cosmic ray events which are simultaneously detected by both Surface Detector and Fluorescence Detector subsystems. The hybrid method and its performances are presented.

Keywords: Cosmic rays
PACS: 13.85.Tp, 95.55.Vj, 95.85.Ry, 96.40.-z, 98.70.Sa

INTRODUCTION

The Pierre Auger Observatory [1] is an international experiment with the goal of exploring with unprecedented statistics the cosmic ray spectrum above 10^{19} eV. Of particular interest are cosmic ray particles with energy $> 10^{20}$ eV. At these energies they interact with cosmic microwave background radiation thus generating a spectrum cutoff, known as GZK effect [2]. This effect attenuates the particle flux except if their sources are in our cosmological neighborhood (< 100 Mpc). Furthermore protons of these energies may point back to the source and open a new kind of astronomy with charged particles.

The extremely low rate, a few particles per km$^2 \cdot$ sr $\cdot$ century, of cosmic rays above the GZK cutoff requires a large area detector. The Auger Southern Observatory, in advanced stage of construction close to the town of Malargüe, Province of Mendoza, Argentina, covers an area of 3000 km^2 (see Figure 1).

Cosmic rays are detected by the Auger Observatory with two different experimental techniques. The Surface Detector (SD), a giant array of 1600 water Cherenkov tanks, placed over the Observatory area with a spacing of 1.5 km, measures the shower particle density and arrival times at ground level. Presently about one half of the SD tanks are taking data.

The Fluorescence Detector (FD), composed by a set of 24 telescopes, simultaneously measures the longitudinal development of the cosmic ray shower in the atmosphere above ground. The telescopes are arranged in four peripheral buildings (Eyes), each housing 6 telescopes, overlooking the SD array. The first three buildings at Los Leones, Coihueco and Los Morados are completed, with their 18 FD telescopes taking data. The Loma Amarilla building is under construction.

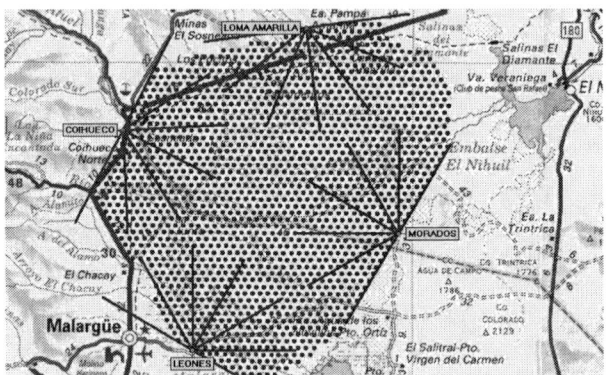

FIGURE 1. A map of the Auger Southern Observatory. Dots represent SD tanks. FD Eyes are shown with their fields of view.

Numerous data are being collected by Auger. Preliminary analyses based on either subsystems show good consistency and comprehension of the whole detector. In this stage, particular attention is devoted to hybrid detection, i.e. with the cosmic ray observed simultaneously by the array tanks and the fluorescence telescopes. This allows high precision energy and arrival direction determination and deeper understanding of the systematics of energy scale of the two techniques. This report will mainly focus on hybrid reconstruction method and its application to preliminary data.

THE SURFACE DETECTOR

The Surface Detector (SD) is made of water Cherenkov tanks. The tanks have 3.6 m diameter and 1.2 m height to contain 12 m^3 of clean water viewed by three 9" photomultiplier tubes (PMT). A solar panel and a buffer

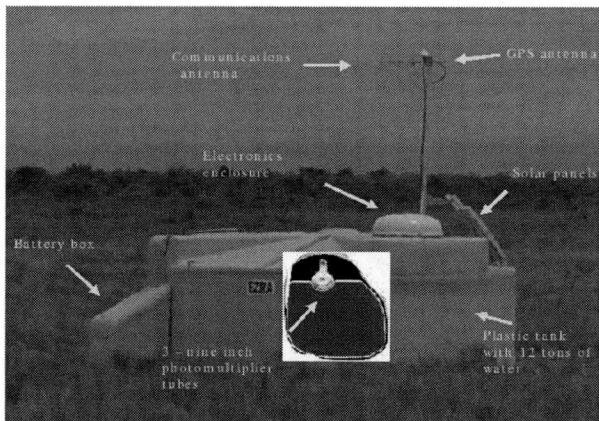

FIGURE 2. A SD tank in the field. The main components of the detector are sketched in the figure.

battery provide electric power for the local intelligent electronics, GPS synchronization system and wireless LAN communication. A picture of one tank in the field, showing its main components, is shown in Figure 2.

Cosmic ray muons produce an essential calibration signal of about 80 photoelectrons in one PMT. The signals are continuously digitised with 16 bit dynamic range at 40 MHz sampling rate and temporarily stored in local memory. The time structure of PMT pulses carries rich information related to the mass of the primary particle. The trigger conditions include a *threshold trigger* (one or more FADC counts above 3.2 Vertical Equivalent Muon [*VEM*] in each of 4 or more tanks) and a *time over threshold trigger* (12 FADC bins exceeding 0.2 VEM in sliding window of 3 μs in each of 3 or more tanks). Detection efficiency will begin around 10^{18} eV and reach 100% at 10^{19} eV.

THE FLUORESCENCE DETECTOR

The Fluorescence Detection method is based on the measurement of fluorescence photons emitted from the shower in its development through the atmosphere. The nitrogen molecules excited by shower particles emit isotropically fluorescence photons, with wavelengths between 300 to 400 nm. The fluorescence yield is ≈ 4 photons per electron traversing one meter of atmosphere, approximately constant as a function of altitude. The fluorescence light is collected by a large mirror and focused on a pixellated surface. The arrival direction of the cosmic ray is then reconstructed from the pixel directions and signal times.

The Fluorescence Detector (FD) consists of 24 wide-angle Schmidt telescopes grouped in four stations. Each telescope has a 30° field of view in azimuth and vertical

FIGURE 3. The FD telescope: on the left the spherical mirror with square shaped segments, on the right the PMT camera. The adjacent telescope's camera is also seen in the background.

angle. The four stations at the perimeter of the surface array consist of six telescopes each for a 180° field of view inward over the array.

A picture of one of the FD telescopes installed in the Los Leones building is shown in Figure 3. Each telescope is formed by segments to obtain a total surface of 12 m^2 on a radius of curvature of 3.40 m. The aperture has a diameter of 2.2 m and is equipped with optical filters and a corrector lens. In the focal surface a photomultiplier camera detects the light on 20×22 pixels. Each pixel covers $1.5° \times 1.5°$ and the total number of photomultipliers in the FD system is 13,200. PMT signals are continuously digitised at 10 MHz sampling rate with 15 bit dynamic range. The FPGA-based trigger system is designed to filter out shower traces from the random background of 100 Hz per PMT.

The absolute calibration of the detector follows an end-to-end approach, based on the uniform illumination of the pixels from a calibrated light source. This is obtained with a "drum" illuminator consisting of a pulsed UV LED embedded in a small cylinder of teflon and illuminating the interior of a 2.5 m diameter cylindrical drum, 1.4 m deep. The drum is positioned at the entrance aperture of the telescope under calibration. Relative optical calibration is also used to monitor time variations in the telescopes calibration during the periods of data taking.

Finally, attention is given to atmospheric monitoring, making use of laser beams, LIDAR's, calibrated light sources and continuous recording of weather conditions. Special efforts are being made to determine the air fluorescence efficiency and its dependence on relevant conditions.

HYBRID RECONSTRUCTION

In the Fluorescence Detector, the cosmic ray shower is seen as a sequence of triggered pixels with typical space-time characteristics, as shown in Figure 4 for one of the hybrid events collected during this first period of data taking. Each of the FD pixels views the light emitted from a given direction in the sky.

This specific event will be used in the following to illustrate the hybrid reconstruction procedure.

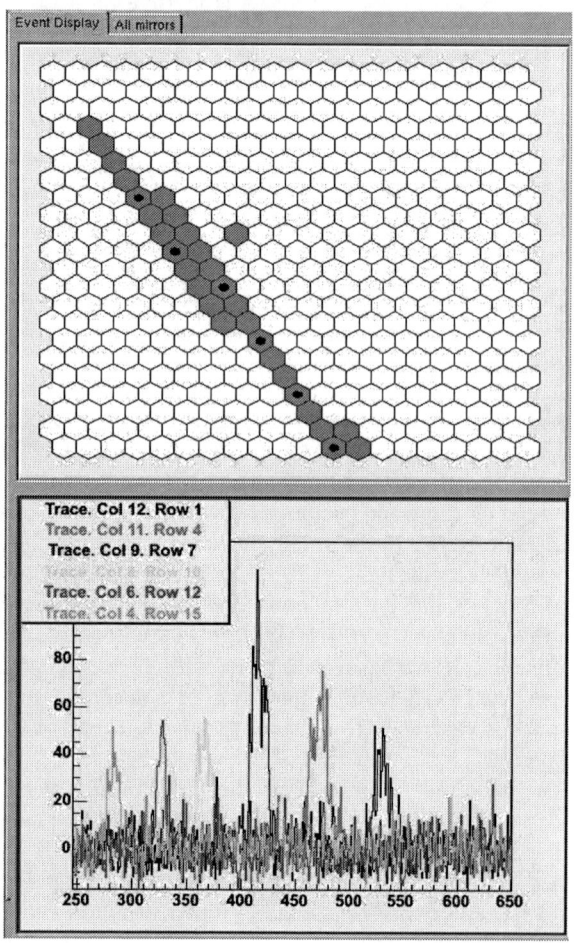

FIGURE 4. A hybrid event as seen by the Los Leones FD camera: pixels with first level trigger, in green (up); FADC time signals, in 100 ns units, for selected pixels along the track (down), showing the time evolution of the shower signal.

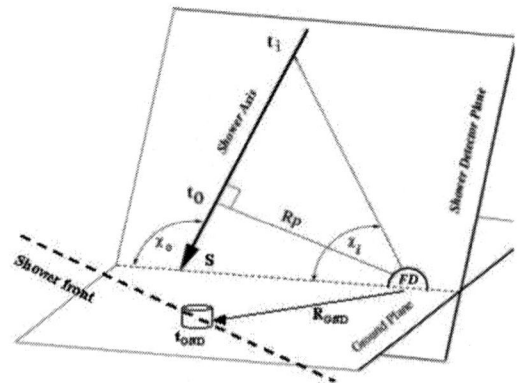

FIGURE 5. Observables for the geometrical reconstruction of a cosmic ray shower.

Geometry reconstruction

FD shower axis reconstruction proceeds in two stages. First, the shower detector plane (SDP) is derived from the angular pattern of hit FD pixels. The SDP is the plane containing the shower axis and the FD. Second, the shower spot angular motion is used to determine the orientation and location of the shower axis in the SDP. The axis is most conveniently described in terms of the SDP normal $\hat{n}_{SDP}$, the perpendicular distance from the FD to shower axis R_p, the angle χ_0 that the axis makes with the horizontal, in the SDP, and t_0 the time at which the shower passes closest to the FD. We call this the "mono" geometric reconstruction method because it relies on information from a single FD telescope. The time information of the pixels is then used for the reconstruction of the shower axis within the SDP. For a given geometry, the arrival time of light onto the pixel i, at angle χ_i, is given by the following expression (see Figure 5):

$$t_i = t_0 + \frac{R_p}{c} tan[(\chi_0 - \chi_i)/2], \qquad (1)$$

The mono procedure suffers from having to find three geometric parameters (R_p, χ_0 and t_0) from a nearly linear relationship between spot position and time. The Auger hybrid procedure solves this problem by exploiting the time of shower front arrival at one or more SD stations. An effective reduction of the fit parameters is accomplished by specifying the time t_0 at which the shower front reaches the position of closest approach, which is related to the ground array tank time t_{GND}, its position $\vec{R}_{GND}$ and the shower axis unit vector $\hat{S}$ by the equation:

$$t_0 = t_{GND} - (\vec{R}_{GND} \cdot \hat{S})/c. \qquad (2)$$

Fitting strategies which use only the timing information from the highest signal tank have been studied, as

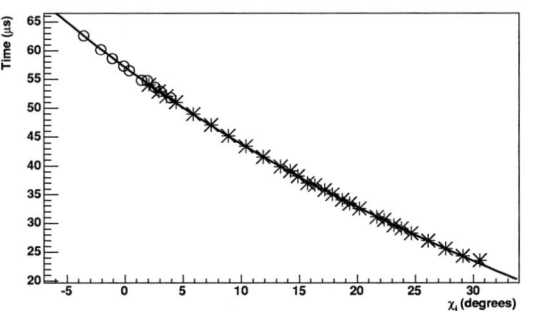

FIGURE 6. Time vs angle correlation for a sample hybrid event: stars represent FD data, open dots SD data, and the line is the results of the best fit to the hybrid geometry.

well as others using all the tanks in the event. In Figure 6, the time as a function of χ_i is shown for the sample hybrid event, with stars representing FD data, and the full line representing the best fit to the hybrid hypothesis. The tanks data are also represented on the same figure as open circles. The event is reconstructed as having a zenith angle of 35°, and core location at 15 km from the Los Leones telescope. The same event as seen by the array is shown in Figure 7. Ten tanks, represented as circles of radius proportional to the logarithm of the energy deposited, are included in the reconstruction. This event was also partly seen by a Coihueco telescope. In this case, the intersection of the SDP's provides an independent, purely geometrical ("stereo") determination of the shower direction and core position. The hybrid reconstruction and the "stereo" one are compatible within a few tenths of a degree, consistent with the expected hybrid accuracy on the shower direction reconstruction.

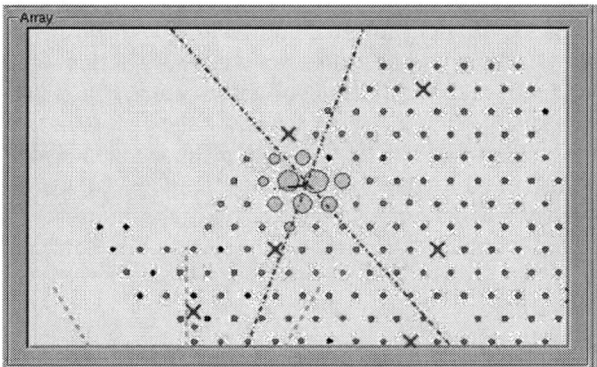

FIGURE 7. The hybrid event shown in Figures 4 and 6, as detected by the array. Tanks are represented by circles of radius proportional to the logarithm of the energy deposited. Dashed-dotted lines represent the intersection between the ground plane and the shower detector planes reconstructed by the telescopes.

Particular attention is being paid to the systematic uncertainties associated to the basic detector measured quantities. The determination of the SDP relies on the FD pixels directions, and it is thus affected by possible misalignments of the telescopes with respect to their nominal direction. Detailed analyses of the reconstructed path of stars on the FD cameras, as well as of the reconstruction of laser shots fired in a known direction, indicate that the alignment is better than 0.1°. The synchronization of the SD stations is given by the GPS receivers installed in each tank. A direct measurement of the SD/FD synchronization is provided by the CLF (Central Laser Facility) which fires periodically the FD with shower-like tracks and injects laser pulses into a tank placed next to the laser facility for this purpose. The tank's measurement of the pulse emission time can then be compared with the FD determination of the same pulse emission time. Differences of only a few hundreds of ns were found, and are corrected for in the data analysis.

Shower energy

In this phase of the analysis, shower energy estimation follows the independent procedures used in SD and FD detection. Nonetheless hybrid detection provides a remarkable constraint through the common geometry obtained with the hybrid fitting method described above.

In the Surface Detector energy is estimated from the lateral distribution function (LDF). This can be obtained from the measured signals at the water tanks as a function of the distance to the shower axis. The measured LDF of the hybrid event is shown in Figure 8. The value of the LDF at 1000 m (S_{1000}) is correlated to the primary cosmic ray energy, giving about $2 \cdot 10^{19}$ eV.

The FD profile reconstruction procedure uses as input the calibrated ADC traces in all pixels. Using the reconstructed geometry of the shower axis and a model of

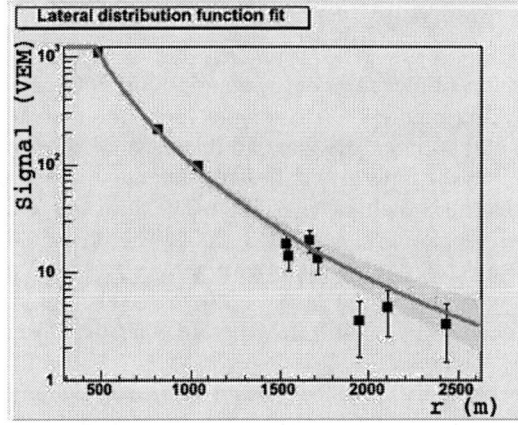

FIGURE 8. Reconstructed LDF for the sample hybrid event: the signal measured for each tank, expressed in Vertical Equivalent Muon (VEM) is plotted as a function of the tank distance to the shower core.

aerosol scattering in the atmosphere is then possible to transform the light received at the detector to the light emitted from the shower axis as a function of slant depth (in g/cm^2). The Cherenkov light contamination, both direct and (Rayleigh and Mie) scattered, is subtracted from the profile [1]. The fluorescence light emitted from a volume of air is proportional to the energy dissipated by the shower particles in that volume. The integral of the light signal is thus proportional to the shower energy. The reconstructed longitudinal profile of the hybrid event, together with a fit to a Gaisser-Hillas function, is shown in Figure 9. The estimated energy from the FD profile reconstruction for this hybrid event was also about $2 \cdot 10^{19}$ eV, in agreement with the S_{1000} determination.

Several hundreds of good quality hybrid events have been collected so far, and their number is continuously increasing with the progress of tanks and telescopes installation.

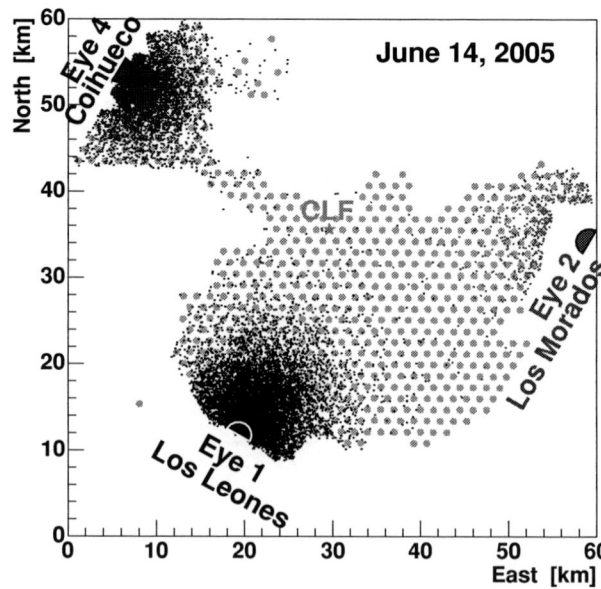

FIGURE 10. Status of the Pierre Auger Observatory on June 14, 2005. Each gray full circle represents a surface station, each color dot is the core location of a reconstructed air shower (where the color indicates which eye saw that event), and each fluorescence site is shown as a colored semi-circle. The Central Laser Facility is shown as a green full star.

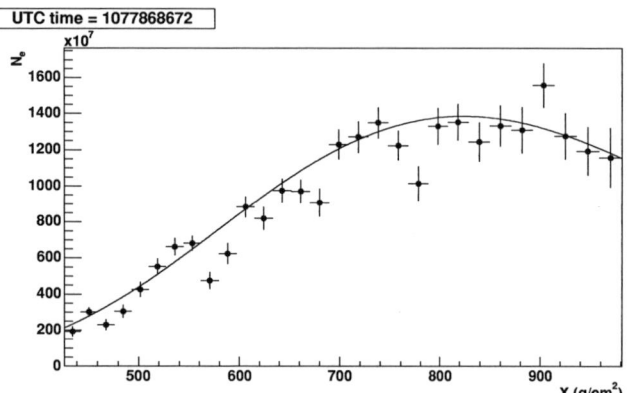

FIGURE 9. The reconstructed longitudinal profile of the hybrid event: the number of shower particles as a function of the slant depth.

OUTLOOK

The status of the Observatory on mid June, 2005 is summarized in Figure 10. There were 740 fully operational tanks at that time. The first two fluorescence sites (eyes 1 and 4 in the figure) were fully operational, i.e. running six telescopes each, in June, 2004. The third site (eye 2) started operation on March 18, 2005. The present average rate is 50 hybrid events per night per eye, for a total of 16090 events up to June 14, 2005. At this rate, 4000 hybrid events per month are expected when the Observatory is completed.

The combination of the air fluorescence measurements and particle detections on the ground provides an energy measurement almost independent of air shower simulations. The fluorescence measurements determine the longitudinal development of the shower, whose integral is proportional to the total energy of the electromagnetic particle cascade. At the same time, the particle density at any given distance from the core can be evaluated with the ground array. The conversion from particle density to the energy of the shower is where the fluorescence measurements become important. Hybrid events that can be independently reconstructed with both techniques are used to establish an empirical rule for the energy conversion.

It is important to note that both techniques have different systematics, and results are preliminary at this stage while the Observatory is under construction. The possibility of studying the same set of air showers with two independent methods is valuable in understanding the strengths and limitations of each technique. The hybrid analysis benefits from the calorimetry of the fluorescence technique and the uniformity of the surface detector aperture.

REFERENCES

1. The Pierre Auger Collaboration, "Properties and performances of the prototype instrument for the Pierre Auger Observatory", Nucl. Instr. Meth. A523 (2004) 50-95
2. K. Greisen, Phys. Rev. Lett. 16 (1966) 748;
   G.T. Zatsepin, V.A. Kuzmin, Sov. Phys. JETP Lett. 4 (1966) 78.

Parallel Session I:

Standard Model

Conveners:

V. Ciulli
University of Firenze and INFN, Italy

M. Cobal
University of Udine and INFN, Italy

S. Moretti
University of Southampton, United Kingdom

The Measure of α_{QED} in e^+e^-: An Alternative Approach[1]

Luca Trentadue

Dipartimento di Fisica, Università di Parma
and
INFN, Gruppo Collegato di Parma, I-43100 Parma, Italy

Abstract. We propose a method to determine the running of α_{QED} from a measurement of small angle Bhabha scattering. The method is suited to high statistics experiments at e^+e^- colliders equipped with luminometers in the appropriate angular region. A new simulation code predicting small angle Bhabha scattering is also presented

Keywords: e^+e^-, Radiative Corrections, Bhabha scattering, α_{QED}
PACS: 11.15.-q, 12.15.Lk, 12.20.-m

The method proposed here deals with an alternative approach for a detailed accurate measurement of a fundamental constant within the Standard Model of Electroweak interactions i.e. α_{QED}. The electroweak Standard Model $SU(2) \otimes U(1)$ contains as a constitutive part Quantum Electrodynamics (QED). The running of the electromagnetic coupling α is determined by the theory

$$\alpha(q^2) = \frac{\alpha(0)}{1 - \Delta\alpha(q^2)} \quad (1)$$

where $\alpha(0) = \alpha_0$ is the Sommerfeld fine structure constant, which has been measured to a precision of $3.7 \cdot 10^{-9}$ [1]. $\Delta\alpha(q^2)$ positive, arises from loop contributions to the photon propagator. The numerical prediction of electroweak observables involves the knowledge of $\alpha(q^2)$, usually for $q^2 \neq 0$. For instance, the knowledge of $\alpha(m_Z^2)$ is relevant for the evaluation of quantities measured by the LEP experiments. This is achieved by evolving α from $q^2=0$ up to the Z mass scale $q^2 = m_Z^2$. The evolution expressed by the quantity $\Delta\alpha$ receives contributions from leptons, hadrons and the gauge bosons. The hadronic contribution to the vacuum polarisation, which cannot be calculated from first principles, is estimated with the help of a dispersion integral and evaluated[2] by using total cross section measurements of $e^+e^- \to hadrons$ at low energies. Therefore, any evolved value $\alpha(q^2)$, particularly for $|q^2| > 4m_\pi^2$, is affected by uncertainties originating from hadronic contributions. The uncertainty on $\alpha(m_Z^2)^{-1}$ induced by these data is as small as $\pm$ 0.09 [2], nevertheless it turned out[3] that it limits the accurate prediction of electroweak quantities within the Standard Model, particularly for the prediction of the Higgs mass. While waiting for improved measurements from BEPC, VEPP-4M and DAFNE as input to the dispersion integral, intense efforts are made to improve on estimating the hadronic shift $\Delta\alpha_{had}$, as for instance[4, 5, 6, 7], and to find alternative ways of measuring α itself. Attempts have been made to measure $\alpha(q^2)$ directly using e^+e^--data at various energies, such as measuring the ratio of $e^+e^-\gamma/e^+e^-$ [8] or more directly the angular distribution of *Bhabha* scattering[9]. We propose[10] the running of α by using small angle *Bhabha* scattering. This process provides unique information on the QED coupling constant α at low *spacelike* momentum transfer $t = -|q^2|$, where $t = -\frac{1}{2} s (1 - \cos\theta)$ is related to the total invariant energy $\sqrt{s}$ and the scattering angle θ of the final state electron. The small angle region has the virtue of giving access to values of $\alpha(q^2)$ without being affected by weak contributions. The cross section can be theoretically calculated with precision at the per mille level. It is dominated by the photonic t-channel exchange and the non-QED contributions have been computed[11] and are on the order of 10^{-4}, in particular contributions from boxes with two weak bosons are safely negligible. In general, the *Bhabha*-cross section is computed from the entire set of gauge invariant amplitudes in the s- and t-channel. The s-channel contribution gives only a negligible contribution[10]. Thus, the measurement of the angular distribution allows indeed to verify directly the running of the coupling $\alpha(t)$. Such a measurement constitutes a genuine test of QED alone. In fact, QED - as part of the electroweak theory - is valid as a consistent theory by itself, since for the applications considered here the conditions $|q^2| \ll m_W^2, m_Z^2$ are fulfilled. Furthermore, for the actual calculations $\theta \gg m_e/E_{beam}$ and $E_{beam} \gg m_e$ must be satisfied. Obviously, in order to manifest the running, the experimental precision must be adequate. This idea can be realized by high statistics experiments at e^+e^--colliders equipped with finely segmented lumi-

---

[1] Work done in collaboration with A.B. Arbuzov, D. Haidt, C. Matteuzzi and M. Paganoni

nometers, in particular by the LEP experiments given their large event samples, by SLC and future Linear Colliders. The relevant luminometers cover the t-range from a few GeV^2 to order 100 GeV^2. The t-dependence of the quantity $\Delta\alpha(t)$ at small values of t may be obtained using the program *alphaQED* by Jegerlehner[2]. At low energies is dominated by the contribution from the leptons, while with increasing energy also the contribution from loops due to hadrons gets relevant. The region where hadronic corrections are critical is contained in the considered t-range. The experimental determination of the angular distribution of the *Bhabha* cross section requires the precise definition of a *Bhabha* event in the detector. The analysis follows closely the procedure adopted in the luminosity measurement which is described in detail, for instance in ref.[14](YR), and elaborates on the additional aspect related to the measurement of a differential quantity. To this aim the luminosity detector must have a sufficiently large angular acceptance and adequate fine segmentation. The variable t is reconstructed on an event-by-event basis.

The method to measure the running of α exploits the fact that the cross section for the process $e^+e^- \to e^+e^-$ can be conveniently decomposed into three factors:

$$\frac{d\sigma}{dt} = \frac{d\sigma^0}{dt} \left(\frac{\alpha(t)}{\alpha(0)}\right)^2 (1 + \Delta r(t)). \quad (2)$$

All three factors are predicted to a precision of 0.1 % or better. The first factor on the right hand side refers to the *Bhabha* Born cross section including soft and virtual photons according to ref.[11], which is precisely known, and accounts for the strongest dependence on t. The vacuum polarization effect in the leading photon t-channel exchange is incorporated in the running of α and gives rise to the squared factor in eq.2. The third factor $\Delta r(t)$ collects all the remaining real (in particular collinear) and virtual radiative effects not incorporated in the running of α[11, 12]. The experimental data after correction for detector effects is to be compared with eq.2. This goal is achieved by using a newly developped program based on the already existing semi-analytical code NLLBHA[11, 15] called SAMBHA [10]. The two-point functions $\Pi(t) = \Delta\alpha(t)$ and $\Pi(s) = \Delta\alpha(s)$ are responsible for the running of α in the space-like and time-like regions. In the language of Feynman diagrams the effect arises from fermion loop insertions into the virtual photon lines. Anticipating the application of the proposed method to measure the t-dependence of $\alpha(t)$ to the data of a real experiment, a Monte Carlo simulation is carried out [10]. Electrons, positrons and photons are observed as clusters. Their reconstruction is based on a cluster algorithm. By applying the selection criteria the event sample is divided in clusters which areattibuted to various rings in the luminometer[10]. The hadronic contribution may be deduced by subtracting the leptonic contribution which is theoretically precisely known[10]. The extraction of the hadronic contribution is only limited by the experimental precision (see the talk of G. Abbiendi at this conference[18]). To conclude a novel experimental approach to access directly the running of α in the t-channel is proposed. It consists in analysing small angle *Bhabha* scattering. Depending on the particular angular detector coverage and on the energy of the beams, it allows to cover a sizeable range of the t-variable. The information obtained in the t-channel can be compared with the existing results of the s-channel measurements. This represents a complementary approach which is direct, transparent and based only on QED interactions and furthermore free of some of the drawbacks inherent in the s channel methods. The method outlined can be readily applied to the experiments at LEP[18] and SLC. It can also be exploited by future e^+e^- colliders as well as by existing lower energy machines. An exceedingly precise measurement of the QED running coupling $\Delta\alpha(t)$ for small values of t may be possibly envisaged with a dedicated luminometer even at low machine energies.

ACKNOWLEDGEMENTS

I would like to thank A. Tricomi, S. Albergo e M. Chiorboli for organizing such an excellent meeting.

REFERENCES

1. D.E. Groom et.al.: Eur. Phys. J. **C15** (2000) 1
2. S. Eidelman and F. Jegerlehner: Z. Phys. **C67** (1995) 602
3. K. Hagiwara, D. Haidt, S. Matsumoto and C.S. Kim : Z.Phys. **C64** (1994) 559; Z.Phys. **C68** (1995) 353; K. Hagiwara, D. Haidt and S. Matsumoto : Eur. Phys. J. **C2** (1998) 95
4. F. Jegerlehner: hep-ph/0308117
5. M. Davier and A. Höcker : Phys. Lett. **B435** (1998) 427; M. Davier, S. Eidelman, A. Höcker, Z. Zhang, Eur. Phys. J. **C27**(2003) 497-521
6. D.Karlen and H.Burkhardt, Eur. Phys. J. **C22** (2001) 39; hep-ex/0105065 (2001)
7. A.A. Pankov, N. Paver, Eur. Phys. J. **C29**(2003) 313-323
8. TOPAZ Collaboration, I. Levine et al.: Phys. Rev. Lett. **78** (1997) 424
9. L3 Collaboration : M. Acciarri *et al*, Phys. Lett. **B476** (2000) 48
10. A.B.Arbuzov, D.Haidt, C.Matteuzzi, M.Paganoni and L. Trentadue, European Physics Journal C35, 267 (2004) . hep-ph/0402211.
11. A.B. Arbuzov, V.S. Fadin, E.A. Kuraev, L.N. Lipatov, N.P. Merenkov and L. Trentadue, Nucl. Phys. **B485** (1997) 457. ro
12. E.A. Kuraev and V.S. Fadin, Sov. J. Nucl. Phys. **41** (1985) 466. O. Nicrosini and L. Trentadue, Phys.Lett. **B196** (1987)

551; Z. Phys. **C39** (1988) 479.
13. S.J. Brodsky, G.P. Lepage and P.B.Mackenzie, Phys. Rev. **D28** (1983) 228
14. LEP Working Group, Yellow Report CERN 96-01, Event Generators for Bhabha scattering, Convenors : S.Jadach and O.Nicrosini, p.229
15. Yellow Report CERN 96-01 : *description of NLLBHA*
16. S. Eidelman and F. Jegerlehner, Z. Phys. **C67** (1995) 585; hep-ph/9502298.
17. A. Arbuzov *et al.*, Phys. Lett. **B383** (1996) 238; hep-ph/9605239.
18. OPAL collaboration preprint 2005

Multiple Photon Corrections to W and Z Boson Production at Hadron Colliders

C. M. Carloni Calame[*,†], G. Montagna[†,*], O. Nicrosini[*,†] and M. Treccani[†]

[*]*Istituto Nazionale di Fisica Nucleare, Sezione di Pavia, via A. Bassi 6, 27100, Pavia, Italy*
[†]*Dipartimento di Fisica Nucleare e Teorica, Università di Pavia, via A. Bassi 6, 27100, Pavia, Italy*

Abstract. The W and Z production processes play an important role in the physics program of hadron colliders. The expected accuracy at the Tevatron Run II and the LHC requires progress in the calculation of electroweak radiative corrections to these processes. To this end, higher order QED corrections to the Drell-Yan-like W and Z production, due to final state multiple photon radiation, are calculated. Particular attention is paid to the effects induced by such corrections on the experimental observables which are relevant for high precision measurements of the W boson mass. The calculation is implemented in the Monte Carlo event generator HORACE, which is available for data analysis.

Keywords: hadron collider, Drell-Yan processes, radiative corrections
PACS: 12.15.Lk, 13.40.Ks, 14.70.Fm, 14.70.Hp

INTRODUCTION

The experiments at the high energy hadron colliders Tevatron RunII and LHC [1, 2] are expected to continue the program of precision physics successfully carried out at LEP, SLC and the Tevatron itself. In particular, a very precise determination of the W boson mass M_W is important because, together with an improved measurement of the top quark mass, it will allow to put more severe indirect bounds on the mass of the Higgs boson.

For the neutral current Drell-Yan process, it is understood that the hadronic measurement of the Z line shape cannot be competitive with the high precision measurements of the LEP collider. However, in order to improve the accuracy in the charged current Drell-Yan process, it is necessary to fully control the uncertainties involved in the Z production process, since in a hadron environment these two kinds of measurements are deeply connected. Most of the uncertainties involved in a hadronic W mass measurement, such as the lepton scale, the lepton resolution and the modeling of p_t^W, can be estimated through the neutral current Drell-Yan process. The latter can also be used to determine the W mass from the ratio of the transverse mass distributions of the W and Z boson (especially at high luminosity), as well as to extract the effective weak mixing angle from the forward-backward asymmetry. Furthermore, both processes are backgrounds to many new physics searches and can also be useful to monitor the collider luminosity and measure the Parton Distribution Functions (PDFs).

At the LHC, the designed luminosity will allow to collect large samples of single Z events. Therefore, since the previous systematics scale as the Z control sample size, they will become almost negligible. However, theoretical systematics are also present in addition to experimental uncertainties. Since the goal at the LHC is to reach a precision of the order of 15 MeV in the W mass determination [1, 2], these systematics will dominate over the statistical error.

Having in mind the precision anticipated for M_W and the need of a precise luminosity monitoring, accurate theoretical predictions, including QCD and electroweak radiative corrections, are necessary for the Drell-Yan-like W and Z production. In the following, we summarize the status of high precision calculations, paying particular attention to multiple photon effects due to final state radiation, as recently computed in [3, 4, 5, 6].

RADIATIVE CORRECTIONS

The α_s corrections to Drell-Yan-like processes were calculated long time ago in [7, 8]. Refined calculations at α_s^2 (NNLO) accuracy have been performed in [9, 10] at the total cross section level. Recently, the NNLO corrected W and Z rapidity distributions have been calculated in [11].

At this level of accuracy, the pure electroweak first order corrections could have *a priori* an impact of the same order of the QCD NNLO effects and therefore the former need to be studied in detail.

Actually, $\mathcal{O}(\alpha)$ electroweak corrections to W and Z production processes are known to contribute at some percent level, as shown in the calculations for W [12, 13, 14] and Z [15, 16] production. As noted first in [17], QED final state radiative corrections result to be the dominant contributions in the electroweak sector. It is therefore natural to argue what is the impact of

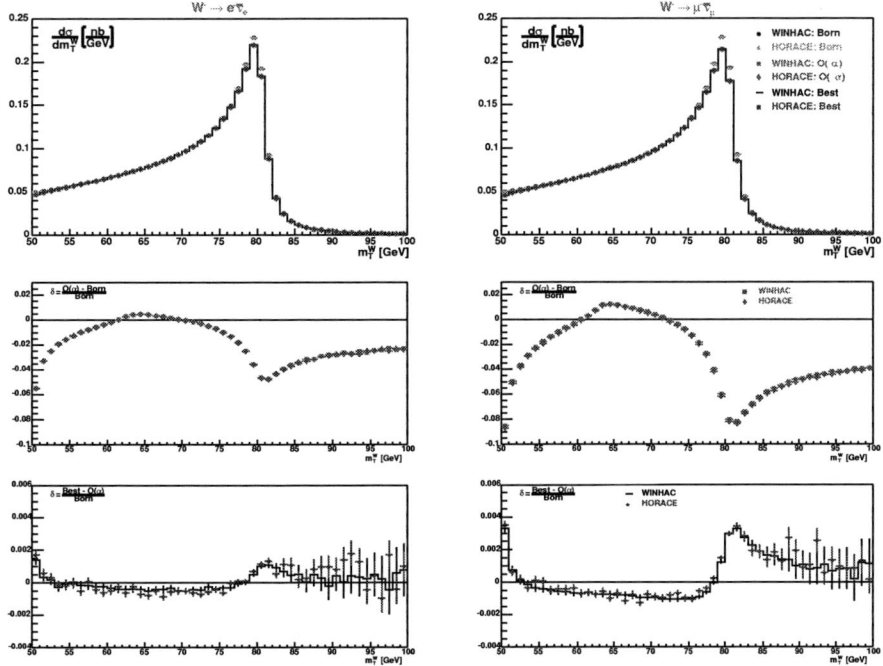

FIGURE 1. M_T^W distribution as obtained by HORACE and WINHAC. The relative effects of $\mathcal{O}(\alpha)$ and higher order QED corrections are shown. From [5].

the multiphoton emission from the final state leptons. The real plus virtual corrections due to multi-photon radiation can be computed in the leading-log approximation using the QED structure function approach. The corrections can be calculated by solving the QED DGLAP equation by means of the QED Parton Shower algorithm developed in [18, 19]. This formulation is implemented into a Monte Carlo generator, HORACE [3], which can incorporate also lepton identification criteria and detector resolution effects, in order to perform simulations for the hadronic processes as realistic as possible. HORACE calculates QED corrections to all orders and at $\mathcal{O}(\alpha)$, in order to disentangle the effect of higher order contributions and to compare with the available $\mathcal{O}(\alpha)$ programs.

MULTI-PHOTON EFFECTS

The preferred quantity to determine M_W at hadron colliders is the transverse mass spectrum M_T^W, because it is less sensitive than the lepton p_T distribution to the W transverse motion, which is difficult to model and to measure.

In Fig. 1, the predictions of $\mathcal{O}(\alpha)$ and higher order QED corrected M_T^W distribution from the programs HORACE and WINHAC [4] are shown. The results of the two generators are in good agreement and in particular the $\mathcal{O}(\alpha)$ corrections around the Jacobian peak amount to about 5% and about 10% for the electron and muon channel, respectively, while higher order effects vary from 0.2% to 0.5%.

In the neutral current process, various observables are exploited to calibrate the calorimeter and also to infer information about the p_T^W distribution. The invariant mass distribution and the Z transverse mass distribution are two observables of particular interest.

Fig. 2 shows the effect of multi-photon radiation on the invariant mass distribution with respect to the fixed $\mathcal{O}(\alpha)$ prediction, when using the cuts and lepton identification requirements of [6]. Higher order corrections pull back the distribution towards the Born level one and change sign when passing from below to above the Z peak. The relative correction reaches the order of 10% for bare electrons and of 1% for muons, because of different lepton masses involved. After photon recombination, the higher order correction for calorimetric electrons becomes flat and is reduced well below the 1% level, because of the partial disappearance of the lepton mass logarithms. The same results hold also in the transverse mass distribution, while for non strongly varying distributions such as lepton and Z rapidity and the forward-backward asymmetry, the higher order corrections are at a few 0.1% level.

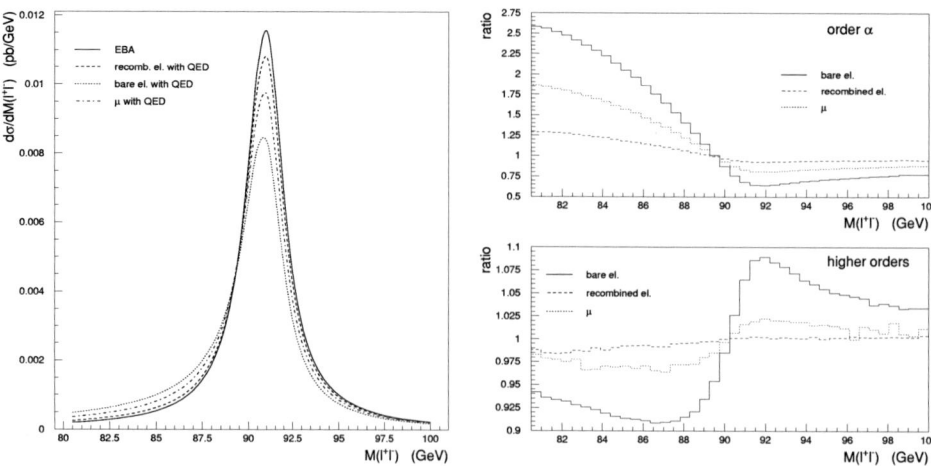

FIGURE 2. The invariant mass distribution $M_{l^+l^-}$ (left panel) and relative effect of $\mathcal{O}(\alpha)$ (right panel, up) and higher order QED corrections (right panel, down) for the Z lepton decays $Z \to e^+e^-, \mu^+\mu^-$ at $\sqrt{s} = 2$ TeV, according to the cuts and lepton identification criteria used in [6].

W AND Z MASS SHIFTS

In order to complete the phenomenological analysis, we investigated the shift induced by multiple QED radiation on the fitted boson masses. The strategy followed by the CDF and DØ collaborations to extract the $W(Z)$ mass from the data is to perform a fit to the transverse mass(invariant mass) distribution. In our study, we perform binned χ^2 fits to the M_T^W (M_{inv}) distribution, in analogy with the experimental procedure. According to this procedure, we generate a sample of "pseudo-data" for the distribution of interest, calculating the distribution at the lowest order (LO) in terms of an input boson mass $M^{\text{LO,input}}$, with high numerical precision. Next, we compute the same distribution including $\mathcal{O}(\alpha)$ radiative corrections for a number of boson mass values and we calculate, for each M value, the χ^2 as

$$\chi^2 = \sum_i (\sigma_{i,\alpha} - \sigma_{i,\text{LO}})^2 / (\Delta\sigma_{i,\alpha}^2 + \Delta\sigma_{i,\text{LO}}^2) \quad (1)$$

where $\sigma_{i,\text{LO}}$ and $\sigma_{i,\alpha}$ are the Monte Carlo predictions at the LO and $\mathcal{O}(\alpha)$ level, respectively, for bin i and $\Delta\sigma_{i,\text{LO}}, \Delta\sigma_{i,\alpha}$ the corresponding statistical errors due to numerical integration. From the minimum of the χ^2 distribution, we derive the fitted boson mass value M^α_{fitted} and we quantify the mass shift due to $\mathcal{O}(\alpha)$ corrections as $\Delta M^\alpha \equiv M^{\text{LO}}_{\text{input}} - M^\alpha_{\text{fitted}}$. The shift due to higher order corrections is derived according to the same procedure, by generating a sample of "pseudo-data" for the distribution of interest at $\mathcal{O}(\alpha)$ with an input boson mass value M^α_{input} and fitting them in terms of templates of the transverse or invariant mass distribution obtained by including higher order corrections for different boson mass values. The mass shift induced by higher order corrections is obtained as $\Delta M^{\text{h.o.}} \equiv M^\alpha_{\text{input}} - M^{\text{h.o.}}_{\text{fitted}}$, where $M^{\text{h.o.}}_{\text{fitted}}$ is the boson mass value returned by the fit.

In order to properly simulate the finite resolution of the calorimeter, a Gaussian smearing has been applied to the particle four-momenta, and its effect is described in the second column of Tab.1. The $\mathcal{O}(\alpha)$ mass shifts inferred by means of this techinque are in reasonable agreement with the results of the CDF and DØ collaborations, even in the absence of a complete detector simulation. The mass shift due to higher order effects is about -10 MeV for the muon channel and a few MeV for the electron channel. The same conclusion holds also in the LHC case. In the Z case, the higher order contributions determine a relative shift of -44 MeV and -6 MeV in the muon and electron channel. Moreover, the negative sign of the shift due to higher order corrections confirms that they tend to mitigate the $\mathcal{O}(\alpha)$ corrections. It is important to notice that the numerical value of these mass shifts depend in a crucial way from the lepton identification criteria and the detector uncertainties. Therefore, the importance of these results, rather than in its absolute value, is in the relative shift due to the higher order corrections with respect to the shift due to $\mathcal{O}(\alpha)$ contributions, which turns out to be about 10% of the latter.

CONCLUSIONS

Recent effort in the calculation of higher order QED corrections to single W and Z production at hadron colliders has been reviewed, and the impact of final state multiphoton corrections on several distributions of experimental

TABLE 1. The Z (W) mass shifts due to $\mathcal{O}(\alpha)$ ΔM_Z^α (ΔM_W^α) and higher order QED corrections $\Delta M_Z^{h.o.}$ ($\Delta M_W^{h.o.}$), according to the different experimental conditions discussed in [6].

Particle	Smearing	Lepton ID	ΔM_Z^α (MeV)	$\Delta M_Z^{h.o.}$ (MeV)	ΔM_W^α (MeV)	$\Delta M_W^{h.o.}$ (MeV)
e	no	no	595	-135		
μ	no	no	270	-31		
e	no	yes	75	-5		
μ	no	yes	215	-28		
e	yes	no	780	-159	400	-40
μ	yes	no	565	-49	220	-10
e	yes	yes	105	-6	20	-2
μ	yes	yes	420	-44	110	-10

interest has been evaluated.

For the neutral current process, and in the presence of lepton identification requirements, multiple photon corrections to the lepton pair invariant mass and to the Z transverse mass distribution were found to modify the distributions around the Z peak by about 0.1-1%. Smaller effects are observed for quantities of interest for luminosity, such as the lepton and Z rapidity distributions, and for the forward-backward asymmetry.

In the W case, the predictions of **HORACE** have been compared with those of the independent program WIN-HAC. For the W transverse mass distribution, the two codes agree in estimating the impact of higher order QED corrections at the 0.2% level in the electron decay channel and at 0.5% in the muon decay channel.

In both W and Z production, the higher order corrections soften the impact of the $\mathcal{O}(\alpha)$ ones, moving the distributions shape closer to the Born level one. These results clearly indicate that higher order QED effects have a relevant effect on most of the distributions adopted in W mass determination, calorimeter calibration as well as PDFs and parton luminosity measurements at the LHC and therefore they need to be properly taken into account in the present and future experimental analysis. As a final step, the shifts induced by higher order contributions in the measured boson masses have been estimated by means of a simulated experiment, finding that, in view of the expected precision at the Tevatron Run II and the LHC, the impact of multiple QED radiation on the vector boson masses is non-negligible in both the decay channels.

ACKNOWLEDGMENTS

One of us (M. T.) thanks S. Moretti for the kind invitation.

REFERENCES

1. R. Brock et al., Report of the working group on precision measurements, in Proceedings of the workshop QCD and weak boson physics in Run II, Fermilab 1999
2. S. Haywood et al., Electroweak Physics, in Proceedings of the Workshop on Standard Model Physics (and more) at the LHC, G. Altarelli and M.L. Mangano eds., CERN Report 2000-04, p. 117
3. C.M. Carloni Calame, G. Montagna, O. Nicrosini, M. Treccani, Phys. Rev. D **69** (2004) 037301
4. W. Płaczek and S. Jadach, Eur. Phys. J. C **29**, (2003) 325
5. C.M. Carloni Calame, S. Jadach, G. Montagna, O. Nicrosini, W. Płaczek, Acta Phys. Polon. B **35** (2004) 1643
6. C. M. Carloni Calame, G. Montagna, O. Nicrosini and M. Treccani, JHEP **0505** (2005) 019
7. G. Altarelli, R. K. Ellis and G. Martinelli, Nucl. Phys. B **157**, (1979) 461
8. B. Humpert and W. L. Van Neerven, Phys. Lett. B **85**, (1979) 293
9. R. Hamberg, W. L. van Neerven and T. Matsuura, Nucl. Phys. B **359** (1991) 343 [Erratum, ibid. B **644** (2002) 403]
10. R. V. Harlander and W. B. Kilgore, Phys. Rev. Lett. **88** (2002) 201801
11. C. Anastasiou, L. J. Dixon, K. Melnikov and F. Petriello, Phys. Rev. D **69**, (2004) 094008
12. U. Baur, S. Keller and D. Wackeroth, Phys. Rev. D **59** (1999) 013002
13. S. Dittmaier and M. Krämer, Phys. Rev. D **65** (2002) 073007
14. U. Baur and D. Wackeroth, Phys. Rev. D **70** (2004) 073015
15. U. Baur, S. Keller and W. K. Sakumoto, Phys. Rev. D **57**, (1998) 199
16. U. Baur, O. Brein, W. Hollik, C. Schappacher and D. Wackeroth, Phys. Rev. D **65**, (2002) 033007
17. F. A. Berends and R. Kleiss, Z. Phys. C **27**, (1985) 365
18. C.M. Carloni Calame, C. Lunardini, G. Montagna, O. Nicrosini, F. Piccinini, Nucl. Phys. B **584** (2000) 459
19. C.M. Carloni Calame, Phys. Lett. B **520** (2001) 16

Electroweak Physics at the Tevatron

Sandra Leone [1]

INFN Sezione di Pisa, Largo B. Pontecorvo 3, 56100 Pisa, Italy

Abstract. We present some recent measurements on electroweak physics obtained by the CDF and DØ experiments analyzing data collected at the Tevatron proton anti–proton Collider in Run II.

Keywords: Electroweak, Tevatron, CDF, DØ, W boson, Z boson
PACS: 12.15.-y, 12.15.Ji, 12.38.Qk, 13.38.Be, 13.38.Dg, 14.70.Fm, 14.70.Hp

INTRODUCTION

Electroweak measurements at the Tevatron are complementary to those performed at e^+e^- machines. Over the next few years the Tevatron is the only accelerator which can produce W and Z bosons. Measuring their properties is an important test of the standard model of elementary particles (SM). Significant deviations from SM predictions could indicate the presence of new physics.

Both CDF and DØ are multipurpose detectors. They have been extensively described elsewhere [1], [2].

W AND Z INCLUSIVE CROSS SECTION MEASUREMENTS

W and Z bosons are identified by their leptonic decay into electrons, muons and taus. Inclusive cross sections of both W and Z have been measured in all the three lepton channels [3]. Figure 1 summarizes the CDF and DØ cross section measurements. All measurements are in agreement with the NNLO calculations, represented by the vertical band [4].

The ratio R of the cross section measurements for W and Z can be used to indirectly extract the total width of the W boson. R can be expressed as:

$$R = \frac{\sigma(p\bar{p} \to W)}{\sigma(p\bar{p} \to Z)} \frac{\Gamma(W \to \ell\nu)}{\Gamma(Z \to \ell\ell)} \frac{\Gamma(Z)}{\Gamma(W)}.$$

Inserting the SM predictions for the total cross sections and $\Gamma(W \to \ell\nu)$ and using the experimental Z total and partial width from LEP one can extract $\Gamma(W)$. CDF measured the following values:
$\Gamma(W)$ = 2079±41 MeV $e + \mu$ channel, $\int \mathcal{L}\, dt$ = 72 pb^{-1},
$\Gamma(W)$ = 2056±44 MeV μ channel, $\int \mathcal{L}\, dt$ = 194 pb^{-1},
in agreement with both the PDG world average [5] and the SM prediction (2091.1 ± 2.5 MeV).

DIRECT W WIDTH MEASUREMENT

DØ measured directly the W width in the electron channel using the transverse mass distribution, defined as:

$$M_T = \sqrt{2 p_T^\ell p_T^\nu (1 - \cos\Delta\phi)},$$

where p_T is the lepton transverse momentum and $\Delta\phi$ is the difference in azimuthal angle between the two leptons [6].

The width is determined by normalizing the signal and background M_T distribution in the region of $50 < M_T < 100$ GeV/c^2 and then fitting the predicted shape of the candidate events in the tail region $100 < M_T < 200$ GeV/c^2 which is most sensitive to the width. Figure 2 shows the M_T distribution. The measurement, obtained using 177 pb^{-1} of data, yields: $\Gamma(W)$ = 2011 ± 93(stat) ± 107(syst) MeV. The uncertainty is already lower than DØ run 1 result.

W CHARGE ASYMMETRY

W bosons at the Tevatron are primarily produced through annihilation of valence u (d) and anti–d (anti–u) quarks for W^+ (W^-). Since u quarks carry, on average, a higher fraction of the proton momentum than d quarks, a W^+ tends to be boosted in the proton direction, while a W^- in the anti–proton direction. This results in a charge asymmetry defined as:

$$A_{y_W} = \frac{d\sigma(W^+)/dy_W - d\sigma(W^-)/dy_W}{d\sigma(W^+)/dy_W + d\sigma(W^-)/dy_W},$$

where y_W is the W rapidity and $d\sigma(W^\pm)/dy_W$ is the differential cross section for $W^\pm$ production. A measurement of the charge asymmetry is sensitive to the ratio of u and d quark components of parton distribution functions (PDF). However, since the longitudinal component of the neutrino momentum is not measured, y_W cannot

---

[1] on behalf of the CDF and DØ Collaborations

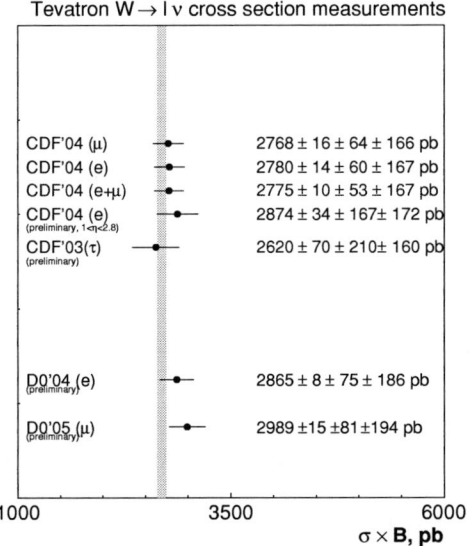
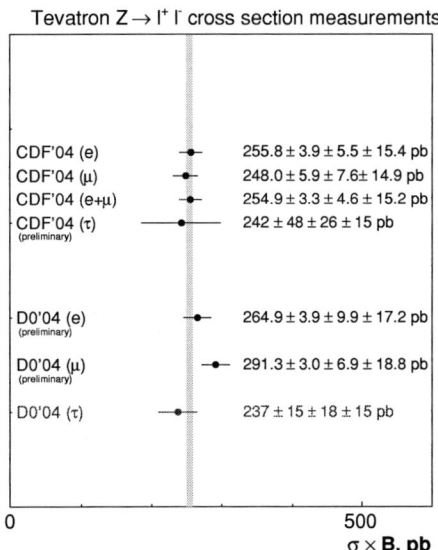

FIGURE 1. $\sigma(p\bar{p} \to W) \times BR(W \to \ell\nu_\ell)$ (left) and $\sigma(p\bar{p} \to Z) \times BR(Z \to \ell\ell)$ (right) measured at CDF and DØ. The uncertainties are listed in the following order: statistical, systematic and luminosity. The vertical band indicates the theoretical (NNLO) predictions.

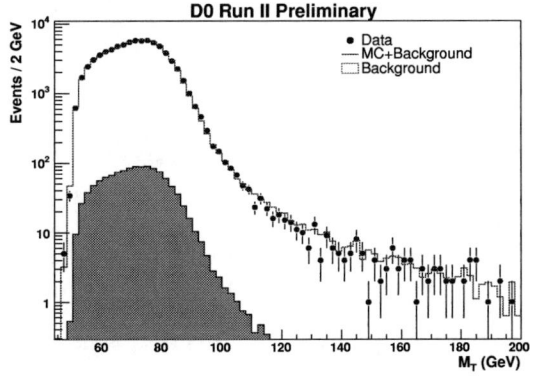

FIGURE 2. Transverse mass distribution M_T used by DØ for direct W width determination.

be determined. Therefore we measure the electron charge asymmetry defined as:

$$A(\eta_e) = \frac{d\sigma(e^+)/d\eta_e - d\sigma(e^-)/d\eta_e}{d\sigma(e^+)/d\eta_e + d\sigma(e^-)/d\eta_e},$$

where η_e is the electron pseudorapidity. The observed asymmetry is a convolution of the W production charge asymmetry and the $V-A$ asymmetry of the W decay. CDF made this measurement on a data sample of 170 pb^{-1}. Figure 3 shows the measured asymmetry corrected for the effect of charge misidentification and background contributions [7]. This measurement will provide an important input for the next calculation of PDF's.

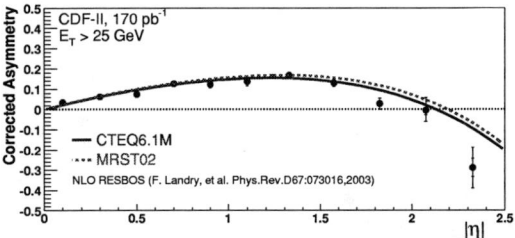

FIGURE 3. W charge asymmetry as a function of the electron η. The predictions using different PDF's are shown.

Z RAPIDITY DISTRIBUTION

Measurement of the rapidity distribution of Drell–Yan pairs in the Z boson mass region provides a test of PDF's since the momentum fraction carried by the parton is directly related to the rapidity of the Z boson. DØ measured $d\sigma/dy_Z$ of the Drell–Yan process in the dielectron's mass range $71 < M_{ee} < 111$ GeV/c^2, using 337 pb^{-1} of data. Figure 4 shows the $d\sigma/dy_Z$ measured distribution [8].

W MASS

The W mass (M_W) measurement, together with the top quark mass measurement, constraints, in the framework of the SM, the mass of the unobserved Higgs boson. At

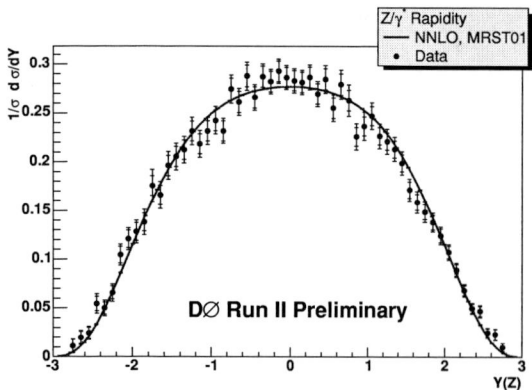

FIGURE 4. $d\sigma/dy_Z$ measurement compared to NNLO prediction based on MRST01 PDF.

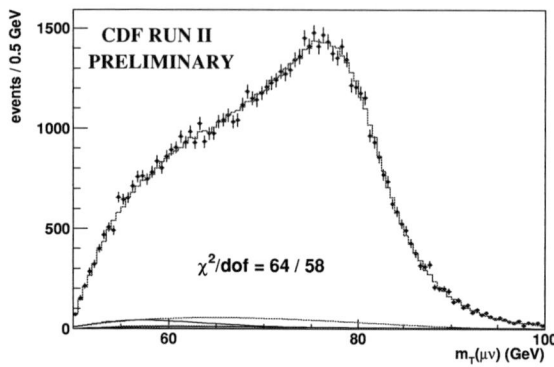

FIGURE 5. M_T spectrum for $W \to \mu\nu$ candidate events at CDF. The histogram represents the simulation with background contribution added. The region between 60–90 GeV/c^2 is used to fit the M_W.

hadron colliders, it is measured from a maximum likelihood fit to the transverse mass spectrum in the $e\nu$ and $\mu\nu$ channels. There are two main components leading to a precise M_W measurement: calibration of the detector to the highest possible precision and simulation of the M_T spectrum. CDF estimated the W mass uncertainty analyzing 200 pb^{-1} of Run II data. Figure 5 shows the M_T distribution in the muon channel compared to the simulation and the expected background contribution. Table 1 summarizes the uncertainties associated with the CDF M_W measurement. The current overall uncertainty is 76 MeV/c^2, already lower than CDF run 1 result [9]. The M_W fit results are currently blinded with a constant offset which will be removed when further cross checks will be completed.

DIBOSON PRODUCTION

The structure of the SM implies that the electroweak gauge bosons W and Z can interact with one another through trilinear and quartic gauge boson vertices. Study of events containing pairs of vector bosons provides a sensitive test of the SM since physics beyond the SM could alter the cross sections and the production kinematics.

$W\gamma$ and $Z\gamma$

The $W\gamma$ production can be used to study the $WW\gamma$ vertex. The final state used in this study is $p\bar{p} \to \ell\nu\gamma$ ($\ell = e$ or μ).

The trilinear gauge couplings of the Z boson to the photon are zero. Therefore evidence of such an interaction would indicate new physics. The final state used in this study is $p\bar{p} \to \ell\ell\gamma$.

The main background for these processes is $V+$ jet production ($V = W, Z$) where the jet is misidentified as a photon. Both CDF and DØ studied these channels. Figure 6 shows on the left the photon E_T spectrum for $W\gamma$ candidate events at DØ and on the right the photon E_T spectrum for $Z\gamma$ candidate events at CDF.

For the $W\gamma$ process CDF measures: $\sigma \times BR(W\gamma \to \ell\nu\gamma) = 18.1 \pm 3.1$ pb for the kinematic region $E_T^\gamma > 7$ GeV and $\Delta R(\gamma\ell) > 0.7$ [10]. DØ measures $\sigma(\ell\nu\gamma) = 14.8 \pm 1.0$(stat)$\pm 1.0$(syst)$\pm 1.2$(lum) pb requiring $E_T^\gamma > 8$ GeV and $\Delta R(\gamma\ell) > 0.7$ [11].

For the $Z\gamma$ process CDF measures: $\sigma \times BR(Z\gamma \to \ell\ell\gamma) = 4.6 \pm 0.6$ pb [10]. DØ measures $\sigma(\ell\ell\gamma) = 4.2 \pm 0.4$(stat+syst)$\pm 0.3$(lum) pb [12].

Anomalous gauge couplings would increase the high energy tail of the photon E_T. DØ sets one and two dimensional limits on the coupling parameters Δk_γ and λ_γ (for the $W\gamma$ process) and on the CP–violating and CP–conserving anomalous couplings h_i^Z ($i = 1, 2, 3, 4$) for the $Z\gamma$ process.

WW Production

The first evidence of W pair production was found in $p\bar{p}$ collisions by CDF [13]. The properties of W pairs have been extensively studied at LEP [14]. The Tevatron offers the possibility to probe much higher center of mass energies. CDF and DØ measured the WW production cross section in the $W^+W^- \to \ell^+\nu\ell^-\nu$ channel:

$$\sigma_{CDF} = 14.6^{+5.8}_{-5.1}(\text{stat})^{+1.8}_{-3.0}(\text{syst}) \pm 0.9(\text{lum}) \text{ pb [15]}$$
$$\sigma_{D\emptyset} = 13.8^{+4.3}_{-3.8}(\text{stat})^{+1.2}_{-0.9}(\text{syst}) \pm 0.9(\text{lum}) \text{ pb [16]}.$$

TABLE 1. The uncertainty on the M_W measurement in MeV/c² obtained from 200 pb⁻¹ of CDF run II data. The CDF run 1b uncertainties are shown for comparison [9].

Systematic uncertainty	Electrons (Run 1b)	Muons (Run 1b)
Production and decay model	30 (30)	30 (30)
Lepton energy scale and resolution	70 (80)	30 (87)
Recoil energy and resolution	50 (37)	50 (35)
Backgrounds	20 (5)	20 (25)
Statistics	45 (65)	50 (100)
Total:	105 (110)	85 (140)

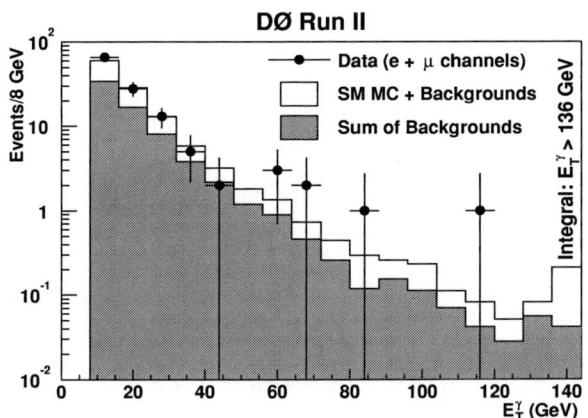 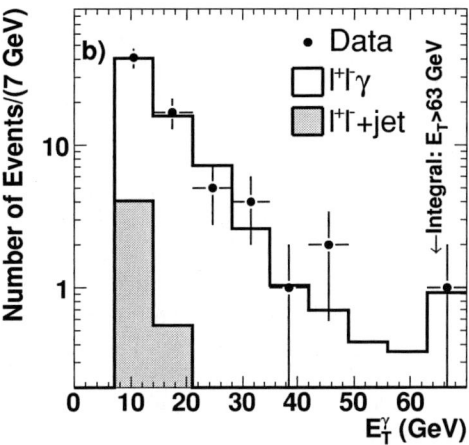

FIGURE 6. Photon E_T spectrum for $W\gamma$ candidates at DØ (left) and $Z\gamma$ candidates at CDF (right).

These values are in good agreement with the NLO calculation at $\sqrt{s} = 1.96$ TeV.

WZ and *ZZ* Production

DØ has established a 95% C.L. upper limit on the *WZ* cross section of 13.3 pb [17]. CDF performed a combined analysis of *WZ* and *ZZ* possible final states, putting a 95% C.L. upper limit on the combined cross section of 15.2 pb [18].

CONCLUSIONS

The Run II of the Tevatron is well underway. Both CDF and DØ are producing interesting results in the electroweak sector. DØ has made the first Run II direct measurement of the *W* width. CDF has determined the uncertainty on the *W* mass with the first 200 pb⁻¹ of Run II data to be 76 MeV/c². Run II should provide the world's highest precision measurement of the *W* boson mass (30 MeV/c² for 2 fb⁻¹).

REFERENCES

1. F. Abe et al., *Nucl. Instrum. Methods A*, **271**, 387 (1988).
2. S. Abachi et al., *Nucl. Instrum. Methods A*, **338**, 185 (1994).
3. D. Acosta et al., *Phys. Rev. Lett.* **94**, 091803 (2005).
4. C. Anastasiou et al., *Phys. Rev. D* **69**, 094008 (2004).
5. S. Eidelman et al., *Physics Letters B* **592**, 1 (2004)
6. DØ note 4563–CONF (2004).
7. D. Acosta et al., *Phys. Rev. D* **71**, 051104 (2005).
8. DØ note http://www-d0.fnal.gov/Run2Physics/WWW/results/prelim/EW/E12/E12.pdf
9. T. Affolder et al., *Phys. Rev. D* **64**, 052001 (2001).
10. D. Acosta et al., *Phys. Rev. Lett.* **94**, 041803 (2005).
11. V.M. Abazov et al., *Phys. Rev. D* **71**, 091108 (2005).
12. V.M. Abazov et al., hep-ex/0502036.
13. F. Abe et al., *Phys. Rev. Lett.* **78**, 4536 (1997).
14. LEP Collaborations, hep-ex/0312023.
15. D. Acosta et al., *Phys. Rev. Lett.* **94**, 211801 (2005).
16. V.M. Abazov et al., *Phys. Rev. Lett.* **94**, 151801 (2005).
17. V.M. Abazov et al., hep-ex/0505019.
18. D. Acosta et al., *Phys. Rev. D* **71**, 091105 (2005).

Top Quark Measurements at Tevatron Run II

P. Azzi

I.N.F.N.-Sezione di Padova, Via Marzolo 8, 35100 Padova, Italy
E-mail: azzi@pd.infn.it

Abstract. The top quark, discovered at the Tevatron in 1995[1, 2, 3], is a very interesting particle. Precise measurement of the top properties using large data samples will allow stringent tests of the Standard Model and offer a unique window on new physics. This report contains a review of the latest measurement of top quark properties using approximately 200 pb^{-1} data collected with the CDF and D0 detectors at the Fermilab Tevatron $p\bar{p}$ collider at a center of mass energy of $\sqrt{s} = 1.96$ TeV.

Keywords: Particle physics, hadron colliders; top quark; electroweak physics; heavy quark
PACS: 14.65.Ha

INTRODUCTION

The top quark is one of the building blocks of the Standard Model of Electroweak interactions as the partner of the bottom quark in the SU(2) isospin doublet of the third family of quarks. The top quark mass is a fundamental Standard Model parameter that needs to be measured with the greatest possible precision. It is needed to determine the strength of the ttH coupling and it has a substantial effect on radiative corrections. In fact, an uncertainty of 2 GeV/c^2 on the top mass would constrain the Higgs mass to 35%. It is not an easy task to achieve such a small uncertainty, but several experimental handles are available in Run II. On one hand the increased detector acceptance and large data sample will allow one to select purer samples less sensitive to systematic uncertainties: for instance requiring events with well measured jets (lowers the energy scale uncertainty) and two b-tagged jets (lowers the overall background). On the other hand, since most of the systematics are data driven, their uncertainty will scale approximately with $1/\sqrt{N}$, N being the number of events of the control samples themselves. Right now the top mass world average from the measurements performed in Run I with a luminosity of about 100 pb^{-1} is 178 ± 4.3 GeV/c^2[4] giving an indirect limit on the SM Higgs mass of $M_H = 126^{+73}_{-48}$ GeV/c^2[5].

The Run II analyses presented here on an accumulated statistics of about 200 pb^{-1} use newer algorithms and techniques not only to lower the statistical and systematical uncertainties but also to study the top samples in a more inclusive fashion.

TOP QUARK PRODUCTION AND DECAY

At the Tevatron the top quark is produced mainly in pairs through the process $qq, gg \to t\bar{t}$ with a cross section of about 6.7 pb[6] for $m_t = 175$ GeV/c^2 and $\sqrt{s} = 1.96$ TeV. Within the Standard Model framework the top decays almost 100% of the time via $t \to Wb$. Therefore it is customary to classify the $t\bar{t}$ final states based on the W decays modes: dileptons ($\ell = e, \mu$), lepton+jets ($\ell = e, \mu$), all-jets and inclusive τ (hadronically decaying) final-state events.

Lepton Plus Jets Channel

The lepton plus jets signature is characterized by the presence of one high p_T lepton and large $\slashed{E}_T$ due to the leptonic W decay plus three or more jets from the hadronically decaying W and the b jets. This final state suffers from large $W + jets$ background. However, kinematical and topological properties of the $t\bar{t}$ signal or its heavy flavor content (or both) provide a good separation from the background processes. The significant branching ratio for this final state of about 30% and the good purity achievable after selection make the lepton+jets sample the best one to measure other top properties with small uncertainties (such as the top mass) and to search for new production processes (single top) or exotic particles (t').

Cross Section Measurement

In the kinematical approach, after the basic event selection, variables such as H_T, the scalar sum of all the objects' transverse energies in the event, or the aplanarity $\mathscr{A}$, a measure of the event shape, are found to be the most discriminant. A complementary approach is to exploit the heavy flavor content of signal events and the large b-tagging efficiency compared to the low fake rate. There are several b identification algorithms available at the moment, some employ the silicon vertex detector information, while another category focuses on the peculiar properties of leptons from b semileptonic decays. Due to the increased statistics of the data sets and the improved efficiency of the b-tagging algorithms, both CDF and D0 have been able to perform a measurement of the $t\bar{t}$ cross section in lepton+jets events with two identified b-jets. The most recent cross section measurements in this channel are summarized in Tab. 1.

Top mass Measurement

One challenge of the top mass measurement common to all analyses is to solve the combinatorics problem in the event reconstruction. For example in the lepton+jets channel $t\bar{t} \to \ell\nu q\bar{q}'b\bar{b}$ there are 12 possible jet-parton assignement. The problem is simplified by identifying the b jets with specific algorithms. Another limitation of the measurement is the large uncertainty in the modelling of the jet energy response, referred to as the jet energy scale uncertainty (JES) that will be discussed below.

Several techniques are employed in the event reconstruction. In one case, templates of reconstructed top quark mass are created in Monte Carlo events as a function of the true top quark mass and compared with the data. Both CDF and D0 have performed measurement using this approach on various samples of lepton+jets events with different signal to background ratios: with 0, 1 or 2 b-tagged jets, or kinematically selected.

A different class of analyses uses the matrix-element method that consists in computing a probability for each event to be signal with a given top quark mass. The probability is then computed using the full SM production and decay matrix-elements. This method has been shown to be very powerful statistically, as demonstrated by the best measurement from Run I obtained by D0 [7]. The CDF collaboration has also presented a measurement using this method on Run II data. The result is in very good agreement with the template analysis above. The latest measurement are summarized in Tab. 3.

Improving the Top mass Measurement

A precise measurement of the top mass combines cutting edge theoretical knowledge with state-of-the-art detector calibration. The highest contribution to the systematic uncertainty in the top mass measurement comes from the jet energy scale. The current value for this systematic is about 3 GeV/c^2 for CDF and 5-6 GeV/c^2 for D0.

Currently the best calibration sample for the jet energy scale consists of events where a jet is recoiling against a well measured photon, however in the near future the hadronic W in lepton plus jets events will provide an in situ calibration for the light quark jets. With larger statistics the $Z \to b\bar{b}$ signal will become useful for the calibration of heavy flavor jets. Moreover a large amount of data will allow one to not only reduce the systematics above but also to pick the best measured event categories with smaller backgrounds and that are less sensitive to systematics uncertainties. Finally to achieve the ultimate precision excellent Monte Carlo generators implementing the latest theory knowledge and understanding of all the various effects (ISR, FSR, PDF's) plus an accurate detector simulation are essential.

Search for Single Top and t'

In addition to pair production, single top quarks can be produced by weak interaction by a virtual W or through Wg fusion, with a total cross section of about $\sigma_{tX} = 2.9$ pb.[8] Single top production is interesting in its own right: a precise measurement of the cross section would provide a direct determination of $|V_{tb}|$ with a 14% uncertainty expected for $\mathscr{L}_{int} = 2\,\text{fb}^{-1}$ of data. Moreover single top events have the same final-state experimental signature as the Standard Model Higgs associated production process ($HW \to b\bar{b}\ell\nu_\ell$). The extraction of a single top signal is more challenging than the pair produced case since there are fewer objects in the final state and the overall event properties are less distinct from the $W+jets$ background. In Run II searches for single top production in the s- and t-channels separately and combined have been performed by both CDF and D0. The current best limit has been obtained by D0 on 230 pb^{-1} of data using a neural network technique and amounts to: $\sigma_{tX}(s-channel) < 6.4$ pb @ 95%CL and $\sigma_{tX}(t-channel) < 5.0$ pb @ 95%CL.

The lepton+jets sample has also been used for a search for a higher mass fourth generation t' particle that would decay in the same final state as the SM top. Employing the same kinematical selection already used for the cross section measurement CDF obtains a limit of $m_{t'} > 225$ GeV/c^2 at 95% C.L..

Dilepton Channel

This final state is characterized by the presence of two high p_T leptons (a reconstructed e, μ or an isolated track), large missing transverse energy from the missing neutrinos and two or more central jets. The main sources of background for this channel come from other Stan-

TABLE 1. Summary of $t\bar{t}$ cross section measurements in the lepton+jets channel for $m_t = 175\,\text{GeV}/c^2$

Experiment	Method	L(pb^{-1})	σ(pb)
CDF	Kinematic	193	$4.7^{+1.6+1.8}_{-1.6-1.8}$
CDF	Kinematic NN	193	$6.7^{+1.1+1.6}_{-1.1-1.6}$
CDF	Vertex Tag + Kin.	162	$6.0^{+1.6+1.2}_{-1.6-1.2}$
CDF	Vertex Tag	162	$5.6^{+1.2+0.9}_{-1.1-0.6}$
CDF	Double Vertex Tag	162	$5.0^{+2.4+1.1}_{-1.9-0.8}$
CDF	Jet Prob. Tag	162	$5.8^{+1.3+1.3}_{-1.2-1.3}$
CDF	Soft Muon Tag	193	$5.2^{+2.9+1.3}_{-1.9-1.0}$
D0	Topological	143	$7.2^{+2.6+1.6}_{-2.4-1.7}$
D0	Soft Muon Tag	93	$11.4^{+4.1+2.0}_{-3.5-1.8}$
D0	Impact Parameter	164	$7.2^{+1.3+1.9}_{-1.2-1.4}$
D0	Vertex Tag	164	$8.2^{+1.3+1.9}_{-1.3-1.6}$

TABLE 2. Summary of $t\bar{t}$ cross section measurements in the dilepton channel for $m_t = 175\,\text{GeV}/c^2$

Experiment	Method	L(pb^{-1})	σ(pb)
CDF	Combined	200	$7.0^{+2.4+1.7}_{-2.1-1.2}$
CDF	$\not{E}_T$, # jets	193	$8.6^{+2.5+1.1}_{-2.4-1.1}$
D0	Combined	146	$14.3^{+5.1+2.5}_{-4.3-1.9}$
D0	Vertex Tag	158	$11.1^{+5.8+1.4}_{-4.3-1.4}$

dard Model processes with similar signatures (Drell-Yan $\gamma^*/Z^0 \to e^+e^-, \mu^+\mu^-, Z^0 \to \tau\tau, W^+W^-/W^\pm Z^0$) and processes with one real lepton and another object that fakes the second lepton.

Cross Section Measurements

The dilepton event selection starts with two oppositely charged high p_T leptons, asking that one or both are well isolated from nearby track activity. Different techniques are employed in order to reduce the contribution from Z^0 events without reducing the signal acceptance. $\not{E}_T$ is required to be large given the two neutrinos from W decay and the presence of two central jets accounts for the two b's from the top decay. Other kinematical and topological cuts (H_T, total energy of all the object in the event, $\Delta\phi(\not{E}_T, object)$) are finally employed in order to reduce the remaining backgrounds. The increased statistics of this dataset and the high purity achievable allow new and more inclusive analyses to be performed for instance applying simultaneous fits to kinematical distribution of various SM processes. The $t\bar{t}$ production cross section has also been measured on dilepton samples of different purity obtained requiring only one or two tight leptons, and with one identified b jet. The latest cross section measurements in this channel are reported in Tab.2.

Top Mass Measurement

The top mass can also be measured in the dilepton channel but with larger statistical uncertainties due to the smaller branching ratio and the two neutrinos in the final state that complicate the event reconstruction. As in the lepton+jets case there are different strategies to solve the problem (template and matrix-element methods): between CDF and D0 three independent mass analyses have been presented here with consistent results, see Tab. 3 for the latest best measurements.

All Hadronic Channel

The all-jets final state where both W decay hadronically is a very challenging signature of six central and energetic jets, swamped by a QCD multijet background of several orders of magnitude bigger than the $t\bar{t}$ signal. In order to improve the signal to background at a level where a cross section measurement is significant both b-tagging and a strict kinematical and topological selection have to be employed. Both experiments have measured the $t\bar{t}$ cross section in this channel at $\sqrt{s} = 1.96\,\text{TeV}$ to be $\sigma = 7.8^{+2.5+4.7}_{-2.5-2.3}$ (CDF, L=165 pb^{-1}) and $\sigma = 7.7^{+3.4+4.7}_{-3.3-3.8}$ (D0, L=162 pb^{-1}).

CONCLUSIONS

The top is still a fairly young particle and most of our current knowledge about its properties still comes from the Run I Tevatron measurements. In Run II a data sample about 50 times the Run I statistics will be collected and it will be possible to achieve better precision in the measurements and perform significant tests of the Standard Model expectations. The preliminary Run II mea-

TABLE 3. Summary of the best Run II top mass measurements in each final state presented at this Conference

Experiment	Channel	L(pb^{-1})	σ(pb)
D0	Dilepton	230	$155^{+14.0}_{-13.0} \pm 7.0$
CDF	Dilepton	200	$168.1^{+11.0}_{-9.8} \pm 8.6$
CDF	Lepton+Jets	162	$177.8^{+4.5}_{-5.0} \pm 6.2$
D0	Lepton+Jets	230	$170.6^{+4.2}_{-4.2} \pm 6.0$

surements of the production cross section and mass presented here show that the the uncertainties are rapidly dropping to the Run I level or better.

REFERENCES

1. F.Abe *et al.*, *Phys. Rev. Lett.* 732251994.
2. F.Abe *et al.*, *Phys. Rev. Lett* 7426261995.
3. S.Abachi *et al.*, *Phys. Rev. Lett.* 7426321995.
4. CDF and D0 Collaborations, hep-ex/0404010.
5. LEP Electroweak Working Group, http://lepewwg.web.cern.ch/LEPEWWG/, to be published.
6. M. Cacciari *et al.*, JHEP 0404, 068(2004); N.Kidonakis and R.Vogt, *Phys.Rev.* D 68 1140142003.
7. Abazov *et al.*, Nature 429, 638 (2004).
8. B.W.Harris *et al.*, *Phys. Rev.* D 660540242002.

QCD Physics at the Tevatron Collider

Andrea Messina

INFN Sezione di Roma

Abstract. In this contribution some of the prominent QCD physics results from CDF and D0 experiments in Run II are presented. The cross sections and the properties of jets are discussed for both the inclusive and the b-jet production. Results on the associate production of light and heavy flavour jets together with vector bosons are also reported.

Keywords: QCD, Collider Physics, Tevatron
PACS: 12.38.Qk, 13.87.Ce, 13.87.Fh

INTRODUCTION

The Run II at the Tevatron $p\bar{p}$ collider will explore extended kinematic regions in many QCD processes. As a matter of fact the increase in the instantaneous luminosity, center-of-mass energy (from 1.8 to 1.96 TeV) and the improvement in both CDF [1] and D0 [2] detectors will allow stringent tests of QCD in hadronic collisions. In these collisions the final state is characterized by a hard interaction between the hadron constituents overlaid to a soft contribution (underlying event) from the initial state radiation and the multiple parton interaction between the remnants. In order to make a proper comparison with theory predictions, all these contributions have to be understood and correctly modeled. This will lead to improved QCD Monte Carlo models and to more precise parton distribution functions. For this respect the Tevatron QCD physics program is spanning a wide range of processes to be able to characterize the soft and hard contributions. This talk will briefly present some of the most relevant analyses contributing to the understanding of hadron collisions. The study of the underlying event and the jet fragmentation properties will be introduced, then the first section will continue with the discussion of the jet inclusive and the b-jet cross sections. In the second section the properties of jets produced in association with a vector boson will be described.

JET PHYSICS

Jets are experimentally observed by adding the energy measured in each calorimetry cell associated to a cluster by a defined jet algorithm [3]. Currently, both CDF and D0, are exploring different jet algorithms, alternative to the cone-based algorithm used in Run I (JetClu), that are not collinear and infrared safe. Following the theoretical work, the first measurement of the inclusive jet cross section have been performed using the K_T algorithm [4]. However this new algorithm suffers from the presence of the underlying event, and it has yet to be proved to work correctly in a hadronic environment as the cone-base does. MidPoint, a generalization of JetClu, has been also used. This algorithm has the advantage of overcoming many of the theoretical limitations of JetClu, still showing a well understood experimental behaviour.

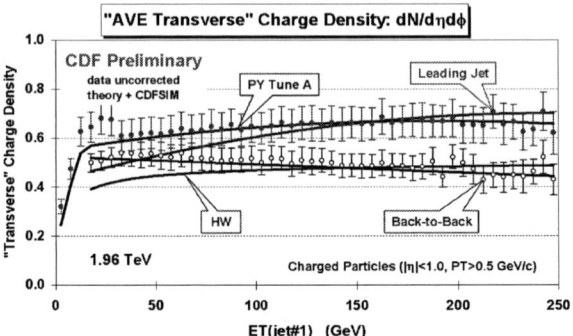

FIGURE 1. Measured average track density in the transverse region as a function of the E_T of the leading jet.

Study of the underlying event

CDF studied the underlying event using the direction of the leading calorimeter jet (JetClu, R=0.7) to isolate regions of $\eta - \phi$ space that are sensitive to the underlying event. Looking in the perpendicular plane of the scattering, the contribution of the underlying event is better visible in the ϕ region opposite ($60 \leq \Delta\phi \leq 120$) to the jet (transverse region). The analysis of the transverse region is restricted to charged particles in the range $p_T > 0.5$ GeV/c and $|\eta| < 1$, while events are required to have at least a jet within $|\eta| < 2$ ("leading jet" events). A subset of these events ("back-to-back" events) is selected by requiring events with back-to-back jets in order to

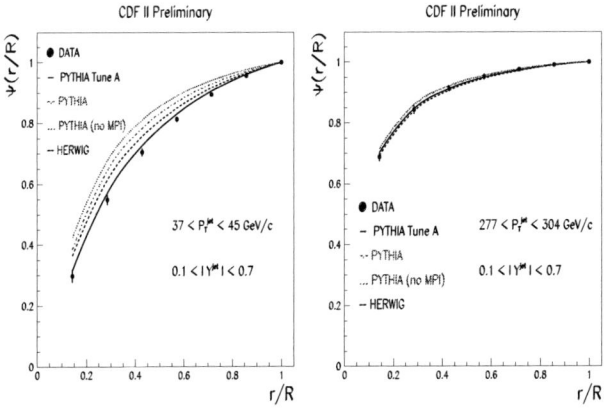

FIGURE 2. The measured integrated jet shape compared to different MC models.

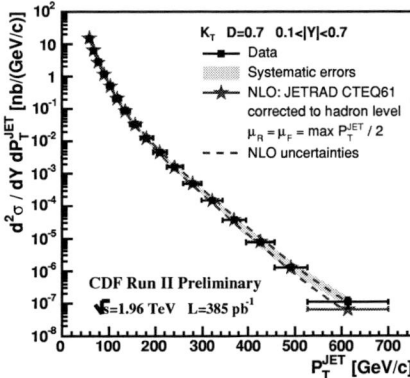

FIGURE 3. The measured inclusive jet cross section from CDF compared to NLO predictions. Jets are searched for using the K_T algorithm.

reduce the extra hard-gluon radiation. Fig. 1 compares PYTHIA [5] and HERWIG [6] with the "leading jet" and the "back-to-back" data on the average P_T track density in the transverse region as a function of the jet E_T. The multiple interaction and the soft gluon initial state radiation parameters of PYTHIA were adjusted to agree with Run I data (Tune A). The "back-to-back" data show a slight decrease in the transverse density with increasing E_T which is described by PYTHIA Tune A but not by Herwig (without multiple parton interaction).

Study of jet shapes

The internal structure of the jets is dominated by the gluon radiation of the primary final-state parton. It is sensitive to the quark and gluon jet fraction and to the color structure of the partonic event, it receives contributions from the initial state radiation as well as from the beam remnant interactions. Study of jet shape constitutes a stringent test of the jet hadronization processes and of their current implementation in the Monte Carlo models. The CDF experiment obtained results on jet shapes for central jets reconstructed with the MidPoint algorithm (R=0.7). The integrated jet shape variable $\Psi(r)$ is defined as:

$$\Psi(r) = \frac{1}{N_{jets}} \Sigma_{jets} \frac{P_T(0,r)}{P_T(0,R)}, \quad 0 \leq r \leq R \quad (1)$$

where N_{jets} denotes the number of jets. The measured jet shape have been compared to the prediction from PYTHIA tune A and HERWIG (Fig. 2). To illustrate the importance of a proper description of the underlying event another PYTHIA sample without the multiple parton interaction have been used.

Inclusive jet production at Tevatron

The measurement of the inclusive jet cross section provides a powerful test of perturbative QCD and a sensitive probe of quark substructure down to a scale of about 10^{-17} cm. The increase of center-of-mass energy in RunII from 1.8 to 1.96 TeV and the increased luminosity result in a larger kinematic range up to jet $E_T <550$ GeV (to be compared to 400 GeV limit in RunI). In order to have a good measurement of jet energy, only jets reconstructed in the central region of the calorimeter are considered. Fig. 3 shows the measured inclusive jet cross section by CDF using the K_T algorithm and based on the first 145 pb^{-1} of Run II data. Measurement have been performed using values of the D parameter in the K_T expression,

$$K_{ij} = min(p_{T,i}^2, p_{T,j}^2) \frac{(y_i - y_j)^2 + (\phi_i - \phi_j)^2}{D} \quad (2)$$

equal to 0.5, 0.7, 1.0. The measurement are compared to pQCD NLO calculation using CTEQ6 [7] parton density function (PDF) and the renormalization and factorization scale set to $p_T^{max}/2$. The measured cross section is fairly well described by the prediction for $P_T > 150$ GeV within the uncertainties. The systematic errors on the data are dominated by the uncertainty on the jet energy scale while the theoretical predictions suffers from the limited knowledge of the gluon distribution at high x. At low P_T, the data are systematically above the theory and the effect increases as D increases. This indicates the effect of the underlying event which is not taken into account yet. Fig. 4 shows the jet inclusive cross section measured by D0 based on the first 143 pb^{-1} of Run II data. The new MidPoint algorithm has been used with a cone radius of 0.7. Data are in good agreement with pQCD NLO prediction using CTEQ6 PDF. However the

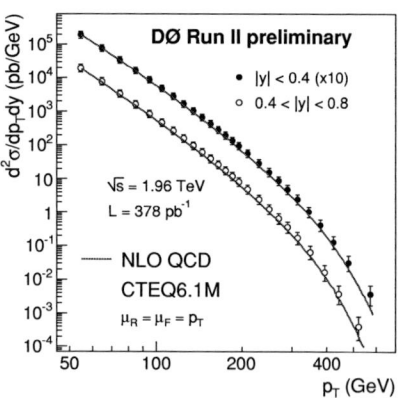

FIGURE 4. The measured inclusive jet cross section from D0 compared to NLO predictions. Jets are searched for using the MidPoint algorithm.

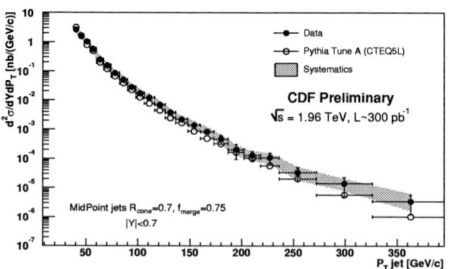

FIGURE 5. The b-jet inclusive cross section measurement from CDF.

measurement is dominated by a relatively large uncertainty on the jet energy scale.

B-jet cross section

Measurements of the b-jet production cross section at $p\bar{p}$ colliders provide an important quantitative test of QCD. The mass of the b quark is considered large enough to justify perturbative expansions in the strong coupling constant α_S. Consequently, data on b-jet production are expected to be adequately described by calculations at the NLO order in α_S.

In this analysis b-jets are identified by studying the invariant mass of the tracks pointing to a common vertex (secondary vertex) displaced from the beam direction. The transverse distance between the secondary and the primary vertex is due to the long lifetime of the heavy b-quark. The b-jet fraction is determined by fitting the secondary vertex invariant mass to Monte Carlo templates as a function of the p_T of the b-jet.

Fig. 5 shows the inclusive b-jet cross section measured by CDF using the first 300 pb^{-1} of Run II data. Jets are searched for using the MidPoint algorithm with a cone size of 0.7. Only jets in the central rapidity region are considered and their energy is corrected, using Monte Carlo simulation, to the hadronic level. The uncertainty on this correction accounts for the largest contribution to the systematic in the low energy bins, while the b-identifcation related uncertainties dominate the highest transverse energy bin systematic uncertainty 5. The b-jet cross section is in a good agreement with the LO prediction from PYTHIA tune A using the CTEQ5L PDF.

BOSON + JETS PRODUCTION

A detailed study of hard processes involving the associated production of vector bosons and a given number of jets in the final state has a prominent importance in the Tevatron Run II physics program. This processes constitute a relevant background to a number of analyses. The W + jets process is the biggest background to the Top and Higgs production in hadron collisions as well as the γ + jet is the background to many new physics searches. The production of bosons accompanied with c/b-jets probes the heavy flavours content in the proton. Moreover the presence of a massive boson ensures high p_T interactions calculable in pQCD. The presence of electroweak interacting particles limits also the number of diagrams contributing to the process compared to a pure QCD event, allowing theoretical predictions up to high jet multiplicity in the final state. During the last years a number of new boson + N_{jet} LO programs have become available [8] which include large jet multiplicity in the final state, in addition to NLO calculation. These programs have been interfaced to parton-showering models in order to produce a realistic final state. Many procedures to avoid double counting in the parton radiation have also been proposed, the next paragraf constitue one of the first such a test.

W + jet cross section

The CDF experiment measured the inclusive cross section for $W+ \geq N_{jet}$ production using $127 pb^{-1}$ of Run II data. Jet with an $E_T \geq 15 GeV$ and $|\eta| < 2.4$ reconstructed using the JetClu algorithm (R=0.4) are considered. The measurement have been compared to similar results form Run I and pQCD LO as implemented in Alpgen interfaced to the parton cascade and fragmentation from HERWIG. The increase in the center-of-mass energy of Run II reflects into an higher cross section with respect to Run I. The pQCD prediction describes fairly

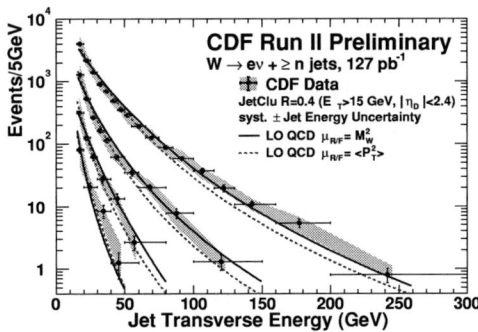

FIGURE 6. E_T^{jet} spectrum for the Nth jet in $W + \geq N_{jet}$ production as measured by CDF compared to LO prediction from Alpgen interfaced to HERWIG.

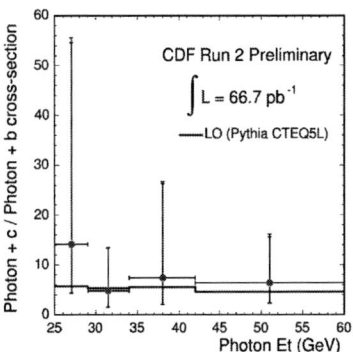

FIGURE 7. Ratio of γ + b-jet to γ + c-jet as measured by CDF compared to the PYTHIA MC.

well the data but suffer form large uncertainties due to the dependence on the renormalization scale in the calculation. Fig. 6 shows the measured E_T^{jet} spectrum for the N^{jet} in $W+ \geq N^{jet}$ production. These spectra are sensitive to the details of the interface between the pQCD LO calculation and the parton shower evolution. The measured spectra are in agreement with the prediction from Alpgen+HERWIG within the present uncertainties. In the data the systematic error are dominated by the jet energy scale while the LO theoretical prediction shows a strong dependence on the renormalization scale.

Z + b-jet production

Inclusive Z+b-jet production is expected to be a major background to Higgs production in $p\bar{p} \to ZH$ channel, with Higgs decaying in $b\bar{b}$. The D0 experiment has measured the ratio of production cross sections of inclusive Z+b jets to Z+jets using $180pb^{-1}$. Jets are reconstructed with a cone algorithm of cone size 0.5, and the Z boson is reconstructed using the leptonic ($ee, \mu\mu$) decay mode. b-jets are identified with a secondary vertex tagging algorithm by searching for tracks inside the jet pointing to a displaced vertex with respect to the beam axis.

This ratio, which benefits from cancellations of many systematic uncertainties, is measured as $0.021 \pm 0.004(stat)^{+0.002}_{-0.003}(syst)$. Which is in good agreement with the NLO prediction of 0.018 ± 0.004 using to CTEQ6 PDF.

γ + heavy flavour production

Events containing an isolated, high E_T photon and a heavy flavour jet are sensitive to the c quark PDF of the proton. The γ + b/c-jet cross-sections are measured both as a function of photon E_T, in order to test QCD predictions at different energy scales, and for all photons with an $E_T \geq 25$ GeV to gain maximal statistical sensitivity to deviations that could signal new physics production. Heavy flavour jets are searched for by looking at the invariant mass of the secondary vertex of the jet, as described for the b-jet cross section measurement. The CDF experiment has measured the cross section of γ+c-jet and γ+b-jet as a function of the photon E_T. In Fig. 7 the ratio between γ + c-jet and b-jet cross section are shown compared with PYTHIA LO predictions. No evidence was found so far of new physics production.

ACKNOWLEDGMENTS

I thank the IFAE 2005 organizers for their kind invitation and ospitality. I would like also to acknowledge the CDF and D0 members for their work to achieve the results reported in this contribution.

REFERENCES

1. CDF Collaboration, D. Acosta et al., Phys. Rev. **D 71** 032001 (2005).
2. D0 Collaboration, V. Abazov et al., T. LeCompte and H.T. Diehl, Annu. Rev. Nucl. Part. Sci. **50**, 71 (2000).
3. G.C. Blazey et al., arXiv:hep-ex/0005012.
4. S. D. Ellis, D. E. Soper, Phys. Rev. **D 48** 3160-3166 (1993).
5. H.-U. Bengtsson and T.Sjostrad, Comp. Phyis. Comm. **46** 43 (1987).
6. G. Marchesini et al., Comp. Phys. Comm. **67** 456 (1992).
7. J. Pumplin et al., JHEP **0207** 012 (2002).
8. M. Mangano et al., JHEP **07** 001 (2003).
   E. E. Boos et al., preprint hep-ph/9503280.
   S.Tsuno et al., Comp. Phys. Comm. **151** 216 (2003).
   F. Maltoni, T. Stelzer., JHEP **0302** 027 (2003).
   J. Campbell, R.K. Ellis, Phys. Rev **D 65** 113007 (2002).

The Muon g-2 in the Standard Model

Massimo Passera

Università di Padova and INFN, Padova, Italy

Abstract. The Standard Model prediction of the anomalous magnetic moment of the muon is reviewed. QED, electroweak and hadronic contributions are presented, and open questions discussed.

Keywords: Magnetic moments, Muons
PACS: 13.40.Em, 14.60.Ef

INTRODUCTION

Schwinger's 1948 calculation [1] of the leading contribution of the Standard Model (SM) prediction for the anomalous magnetic moment of the muon $a_\mu \equiv (g_\mu - 2)/2$, equal to the one of the electron, was one of the very first results of QED. Its agreement with the experimental value of the anomalous magnetic moment of the electron, a_e, provided one of the early confirmations of this theory.

While a_e is rather insensitive to strong and weak interactions, hence providing a stringent test of QED and leading to the most precise determination to date of the fine-structure constant α, a_μ allows to test the entire SM, as each of its sectors contribute in a significant way to the total prediction. Compared with a_e, a_μ is also much better suited to unveil or constrain "new physics" effects [2].

In a sequence of increasingly more precise measurements [3, 4], the E821 Collaboration at Brookhaven has reached a fabulous relative precision of 0.5 parts per million (ppm) in the determination of a_μ. This note provides a brief summary of the present status of the three contributions into which the SM prediction a_μ^{SM} is usually split – QED, electroweak and hadronic – and a comparison with the current experimental value. Extensive reviews can be found in [5, 6, 7].

THE QED CONTRIBUTION

The QED contribution to the anomalous magnetic moment of the muon arises from the subset of SM diagrams containing only leptons (e, μ, τ) and photons. Only one diagram is involved in the evaluation of the lowest-order contribution (second-order in the electric charge); it provides the famous result by Schwinger [1]: $a_\mu^{QED}(1 \text{ loop}) = \alpha/(2\pi)$. Also the fourth- and sixth-order QED terms are known analytically. I refer the reader to [5] for an updated review of these contributions. Reference [5] also provides the numerical re-evaluation of these two- and three-loop contributions employing the updated CODATA [8] and PDG [9] values for the lepton masses. The eighth-order term has thus far been evaluated only numerically. This formidable task was first accomplished by Kinoshita and his collaborators in the early 1980s [10]. The latest analysis appeared in [11]. Note that this eighth-order QED contribution is about six times larger than the present experimental uncertainty of a_μ. The evaluation of the five-loop QED contribution is in progress [12].

Adding up the above contributions and using the latest CODATA recommended value for the fine-structure constant [8], $\alpha^{-1} = 137.03599911(46)$, known to 3.3 ppb, one obtains [5] $a_\mu^{QED} = 116584718.8(0.3)(0.4) \times 10^{-11}$. The first error is due to the uncertainties of the $O(\alpha^2)$, $O(\alpha^4)$ and $O(\alpha^5)$ terms, and is strongly dominated by the last of them. (The uncertainty of the $O(\alpha^3)$ term is negligible.) The second error is caused by the 3.3 ppb uncertainty of the fine-structure constant α.

THE ELECTROWEAK CONTRIBUTION

The electroweak (EW) contribution to the anomalous magnetic moment of the muon is suppressed by a factor $(m_\mu/M_W)^2$ with respect to the QED effects. The one-loop part was computed in 1972 by several authors [13]:

$$a_\mu^{EW}(1 \text{ loop}) = \frac{5 G_\mu m_\mu^2}{24\sqrt{2}\pi^2}\left[1 + \tfrac{1}{5}\left(1 - 4\sin^2\theta_W\right)^2 + \ldots\right],$$

where the ellipsis indicate terms of $O(m_\mu^2/M_{Z,W,H}^2)$, $G_\mu = 1.16637(1) \times 10^{-5}$ GeV^{-2}, M_Z, M_W and M_H are the masses of the Z, W and Higgs bosons, and θ_W is the weak mixing angle. We can employ the on-shell definition $\sin^2\theta_W = 1 - M_W^2/M_Z^2$ [14], where $M_Z = 91.1875(21)$ GeV and M_W is the theoretical SM prediction of the W mass. The latter can be easily derived from the simple analytic formulae of ref. [15] (on-shell scheme II with $\Delta\alpha_h^{(5)} = 0.02761(36)$, $\alpha_s(M_Z) = 0.118(2)$ and $M_{top} = 178.0(4.3)$ GeV), leading to $M_W = 80.383$ GeV for

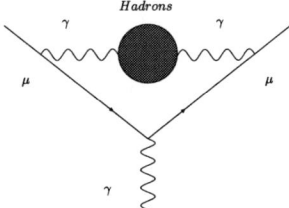

FIGURE 1. Leading hadronic contribution to a_μ.

$M_H = 150\,\text{GeV}$, compared with the direct experimental value $M_W = 80.425(38)\,\text{GeV}$ [9], which corresponds to a small M_H [16]. For $M_H = 150\,\text{GeV}$ we thus obtain $a_\mu^{EW}(1\,\text{loop}) = 194.8 \times 10^{-11}$.

The two-loop EW contribution to a_μ was computed in 1995 [17]. Naïvely one would expect the two-loop EW contribution $a_\mu^{EW}(2\,\text{loop})$ to be negligible, but this turns out not to be so because of the appearance of terms enhanced by a factor of $\ln(M_{Z,W}/m_f)$, where m_f is a fermion mass scale much smaller than M_W [18]. The question of how to treat properly the contribution of the light quarks was addressed in [19, 20]. These refinements significantly improved the reliability of the fermionic part (that containing closed fermion loops) of $a_\mu^{EW}(\text{two loop})$ leading, for $M_H = 150\,\text{GeV}$, to $a_\mu^{EW} = 154(1)(2) \times 10^{-11}$ [20], where the first error corresponds to hadronic loop uncertainties and the second to an allowed Higgs mass range of $114\,\text{GeV} < M_H < 250\,\text{GeV}$, the current top mass uncertainty and unknown three-loop effects. The leading-logarithm three-loop contribution to a_μ^{EW} is extremely small (indeed, consistent with zero to a level of accuracy of 10^{-12}) [20, 21]. The recent calculation of [22] for the two-loop bosonic part of a_μ^{EW}, performed without the approximation of large Higgs mass previously employed, agrees with the result of [17]. Work is also in progress for an independent recalculation based on the numerical methods of refs. [23].

THE HADRONIC CONTRIBUTION

Leading-order Contribution. The leading hadronic contribution a_μ^{HLO} arises from the diagram in fig. 1, whose evaluation involves long-distance QCD for which perturbation theory cannot be employed. However, using analyticity and unitarity, it was shown long ago [24] that this contribution can be computed from hadronic e^+e^- annihilation data via the dispersion integral [24, 25]

$$a_\mu^{HLO} = \frac{1}{4\pi^3}\int_{s_{thr}}^\infty ds\, K(s)\sigma^{(0)}(s) = \frac{\alpha^2}{3\pi^2}\int_{s_{thr}}^\infty \frac{ds}{s} K(s)R(s), \quad (1)$$

where $\sigma^{(0)}(s)$ is the total cross section for e^+e^- annihilation into any hadronic state, with extraneous QED radiative corrections subtracted off, and $R(s) = \sigma^{(0)}(s)/(4\pi\alpha^2/3s)$. The kernel function $K(s)$ decreases monotonically for increasing s and for large s behaves as $m_\mu^2/3s$ to a good approximation. For this reason the low-energy region of the dispersive integral is enhanced by $\sim 1/s^2$.

A prominent role among all data sets is played by the precise measurements by the CMD-2 detector at the VEPP-2M collider in Novosibirsk [26, 27] of the cross section for $e^+e^- \to \pi^+\pi^-$ at values of s between 0.37 and 0.93 GeV^2. The quoted systematic error of these data is 0.6% [27], dominated by the uncertainties in the radiative corrections (0.4%). In July 2004, also the KLOE experiment at the DAΦNE collider presented the final analysis [28] of the 2001 data for the precise measurement of $\sigma(e^+e^- \to \pi^+\pi^-)$ via the radiative return method from the ϕ resonance. In [28] the cross section $\sigma(e^+e^- \to \pi^+\pi^-)$ was extracted for the range $s \in [0.35, 0.95]\,\text{GeV}^2$ with a systematic error of 1.3% and a negligible statistical one. KLOE's 2004 results are in "fair agreement" [29] with the latest energy scan data from CMD-2. The study of the $e^+e^- \to \pi^+\pi^-$ process via the initial-state radiation method is also in progress at BABAR [30]. On the theoretical side, the properties of analyticity, unitarity and chiral symmetry provide strong constraints for the pion form factor $F_\pi(s)$ in the low-energy region [31].

The evaluations of the dispersive integral in eq. (1) based on the latest CMD-2 reanalysis [27] (which supersedes all their earlier ones) are in very good agreement:

$$[29] \quad a_\mu^{HLO} = 6934(53)_{exp}(35)_{rad} \times 10^{-11}, \quad (2)$$

$$[32] \quad a_\mu^{HLO} = 6948(86) \times 10^{-11}, \quad (3)$$

$$[33] \quad a_\mu^{HLO} = 6934(92) \times 10^{-11}, \quad (4)$$

$$[34] \quad a_\mu^{HLO} = 6924(59)_{exp}(24)_{rad} \times 10^{-11}, \quad (5)$$

$$[35] \quad a_\mu^{HLO} = 6944(48)_{exp}(10)_{rad} \times 10^{-11}. \quad (6)$$

The preliminary result in eq. (2) already includes KLOE's 2004 data analysis and updates the one of [36].

The authors of ref. [37] pioneered the idea of using vector spectral functions derived from the study of hadronic τ decays [38] to improve the evaluation of the dispersive integral in eq. (1). Indeed, assuming isospin invariance to hold, the isovector part of the cross section for $e^+e^- \to$ hadrons can be calculated via the Conserved Vector Current (CVC) relations from τ-decay spectra. An updated analysis is presented in [36], where τ spectral functions are obtained from the results of ALEPH, CLEO and OPAL, and isospin-breaking corrections are applied [39]. Their result is $a_\mu^{HLO} = 7110(50)_{exp}(8)_{rad}(28)_{SU(2)} \times 10^{-11}$ [36], where the quoted uncertainties are experimental, missing radiative corrections to some e^+e^- data, and isospin violation. Also an analysis of [35] includes information from τ decay; they obtain $a_\mu^{HLO} = 7027(47)_{exp}(10)_{rad} \times 10^{-11}$.

Although the latest CMD-2 $e^+e^- \to \pi^+\pi^-$ data are consistent with τ data for the energy region below 850 MeV, there is an unexplained discrepancy for larger energies. The $\pi^+\pi^-$ spectral function derived from KLOE's 2004 analysis confirms the discrepancy with the τ data. Among the possible causes of this discrepancy, one may wonder about inconsistencies in the e^+e^- data, in the τ data, or in the isospin-breaking corrections applied to the τ spectral functions. Given the good consistency of the ALEPH and CLEO data sets, and the confirmation by KLOE of the trend exhibited by other e^+e^- data, further careful investigations of the isospin-violating effects are clearly warranted [29, 40].

Higher-order Contributions. The $O(\alpha^3)$ hadronic contribution can be divided into two parts: $a_\mu^{HHO} = a_\mu^{HHO}(\text{vp}) + a_\mu^{HHO}(\text{lbl})$. The first term is the $O(\alpha^3)$ contribution of diagrams containing hadronic vacuum polarization insertions; the second is the light-by-light one. In recent years, $a_\mu^{HHO}(\text{vp})$ was evaluated in [41] and its latest value is $a_\mu^{HHO}(\text{vp}) = -97.9(0.9)_{exp}(0.3)_{rad} \times 10^{-11}$ [34] (obtained using e^+e^- annihilation data; it changes by $\sim -3 \times 10^{-11}$ if hadronic τ-decay data are used instead [6]). The hadronic light-by-light contribution cannot be expressed in terms of experimental observables determined from data, and its evaluation therefore relies on purely theoretical considerations. The estimate of the authors of [42, 43], who uncovered in 2001 a sign error in earlier evaluations, is [42] $a_\mu^{HHO}(\text{lbl}) = +80(40) \times 10^{-11}$. Earlier determinations now agree with this result [44]. At the end of 2003 a higher value was reported in [45], $a_\mu^{HHO}(\text{lbl}) = +136(25) \times 10^{-11}$. It was obtained by including short-distance QCD constraints previously overlooked. Further independent calculations would provide an important check of this result for $a_\mu^{HHO}(\text{lbl})$, a contribution whose uncertainty may become the ultimate limitation of the SM prediction of the muon $g-2$.

SM PREDICTION VS. MEASUREMENT

The first column of table 1 (from ref. [5]) shows the values obtained for $a_\mu^{SM} = a_\mu^{QED} + a_\mu^{EW} + a_\mu^{HLO} + a_\mu^{HHO}(\text{vp}) + a_\mu^{HHO}(\text{lbl})$. The values employed for a_μ^{HLO} are indicated by the reference number. The value $a_\mu^{HHO}(\text{lbl}) = 136(25) \times 10^{-11}$ [45] has been used for the hadronic light-by-light contribution. Errors were added in quadrature.

The present world average experimental value for the muon $g-2$ is $a_\mu^{EXP} = 116592080(60) \times 10^{-11}$ (0.5 ppm) [4]. The comparison of the SM results with the above experimental average gives the discrepancies $(a_\mu^{EXP} - a_\mu^{SM})$ listed in the second column of table 1. The number of standard deviations, shown in the third

TABLE 1. a_μ: SM predictions vs. experiment.

$a_\mu^{SM} \times 10^{11}$		$(a_\mu^{EXP} - a_\mu^{SM}) \times 10^{11}$	σ	
[29]	116591845 (69)	235 (91)	2.6	(3.0)
[32]	116591859 (90)	221 (108)	2.1	(2.5)
[33]	116591845 (95)	235 (113)	2.1	(2.5)
[34]	116591835 (69)	245 (91)	2.7	(3.1)
[35]	116591855 (55)	225 (81)	2.8	(3.2)
[36]	116592018 (63)	62 (87)	0.7	(1.3)
[35]	116591938 (54)	142 (81)	1.8	(2.3)

column, spans a wide range from 0.7 to 2.8. Somewhat higher discrepancies, shown in parentheses also in the third column, are obtained if the hadronic light-by-light contribution $a_\mu^{HHO}(\text{lbl}) = 80(40) \times 10^{-11}$ [42] is used instead of $a_\mu^{HHO}(\text{lbl}) = 136(25) \times 10^{-11}$ [45].

CONCLUSIONS

The discrepancies in table 1 between recent SM predictions and the current world average experimental value vary in a very wide range, from 0.7 to 3.2 standard deviations, according to the values used for the leading-order and light-by-light hadronic contributions. In particular, the contribution of the hadronic vacuum polarization depends on which of the two data sets, e^+e^- collisions or τ decays, are employed. This puzzling discrepancy between the $\pi^+\pi^-$ spectral functions from e^+e^- and isospin-breaking-corrected τ data could be caused by inconsistencies in the e^+e^- or τ data, or in the isospin-breaking corrections applied to the latter. Indeed, the question remains whether all possible isospin-breaking effects have been properly taken into account. Until we reach a better understanding of this problem, it is probably safer to discard information from hadronic τ decays [29]. If e^+e^- annihilation data are used to evaluate the leading hadronic contribution, the SM prediction of the muon $g-2$ deviates from the present experimental value by 2–3 standard deviations.

The measurement of the muon $g-2$ by the E821 experiment, with an impressive relative precision of 0.5 ppm, is still limited by statistical errors rather than systematic ones. A new experiment, E969, has been approved (but not yet funded) at Brookhaven in September 2004 [46]. Its goal would be to reduce the present experimental uncertainty by a factor of 2.5 to about 0.2 ppm. A letter of intent for an even more precise $g-2$ experiment was submitted to J-PARC with the proposal to reach a precision below 0.1 ppm [47]. While the theoretical predictions for the QED and EW contributions appear to be ready to rival these precisions, much effort will be needed in the hadronic sector to test a_μ^{SM} at an accuracy comparable to the experimental one. Such an effort is certainly well motivated by the excellent opportunity the muon $g-2$ is providing us to unveil or constrain "new physics" effects.

ACKNOWLEDGMENTS

I would like to thank the organizers for their kind invitation and for the excellent organization of this meeting.

REFERENCES

1. J.S. Schwinger, Phys. Rev. **73** (1948) 416.
2. A. Czarnecki, W.J. Marciano, Phys. Rev. D **64** (2001) 013014.
3. H.N. Brown *et al.*, Phys. Rev. D **62** (2000) 091101; Phys. Rev. Lett. **86** (2001) 2227; G.W. Bennett *et al.*, Phys. Rev. Lett. **89** (2002) 101804 [Erratum-ibid. **89** (2002) 129903].
4. G.W. Bennett *et al.*, Phys. Rev. Lett. **92** (2004) 161802.
5. M. Passera, J. Phys. G **31** (2005) R75.
6. M. Davier and W.J. Marciano, Annu. Rev. Nucl. Part. Sci. 54 (2004) 115.
7. M. Knecht, arXiv:hep-ph/0307239; K. Melnikov, Int. J. Mod. Phys. A **16** (2001) 4591; W.J. Marciano and B.L. Roberts, arXiv:hep-ph/0105056; V.W. Hughes and T. Kinoshita, Rev. Mod. Phys. **71** (1999) S133; T. Kinoshita and W.J. Marciano, in *Quantum Electrodynamics*, T. Kinoshita ed., World Scientific, pp. 419–478; A. Czarnecki and W.J. Marciano, Nucl. Phys. Proc. Suppl. **76** (1999) 245.
8. P.J. Mohr and B.N. Taylor, Rev. Mod. Phys., **77** (2005) 1.
9. S. Eidelman *et al.*, Phys. Lett. B **592** (2004) 1.
10. T. Kinoshita and W.B. Lindquist, Phys. Rev. Lett. **47** (1981) 1573; T. Kinoshita, B. Nizic and Y. Okamoto, Phys. Rev. Lett. **52** (1984) 717.
11. T. Kinoshita and M. Nio, Phys. Rev. D **70** (2004) 113001.
12. T. Kinoshita, Nucl. Phys. Proc. Suppl. **144** (2005) 206.
13. R. Jackiw and S. Weinberg, Phys. Rev. D **5** (1972) 2396; I. Bars and M. Yoshimura, Phys. Rev. D **6** (1972) 374; G. Altarelli, N. Cabibbo and L. Maiani, Phys. Lett. B **40** (1972) 415; W.A. Bardeen, R. Gastmans and B. Lautrup, Nucl. Phys. B **46** (1972) 319; K. Fujikawa, B.W. Lee and A.I. Sanda, Phys. Rev. D **6** (1972) 2923.
14. A. Sirlin, Phys. Rev. D **22** (1980) 971.
15. A. Ferroglia, G. Ossola, M. Passera and A. Sirlin, Phys. Rev. D **65**, 113002 (2002); G. Degrassi and P. Gambino, Nucl. Phys. B **567** (2000) 3; G. Degrassi, P. Gambino, M. Passera and A. Sirlin, Phys. Lett. B **418** (1998) 209.
16. A. Ferroglia, G. Ossola and A. Sirlin, arXiv:hep-ph/0406334.
17. A. Czarnecki, B. Krause and W.J. Marciano, Phys. Rev. D **52** (1995) 2619; Phys. Rev. Lett. **76** (1996) 3267.
18. T.V. Kukhto *et al.*, Nucl. Phys. B **371** (1992) 567.
19. S. Peris, M. Perrottet and E. de Rafael, Phys. Lett. B **355** (1995) 523; M. Knecht, S. Peris, M. Perrottet and E. de Rafael, JHEP **0211** (2002) 003.
20. A. Czarnecki, W.J. Marciano and A. Vainshtein, Phys. Rev. D **67** (2003) 073006.
21. G. Degrassi, G.F. Giudice, Phys.Rev.D**58** (1998) 053007.
22. S. Heinemeyer, D. Stöckinger and G. Weiglein, Nucl. Phys. B **699** (2004) 103.
23. F.V. Tkachov, Nucl. Instrum. Meth. A **389** (1997) 309; L.N. Bertstein, Funct. Anal. Appl. **6** (1972) 66; G. Passarino, Nucl. Phys. B **619** (2001) 257; G. Passarino and S. Uccirati, Nucl. Phys. B **629** (2002) 97; A. Ferroglia, G. Passarino, M. Passera and S. Uccirati, Nucl. Phys. B **650** (2003) 162; Nucl. Instrum. Meth. A **502** (2003) 391; Nucl. Phys. B **680** (2004) 199; S. Actis, A. Ferroglia, G. Passarino, M. Passera and S. Uccirati, Nucl. Phys. B **703** (2004) 3.
24. C. Bouchiat and L. Michel, J.Phys.Radium 22 (1961) 121.
25. L. Durand, Phys. Rev. **128** (1962) 441; Erratum-ibid. **129** (1963) 2835; S.J. Brodsky and E. de Rafael, Phys. Rev. **168** (1968) 1620; M. Gourdin and E. de Rafael, Nucl. Phys. B **10** (1969) 667.
26. R.R. Akhmetshin *et al.*, arXiv:hep-ex/9904027; Phys. Lett. B **527** (2002) 161.
27. R.R. Akhmetshin *et al.*, Phys. Lett. B **578** (2004) 285.
28. A. Aloisio *et al.*, Phys. Lett. B **606** (2005) 12.
29. A. Höcker, arXiv:hep-ph/0410081.
30. M. Davier, Nucl. Phys. B (Proc. Suppl.) **131** (2004) 82.
31. J. Gasser and U.G. Meissner, Nucl. Phys. B **357** (1991) 90; G. Colangelo, J. Gasser and H. Leutwyler, Nucl. Phys. B **603** (2001) 125; H. Leutwyler, arXiv:hep-ph/0212324; G. Colangelo, Nucl. Phys. B (Proc. Suppl.) **131** (2004) 185; F. Guerrero and A. Pich, Phys. Lett. B **412** (1997) 382; J. Portoles and P.D. Ruiz-Femenia, Nucl. Phys. B (Proc. Suppl.) **131** (2004) 170.
32. F. Jegerlehner, Nucl. Phys. B (Proc. Suppl.) **126** (2004) 325; Nucl. Phys. B (Proc. Suppl.) **131** (2004) 213.
33. O.V. Zenin, private communication, January 2005. Preliminary update of V.V. Ezhela, S.B. Lugovsky and O.V. Zenin, arXiv:hep-ph/0312114.
34. K. Hagiwara *et al.*, Phys. Rev. D **69** (2004) 093003.
35. J.F. de Trocóniz and F.J. Ynduráin, Phys. Rev. D **71** (2005) 073008.
36. M. Davier, S. Eidelman, A. Höcker and Z. Zhang, Eur. Phys. J. C **31** (2003) 503.
37. R. Alemany, M. Davier and A. Höcker, Eur. Phys. J. C **2** (1998) 123.
38. R. Barate *et al.*, Z. Phys. C **76** (1997) 15.
39. W.J. Marciano and A. Sirlin, Phys. Rev. Lett. **61** (1988) 1815; A. Sirlin, Nucl. Phys. B **196** (1982) 83; V. Cirigliano, G. Ecker and H. Neufeld, Phys. Lett. B **513** (2001) 361; JHEP **0208** (2002) 002.
40. S. Ghozzi and F. Jegerlehner, Phys. Lett. B **583**, 222 (2004); M. Davier, Nucl. Phys. B (Proc. Suppl.) **131** (2004) 123; ibid. **131** (2004) 192; W.M. Morse, arXiv:hep-ph/0410062; D.N. Gao, Phys. Rev. D **71** (2005) 051301.
41. B. Krause, Phys. Lett. B **390** (1997) 392.
42. A. Nyffeler, Acta Phys. Polon. B **34** (2003) 5197.
43. M. Knecht and A. Nyffeler, Phys. Rev. D **65** (2002) 073034; M. Knecht, A. Nyffeler, M. Perrottet and E. de Rafael, Phys. Rev. Lett. **88** (2002) 071802.
44. M. Hayakawa and T. Kinoshita, arXiv:hep-ph/0112102; Phys. Rev. D **57** (1998) 465 [Erratum-ibid. D **66** (2002) 019902]; J. Bijnens, E. Pallante and J. Prades, Nucl. Phys. B **626** (2002) 410.
45. K. Melnikov and A. Vainshtein, Phys. Rev. D **70** (2004) 113006.
46. R.M. Carey *et al.*, Proposal of the BNL Experiment E969, 2004; B.L. Roberts, arXiv:hep-ex/0501012.
47. J-PARC Letter of Intent L17, B.L. Roberts contact person.

Bottom-Quark Fragmentation in Higgs Decay

G. Corcella

CERN, Department of Physics, Theory Division, CH-1211 Geneva 23, Switzerland

Abstract. I investigate bottom-quark fragmentation in $H \to b\bar{b}$ processes. I discuss the implementation of collinear and soft resummation, in the next-to-leading logarithmic approximation, and the inclusion of non-perturbative contributions taken from LEP and SLD data.

Keywords: Higgs boson, fragmentation functions, b quarks, B hadrons
PACS: 12.38.Bx, 12.38.Cy, 14.65.Fy

Heavy-quark physics is currently one of the major fields of investigation in experimental and theoretical particle physics. In order to investigate heavy-quark phenomenology, fixed-order calculations are reliable as long as one considers inclusive observables, such as total cross sections or widths. For exclusive quantities, one needs to resum large contributions, which correspond to collinear or soft parton radiation.

In this talk shall discuss bottom-quark and B-hadron production in the Standard Model Higgs decay $H \to b\bar{b}$. In fact, the favourite discovery channel at the Tevatron accelerator is H production in association with a vector boson, followed by the decay $H \to b\bar{b}$. At the LHC, the process $gg(q\bar{q}) \to H \to b\bar{b}$ will be affected by larger backgrounds. However, the decay channel $H \to b\bar{b}$ will still be relevant, especially for Higgs production in association with $t\bar{t}$ pairs [1].

I shall present results for the the b-quark energy fraction x_b in $H \to b\bar{b}(g)$, which is given by:

$$x_b = \frac{2 p_b \cdot p_H}{m_H^2}. \quad (1)$$

The next-to-leading order (NLO) x_b spectrum is given by the following general result:

$$\frac{1}{\Gamma_0}\frac{d\Gamma}{dx_b} = \delta(1-x_b) + \frac{\alpha_S(\mu)}{2\pi}\left[P_{qq}(x_b)\ln\frac{m_H^2}{m_b^2} + A(x_b)\right]$$
$$+ \mathscr{O}\left(\frac{m_b^2}{m_H^2}\right)^p. \quad (2)$$

In Eq. (2), Γ_0 is the width of the LO process, μ is the renormalization scale, $A(x_b)$ is a function independent of m_b, $p \geq 1$, $P_{qq}(x_b)$ is the Altarelli–Parisi splitting function. Equation (2) is formally equal to the bottom energy spectrum in other processes, such as e^+e^- annihilation [2] or top decay [3], after replacing m_H with the centre-of-mass energy $\sqrt{s}$ of the e^+e^- collision or with the top mass m_t. Function $A(x_b)$ is instead process-dependent.

Equation (2) presents a large mass logarithm $\sim \alpha_S \ln(m_H^2/m_b^2)$, which needs to be resummed to all orders to extend the validity of the perturbative calculation. We follow the approach of perturbative fragmentation functions [2], which expresses the energy spectrum of a heavy quark as the convolution of a coefficient function, describing the emission of a massless parton, and a perturbative fragmentation function $D(m_b, \mu_F)$, associated with the transition of a massless parton into a massive quark:

$$\frac{1}{\Gamma_0}\frac{d\Gamma_b}{dx_b} = \sum_i \int_{x_b}^1 \frac{dz}{z}\left[\frac{1}{\Gamma_0}\frac{d\hat{\Gamma}_i}{dz}(z,m_H,\mu,\mu_F)\right]^{\overline{\text{MS}}}$$
$$\times\; D_i^{\overline{\text{MS}}}\left(\frac{x_b}{z}, \mu_F, m_b\right) + \mathscr{O}\left[\left(\frac{m_b^2}{m_H^2}\right)^p\right] \quad (3)$$

In Eq. (3), $d\hat{\Gamma}_i/dz$ is the differential width for the production of a massless parton i in Higgs decay with an energy fraction z; $D_i(x, \mu_F, m_b)$ is the perturbative fragmentation function for a parton i to fragment into a massive b quark, μ_F is the factorization scale. Neglecting $g \to b\bar{b}$ splitting, $i = b$ on the right-hand side of Eq. (3). The coefficient function for $H \to b\bar{b}$ processes was computed in [4] in the $\overline{\text{MS}}$ factorization scheme.

The perturbative fragmentation function follows the Dokshitzer–Gribov–Altarelli–Parisi (DGLAP) evolution equations. Its value at a given scale μ_F can be obtained once an initial condition is given. In [2] the initial condition $D_b^{\text{ini}}(x_b, \mu_{0F}, m_b)$ was calculated and its process-independence was established on more general grounds in [5]. It is given at NLO by:

$$D_b^{\text{ini}}(x_b, m_b) = \delta(1-x_b) + \frac{\alpha_S(\mu_0)C_F}{2\pi}d(x_b, m_b), \quad (4)$$

with

$$d(x_b, m_b) = \left[\frac{1+x_b^2}{1-x_b}\left(\ln\frac{\mu_{0F}^2}{m_b^2} - 2\ln(1-x_b) - 1\right)\right]_+. \quad (5)$$

As discussed in [2], solving the DGLAP equations for an evolution from μ_{0F} to μ_F, with a NLO kernel, allows one to resum leading (LL) $\alpha_S^n \ln^n(\mu_F^2/\mu_{0F}^2)$ and next-to-leading (NLL) $\alpha_S^n \ln^{n-1}(\mu_F^2/\mu_{0F}^2)$ logarithms (collinear resummation). In $H \to b\bar{b}$ processes, for an evolution from $\mu_{0F} \simeq m_b$ to $\mu_F \simeq m_H$, the large logarithms $\ln(m_H^2/m_b^2)$ appearing in the massive calculation (2) are resummed with NLL accuracy. Furthermore, both the coefficient function in Ref. [4] and the initial condition (4) present terms that become large once the b-quark energy fraction x_b approaches 1, which corresponds to soft-gluon radiation. Soft contributions to the initial condition are process-independent and were resummed in [5] in the NLL approximation. NLL soft-gluon resummation in the coefficient function of $H \to b\bar{b}$ was implemented in [4].

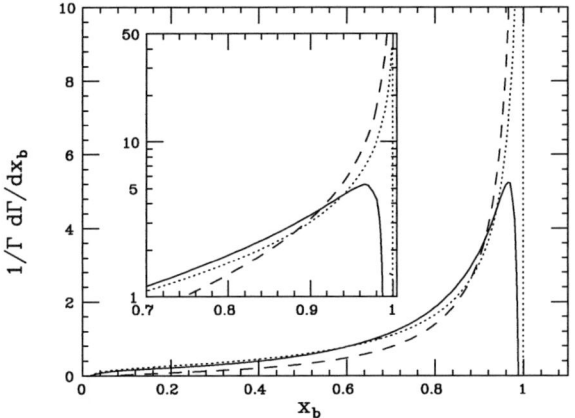

FIGURE 1. b spectrum in Higgs decay according to the NLO massive calculation (dashed), including NLL collinear resummation (dotted), and both NLL collinear and soft resummations (solid). In the inset figure, the same curves are shown at large x_b, on a logarithmic scale.

In Fig. 1, the bottom energy distribution in Higgs decay is shown. I have plotted the NLO massive calculation (dashed), including NLL collinear resummation (dotted), and with both NLL collinear and soft resummations (solid). I have set $\mu = \mu_F = m_H$, $\mu_0 = \mu_{0F} = m_b$, $m_H = 120$ GeV, $m_b = 5$ GeV. The fixed-order calculation lies below the resummed ones and grows as $x_b \to 1$, since $\sim 1/(1-x_b)_+$; the collinear-resummed spectrum exhibits instead a sharp peak at large x_b. Soft resummation is relevant at $x_b > 0.6$: the distribution is further smoothed and shows the Sudakov peak at $x_b \simeq 0.97$.

I would like to present results for b-flavoured B hadron production in $H \to b\bar{b}$ processes. The B spectrum will be given by the convolution of the b-energy distribution with a non-perturbative fragmentation function associated with the hadronization. As in [3, 4, 6], I consider a power law with two tunable parameters

$$D^{np}(x;\alpha,\beta) = \frac{1}{B(\beta+1,\alpha+1)}(1-x)^\alpha x^\beta, \quad (6)$$

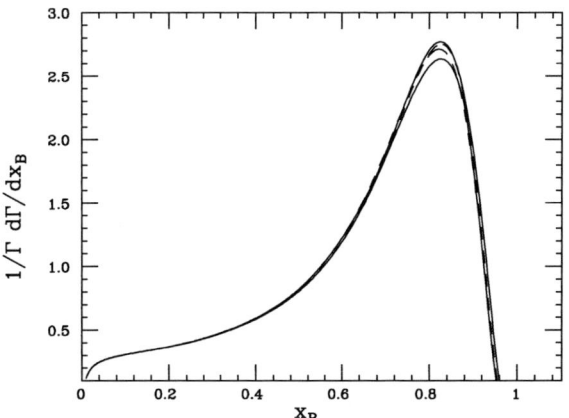

FIGURE 2. B-hadron spectrum in Higgs decay: the hadronization is described according to the power law (6) (solid) and the Kartvelishvili model (7) (dashes).

the model of Kartvelishvili et al. [7]

$$D^{np}(x;\delta) = (1+\delta)(2+\delta)(1-x)x^\delta, \quad (7)$$

and the Peterson model [8]

$$D^{np}(x;\varepsilon) = \frac{A}{x[1-1/x-\varepsilon/(1-x)]^2}. \quad (8)$$

In Eq. (6), $B(x,y)$ is the Euler beta function; in (8) A is a normalization constant. The parameters α, β, δ and ε can be obtained after fitting models (6)–(8) to x_B data in e^+e^- collisions, where x_B is the B energy fraction. As in [4], I consider data from ALEPH [9] and SLD [10] Collaborations in the range $0.18 \leq x_B \leq 0.94$. The best-fit parameters, along with the χ^2 per degree of freedom, are listed in Table 1.

TABLE 1. Results of fits to $e^+e^- \to b\bar{b}$ data from the ALEPH and SLD Collaborations, I have set $\Lambda = 200$ MeV, $\mu_{0F} = \mu_0 = m_b = 5$ GeV and $\mu_F = \mu = \sqrt{s} = 91.2$ GeV.

α	0.90 ± 0.15
β	16.23 ± 1.37
$\chi^2(\alpha,\beta)/\text{dof}$	$33.42/31$
δ	17.07 ± 0.39
$\chi^2(\delta)/\text{dof}$	$33.80/32$
ε	$(1.71 \pm 0.09) \times 10^{-3}$
$\chi^2(\varepsilon)/\text{dof}$	$166.36/32$

The power law (6) and the Kartvelishvili model (7) yield very good fits, while the Peterson model (8) is unable to reproduce the data. In Fig. 2 we show the x_B distribution in Higgs decay according to models (6) and (7), using the results listed in Table 1. Plotted are

TABLE 2. Experimental data for the moments σ_N^B from DELPHI [11], the extracted non-perturbative contribution D_N^{np}, and the predicted moments in $H \to b\bar{b}$ processes.

	$\langle x \rangle$	$\langle x^2 \rangle$	$\langle x^3 \rangle$	$\langle x^4 \rangle$
e^+e^- data σ_N^B	0.7153±0.0052	0.5401±0.0064	0.4236±0.0065	0.3406±0.0064
e^+e^- NLL σ_N^b	0.7801	0.6436	0.5479	0.4755
D_N^{np} [B]	0.9169	0.8392	0.7731	0.7163
H-decay NLL Γ_N^b	0.7578	0.6162	0.5193	0.4473
H-decay Γ_N^B	0.6948	0.5171	0.4015	0.3204

the edges of bands at one-standard-deviation confidence level: we can see that the predictions yielded by the two models are in statistical agreement.

Finally, we can use the moments on B production in e^+e^- annihilation from the DELPHI Collaboration [11] to predict the same moments Higgs decay. In N-space we have: $\sigma_N^B = \sigma_n^b D_N^{np}$, $\Gamma_N^B = \Gamma_N^b D_N^{np} = \Gamma_N^b \sigma_N^B / \sigma_N^b$, where $\sigma_N^{b,B}$ and $\Gamma_N^{b,B}$ are the moments of the e^+e^--annihilation cross section and of the $H \to b\bar{b}$ width, at hadron or parton level. Fitting directly the moments σ_N^B, we avoid the problems exhibited by x-space spectra, which become negative at very small and large x. The results of the N-space fits and the predictions for the $H \to b\bar{b}$ moments are shown in Table 2.

In summary, I have reviewed recent results on bottom-quark fragmentation in $H \to b\bar{b}$ decays. Parton-level spectra exhibit a relevant impact of NLL collinear and soft resummation; using information from LEP and SLD experiments, B-hadron energy distributions have been predicted in x and moment spaces. It will be now very interesting to compare the results yielded by the perturbative fragmentation approach with the ones obtained using Monte Carlo event generators. This is in progress [12].

REFERENCES

1. V. Drollinger, Th. Müller and D. Denegri, CMS Note 2001/054, hep-ph/0111312.
2. B. Mele and P. Nason, *Nucl. Phys. B* **361** (1991) 626.
3. G. Corcella and A.D. Mitov, *Nucl. Phys. B* **623** (2002) 247.
4. G. Corcella, *Nucl. Phys. B* **705** (2005) 363 [Erratum-ibid. B **715** (2005) 609].
5. M. Cacciari and S. Catani, *Nucl. Phys. B* **617** (2001) 253.
6. M. Cacciari, G. Corcella and A.D. Mitov, *JHEP* **0212** (2002) 015.
7. V.G. Kartvelishvili, A.K. Likehoded and V.A. Petrov, *Phys. Lett. B* **78** (1978) 615.
8. C. Peterson, D. Schlatter, I. Schmitt and P.M. Zerwas, *Phys. Rev. D* **27** (1983) 105.
9. ALEPH Collaboration, A. Heister et al., *Phys. Lett. B* **512** (2001) 30.
10. SLD Collaboration, K. Abe et al., *Phys. Rev Lett.* **84** (2000) 4300.
11. DELPHI Collaboration, ICHEP 2002 Note, DELPHI 2002-069 CONF 603.
12. G. Corcella and V. Drollinger, in preparation.

Electroweak Physics at LHC

Andrea Giammanco

Scuola Normale Superiore
piazza dei Cavalieri, 7
56126 Pisa, Italy
andrea.giammanco@cern.ch

Abstract. The Large Hadron Collider experiments will be able to perform precision electroweak studies in new kinematic regions, due to the high collision energy and the huge cross sections for many Standard Model processes. This article describes how the two multipurpose detectors ATLAS and CMS will allow a significant improvement of the precision on several Standard Model parameters. The selected items include the determination of parton distribution functions, tests of QCD, the determination of the W mass, studies of the Drell-Yan lepton pairs production and of di-boson production.

Keywords: Standard Model, LHC

INTRODUCTION

The Large Hadron Collider (LHC) is designed to collide protons at centre-of-mass energy of 14 TeV and luminosities in the range from 10^{33} cm^{-2}s^{-1} in the initial phase up to 10^{34} cm^{-2}s^{-1} at a later stage.

Already during the low luminosity phase the very high centre-of-mass energy will result in large production rates for several Standard Model particles: approximately 200 W and 50 Z bosons will be produced every second. Even after tight selection cuts in specific clean final states (e and μ decay channels), large data samples will be available very shortly after the start-up of the collider, allowing many precision measurements and also the possibility to calibrate the detectors with known resonances.

PARTON DISTRIBUTION FUNCTIONS

Most part of the LHC physics program concerns hard scattering processes, whose description involves elementary interactions between the partons inside the protons. Therefore the accurate knowledge of the parton distribution functions (PDF) of quarks, anti-quarks and gluons is of paramount importance for the understanding of data.

The current knowledge on the PDFs largely relies on the measurements carried on at the HERA electron-proton collider, and theoretical models like the DGLAP evolution [1] are required to extrapolate from the experimentally tested regions at low Q^2 to the higher momentum transfers relevant for the LHC physics. A great improvement will come from using PDFs determined directly from LHC data.

The differential distributions of transverse momentum p_T and pseudo-rapidity η for heavy particles can help constrain the PDF models. For example, the cross section ratio of W^+ to W^- bosons is sensitive to the ratio of u- and d- quarks in the proton. From the experimental point of view this measurement consists in measuring the ratio of positively to negatively charged electrons and muons as a function of rapidity from selected $W^\pm$ decays. As an example, Fig. 1 shows this ratio for two PDF models with small differences in the sea quark distribution [2]. The statistics corresponding to 0.1 fb^{-1} integrated luminosity (collected in a few days during the initial low luminosity phase) is sufficient to distinguish between these two models.

Examples of other "probes" to the PDFs are the quark-gluon scattering processes $gb \to b\gamma$, $gc \to c\gamma$ and $gs \to cW$, exploiting the clean signatures of isolated photons or W bosons (decaying leptonically) and tagging the jets from b- and c- quarks through displaced vertices and non isolated leptons [3].

QCD AND JET PHYSICS

Being a hadron collider, LHC is an ideal environment to test strong interaction physics. The transverse jet energy spectrum can be studied up to several TeV's, as illustrated by Fig. 2. Of particular interest is the highest part of the spectrum, corresponding to small distance scales, where new physics may appear (for example a possible quark substructure). For an integrated luminosity of 300 fb^{-1}, corresponding to a few years of operation during the high luminosity phase, about 4 million jets with $E_T > 1$ TeV and 400 with $E_T > 3$ TeV are expected per experiment [4].

The jet energy spectrum can be used to extract the

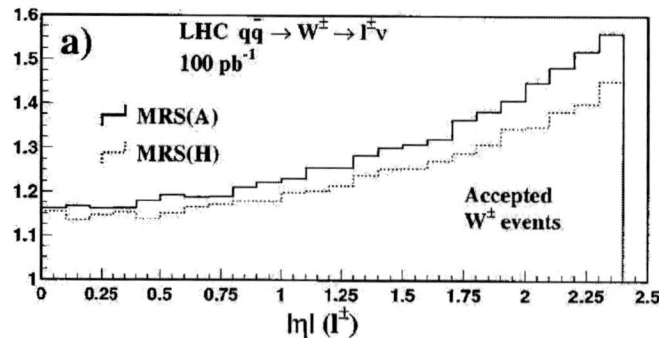

FIGURE 1. Ratio of positive to negative leptons (e, μ) from W boson decays, as a function of absolute rapidity $|\eta|$.

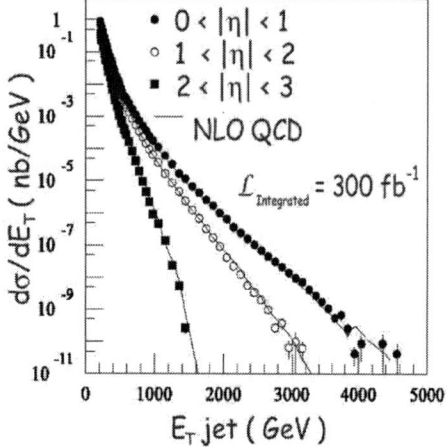

FIGURE 2. Differential cross section of the transverse jet energy in three different rapidity ranges.

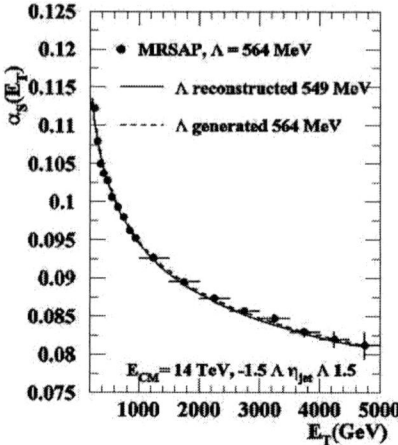

FIGURE 3. Running of α_s determined from the spectrum in Fig. 2.

strong coupling constant α_s, demonstrating its running from the energy scale of about 100 GeV (investigated at previous colliders) up to a few TeV, as illustrated in Fig. 3. However, the absolute precision on $\alpha_s(M_Z)$, limited to around $\pm 10\%$ mainly by the uncertainties on PDFs and on the renormalization and factorization scale [4], will not be competitive with the existing measurements.

THE W BOSON MASS

The W boson mass is related to the QED fine structure constant α, the Fermi constant G_F and the Weinberg angle $\sin \theta_W$ by Eq. (1),

$$M_W = \sqrt{\frac{\pi \alpha}{G_F \sqrt{2}}} \frac{1}{\sin \theta_W \sqrt{1 - \Delta r}}, \quad (1)$$

where Δr accounts for the electroweak corrections, which amount to about 4% [5] in the Standard Model and depend on the top quark mass (as M_t^2) and on the Higgs boson mass (as $\log M_H$). Therefore, precise measurements of both the W boson mass and the top quark mass give an indirect constraint to the SM Higgs boson mass.

The cross section expected for the $pp \to W(+X)$ process, times the branching ratio for $W \to e, \mu$, is 30 nb. For 10 fb^{-1} this translates in about 300 million events, resulting in a statistical uncertainty on M_W of less than 2 MeV/c^2 [6]. Therefore, the precision of the measurement will be dominated by the systematic uncertainties, both from detector and physics modeling.

The W boson mass is extracted by fitting the experimental distribution of the transverse mass (shown in Fig. 4 on simulated data both with and without detector effects). The sources of error and their contribution are shown in table 1 for the ATLAS case.

The quoted figures for most of these uncertainties assume that they will be constrained in situ by using data samples such as leptonic Z decays to determine the lepton energy scale and measure the detector resolution (by

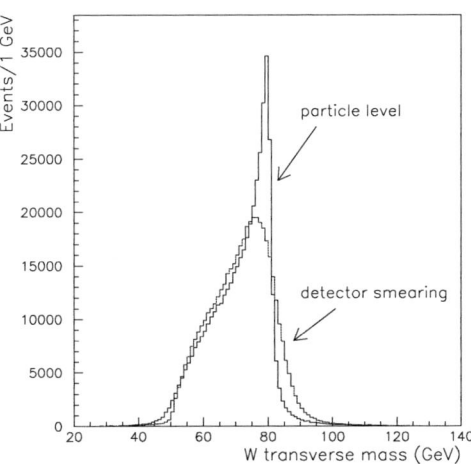

FIGURE 4. W boson transverse mass distribution including expected detector resolution.

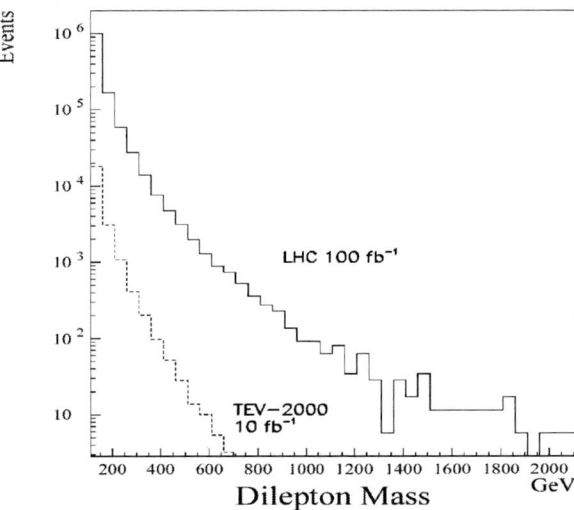

FIGURE 5. Expected number of di-lepton events per channel and per experiment as a function of the invariant mass of the pair, at LHC with 100 fb^{-1} and Tevatron with 10 fb^{-1}.

TABLE 1. Sources of error to M_W, and their individual contributions, for ATLAS with 10 fb^{-1}.

Source	ΔM_W (MeV/c^2)
Statistics	< 2
Lepton E-p scale	15
Lepton energy resolution	5
Recoil model	5
Lepton identification	5
p_T^W spectrum modeling	5
PDFs	< 10
W width	7
Radiative decays	10
Background	5
Total	**< 25**

exploiting the precise knowledge of the Z mass from the LEP and SLD measurements [7]), as well as model the p_T spectrum and the detector response to the hadronic system recoiling against the W boson (by using $Z + jets$ events with $Z \to l^+l^-$, since their kinematic is very similar to the $W + jets$ signal events).

With 10 fb^{-1} data, using only one lepton species, the ATLAS experiment expects an overall accuracy of better than 25 MeV/c^2. By combining the lepton channels (taking properly into account the correlations among systematics) this reduces to better than 20 MeV/c^2, and assuming the same systematics for the CMS case this would result in better than 15 MeV/c^2 for the LHC combination. This would reduce the uncertainty on $log\,M_H$ by a factor of 2, from 0.2 to 0.1.

DRELL-YAN LEPTON PAIRS PRODUCTION

The so called Drell-Yan (DY) process $q\bar{q} \to l^+l^-$ ($l = e, \mu$) constitutes in some sense the equivalent of the $e^+e^- \to hadrons$ which was extensively studied at LEP and SLC, and, like it, at the Z mass peak may provide a measurement of $sin\,\theta_W$ from the observation of a forward-backward asymmetry A_{FB}. The ambiguity in the direction of the incoming quark and anti-quark is resolved by assuming that the anti-quark, being a seaquark, is usually the parton with the lower x value. Preliminary studies show that, provided a reliable identification of leptons in the forward region is available, a statistical error of $\Delta sin^2\,\theta_W \approx 1.4 \times 10^{-4}$ is expected for 100 fb^{-1}, comparable with the one from electron-positron colliders. Further studies are needed to estimate if systematic errors from the limited knowledge of the PDFs, of the lepton acceptance and of radiative corrections can be controlled to the required level such that this measurement will become significant.

Reversing the logic, DY events may be used to constrain the PDFs, both at the Z peak and at higher invariant masses. Furthermore, the invariant mass spectrum of the lepton pair can show new heavy particles decaying into leptons, e.g. a Z' boson.

Fig. 5 shows the SM expectations for the number of events per di-lepton channel and per experiment as a function of the mass for a total luminosity of 100 fb^{-1} in comparison to the expectation at the Tevatron Run II (with 10 fb^{-1}) [8, 9]

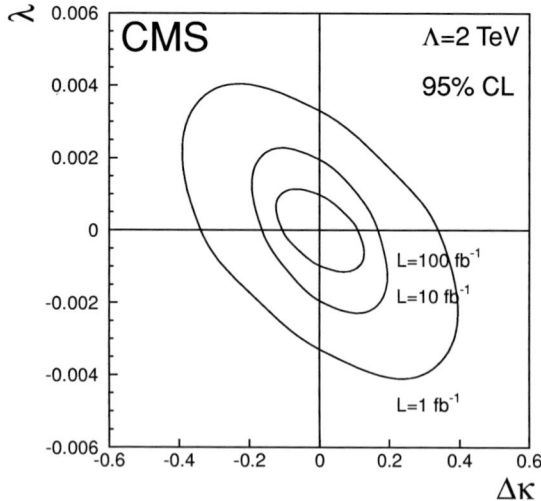

FIGURE 6. Limits in the plane $\lambda - \Delta\kappa$ obtainable at LHC.

TABLE 2. Coupling parameters tested by di-boson production measurements at hadron colliders, their Standard Model values and their dependence on s.

Coupling parameter	SM value	Scaling with s)
$\kappa_{\gamma,Z}$	1	$\sqrt{s}$
$\lambda_{\gamma,Z}$	0	s
g_1^Z	1	$\sqrt{s}$
$h_1^{\gamma,Z}$	0	$s^{3/2}$
$h_2^{\gamma,Z}$	0	$s^{5/2}$
$h_3^{\gamma,Z}$	0	$s^{3/2}$
$h_4^{\gamma,Z}$	0	$s^{5/2}$
$f_4^{\gamma,Z}$	0	$s^{3/2}$
$f_5^{\gamma,Z}$	0	$s^{3/2}$

DI-BOSON PRODUCTION

The high centre-of-mass energy makes the LHC also an ideal place to study di-boson production, and consequently the Triple Gauge Boson (TGC) couplings, providing a powerful test of the Standard Model. In the SM, in fact, the WWγ and WWZ vertices exist whereas all others, in particular those involving only the neutral bosons γ and Z, do not.

The WWγ coupling is tested in Wγ final states [10, 11, 12], and the transverse momentum distribution of the photon is sensitive to deviations from the Standard Model, which can be described by the parameters λ and $\Delta\kappa$ (found by LEP to be consistent with zero at the percent level). Fig. 6 shows the expected limits on the coupling parameters at LHC. The sensitivity to λ is much higher than to $\Delta\kappa$ at LHC, because the deviations from the SM induced by the λ coupling scale with s, against $\sqrt{s}$ for $\Delta\kappa$ (see Table 2).

Likewise, $Z\gamma$ final states will be used to search for anomalous $ZZ\gamma$ and $Z\gamma\gamma$ vertexes [10, 13], whose observable effects grow even stronger, as $s^{3/2}$ or $s^{5/2}$ (see Table 2).

CONCLUSIONS

The two LHC general purpose experiments ATLAS and CMS will allow precise tests of the SM, thanks to the high energy and luminosity that will open unexplored regions of phase space and provide very high statistics making LHC a "W and Z factory". Most of the SM analyses will be limited by systematic uncertainties, both from physics and detector, and their reduction for the achievement of optimal precision is a challenge for the experimenters, inducing them to devise sophisticated analysis techniques. LHC data themselves may help to pin down these systematics, by studying the physics behind them (e.g. PDFs) and by using the known SM particles as "standard candles" for detector calibration.

ACKNOWLEDGMENTS

I wish to thank Joachim Mnich and Roberto Chierici for the help in preparing my presentation and in understanding some of the experimental issues described.

REFERENCES

1. G. Altarelli and G. Parisi, Nucl. Phys. 126 (1977) 297; V.N. Gribov and L.N. Lipatov, Sov. J. Nucl. Phys. 15 (1972) 438 and 675; Yu.L. Dokshitzer, Sov. Phys. JETP 46 (1977) 641.
2. M. Dittmar, F. Pauss, D. Zürcher, Phys. Rev. D56 (1997), 7284.
3. M. Dittmar, K. Mazumdar, CMS-NOTE-2001/002.
4. H. Stenzel, ATLAS, ATL-PHYS-2001-003.
5. G. Altarelli et al., CERN/LHCC 2000-004.
6. ATLAS Detector and Physics Performance, Technical Design Report, CERN/LHCC 99-14.
7. The LEP Electroweak Working Group, CERN-PH-EP/2004-069.
8. Workshop on Standard Model Physics (and more) at the LHC, CERN Yellow Report 2000-004.
9. D. Bourilkov, Proc. 30th Int. Conf. on High Energy Physics, Osaka (2000), eds. C.S. Lim and T. Yamanaka.
10. T. Mueller et al., CMS-NOTE-2000/017.
11. C. Mackay, P. Hobson., CMS-NOTE-2001/052.
12. C. Mackay, P. Hobson., CMS-NOTE-2001/056.
13. C. Mackay, P. Hobson., CMS-NOTE-2002/028.

Top Physics at ATLAS

M. Barisonzi

University of Twente, Postbus 217, 7500 AE Enschede, The Netherlands
NIKHEF, Kruislaan 409, 1098 SJ Amsterdam, The Netherlands

Abstract. The Large Hadron Collider **LHC** is a top quark factory: due to its high design luminosity, LHC will produce about 200 millions of top quarks per year of operation. The large amount of data will allow to study with great precision the properties of the top quark, most notably cross-section, mass and spin. The Top Physics Working Group has been set up at the ATLAS experiment, to evaluate the precision reach of physics measurements in the top sector, and to study the systematic effects of the ATLAS detector on such measurements. This reports give an overview of the main activities of the ATLAS Top Physics Working Group in 2004.

Keywords: top quark physics, top spin correlation, b-jet tagging, ATLAS
PACS: 14.65.Ha Top quarks

OVERVIEW OF THE TOP QUARK

The top quark is the fermion with the largest mass in the Standard Model. In the Higgs mechanisms, this means that the top quark has a large coupling strength with the Higgs boson. This property can be exploited to constrain the expected mass of the Higgs boson. A NLO calculation of the mass of the W boson contains corrections term dependent on the top mass and the Higgs mass. By accurately measuring M_W and m_t, it is possible to obtain an estimate for the M_H. Recent Tevatron measurements for the top mass of 178.0±4.3 GeV result in a Higgs mass prediction of 126^{+73}_{-48} GeV.

The achievement of a high top mass resolution is instrumental for a precise prediction of M_H.

Production processes

Top quarks are produced at LHC by two types of processes: QCD and Electroweak production. In QCD production, top quarks are created in $t\bar{t}$ pairs via the processes $q\bar{q} \to t\bar{t}$ and $gg \to t\bar{t}$. The combined NLO cross-section for these two processes is ~825 pb; 90% of the contribution comes from the gluon-gluon fusion process, the remaining 10% comes from quark-antiquark annihilation.

Electroweak production is composed of three channels: the s-channel, the t-channel (also referred as W-gluon fusion) and associated Wt production. The three processes have cross sections of 11 pb, 60 pb, and 247 pb respectively.The cross-section for EW processes is directly proportional to the CKM matrix element V_{tb}: a deviation from the SM prediction may point to the existence of a fourth quark generation.

Experimental signatures

In the Standard Model, the top quark decays predominantly into a W and a b-quark with a branching ratio of 0.998. Because of fermion universality in electroweak interactions, the W boson decays 1/3 of the time into a lepton/neutrino pair and 2/3 of the time into a $q\bar{q}$ pair. Since in $t\bar{t}$ events two real W bosons are present, the signatures of the events are classified according to the decays of the two W bosons:

all jets channel In this channel both Ws decay into a quark/anti-quark pair. The event has at least six high-p_T jets, two of which have to be b-tagged. Despite having the highest branching ratio (44%), this decay channel suffers heavily from QCD background and ambiguities in the assignment of jets to the originating Ws.

lepton+jets channel In this decay channel one W decays into a lepton–neutrino pair, the other W into a quark/anti-quark pair. Thus, the top and anti-top quarks are called "leptonic" or "hadronic" accordingly with the decay mode of their daughter W. The event signature is characterised by one isolated lepton one b-jet and missing E_T in the leptonic branch, two light jets and one b-jet in the hadronic branch. The branching ratio for this channel is about 30%.

di-lepton channel In this decay channel, both Ws decay into a lepton–neutrino pair. For practical purposes, only e, μ are considered, since τ decays are difficult to distinguish from the QCD background. The events have two high-p_T leptons, two jets (at least one of which is b-tagged) and missing energy due to the neutrinos. This signature is quite clear, being affected mainly by electroweak background. The only drawback of this decay channel is its low branching ratio (5%).

The lifetime of the top quark is much shorter than the time constant for spin-flip via gluon emission. Thus, decay products retain the spin information of the original top quark. By measuring the spin of the decay products, it is possible to verify the $V-A$ coupling of the Wtb vertex or discover a non-SM right-handed coupling.

MASS MEASUREMENTS

The lepton+jets $t\bar{t}$ channel is the most promising for top mass measurement; however the channel is affected by combinatorial background. The assignment of the two b-jets to the hadronic and leptonic branches is not unique, and can spoil the resolution of the top mass measurement.

A study was performed to evaluate the best strategy for the assignment of b-jets. The study was performed on a MC sample of 200,000 $t\bar{t}$ events with full detector simulation. The selection cuts required at least 4 Jets of p_T>20 GeV and $|\eta|$<2.5, E_T^{miss}>20 GeV and only one isolated lepton of p_T>20 GeV. In addition, two b-tagged jets were required with varying p_T thresholds.

The assignment algorithm proceeded as follows: the hadronic branch is assigned to the b-jet with the minimum distance ΔR from the hadronic W, while the leptonic branch is assigned to the b-jet with the minimum azimuthal angle ϕ from the leptonic W. If the algorithm produces conflicting assignment, then the algorithm chooses the basis of another variable: the maximum ΔR between the b-jet and the lepton, the combination that maximizes $p_T(Top_{had}) + p_T(Top_{lep})$ or the combination that minimizes $\phi(b_i,W_{lep}) + \phi(b_j,W_{had})$ or one of the two angles $\phi(b_i,W_{lep})$, $\phi(b_j,W_{had})$.

For each of the variables used in the assignment algorithm, the purity of the assignment was computed. The purity varies according to the cut on the minimum p_T of the b-jets: the higher the threshold, the better the purity achieved (Figure 1). On the other hand, a high threshold biases the mass of the reconstructed top quark towards higher value. The best compromise is to choose a p_T threshold for b-jets of 40 GeV, which results in an assignment purity of 70-80%.

SPIN CORRELATIONS

Top/anti-top quark pairs produced in QCD processes are not polarized; however, the top and anti-top spins are correlated. For QCD processes close to production threshold, the $t\bar{t}$ system is produeced in a $^3S_1$ state for $q\bar{q}$ annihilation, or in a $^1S_0$ state for gluon-gluon fusion. Hence, in the first case the top and the anti-top have

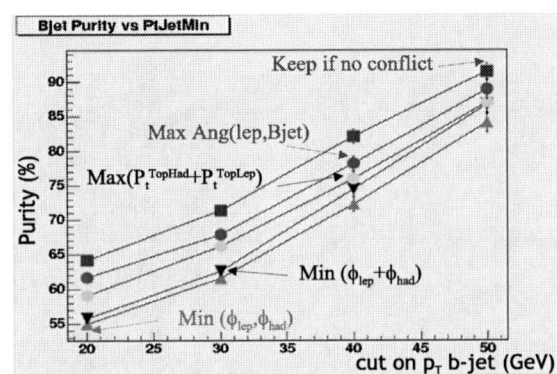

FIGURE 1. Purity in the $W-b$ assignment for various assignment methods over varying p_T threshold for the b-jets.

parallel spins, while in the second case the spins are antiparallel. Since at LHC the gluon-fusion process has a much larger cross-section than $q\bar{q}$ annihilation, NLO calculations predict an excess of top pairs with opposite spins. The spin of the quarks can be evaluated in the helicity basis, which corresponds to the top (anti-top) direction of flight in the $t\bar{t}$ system. In this basis, the asymmetry parameter — that expresses the excess of same-helicity pairs — is given by:

$$A = \frac{\sigma(t_L\bar{t}_L) + \sigma(t_R\bar{t}_R) - \sigma(t_L\bar{t}_R) - \sigma(t_R\bar{t}_L)}{\sigma(t_L\bar{t}_L) + \sigma(t_R\bar{t}_R) + \sigma(t_L\bar{t}_R) + \sigma(t_R\bar{t}_L)}$$
$$= 0.326^{+0.003}_{-0.002}(\mu_{r,f})^{+0.013}_{-0.001}(PDF)$$

For $t\bar{t}$ pairs produced with total invariant mass much larger than production threshold, the asymmetry is diluted because of the presence of higher spin $t\bar{t}$ pairs. Hence it is useful to introduce a cut in the total invariant mass $m_{t\bar{t}} < 550$ GeV to maximize the power of the experimental analysis. A significant deviation of the measured asymmetry parameter from the theoretical value may point to non-SM physics, such as right-handed weak interactions.

The asymmetry can be evaluated by studing the angular distributions of the top and anti-top decay products (spin analyzers):

$$\frac{1}{N}\frac{d^2N}{d\cos\theta_1 d\cos\theta_2} = \frac{1}{4}(1 - C\cos\theta_1\cos\theta_2)$$
$$\frac{1}{N}\frac{dN}{d\cos\phi} = \frac{1}{2}(1 - D\cos\phi)$$

where θ_1 (θ_2) indicates the the angle between the decay product of t ($\bar{t}$) in the t ($\bar{t}$) rest frame and the t ($\bar{t}$) direction in the $t\bar{t}$ frame, while ϕ indicates the angle between the decay products of t and $\bar{t}$ in the respective rest frames [3].

The parameters C and D are the spin correlation variables, which can be evaluated using the following unbiased estimators:

$$C = -9\langle \cos\theta_1 \cos\theta_2 \rangle$$
$$D = -3\langle \cos\phi \rangle$$

The spin correlation parameters have been measured on MonteCarlo data in the dileptonic and the lepton+jets decay channels. For the dileptonic channel, the natural choice for spin analyzers is the two leptons from the decay of the Ws, since leptons are 100% polarized with respect to the top spin. In the lepton+jets channel, the best choice for spin analyzers would be a lepton and a d-type quark, since d have 100% polarization as well. However, since it is experimentally impossible to distinguish between light quark flavours, the second spin analyzer is the least energetic jet in the t ($\bar{t}$) rest frame, which is polarised at 51%.

The $t\bar{t}$ sample was created using the event generator TopReX 4.05 [4] in combination with the hadronisation software Pythia 6.2 [5] and the fast detector simulation ATLFAST [6]. The background sample included $t\bar{t}$ events decaying to all-hadrons and τ+jets, W+jets, Z+jets, $Wb\bar{b}$, ZZ, WW, WZ and single top production. The total generated sample corresponds to 10 fb^{-1}, or one year at LHC at low luminosity; the sample was generated with varying production parameters to study the systematic effects of:

- Q-scales;
- Parton Distribution Functions;
- Initial and Final State Radiation;
- b-fragmentation and hadronisation scheme;
- b-tagging and b-jet calibration;
- generated top mass.

The results of the spin correlation measurements in the two channels are summarised in Table 1; the spin correlation parameter D resulted more suitable for measurements, since it is less sensitive to systematic effects.

DETECTOR COMMISSIONING STUDIES

The top pair production process is valuable for the in-situ calibration of ATLAS in the detector commissioning stage: the large cross section and the large S/B ratio for the lepton+jets $t\bar{t}$ channel allow to produce high purity samples with large statistics in a short time period. The experimental signature for top events — high energy leptons and jets, b-jets, missing E_T — involves most parts of the ATLAS detector, so top samples can play

TABLE 1. Comparison between the theoretical predictions and the MC measurements for spin correlation parameters C and D in the di-leptonic and the lepton+jets sample. The statistical error is written in italics, while the systematic error is written in plain typeface.

Parameter	Di-lepton	Lepton+jets
C (theory)	0.42	0.21
C (MC)	0.40±*0.02*±0.02	0.21±*0.01*±0.04
Precision	8%	20%
D (theory)	-0.29	-0.15
D (MC)	-0.29±*0.01*±0.01	-0.15±*0.006*±0.02
Precision	5%	13%

an important role in the calibration of the detector in the initial phases.

Top analysis without b-tagging

The experimental signature for top events include one or more b-jets, arising from the top decay. Thus, the b-tagging performance of ATLAS has an important role for top analyses. However, efficient b-tagging needs precise alignment of the trackers of the Inner Detector, which will be reached only after few months of data taking. Will it be possible to perform top physics in the commissioning stage? A study has explored the possibility of reconstructing top events in $t\bar{t}$ production by assuming the absence of b-tagging.

The kinematic cuts applied to select the top-enriched sample require exeactly one isolated lepton with p_T>20 GeV and exactly four jets — reconstructed with a cone algorithm of size ΔR=0.4 — with p_T>40 GeV. The analysis reconstructs exclusively the hadronic branch of the $t\bar{t}$ event: the combination of 3 out of 4 jets with the highest vector p_T sum is assumed to belong to the same top (antitop) branch. The invariant mass of the 3-jets combination is an estimate of the top mass.

The analysis was performed on a generated data sample which included the $t\bar{t}$ signal plus a W+4jets background (leptonic decay of W with 4 extra light jets). The invariant mass distribution (Figure 2) is fitted with a gaussian curve — describing the top peak mass — plus a polynomial curve — which accounts for the W+jets signal plus the combinatorial background. The fitted gaussian curve peaks at 167.0 GeV, with a RMS of 12 GeV.

The generated sample corresponds to 150 pb^{-1}, i.e. a few days of data taking at initial luminosity. This study shows that top reconstruction can be performed in absence of b-tagging.

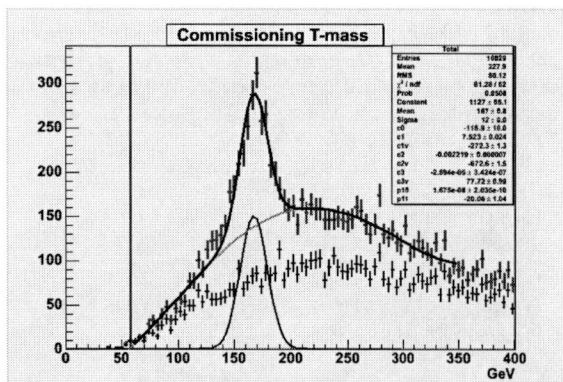

FIGURE 2. Invariant mass distribution of the 3-jets combinantion with the highest vector p_T sum. The full curve fits the signal+background (W+jets and combinatorial), while the lower distribution shows the contribution of W+jets only.

W calibration from top decays

In order to obtain a precise measurement of the mass of the top quark, it is necessary to know the absolute energy scale of hadronic jets. Miscalibrations can arise from detector effects (dead channels, imprecise cell weighing), physics effects (final state radiation, pile-up), cone algorithms effects (out-of-cone energy, jet overlap). The best method to achieve a reliable calibration is to use the physics data to reconstruct particles of known properties. One of this techniques proposes to reconstruct the hadronic W from the lepton+jets $t\bar{t}$ sample. The kinematic cuts used to select the sample are the same as the "commissioning top" cuts described in the previous section. In addition, a quality cut is applied: the 4-vectors of the fourth jet, the lepton and the E_T^{miss} are summed, and the invariant mass is required to be in the range 140 GeV$<M_{j_4 \ell \nu}<$200 GeV. With this set of cuts, the purity for W in the selected sample is $\sim$85%.

Since the mass of the W is known to a precision of a few MeV, M_W can be used to obtain calibration factors for the energies of the jets originating from the decay of the W. For the jets j_1, j_2 from the W decay, momentum conservation implies:

$$M_W^2 = 2E_{j_1}E_{j_2}(1-\cos\theta_{j_1 j_2})$$

where E_{j_i} indicates the energy of jet i and $\theta_{j_1 j_2}$ indicates the angle between the two jets. Thus, the mass measurement is influenced both by the resolution on the jet energy scale and the angular measurement. To disentangle the energy and the angular contributions, the invariant mass of the two-jets system, the W nominal mass and the kinematic properties of the jets are used as input in a constrained χ^2 fit:

$$\chi^2 = \left(\frac{m_{jj}-M_W}{\sigma_{M_W}}\right)^2 + \sum_{i,X}\left[\frac{X_i - \alpha_E^i X_i}{\sigma_X}\right]^2$$

where $X = E, \eta, \phi$. The result of the fitting procedure is the correction factor α_E, which is assumed to depend on the jet energy E. By plotting α_E versus the jet energy, a correction function $\alpha_E = f(E)$ is obtained. Applying the correction factor allows to reconstruct the W mass with a 1% precision (Figure 3). To obtain such a precision level, only 10000 $t\bar{t}$ events are necessary — that means that the calibration will be achieved within one or two months of data taking. The calibration factors have been applied to a Z+jets sample, allowing the same level of precision and showing that the results of the calibration can be used for other physics signatures involving light jets.

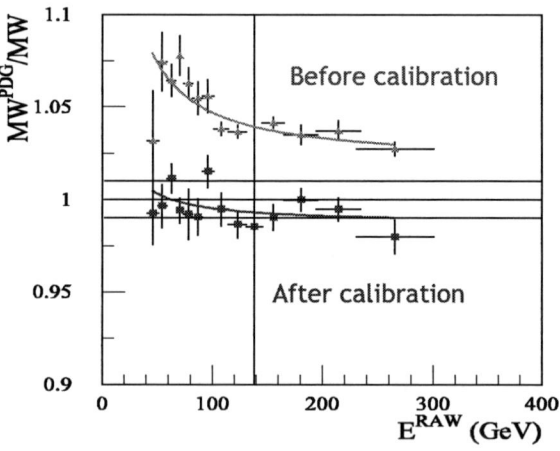

FIGURE 3. Ratio between the nominal and reconstructed W mass, as a function of the uncalibrated W energy, before and after the jet scale calibration.

REFERENCES

1. D.Chakraborty *et al.*, *Review of Top Quark Physics* Annu. Rev. Nucl. Part. Sci. 2003 63
2. I.Borjanović *et al.*, *Investigation of top mass measurements with the ATLAS detector at LHC*, SN-ATLAS-2004-040
3. F.Hubaut *et al.*, *ATLAS sensitivity to ttbar spin correlation*, ATL-PHYS-PUB-2005-001
4. S.R.Slabospitsky, L.Sonnenschein, *TopReX generator (version 3.25) Short Manual* Comput. Phys. Commun. **148** (2002), 87, hep-ph/0201292
5. T.Sjöstrand *et al.*, *PYTHIA 6.2: Physics and Manual*, LU-TP-01-21, hep-ph/0108264
6. E.Richter-Was *et al.*, *ATLFAST 2.0 a fast simulation package for ATLAS*, ATL-PHYS-98-131

Measurement of $\Delta\Gamma_s$ in $B_s^0 \to J/\psi\phi$ with CMS

Nicolò Magini

Università degli Studi di Firenze and INFN Sezione di Firenze, Italy

Abstract. The angular distributions of the products of the $B_s^0 \to J/\psi\phi \to \mu^+\mu^- K^+K^-$ decay provide information about the decay width difference $\Delta\Gamma_s$ between the two states that compose the system of B_s^0 mesons. The performance of the CMS detector in the reconstruction of this decay is presented. A trigger selection based on the muon detectors al Level-1 and on a full decay chain reconstruction with the inner silicon tracker at High-Level Trigger is described. The strategy for offline reconstruction and analysis is presented, and the potential of the CMS detector for the measurement of $\Delta\Gamma_s$ is estimated.

Keywords: CMS, B Physics
PACS: 13.25.Hw, 14.40.Nd

INTRODUCTION

In the year 2007 the LHC accelerator will begin operations at CERN, providing proton-proton collisions at a centre-of-mass energy of 14 TeV and with a starting low luminosity of 2×10^{33} cm^{-2}s^{-1}. Among many physical processes of interest, the LHC collider will provide about 10^{13} b quark pairs every year during the initial low luminosity phase, allowing the detailed study of the properties of B mesons. The angular distributions of the products of the $B_s^0 \to J/\psi\phi$ decay, in particular, provide information about physical parameters of B_s^0 mesons, including the decay width difference $\Delta\Gamma_s$ between the two states that compose the system.

One of the four experiments that will be installed at LHC is CMS, whose main feature is the 4 Tesla superconducting solenoid magnet, allowing a precise measurement of particle momentum with the inner silicon microstrip tracking detector.

One of the main issues for LHC experiments is the trigger. The CMS trigger selection is divided in two levels: a Level-1 hardware trigger using calorimeter and muon chamber information, and a software High-Level Trigger (HLT) running in thousands of CPU units.

In the following, the trigger selection for the $B_s^0 \to J/\psi\phi$ decay, which allows to store 180000 signal events per year at low luminosity, will be described. The offline analyis strategy will be introduced; the expected precision on the measurement of $\Delta\Gamma_s/\Gamma_s$ after three years is 0.018.

MONTE CARLO SAMPLES

A signal dataset of 200000 signal events of the decay chain $B_s^0/\bar{B}_s^0 \to J/\psi\phi \to \mu^+\mu^- K^+K^-$ was produced with SIMUB [1], a generator dedicated to the production of B physics events which fully reproduces the angular distributions of the final decay products of B mesons. The following kinematical cuts were used in the generation:

$$p_T(\mu) > 2 \text{ GeV}/c$$
$$p_T(K) > 0.5 \text{ GeV}/c$$
$$|\eta(\mu,K)| < 2.5$$

The predicted cross section for this process, taking into account kinematical cuts, is:

$$\sigma(B_s^0 \to J/\psi\phi \to \mu^+\mu^- K^+K^-) = 167 \text{ pb} \quad (1)$$

The main background considered was an inclusive sample of 200000 B hadron decays to states with a J/ψ, followed by $J/\psi \to \mu^+\mu^-$ with the same kinematical cuts used for the signal; for this sample, no detailed simulation of angular distributions of the final decay products was performed. The cross section for this decay chain at LHC energies, including also the kinematic cuts, is:

$$\sigma(b \to J/\psi X \to \mu^+\mu^- X) = 51.4 \text{ nb} \quad (2)$$

Another important background at trigger level is the direct production of prompt J/ψ mesons in strong interactions. A modified version [2] of PYTHIA including color-octet processes has been used to simulate a sample of J/ψ decaying to two muons, with the same kinematical cuts used for the signal. The estimated cross section for this process at LHC energies, including also the kinematic cuts, is:

$$\sigma(pp \to J/\psi X \to \mu^+\mu^- X) = 310 \text{ nb}. \quad (3)$$

All the generated events were processed with a full simulation of detector response.

LEVEL-1 TRIGGER OF $B_s^0 \to J/\psi\phi$

The $B_s^0 \to J/\psi\phi \to \mu^+\mu^- K^+ K^-$ decay chain is selected at Level-1 with the outer muon tracking system by the di-muon trigger stream. At low luminosity it is foreseen [3] that it will be possible to use a symmetric threshold of 3 GeV/c on the transverse momentum of the two muons, still keeping a low bandwidth occupancy of 0.9 kHz. Such a low p_T threshold ensures a very high selection efficiency on this channel.

The overall efficiency of the Level-1 di-muon Trigger selection (with the additional requirement that the triggering muons have opposite charge) on the Monte Carlo sample of the signal channel is reported in Tab. 1.

HIGH-LEVEL TRIGGER RECONSTRUCTION WITH THE TRACKER

The High-Level Trigger selection of this decay is performed in two consecutive steps: at Level-2 the muon tracks from the decay of the J/ψ resonance are reconstructed with the tracker, checking that they are produced in a displaced secondary vertex; then, at Level-3 the kaon tracks from the ϕ decay are reconstructed with the tracker and the B_s^0 meson is identified.

At first, primary vertex identification is performed with a dedicated algorithm which uses hits in the Pixel detector, described in [4]. For each of the Level-1 muon candidates, a region of $\Delta\eta < 0.3$, $\Delta\varphi < 0.5$ around its direction and centered in the identified primary vertices is inspected. A regional track reconstruction is performed inside those tracking regions, using the Pixel detector and the two inner layers of silicon microstrip detectors in the Tracker. After assigning muon mass to the reconstructed tracks, the invariant mass is calculated; a resolution of 53 MeV/c^2 is obtained with 5 hits per track. Mass distribution is shown in Fig. 1.

A J/ψ candidate is then accepted if it lies within 150 MeV/c^2 from the nominal J/ψ mass. The dominant background contribution at Level-2 is prompt J/ψ production. In order to reject these events it is necessary to reconstruct the decay vertex of the J/ψ resonance and check for its compatibility with the beam axis. A Kalman vertex fit [5] is performed on the two muon tracks, and events where the reduced χ^2 of the fit is less than 10 are retained.

Resolution obtained on the decay length L_T (in the transverse plane) travelled by the B_s^0 meson before decaying is 111 μm. The distributions for signal and background of the significance of the transverse decay length (defined as the transverse decay length itself divided by its error) are reported in Fig. 2.

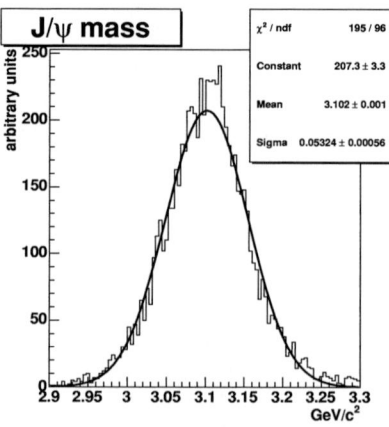

FIGURE 1. Mass distribution for the J/ψ with the Level-2 reconstruction, stopping track propagation after 5 hits have been added to the tracks.

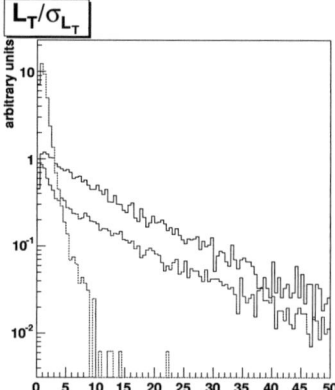

FIGURE 2. Distribution of the significance of the transverse decay length L_T/σ_{L_T} with the Level-2 reconstruction for signal events (in black), direct J/ψ background (in red) and for $b \to J/\psi$ background (in blue). Signal events are rescaled by a factor 10^3.

A threshold of $L_T/\sigma_{L_T} > 3$ is chosen as the working point for the High-Level Trigger selection. After the Level-2 trigger selection, $b \to J/\psi X$ events are the dominant contribution to the trigger rate. This selection is therefore well suited for an inclusive selection of decays of B hadrons to J/ψ, which may be used for many different B Physics studies. Average CPU time spent on Level-2 reconstruction (normalized to a 1 GHz PIII processor) is about 190 ms, which is compatible with the constraints imposed by operation at the HLT stage.

An additional trigger rate reduction may then be achieved by performing a Level-3 reconstruction of the full decay chain, inspecting the region ($|\Delta\eta| < 1.0, |\Delta\varphi| < 1.0$) around the J/ψ direction in order to reconstruct kaons from the ϕ decay.

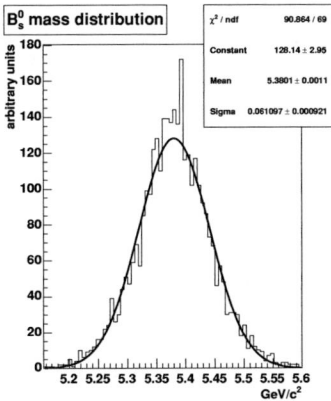

FIGURE 3. Mass distribution for the B_s^0 with the Level-3 reconstruction, stopping track propagation after 5 hits have been added to the tracks.

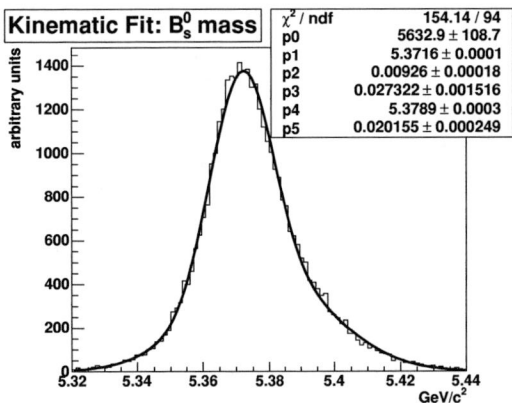

FIGURE 4. Mass distribution for B_s^0 after kinematic fitting. The distribution is fitted with a double Gaussian, with σ_1=9 MeV/c^2 and σ_2=20 MeV/c^2.

All tracks with transverse momentum greater than 0.5 GeV/c in this tracking region are reconstructed, stopping track reconstruction after 5 hits have been added.

Tracks reconstructed in this way are assigned K mass, and their invariant mass is calculated; compatibility of the kaon tracks with the previously identified decay vertex is also checked.

The final resolution on the mass of the B_s^0 meson obtained with the trigger reconstruction is shown in Fig. 3 and is 61 MeV/c^2. The following cuts are imposed on the reconstructed masses:

- $|\Delta m_\phi| < 0.02$ GeV/c^2
- $|\Delta m_{B_s^0}| < 0.2$ GeV/c^2

Tab. 1 shows the number of signal events selected by the complete High Level Trigger selection and the remaining background rate for each of the backgrounds which have been simulated.

The total rate for this High Level Trigger selection is well below 1 Hz, which is less than one percent of the maximum allowed total High-Level Trigger rate of 150 Hz. The number of $B_s^0 \to J/\psi\phi$ decays saved on tape for the offline analysis is approximately 180000 for an integrated luminosity of $\int \mathscr{L} = 20$ fb^{-1}, corresponding to one year of running at low luminosity.

OFFLINE RECONSTRUCTION AND SELECTION

A full track reconstruction in the complete Tracker acceptance ($|\eta| < 2.4$) is performed, followed by a combinatorial reconstruction of B_s^0 candidates. The interaction primary vertices are reconstructed with a *divisive* method: all reconstructed tracks are initially assigned to a single primary vertex cluster, which is then iteratively split into smaller clusters containing subsets of reconstructed tracks. For precise reconstruction of the decay vertex of the B_s^0 candidate, the method of Kinematic fitting [6] is used. The following kinematic constraints on the reconstructed particles are applied:

- The four final state tracks are required to come from a common secondary decay vertex.
- The invariant mass of the muon pair is required to be equal to the known value of the mass of the J/ψ resonance.
- The momentum of the B_s^0 is required to be aligned with the direction between the primary interaction vertex and the secondary decay vertex.

Imposing these constraints on the tracks of the four final state particles improves the resolution on their measured parameters; mass distribution for reconstructed B_s^0 candidates after applying kinematic constraints is shown in Fig. 4.

The position of the secondary vertex is also evaluated with the fit; the decay length is obtained by projecting the distance between the reconstructed interaction and decay vertices on the direction of the momentum of the B_s^0 meson. The decay time of the B_s^0 meson in its reference frame is related to the measured decay length by a Lorentz transformation; resolution on the proper decay time of the B_s^0 meson is 39 μm/c.

The offline selection cuts were optimized in order to minimize the statistical error on the observables of interest, which in the presence of background is proportional to $\sqrt{S+B}/S$. An optimal statistical error is found with the cuts $|\Delta m_{B_s^0}| < 33$ MeV/c^2, $|\Delta m_\phi| < 8$ MeV/c^2. Tab. 2 reports the number of events reconstructed and selected in the offline analysis for the signal and background event samples.

TABLE 1. Efficiency ε (defined with respect to the number of generated events) and rate R for signal and backgrounds of the Level-1, Level-2 and Level-3 $B_s^0 \to J/\psi\phi$ trigger selections at low luminosity. Quoted errors are statistical.

Event sample	$B_s^0 \to J/\psi\phi$	$pp \to J/\psi$	$b \to J/\psi X$
σ (nb)	0.167	310	51.4
L1 ε	$26.0\% \pm 0.1\%$	$21.1\% \pm 0.2\%$	$20.9\% \pm 0.1\%$
L1 R (Hz)	0.0869(3)	131(1)	21.5(1)
L2 ε	$11.8\% \pm 0.2\%$	$0.23\% \pm 0.02\%$	$8.0\% \pm 0.1\%$
L2 R (Hz)	$3.94(7) \cdot 10^{-2}$	1.44(13)	8.2(1)
L3 ε	$5.3\% \pm 0.1\%$	$< 6.1 \cdot 10^{-5}$	$(3.9 \pm 0.7) \cdot 10^{-4}$
L3 R (Hz)	$1.78(3) \cdot 10^{-2}$	$< 3.7 \cdot 10^{-2}$	$4.0(8) \cdot 10^{-2}$

TABLE 2. Efficiency ε (calculated with respect to the number of generated events) and number of events N_{SEL} selected for the offline analysis (for an integrated luminosity of $\int \mathscr{L} = 10$ fb^{-1}), for signal and background events.

Data sample	$B_s^0 \to J/\psi\phi$	$b \to J/\psi X$
σ (nb)	0.167 ± 0.062	51.4 ± 4.8
ε	$3.81\% \pm 0.08\%$	$(9 \pm 3) \cdot 10^{-5}$
N_{SEL}	$(64 \pm 24) \cdot 10^3$	$(45 \pm 13) \cdot 10^3$

The good mass resolution on the ϕ and B_s^0 resonances obtained with kinematic fitting allows to greatly reduce the background from events where the two tracks used to reconstruct the ϕ resonance come from the decay of different resonances such as the K^*; a signal-to-background ratio of $S/B = 4/3$ is obtained.

MEASUREMENT OF $\Delta\Gamma_S$ WITH THE METHOD OF ANGULAR MOMENTS

The extraction from the experimental data of $\Delta\Gamma_s$ was performed with the method of moments [7], evaluating the effect of experimental resolutions and efficiencies on the precision of the measurement. The angular distribution of the final state particles of the $B_s^0 \to J/\psi\phi$ decay is averaged over a set of appropriate weighting functions. The averages have an analytical dependence on the unknown parameters of the distribution; measuring the averages on the data then allows to determine the parameters of interest. With 64000 events, corresponding to an integrated luminosity of $\int\mathscr{L} = 10$ fb^{-1}, it was determined that a statistical error of 0.028 on the measurement of $\Delta\Gamma_s/\Gamma_s$ would be obtained in an ideal case. The influence of experimental effects was estimated; finite detector resolution was found to have a negligible effect on the error, while background contamination and inefficiency on signal events at low proper time (caused by the decay length selection at the High-Level Trigger stage) were found to increase the statistical error from 0.028 to 0.043. The statistical error that may be ultimately achieved depends on the total integrated luminosity collected during the low luminosity phase; the final result is then 0.018 with an integrated luminosity of $\int\mathscr{L} = 60$ fb^{-1} corresponding to three years of running.

ACKNOWLEDGMENTS

I would like to thank Vitaliano Ciulli for guidance and support in the development of this work. Sincere thanks also go to Thomas Speer, Kirill Prokofiev and Sergei Shulga for a long and fruitful collaboration.

REFERENCES

1. A. Belkov, S. Shulga, "Studies of angular correlations in the decays $B_s^0 \to J/\psi\phi$ by using the SIMUB generator", *Comput. Phys. Commun.*, **156** (2004) 221.
2. B. Cano-Coloma and M.A. Sanchis-Lozano, *Nucl. Phys.*, **B508** (1997) 753.
3. CMS Collaboration, "CMS: The Trigger and Data Acquisition Project, Volume II: Data Acquisition and High-Level Trigger Technical Design Report", CERN/LHCC 02-26, CMS TDR 6.2, 15 December 2002.
4. S. Cucciarelli, M. Konecki, D. Kotlinski, T.Todorov, CMS NOTE-2003/026, "Track Parameter Evaluation and Primary Vertex Finding with the Pixel Detector".
5. R. Fruhwirth, P. Kubinec, W. Mitaroff, M. Regler *Comput. Phys. Commun.*, **96** (1996) 189-208, "Vertex Reconstruction and Track Bundling at the LEP Collider Using Robust Algorithms".
6. P. Avery "Applied Fitting Theory VI: Formulas for Kinematic Fitting" CBX 98-37, June 9, 1998.
7. A.S. Dighe, I. Dunietz, R. Fleischer, "Extracting CKM Phases and $B_s - \bar{B}_s$ Mixing Parameters from Angular Distributions of Non-Leptonic B Decays", *Eur. Phys. J.*, **C6** (1999) 647.

A Preliminary Study of Z+b Production at LHC

M. Verducci*, A. Tonazzo† and S. Diglio**

*CERN and INFN-CNAF, CH-1211 Geneva 23, Switzerland
†INFN and Dipto. di Fisica "E.Amaldi", Università degli Studi di Roma Tre, 00146 Rome,Italy
**Laboratori di Frascati and Dipto. di Fisica, Università degli Studi di Roma Tre, 00146 Rome, Italy

Abstract. The study of Z boson production in association with a single heavy flavor jet ($gQ \rightarrow ZQ(Q = b,c)$) can provide interesting insight on the heavy quark content of the proton. In addition, this process represents a background to the search for the Higgs boson, when it is produced together with a b quark and decays in the same final states ($b\bar{b}, \tau^+\tau^-, \mu^+\mu^-$).

This paper presents a preliminary investigation of the potential to study the channel $Z + b - jet$ with the $Z \rightarrow \mu\mu$ at LHC. The expected statistics and purity of the event sample that could be selected in the data are evaluated considering two different methods for the b tagging.

The analysis has been performed, first, with a fast simulation of the ATLAS detector, and then, using the full simulation and the complete offline reconstruction algorithms in the official ATLAS software framework.

Keywords: Standard Model, pdf b, Z Boson
PACS: 13.38.Dg, 13.85.Qk, 14.60.Ef, 14.70.Hp

INTRODUCTION

The production of electroweak gauge bosons ($W^\pm, Z, \gamma$) together with jets containing heavy-quarks (c, b) is an important process in the hadron colliders environment [1, 2]. The simplest process involves a boson and one heavy-quark jet. The dominant diagrams contributing to Z + Q (where Q represents a heavy quark) in $p-p$ and $p-\bar{p}$ collisions are shown in figure 1.

In this paper we consider the production of a single Z boson with a b jet.

The NLO cross-section for this process at Tevatron and at the LHC is reported in table 1, [1].

Z+b jet production has been recently measured by the D0 experiment [3]. The study of this process should be even more interesting at LHC, as can be understood in the table 1, because the contribution of the leading order process, sensitive to the b content of the proton, is more relevant than at Tevatron, the total cross section is larger by a factor 50, and the relative contribution of background processes, mainly Z+c, is smaller.

In this paper we present a preliminary analysis on the potential of the ATLAS experiment to measure the Z+b-jet production at LHC.

---

[1] Talk given by M.Verducci at IFAE Incontri di Fisica delle Alte Energie, Catania, Italy, Marh 30th - April 2nd

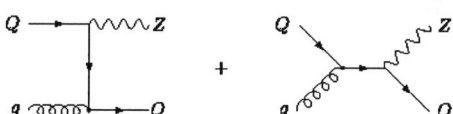

FIGURE 1. Associated production of a Z boson and a single hight-p_T heavy quark (Q=c,b).

TABLE 1. Next-to-leading-order inclusive cross section (pb) for Z Boson production in association with heavy-quark jets at the LHC ($\sqrt{s} = 14TeV\ pp$) and Tevatron ($\sqrt{s} = 1.96TeV\ p\bar{p}$). The calculations are limited to the case of a jet in a range $p_T > 15GeV$ and $|\eta| < 2.5$ (LHC) or $|\eta| < 2.0$ (Tevatron). The labels on the columns mean: ZQ = exactly one jet, which contains a heavy quark, while Zj = exactly one jet, which does not contain a heavy quark, [1].

Cross Section(pb)	Tevatron	LHC
Process		**ZQ inclusive**
$gb \rightarrow Zb$	13.4 ± 0.9 $\pm 0.8 \pm 0.8$	$1040^{+70+70+30}_{-60-100-50}$
$q\bar{q} \rightarrow Zb\bar{b}$	6.83	49.2
$gc \rightarrow Zc$	$20.3^{+1.8}_{-1.5} \pm 0.1^{+1.3}_{-1.2}$	$1390 \pm 100^{+60+40}_{-70-80}$
$q\bar{q} \rightarrow Zc\bar{c}$	13.8	89.7
		Zj inclusive
$gq \rightarrow Zq$		
$q\bar{q} \rightarrow Zg$	$1010^{+44+9+7}_{-40-2-12}$	$15870^{+900+60+300}_{-600-300-500}$

ANALYSIS TOOLS

Z+jet samples have been generated using the PYTHIA [4] Monte Carlo.

The events were been processed with a package providing a fast simulation of the detector response via smearing through parametrization resolution functions (Atlfast) [5].

The generated events were then passed through the full simulation of the ATLAS detector based on GEANT4 [6], using the detector geometry defined as "DC2" layout, [7].

Reconstruction is then performed with the official software package (ATHENA [8]). This package includes dedicated algorithms for the reconstruction of different trigger-objects as such jets, leptons and photons, or B-tagging algorithms. Particularly the reconstruction of single muons uses algorithms that provide the tracks reconstructed in the Muon Spectrometer (MOORE and MUID package [9]), including the corrections due to the multiple scattering and the energy loss in the inert material as described in details in [9], [10] and [11].

EVENT SELECTION CRITERIA

The signal was defined as the sample events containing a Z boson and a b quark with $p_T \geq 15 GeV$ and $|\eta| \leq 2.5$. The background samples containing respectively a c quark within the same cuts, or a jet originating from a light quark or a gluon in the same range, were considered separately. The NLO cross-section computed was used for the signal and for these two classes of background, while the LO cross-section given by PYTHIA was taken for the other types of events [12].

Z+jet Selection

The sample of Z+jet events with the Z boson decays into a couple of muons was selected requiring two high p_T muons satisfying following kinematic cuts:

- two muons of opposite charge,
- two muons of $p_T \geq 20$ GeV,
- di-muons invariant mass in the range $80 \leq M^{\mu\mu} \leq 105$ GeV,

In the case where there are more than two muons satisfying the cuts on the charge and on the p_T, all the possible combinations for the di-muons invariant mass are done, and is chosen the pair with the invariant mass in the range $80 \leq M^{\mu\mu} \leq 105$ GeV. There is a cut on η topology of the muons, according with the coverage of the detector, η must be in the range of $|\eta| \leq 2.7$.

About 64% of signal events are retained after applying these cuts in ATLFAST. The loss is equally due to the η acceptance and to the p_T cut. The full simulation and reconstruction decreases this percentage to about 55% of

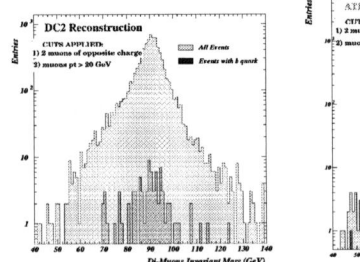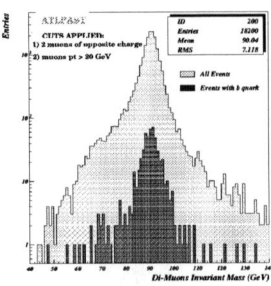

FIGURE 2. On the left the di-muons invariant mass reconstructed with the samples simulated as DC2 layout is shown, while on the right the same mass obtained by ATLFAST analysis.

signal events due to the inefficinecy of the reconstruction algorithm in the $|\eta| \leq 2.0$ region, where different tracking detectors are used.

Then in the selected events, a third muon or a b-jet are searched, always within the eta acceptance of the detector, [12]. Two variables: *efficiency* and *purity* are used to check the goodness of the criteria applied for the identification of b quark in the sample.

Z+b-jet

The selection of events where the jet originates from a b quark was based on two different tagging methods: soft muon tagging and inclusive b-tagging.

The two methods are completely independent one of the other and this allows the possibility to have a direct cross check of the results obtained.

The soft muon tagging method looks for a third non-isolated muons in the event. Hadrons containing a b quark give origin to prompt muons decays in about 12% of the cases. The efficiency of this method, therefore, cannot exceed this value, however the background is also expected to be small. The "third muon", considered to be the muon from the b hadron decay, will in general be softer and closer to a jet than the muons from the Z decay. The algorithm for the reconstruction and identification of the soft muons have been recently developed in the framework of the ATLAS software and good performance has been achieved for the transverse momenta as low as 4 GeV/c. In this case the final efficiency is about 8% for ATLFAST and less than 5% for the full simulation.

The second analysis was based on the inclusive method for the b-tagging, based on the presence of secondary vertexes and of tracks with high impact parameter with respect to the primary vertex, originated from the decay of the long-lived b-hadrons. The ATLFAST package reproduces the ATLAS b-tagging capabilities

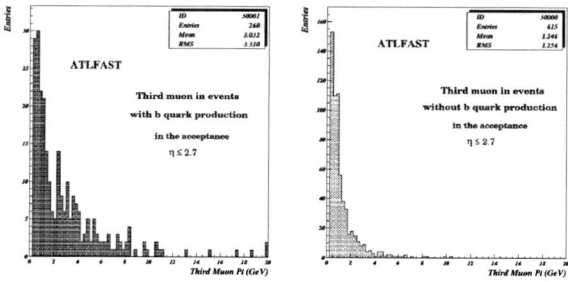

FIGURE 3. (a) Third muons p_T for muons with b quark in events, there is the requirement on the eta acceptance and all the cuts applied on the entire sample. Plots obtained by ATLFAST. (b) Third muons p_T for all muons in events without b quark, there is the requirement on the eta acceptance and all the cuts applied on the entire sample. Plots obtained by ATLFAST.

by applying the tagging efficiency on the b jets and a mis-tag rate on non-b jets on a statistical basis, for the complete reconstruction the b-tagging algorithm uses the cone method with the analysis on the tracks and vertexes to identify the flavor of the jets.

The final efficiency obtained is about 21% for ATL-FAST and about 15% for the full simulation.

Expected Event Samples

The estimation of the number of events is performed using the cross sections at the NLO multiplying by the branching ratio of the Z boson decay, and taking into account the different efficiencies of selection: the one due to the geometric acceptance, the one for the cuts applied for the di-muons invariant mass and finally the one for the selection of the sample with the jet of a particular flavor.

For the soft muon tagging, the efficiency on the signal events, the expected number of the signal and background events with an integrated luminosity of 30 fb^{-1} and expected purity of the selected samples were evaluated for different thresholds on the third muon pt, only the $p_T \geq 4 GeV$ is shown because it provides the best purity value.

With the fast simulation, the soft muon tagging capabilities are quite optimistic, in that full efficiency and no mistag are assumed for the lepton identification, in the full simulation the results are less good due to a more realistic assumptions on the detector layout, this last case is under study. The purities obtained for the full simulation are reported in table 2.

In the case of b-tagging method the purities obtained for the full simulation are reported in table 3.

TABLE 2. The purity and the number of events estimated with the full simulation for the soft muon tagging algorithm. At the third muon is applied a cut on the minimum Pt threshold of about 4 GeV according to the low muon p_T trigger. The number of events are calculated using for 30 fb^{-1} of integrated luminosity. The b,c jets selected satisfied the cuts on p_T and η: $p_T > 15 GeV$ and $|\eta| \leq 2.5$

	N_b	N_c	N_{TOT}	Purity
DC2 Reco	15750	3880	45478	35%

TABLE 3. The purity and the number of events estimated with the full simulation for the b-tagging algorithm. The number of events are calculated using for 30 fb^{-1} of integrated luminosity. The b,c jets selected satisfied the cuts on p_T and η: $p_T > 15 GeV$ and $|\eta| \leq 2.5$

	N_b	N_c	N_{TOT}	Purity
DC2 Reco	116413	31040	220155	53%

CHECK ON SYSTEMATICS

We have investigated the possibility to control the systematic effects directly from data samples.

We focused in particular on the evaluation of b-tagging performance and on the residual background.

The b-tagging efficiency can be checked using b-enriched samples. Based on previous experience at Tevatron and LEP, we can expect a relative uncertainty $\Delta \varepsilon_b / \varepsilon_b = 5\%$.

The background in the selected sample is mainly due to mistagged jets from c and light quarks. To check this we can use W+jet sample where no b-jets are present: such events will be available with large statistics and with jets covering the full p_T range of the signal. We can therefore expect to estimate the background from

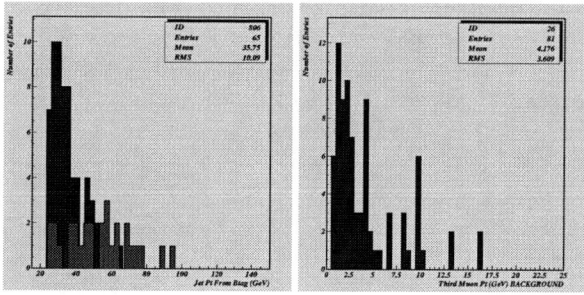

FIGURE 4. On the left is shown the P_T distribution for the jets reconstructed by the full reconstruction chain, divided in jets from b quark (red) and other jets (blue). On the right is presented the distribution in P_T of the third muon after the Soft Muon Tagging in the full reconstruction.

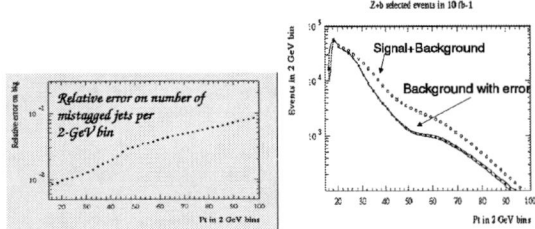

FIGURE 5. The systematics on the mistagging effects for the W+jet sample. The plot on the left defines the threshold on the possible relative error per bins.

mistagging with a relative uncertainty at the level of few percent, see plots in figure 5.

CONCLUSIONS AND OUTLOOK

The fast simulation of the ATLAS detector has shown that this type of events will be observed with very high statistics and good purity at the LHC.

By the fast simulation the purity obtained cutting on third p_T is very high, about 70% with an efficiency of about 8% (remembering that the efficiency is limited by the branching ratio BR($b \rightarrow \mu$) $\approx$ 10%).

Using the complete reconstruction, the efficiencies for the soft muon tagging drop a little, due to the low p_T muon (most muons have not be reconstructed in the Muon Spectrometer). Further analysis on this channel is definitely worthwhile.

Given the large statistics of the sample, the precision of the Z+b cross-section measurement will be limited by systematics effects. The possible sources of systematic uncertainty and their impact on the measurements has been evaluated. The systematics related to the b-tagging have been estimated of the order of few %, both for b-tagging efficiency than mistagging inefficiency. The precision of this measurement will be then dominated by other systematics effects not included in this analysis, as luminosity measurement or jet and energy resolution.

The availability of the large sample opens interesting possibilities for the studies of differential distributions: for instance, measuring the cross-section as a function of η and p_T of the Z boson would allow the measurements of the b PDF as function of the momentum fraction carried by the quark inside the proton. These items are an additional topic for further studies.

ACKNOWLEDGMENTS

Many thanks to the Organizer Committee for the good organization and the stimulating atmosphere during the

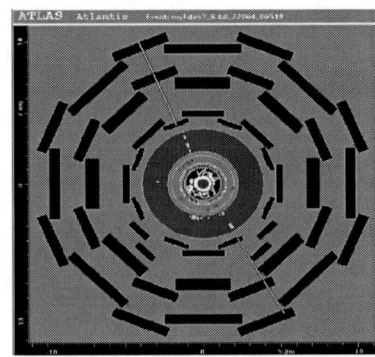

FIGURE 6. Atlas Event Display: ATLANTIS. One reconstructed event is shown, two muons in the Muon Spectrometer (three outside blue rings) and some energy clusters in the Tile Calorimeter (red ring).

IFAE Conference.

REFERENCES

1. J.Campbell, R.K.Ellis, F.Maltoni and S.Willenbrock, Hep-ph/0312024, December 3, 2003
2. J.Campbell, R.K.Ellis, F.Maltoni, S.Willenbrock, Hep-ph/0204093, May 26, 2003
3. DO Note 4388, The DO Collaboration Phys. Rev. Lett. **94** 161801 (2005);
4. Atlas Collaboration Manual, $http://wwwinfo.cern.ch/asdoc/psdir/pythiarus/pythia\_rus.ps.gz$
5. E.Richter-Was et al., ATLAS Internal Note, ATL-PHYS-98-131, 1998
6. S.Agostinelli et al., Nuclear Instruments and Methods in Physics Research, NIM A 506 (2003), 250-303
7. DC1 Task Force, ATL-SOFT-2003-012, Nim publ., 2003
8. Atlas Collaboration Manual, $SOFTWARE/OO/architecture/General/Tech.Doc/Manual/2.0.0-DRAFT/AthenaUserGuide.pdf$
9. D.Adams, K.Assamagan, M.Biglietti, G.Carlino, G.Cataldi, F.Conventi, A.Farilla, Y.Fisyak, S.Goldfarb, E.Gorini, T.Laugori, K.Mairz, L.Merola, A.Nairz, A.Poppleton, M.Primavera, S.Rosati, J.Shank, S.Spagnolo, L.Spogli, G.Stravopoulos, M.Verducci, T.Wenaus, Atlas Internal Note, ATL-SOFT-2003-008, ATL-DAQ-2003-02, 2003
10. D.Adams, K.Assamagan, M.Biglietti, G.Carlino, G.Cataldi, F.Conventi, A.Farilla, Y.Fisyak, S.Goldfarb, E.Gorini, T.Laugori, K.Mairz, L.Merola, A.Nairz, A.Poppleton, M.Primavera, S.Rosati, J.Shank, S.Spagnolo, L.Spogli, G.Stravopoulos, M.Verducci, T.Wenaus, ATLAS Internal Note, ATL-SOFT-2003-007, 2003
11. D.Adams, K.Assamagan, M.Biglietti, G.Carlino, G.Cataldi, F.Conventi, A.Farilla, Y.Fisyak, S.Goldfarb, E.Gorini, T.Laugori, K.Mairz, L.Merola, A.Nairz, A.Poppleton, M.Primavera, S.Rosati, J.Shank, S.Spagnolo, L.Spogli, G.Stravopoulos, M.Verducci, T.Wenaus, ATLAS Internal Note, ATL-CONF-2003-011, 2003
12. S.Diglio, A.Tonazzo, M.Verducci ATLAS Internal Note, ATL-COM-PHYS-2004-078, Novembre 2004

VV-Fusion Process to Investigate EWSB; Interplay LC-LHC

G. Cerminara

University and INFN Torino

Abstract. The VV-fusion is a promising process to investigate the electroweak symmetry breaking (EWSB) mechanism. In fact the cross section for the scattering of longitudinally polarized vector bosons, in absence of a Higgs, violates unitarity at about 1.2TeV. As a consequence, new physics must appear below this threshold. The LHC and the LC will be the machines which will explore the energy region where this new physics is expected. In the following the potential of the two accelerators for the study of this reaction is briefly reviewed and a preliminary study with the CMS detector at LHC is presented.

Keywords: VV-fusion, Electroweak Symmetry Breaking, LHC, LC
PACS: 11.15.Ex, 13.85.Hd

INTRODUCTION

General arguments based on partial wave unitarity imply that, without a Higgs, the electroweak gauge bosons develop strong non-perturbative interactions at an energy scale of about ~ 1.2TeV. Below this threshold new physics is therefore expected: in the Higgs case a resonance will be observed in the VV invariant mass spectrum at m_H; otherwise the cross section will deviate from the Standard Model (SM) prediction.

Partial wave unitarity in the VV scattering

In the SM the scattering of longitudinally polarized vector bosons, without a Higgs, violates the partial wave unitarity [1]. In fact, considering the Feynman diagrams for the W boson scattering not including the Higgs boson (Fig. 1) we would obtain a scattering amplitude

$$\mathcal{M}(L,L;L,L) \sim \sqrt{2} G_F (s+t) \qquad (1)$$

where t is the square of the four momentum transferred in the scattering process and s is the square of the center of mass energy of the boson-boson system. Indeed this amplitude has a bad high energy behavior: the unitarity bound is violated at large s.

If we compute the amplitude for the W scattering considering also the terms from the exchange of a Higgs boson (Fig. 1) than we obtain:

$$\mathcal{M}(L,L;L,L) \sim -\sqrt{2} G_F m_H^2 \left(\frac{s}{s-m_H^2} + \frac{t}{t-m_H^2} \right). \qquad (2)$$

The bad high energy behavior is therefore canceled for $\sqrt{s} \gg m_H$; the Higgs can remove the unitarity violation in the scattering of longitudinally polarized W bosons.

If we consider the unitarity constraint in the limit $m_W^2 \leq s \leq m_H^2$ we can use eq. 2 to find the critical energy $\sqrt{s_c}$ which can be interpreted as an energy scale below which new physics, beyond the SM, must appear. This s_c is the maximum value of s allowed by tree-level unitarity for $m_H \to \infty$. It can be shown that this critical energy is:

$$s_c = \frac{4\pi\sqrt{2}}{G_F} = (1.2 \text{ TeV})^2. \qquad (3)$$

For Higgs masses $m_H \gg \sqrt{s_c}$ the tree level unitarity breaks down and the effect of the Higgs boson on the scattering amplitude vanishes. This scenario is equivalent to a theory without the Higgs boson.

VV scattering as a probe of EWSB

The SM Higgs boson is not the only possibility for recovering tree level unitarity of scattering amplitude, nevertheless the above argument implies that a new sector of physics, associated with EWSB, must be revealed in the VV-fusion process before the threshold presented by eq. 3.

VV-FUSION PROCESS AT LHC

Thanks to its high center of mass energy and high luminosity the Large Hadron Collider (LHC) will be able to explore the high energy region beyond the unitarity violation threshold. The study of the VV-fusion process is therefore extremely promising: if no Higgs is found new physics is expected in this channel. If the Higgs boson exists, the study of this reaction can lead to interesting results. The fusion of longitudinally polarized vector bosons represents the second most important contribution to the cross section for the Higgs production and the WW/ZZ final state is one of the most clear signature for the Higgs decay.

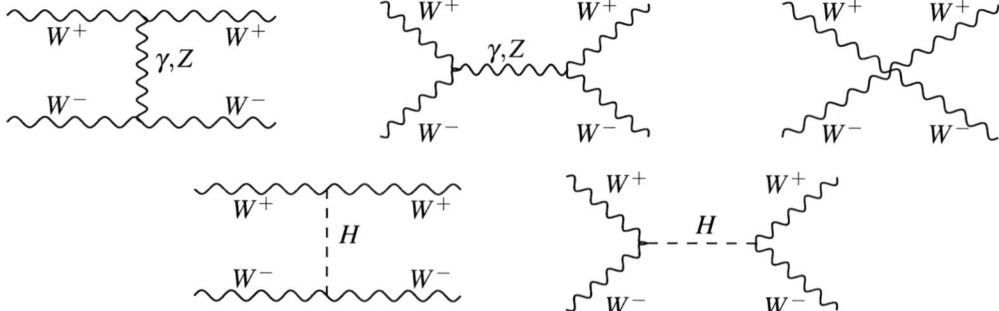

FIGURE 1. Tree-level Feynman diagrams for the W boson scattering without (top) and with (bottom) Higgs boson exchange.

A preliminary study of the VV-fusion process at CMS

An exploratory study [2, 3] has been performed in order to assess the possibility of probing the symmetry breaking mechanism through the VV-fusion process using the Compact Muon Solenoid (CMS) detector [4]. A *model independent* analysis was carried out with no assumptions on the mechanism restoring the unitarity in the scattering amplitude and without any degrees of freedom beyond the SM.

The analysis was focused on the semi-leptonic final state which offers a clear experimental signature, thanks to the presence of high p_t leptons from the W or Z decays together with the highest branching ratio among the other final states which is possible to reconstruct in an hadronic environment. In particular the processes that were studied are:

$$pp \rightarrow qqW_LW_L + X \rightarrow qq\mu\nu qq + X,$$
$$pp \rightarrow qqZ_LW_L + X \rightarrow qq\mu\mu qq + X,$$
$$pp \rightarrow qqZ_LZ_L + X \rightarrow qq\mu\mu qq + X.$$

In order to explore the sensitivity of the analysis method to the entire heavy Higgs mass spectrum, we analyzed Monte Carlo data sets produced using different Higgs masses: $m_H = 500$ GeV, $m_H = 750$ GeV and $m_H = 1000$ GeV. Moreover, we generated a sample with very high Higgs mass ($m_H = 10000$ GeV) which approximates the no-Higgs scenario and allows therefore to consider the VV-fusion cross section in a region where no resonance is present.

All the samples for this preliminary study were generated using PYTHIA [5] and the cross sections after minimal acceptance cuts are shown in Table 1.

Signal and background samples were processed with the CMSJET [6] fast simulation package, which uses a parameterization of CMS acceptance, resolutions and reconstruction efficiencies.

The signal topology (Fig. 2) is characterized by a six fermion final state; the four fermions from the VV decay and the two quarks which radiated the vector bosons. These spectator quarks are scattered predominantly in the high pseudorapidity region and their presence is essential to tag the VV-fusion events (Fig. 3).

TABLE 1. PYTHIA cross sections (fb) for the signal samples after minimal acceptance cuts

	$m_H =$ 500 GeV	$m_H =$ 1000 GeV	No Higgs case
$\sigma_{pp \rightarrow qqWW \rightarrow qq\mu\nu qq}$	64.4	26.9	19.7
$\sigma_{pp \rightarrow qqZW \rightarrow qq\mu\mu qq}$	0.7	1.0	1.5
$\sigma_{pp \rightarrow qqZZ \rightarrow qq\mu\mu qq}$	9.1	3.0	1.7

The most difficult backgrounds to the scattering of longitudinally polarized V bosons are given by processes of the type: $pp \rightarrow V_L V_T + X$ and $pp \rightarrow V_T V_T + X$ with at least one of the two bosons transversely polarized. Contributions to these processes come from the scattering of transversally polarized vector bosons and from processes where the V bosons are emitted by interacting quark lines. At the time of the preliminary work presented here, no appropriate generator was available to simulate the scattering of transverse bosons. Therefore only the production of unpolarized WW and ZV pairs coming from quark anti-quark annihilation was considered. Also the production of $t\bar{t}$ pairs represents a difficult background; the top quark decays into a W boson and a b-quark with a branching ratio of almost 100% giving a final state similar to that produced in VV-fusion events. Also the production of a vector boson in association with a pair of jets has been simulated. In fact this process, although it has a topology quite different from that of the signal, has a huge cross section which must be kept under control.

Preliminary results

The main goal of this exploratory study was to investigate the feasibility of the measurement of the VV-fusion

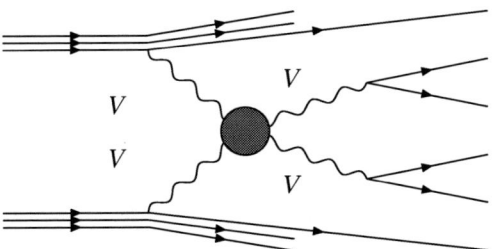

FIGURE 2. Simplified VV-fusion process diagram.

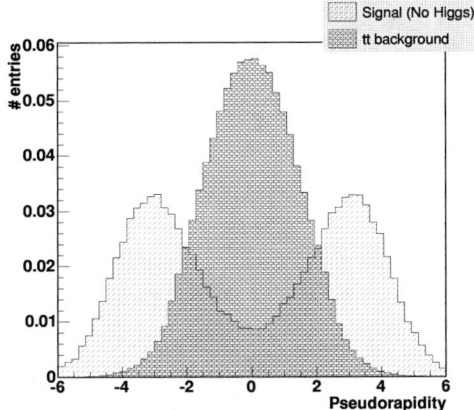

FIGURE 3. Pseudorapidity distribution (arbitrary normalization) for spectator quarks of the signal ($pp \rightarrow qqW_LW_L + X$, no Higgs case) and b-quarks of the $t\bar{t}$ background.

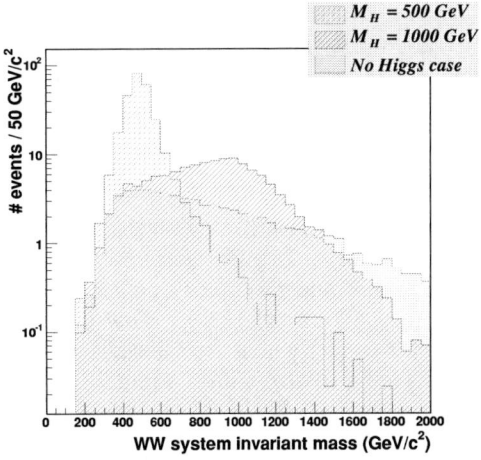

FIGURE 4. Reconstructed VV invariant mass spectra for the samples $pp \rightarrow qqW_LW_L + X \rightarrow qq\mu\nu qq + X$ after 100fb^{-1}.

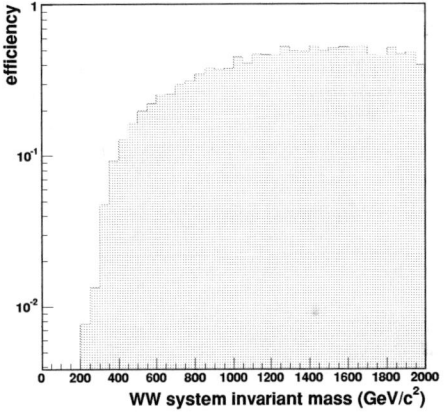

FIGURE 5. Reconstruction efficiency as a function of VV system invariant mass (GeV) for the sample $pp \rightarrow qq\mu\nu qq + X$ in the no Higgs case after all selection cuts.

cross section at CMS. A good sensitivity to all heavy Higgs mass resonances is achieved and also in the no Higgs scenario the VV invariant mass spectrum is pretty well reconstructed (Fig. 4). A good reconstruction efficiency for the region of high VV invariant masses is achieved, as shown in Fig. 5.

Thanks to this reconstruction efficiency the number of events obtained at the end of the selection is quite promising being about 120 events with $M_{WW} > 1$TeV for the $pp \rightarrow qqWW \rightarrow qq\mu\nu qq$ channel (No Higgs) after 100fb^{-1} of integrated luminosity. For the ZZ final state the efficiency at high invariant masses is about 35% and the number of expected events with $M_{ZZ} > 1$TeV is around 20. Also the signal-to-background ratio appears to be encouraging at high invariant masses. However it is worth remembering that the exact estimation of this quantity was far from being the main aim of the study. In fact at this stage only a limited set of backgrounds was considered and important irreducible processes were still missing.

MC generators for six fermion final states and VV-fusion processes

In order to explore the EWSB with the study of the VV-fusion process a precise knowledge of the cross section σ_{VV} on the whole spectrum of M_{VV} is essential. The choice of the MC generator to be used for the simulation is therefore a key aspect for this study. Results presented in the previous section were obtained using PYTHIA event generator. PYTHIA is a parton shower code which uses the Effective Vector Boson Approximation (EVBA)[7, 8] to compute the cross section for VV scattering processes. In this approximation the V bosons are treated as on-shell partons in the quarks: the fermion-fermion cross section is written as a product of the distribution function of the vector boson inside the quark and the cross section for on-shell VV scattering. Moreover PYTHIA can not simulate the scattering of tranverse vector bosons which is the most important irreducible background.

Recently a new matrix element event generator was made available for this kind of studies: PHASE [9], a MC dedicated to SM processes with six fermions in the final state at the LHC. PHASE uses exact leading order matrix elements for the generation of unweighted events. Studies of the VV-fusion at CMS show important differences in the cross sections for the signal with respect to PYTHIA. PHASE also allows a more realistic background treatment since it can generate most of the irreducible backgrounds that were not available in the studies presented in the previous paragraph.

VV-FUSION PROCESS AT LC

A Linear Collider (LC) with a center of mass energy $\sqrt{s} = 500$GeV or $\sqrt{s} = 1000$GeV would be a powerful machine to investigate EWSB in VV scattering. Even if no resonance will be observed in that energy range a precise analysis of the rising of the scattering processes in the sub-threshold region will significantly add information to data obtained at LHC.

At an e^+e^- collider the relevant processes are:

$$e^+e^- \to \nu\bar{\nu}WW \to \nu\bar{\nu}VV,$$
$$e^+e^- \to e^+e^-ZZ \to e^+e^-VV.$$

All the possible W and Z decay channels (exception made for $Z \to \nu\bar{\nu}$) can be detected and the experimental environment of an e^+e^- collider will allow better signal to background ratio than at the LHC. Other clear experimental advantages with respect to pp machines will be the fact that the initial four-momenta of the interacting particles will not depend on the proton PDFs. Furthermore it will be possible to enhance the cross sections of all process involving WW scattering polarizing the initial beams.

From the reactions mentioned above it will be possible to obtain informations about the *anomalous couplings* $\alpha_{4,5,6,7,10}$ which describe deviations form the SM in the four-point interactions of longitudinal gauge bosons in the electroweak effective Lagrangian. In particular, assuming $SU(2)$ invariance, α_4 and α_5 can be constrained with a better precision than what is achievable with LHC collider [10].

At the moment there is no complete study trying to combine the sensitivities of the two machines, but the LC could be important to interpret resonances in the VV-fusion cross section observed at LHC. For example measuring the TGC α_2 and α_3 at $\sqrt{s} = 500$GeV the LC can help distinguish between vector and scalar resonances [11]; if $J = 1$ the measurement of $\alpha_2 + \alpha_3$ provides an indirect estimation of a 1TeV resonance mass with a precision of the order of the GeV. On the other side in case of a $J = 0$ resonance of $m = 1$TeV the LC (with $\sqrt{s} = 800 - 1000$GeV) would be able to measure the mass, through the enhancement in the vector boson scattering cross section, with an accuracy of about 250GeV.

CONCLUSIONS

The VV-fusion will be an important channel to explore the EWSB especially in the case where no light Higgs boson is found. The LHC will be the machine able to explore the high energy region beyond the unitarity violation threshold. In case of new physics the LC can contribute to understand its nature and to improve the resolution on some of its parameters.

REFERENCES

1. H. Haber, "Lectures on Electroweak Symmetry Breaking," in *Testing the Standard Model, Proceedings of the 1990 Theorethical Advanced Study Institute in Elementary Particle Physics*, edited by M. Cvetič, and P. Langacker, 1990.
2. R. Bellan, *Observables in VV-fusion at the CMS Experiment*, Diploma thesis, Università degli Studi di Torino (2004).
3. G. Cerminara, *A Study of the WW-fusion Process at CMS as a Probe of Symmetry Breaking*, Diploma thesis, Università degli Studi di Torino (2003).
4. The CMS Collaboration, *CERN/LHCC 94-38, LHCC/P1* (1994).
5. T. Sjöstrand, P. Edén, C. Friberg, L. Lönnblad, G. Miu, S. Mrenna, and E. Norrbin, **135**, 238–259 (2001).
6. S. Abdullin, A. Khanov, and N. Stepanov, *CMS TN/94-180* (2002).
7. I. Kuss, *Physical Review D*, **55** (1997).
8. I. Kuss, and H. Spierberger, *BI-TP 95/25* (1995).
9. E. Accomando, A. Ballestrero, and E. Maina, hep-ph/0504009 (2005).
10. R. Chierici, S. Rosati, and M. Kobel, *LC-PHSM-2001-038* (2001).
11. The LHC/LC Study Group, hep-ph/0410364 (2004).

Pentaquark Searches in *e-p* Interactions at HERA

A. Polini

I.N.F.N. Bologna, via Irnerio 46, 40126 Bologna, Italy
Email: alessandro.polini@bo.infn.it

Abstract.
Recent results on searches for baryonic resonances compatible with pentaquark states in deep-inelastic collisions at the HERA *ep* collider are reported. Searches and measurements of strange and charmed pentaquarks by the H1 and ZEUS experiments are presented. The analyses were performed in the central rapidity region of inclusive deep-inelastic scattering at an *ep* center-of-mass energy of 300-318 GeV using the full HERA I luminosity. Some conclusions and prospects for future measurements with HERA II are discussed.

Keywords: deep-inelastic scattering, pentaquark
PACS: 12.39.Mk

INTRODUCTION

Interest in hadron spectroscopy has been revived recently by the observation in fixed target experiments of baryonic resonances compatible with the theoretical prediction of pentaquark states, i.e. a bound state of four quarks and an antiquark.

Although all known hadrons can be interpreted in terms of baryons (qqq) or mesons ($q\bar{q}$) the theory of Quantum Chromodynamics does not a priori exclude the existence of other colour-neutral quark combinations such as tetraquarks ($qq\bar{q}\bar{q}$) or pentaquarks ($qqqq\bar{q}$).

While several models allowing such multi-quark bound states are available, the first quantitative predictions have been made only recently [1, 2].

The observation [3, 4] of one such state, the $\theta^+(1530)$ decaying either into $K^0 p$ or $K^+ n$, cannot be explained as a baryon, and requires a quark content of $uudd\bar{s}$.

Subsequently NA49 has reported the observation of manifestly exotic S=-2 baryon states decaying to $\Xi^\pm \pi^\pm$.

In Fig. 1 the minimal pentaquark anti-decuplet with the so-far investigated states is shown.

In fixed-target experiments at low center-of-mass energy, valence quarks from the target nucleons can easily become part of the pentaquark final state. At HERA, collisions of 820–920 GeV protons with 27 GeV electrons or positrons, corresponding to a *ep* center-of-mass energy of 300–318 GeV, allow the investigation of a kinematic region in which hadrons are mainly produced from fragmentation of sea quarks or antiquarks of the struck proton. As a result, hadrons and their antiparticles are expected to be produced in approximately equal proportions and can be both investigated. Furthermore, states of higher mass or composed of heavier quarks, like the θ_c^0 observed by H1, can be searched for.

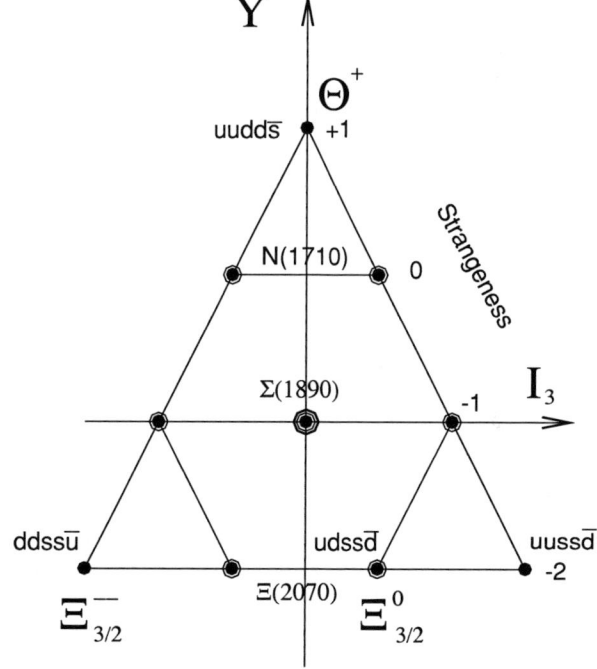

FIGURE 1. Pentaquark multiplets - $\overline{10}_f + 8_f$.

SEARCH FOR $\Theta^+ \to K_S^0 \, p(\bar{p})$

ZEUS has performed a search for pentaquarks in the $K_S^0 \, p(\bar{p})$ decay channel [6]. The data sample corresponds to an integrated luminosity of 121 pb^{-1}. The charged tracks were selected with the central tracking detector ($|\eta| < 1.75$) and (anti)proton-candidates were identified by dE/dx.

Fig. 2 shows the $K_S^0 \, p(\bar{p})$ invariant mass for $Q^2 > 20$ GeV^2, as well as for the $K_S^0 \, p$ and $K_S^0 \, \bar{p}$ samples sep-

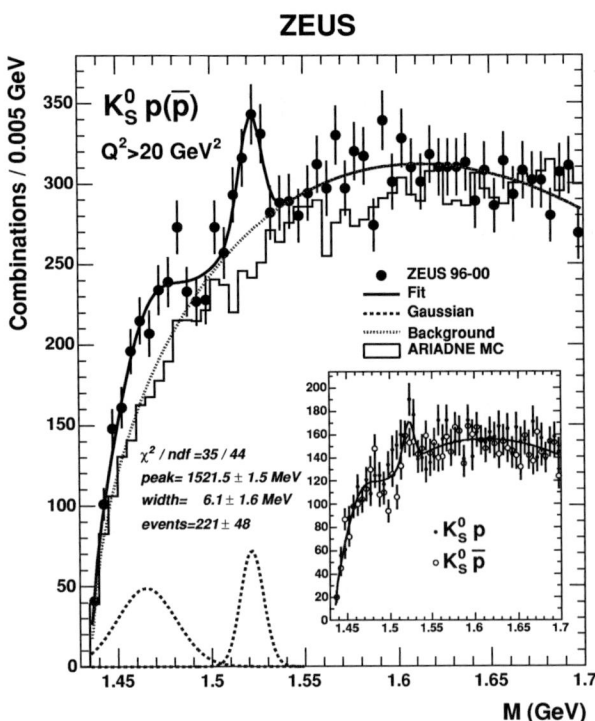
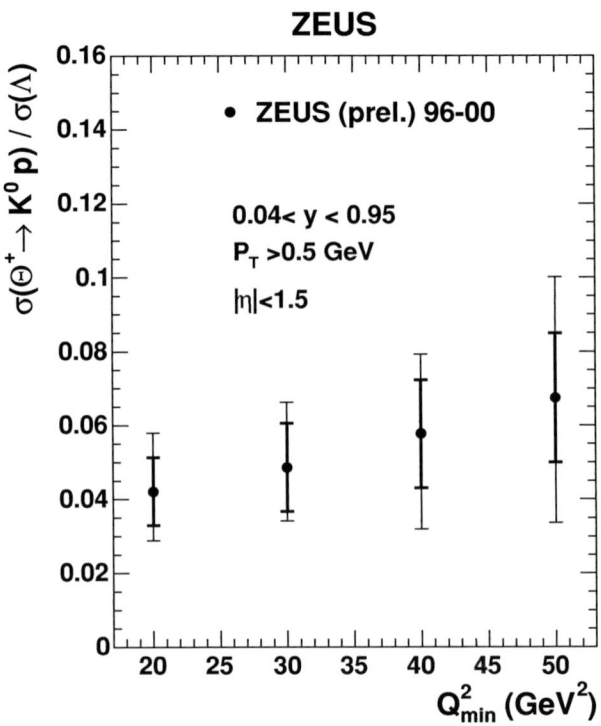

FIGURE 2. Invariant-mass spectrum for the $K_S^0\, p(\bar{p})$ channel for $Q^2 > 20$ GeV^2. The solid line is the result of a fit to the data using the threshold background plus two Gaussians. The dashed lines show the Gaussian components, while the dotted line indicates the background. The prediction of the Monte Carlo simulation is normalized to the data in the mass region above 1650 MeV. The inset shows the $K_S^0\, \bar{p}$ and the $K_S^0\, p$ candidates scaled by a factor 0.5 separately.

FIGURE 3. Ratio of the cross section $\sigma(\Theta^+ \to K_S^0\, p(\bar{p}))$ to the inclusive Λ cross section, $\sigma(\Lambda + \bar{\Lambda})$, measured in the kinematic region given by $0.04 < y < 0.95$, $p_T > 0.5$ GeV and $|\eta| < 1.5$, integrated above Q_{min}^2.

arately. To extract the signal seen at 1522 MeV, a fit was performed using a background function plus two Gaussians. The first Gaussian, which significantly improves the fit at low masses, may correspond to the unestablished PDG $\Sigma(1480)$. The peak position agrees well with the measurements by HERMES, SVD and COSY-TOF for the same decay channel [4]. The statistical significance, estimated from the number of events assigned to the signal by the fit, was 4.6σ. The number of events in the $K_S^0\, \bar{p}$ channel agrees well with the signal extracted for the $K_S^0\, p$ decay mode. If the observed signal corresponds to the pentaquark, this provides the first evidence for its antiparticle with a quark content of $\bar{u}\bar{u}\bar{d}\bar{d}s$.

Furthermore ZEUS has presented the relative cross sections [7] for $\Theta^+ \to K_S^0\, p$ and $\Lambda(1116) \to p\pi^-$ and their charge conjugate. The number of reconstructed Θ^+ events were determined using the fit described previously with the width of the 1522 MeV Gaussian set to 6 MeV. For Λ-reconstruction, a fit with one Gaussian and a second-order polynomial background function was used while the branching ratios for $K^0 \to K_S^0 \to \pi^+\pi^-$ and $\Lambda \to p\pi^-$ were taken from the PDG [11]. Acceptance corrections were calculated using Σ^+-baryons assumed to be Θ^+-baryons with a mass of 1522 MeV and were forced to decay, with 100% branching ratio, to the $K_S^0\, p(\bar{p})$ final state. The acceptance correction for the Θ^+ and Λ's was around 4% and 10%, respectively.

The cross section for $e^\pm p \to e^\pm \Theta^+ X \to e^\pm K^0 p X$ was measured to be $\sigma_{vis} = 125 \pm 27\text{(stat.)}^{+36}_{-28}\text{(syst.)}$ pb in the kinematic region given by $Q^2 \geq 20$ GeV^2, $0.04 < y < 0.95$, $p_T > 0.5$ GeV and $|\eta| < 1.5$.

Fig. 3 shows the ratio of this cross section to the Λ cross section integrated above Q_{min}^2. This ratio, for $Q_{min}^2 = 20$ GeV^2, is $4.2 \pm 0.9\text{(stat.)}^{+1.2}_{-0.9}\text{(syst.)}\%$ and, in the current data, shows no significant dependence on Q_{min}^2. Since the Θ^+ has other decay channels in addition to $\Theta^+ \to K_S^0$, this ratio sets a lower limit on the production rate of the Θ^+ compared to that of the Λ-baryon.

To rule out possible unexpected effects the $K^\mp p(K^\pm \bar{p})$ invariant-mass spectra were investigated in a wide range of minimum Q^2 values. As shown in Fig. 4, no peak is observed near 1522 MeV in the $K^+ p(K^- \bar{p})$ spectra, while a clean signal is seen in the $K^- p(K^+ \bar{p})$ channel at 1518.5 ± 0.6 (stat.) MeV, corresponding to the

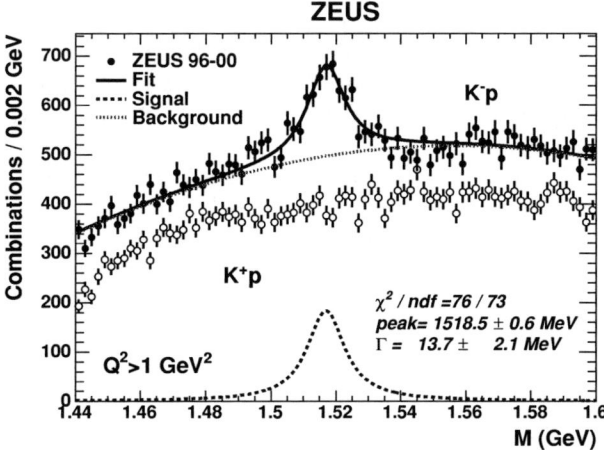

FIGURE 4. $K^{\mp}p(K^{\pm}\bar{p})$ invariant-mass spectra for $Q^2 > 1$ GeV^2.

$\Lambda(1520)D_{03}$ state. The fact that both mass and width are in good agreement with the PDG values verifies that there are no significant reconstruction biases in the pentaquark mass range. The absence of a peak in the like sign spectrum disfavours the isotensor (Θ^0, Θ^+, Θ^{++}) interpretation of the Θ^+ state [8].

Following the results by ZEUS, H1 has recently presented preliminary results [9] on the search for strange Θ^+. The analysis used 71 pb^{-1} of data requiring $0.1 < y < 0.6$, $p_t(k_S^0 p) > 0.5$ GeV in bins of Q^2. Particle identification (K, p) was done via dE/dx with likelihood cut. No signal was found.

In order to compare the result with the one from ZEUS, the dE/dx likelihood cut on the proton was replaced by the visual selection used by ZEUS. Fig. 5 shows the H1 invariant-mass distribution for $20 < Q^2 < 100$ GeV. No appreciable signal was found. Around the mass of 1.522 GeV an upper limit on the cross section of roughly 100 pb was set, which does not anyhow exclude the cross section observed by the ZEUS experiment.

SEARCH FOR S=-2 PENTAQUARKS

Following positive results from NA49 [5], ZEUS has performed a search for pentaquarks decaying to $\Xi^-\pi^\pm$ and their antiparticles using deep inelastic data corresponding to an integrated luminosity of 121 pb^{-1}. Fig. 6 shows the invariant mass for the different channels together with the expectation for the NA49 signal. No pentaquark signal was found and upper limits at 95% C.L. on the ration of $\Xi_{3/2}^{--}(\Xi_{3/2}^0)$ to the clearly seen Ξ^0 were set in the mass range of 1650-2350 MeV. However it has to be noted that the absence of signal in the central-

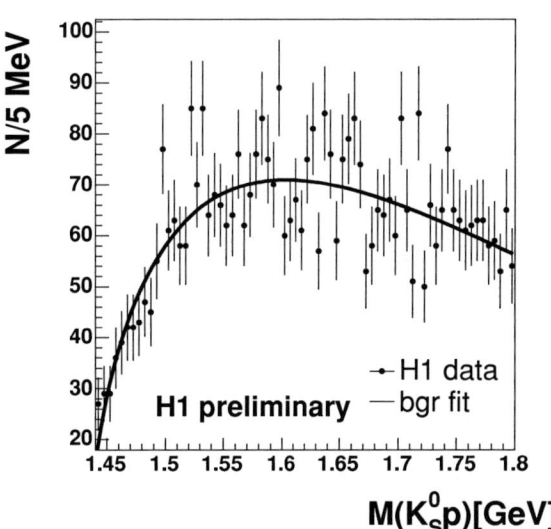

FIGURE 5. Invariant-mass spectrum for the $K_S^0\,p(\bar{p})$ channel for $Q^2 > 20$ GeV^2. The solid line is the result of a fit to the data using the threshold background plus two Gaussians.

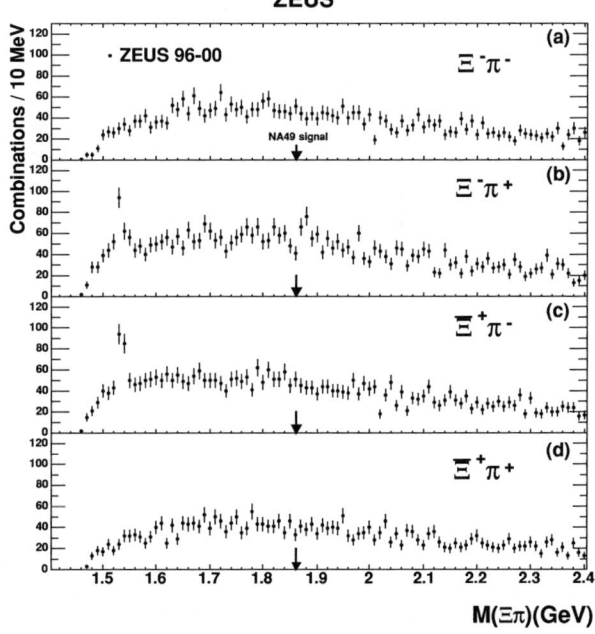

FIGURE 6. The $\Xi\pi$ invariant-mass spectra for each combination reconstructed at $Q^2 > 1$ GeV^2 The arrows show the location of the signal observed by NA49.

fragmentation region in the ZEUS data does not necessarily contradict the observation of NA49, this one being a fixed target experiment which has good acceptance in the forward region.

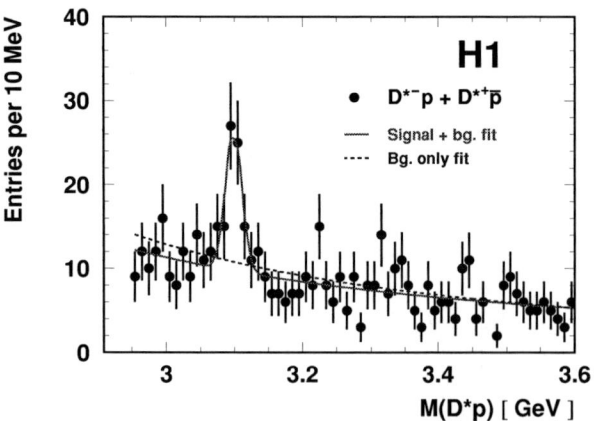

FIGURE 7. ΔM_D^* distribution from the opposite-charge D^*p combinations compared to a result of a fit in which both signal and background components are included (solid line) and with the result of a fit in which only the background components is included (dashed line).

CHARMED PENTAQUARK θ_C^0

Various QCD models imply that, besides the strange pentaquark, heavy pentaquarks, such as $\theta_c^0 = uudd\bar{c}$, should exist.

H1 reported evidence [12] for a narrow resonance in the $D^{*\pm}p^{\mp}$ mass spectrum around 3.1 GeV. The data sample corresponds to an integrated luminosity of 76 pb^{-1}. $D^{*\pm}$ were identified by evaluating the mass difference $\Delta M = M(K^-\pi^+\pi_s^+) - M(K^-\pi^+) + c.c.$ while protons were selected using the dE/dx measurements.

As shown in Fig. 7, the statistical significance of the signal was determined to be 5.4 σ and was observed in both charge coniugate decay channels and in an independent photoproduction sample. In both cases it was reported that about 1% of the $D^{*\pm}$ mesons originate from this state.

ZEUS performed a similar analysis [13] on the 126 pb^{-1}. Although the P-wave charm mesons $D_1^0(2420)$, $D_2^{*0}(2460)$ and $D_{s1}^{\pm}(2536)$ were clearly seen, no resonance structure was observed in the $M(D^{*\pm}p^{\mp})$ spectra The upper limits on the fraction of D^* mesons originating from Θ_c^0 decays was measured to be 0.23% (95% C.L.) and 0.35 % (95% C.L.) for $Q^2 > 1$ GeV^2. These results, not compatible with those from H1, leave the θ_c^0 observation still uncertain.

SUMMARY AND CONCLUSIONS

Recent results from ZEUS and H1 are contributing to the investigation from fixed target experiments in the kinematic region where pentaquarks are expected to be produced via fragmentation.

The Θ^+ observation from ZEUS in deep inelastic scattering, although not confirmed by H1, might indicate a production consistent with a pentaquark state and its antiparticle.

The observation of the charmed pentaquark θ_c^0 by H1 and not seen in ZEUS looks still unclear.

The ongoing high luminosity program of HERA II, aimed at a increase in the total integrated luminosity up to 1 fb^{-1} by end of 2007 will allow to clarify and solve some of the open questions.

REFERENCES

1. D. Diakonov, V. Petrov, M.V. Polyakov, *Z. Phys.* A **359**, 305 (1997); M. Praszalowicz, in *Skyrmions and Anomalies*, eds. M. Jezabek, M. Praszalowicz (World Scientific 1987);
2. R.L. Jaffe, F. Wilczek, *Phys. Rev. Lett.* **91**, 232003 (2003); M. Karliner, H.J. Lipkin, *Phys. Lett.* B **575**, 249 (2003); E. Shuryak, I. Zahed, *Phys. Lett.* B **589**, 21 (2004);
3. LEPS Coll., T. Nakano *et al*, *Phys. Rev. Lett.* **91**, 012002 (2003).
CLAS Coll., V. Kubarovsky *et al*, *Phys. Rev. Lett.* **91**, 252001 (2003) and **92**, 032001 (2004). Erratum; ibid, 049902.
SAPHIR Coll., J. Barth *et al*, *Phys. Lett.* B **572**, 127 (2003).
4. DIANA Coll., V.V. Barmin *et al*, *Phys. Atom. Nucl.* **66**, 1715 (2003).
A.E. Asratyan, A.G. Dolgolenko, M.A. Kubantsev, *Phys. Atom. Nucl.* **67**, 682 (2004); *Yad. Fiz.* **67**, 704 (2004).
SVD Coll., A. Aleev *et al*, Preprint hep-ex/0401024, (2004).
HERMES Coll., A. Airapetian *et al.*, *Phys. Lett.* B **585**, 213 (2004).
COSY-TOF Coll., M. Abdel-Bary *et al.*, *Phys. Lett.* B **595**, 127 (2004).
5. NA49 Coll., C. Alt *et al.*, *Phys. Rev. Lett.* B **92**, 042003 (2004).
6. ZEUS Coll., S. Chekanov *et al.*, *Phys. Lett.* B **591**, 7 (2004).
7. ZEUS Coll., S. Chekanov *et al.*, Submitted Conference Paper, Abstract: 10-0273, ICHEP'04, Beijing.
8. S. Capstick, P.R. Page, W. Roberts, *Phys. Lett.* B **570** 185 (2003).
9. C. Risler, H1 Coll., Submitted Conference Proceedings, DIS 2005, Madison.
10. ZEUS Coll., S. Chekanov *et al*, Preprint hep-ex/0409033 , (2004).
11. Particle Data Group, K. Hagiwara *et al.*, *Phys. Rev.* D **66**, 010001 (2002).
12. H1 Coll., A. Aktas *et al.*, *Phys. Lett.* B **588**, 17 (2004).
13. ZEUS Coll., S. Chekanov *et al.*, Submitted Conference Paper, Abstract: 10-0293, ICHEP'04, Beijing.

Parallel Session II:

SUSY and Beyond the Standard Model

Conveners:

M. Maggi
INFN Bari, Italy

A. Romanino
SISSA Trieste and INFN, Italy

Particle Dark Matter: Searching for new Physics without Accelerators

Nicolao Fornengo

Dipartimento di Fisica Teorica, Università di Torino
Istituto Nazionale di Fisica Nucleare, Sezione di Torino
via P. Giuria 1, I–10125 Torino, Italy

Abstract. Astroparticle physics offers very important, and sometimes unique, tools for studying extensions of the standard model in a way which is complementary to accelerator physics. In this paper we discuss the problem of dark matter in connection with its explanation in terms of relic supersymmetric particles, by analyzing different types of astrophysical signals which can be looked at in order to disentangle the presence of dark matter in the galactic halo.

Keywords: Supersymmetry, Particle Dark Matter, Galactic halo, Cosmology
PACS: 95.35.+d,98.35.Gi,11.30.Pb,12.60.Jv,95.30.Cq

EVIDENCE OF DARKNESS

The presence of large amounts of non–luminous components in the Universe has been identified along the years by different means and on different scales: on the galactic scale, the flatness of the rotational curves of many galaxies indicates a dark component which is presumably distributed as a halo around the galaxies; clusters points toward a sizeable contribution of unseen matter distributed between galaxies; more recently, on cosmological scales, the combination of the results on high–redshift supernovae and on the anisotropies of the cosmic microwave background radiation is pointing toward a flat Universe whose energy density is dominated by a dark vacuum component (cosmological constant, quintessence) together with a sizeable dark component of matter. In terms of the density parameter Ω, the current view can be summarized as follows: the total amount of matter/energy of the Universe is $\Omega_{tot} = 1.02^{+0.06}_{-0.05}$, and this is composed of a matter component $\Omega_M = 0.31^{+0.13}_{-0.12}$ and a vacuum–energy component $\Omega_\Lambda = 0.71 \pm 0.11$. The clear indication of the latest data is therefore that the Universe is strongly dominated by dark (and unknown) components. In fact the numbers above cannot be reconciled with a Universe made only of standard components: primordial nucleosynthesis tells us that baryons can contribute only at the level of $\Omega_b = 0.037 \pm 0.11$, while luminous matter is known to provide only a contribution of order $\Omega_{lum} \sim 0.003$. We are therefore facing the presence of at least *three dark components* in the Universe: dark baryons, dark matter and dark energy. The existence of both dark (relativistic or non–relativistic) exotic matter and dark energy asks for extension of the standard model of fundamental interactions, since no known particle or field can explain either of these components. The amount of non–baryonic dark matter can therefore be summarized as:

$$0.05 \lesssim \Omega_M h^2 \lesssim 0.3 \quad (1)$$

SUPERSYMMETRY AND DARK MATTER: THE NEUTRALINO

The existence of a relic particle in supersymmetric theories arises from the conservation of a symmetry, *R*–parity, which prevents the lightest of all the superpartners from decaying. The nature and the properties of this particle depend on the way supersymmetry is broken. The neutralino can be the dark matter candidate in models where supersymmetry is broken through gravity– (or anomaly–) mediated mechanisms. The actual implementation of a specific susy scheme depends on a number of assumptions on the structure of the model and on the relations among its parameters. This induces a large variability of the phenomenology of neutralino dark matter [1].

The simplest and most direct implementation of supersymmetry is represented by the *minimal supergravity* (mSUGRA) scheme, where, in addition to requiring gauge coupling constant unification at the GUT scale, also all the mass parameters in the susy breaking sector are universal at the same GUT scale. The low–energy sector of the model is obtained by evolving all the parameters through renormalization group equations (RGE): this process also induces the breaking of the electroweak (EW) symmetry in a radiative way (rEWSB). This model is very predictive, since it relies only on four free parameters, but at the same time it has a very constrained phenomenology at low–energy. It also appears to be quite sensitive to some standard model parameters, like the mass of the top and bottom quarks (m_t and m_b) and the strong coupling constant α_s.

A more relaxed implementation of this susy scheme is offered by *non–universal supergravity* (nuSUGRA), where some of the unification conditions at the GUT scale are relaxed: non–universality has been studied in the Higgs, in the sfermion and in the gaugino sectors.

Specific patterns of non–universality may be originated through mechanisms which involve effects of extra–dimensions, like in *D–brane* and *string models*. These models also may be very predictive, with very few free parameters, but the relations among them is different from what is postulated in mSUGRA models.

It has also been realized that unification conditions, both for gauge couplings and/or mass parameters, may occur at scales which are different from the standard GUT scale at about 10^{16} GeV. This unification scale may be lower than the usual GUT scale (*intermediate unification scale models*), and this induces a modification of neutralino phenomenology at low energy.

A different approach is offered by the *low–energy supersymmetric model*, defined directly at the EW scale, which is where the phenomenology of neutralino dark matter is actually studied. Also in this case we have to make assumptions in order to reduce the number of free parameters to a manageable number. These assumptions must be mild enough not to represent an arbitrary over–constraint on the model, and all the relevant parameters at the EW scale must be represented. It is possible in this case to work with six or seven free parameters. Recently it has been put forward the study of a class of low–energy models with a violation of the usual relation between the gaugino masses where the relic neutralino is particle as light as the GeV scale [2]. These kind of models have been shown to have a rich an interesting phenomenology, both from the point of view of Cosmology, and from the point of view of possibility of detection of signals related to these light neutralinos [3, 4].

Other models which have been discussed in the literature in connection with neutralino dark matter are *dilaton domination* models, models with *CP–violation* and *anomaly mediated models*.

DIRECT DETECTION OF RELIC NEUTRALINOS

Direct detection relies on the scattering of dark matter particles off the nuclei of a low–background detector. This method is sensitive to the local properties of the neutralinos in the halo, *i.e.* its local abundance ρ_χ and its local velocity distribution, and depends on the neutralino–nucleus scattering cross section, which is usually dominated by the coherent interaction. The detection rate

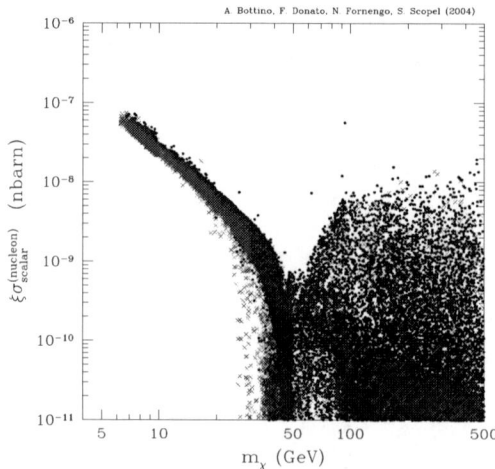

FIGURE 1. Direct detection elastic scattering cross section off a nucleon as a function of the neutralino mass. In this and all the following figures, the scatter plot for neutralino masses lower than 50 GeV refer to gaugino non–universal supersymmetric models, while for higher values of the neutralino mass the supersymmetric model is a low–energy, gaugino–universal model.

is proportional to the product $[\rho_\chi \times \sigma_{\text{scalar}}^{(\text{nucleon})}]$, where $\sigma_{\text{scalar}}^{(\text{nucleon})}$ is the elastic–scattering cross section off a nucleon, which is therefore the relevant quantity in direct detection.

The predictions for the elastic cross section in low–energy supersymmetric models are shown in Fig. 1. We notice that for low neutralino masses (which correspond to supersymmtric models where the universality in the gaugino–sector is not assumed) the theoretical predictions are sizeable and very constrained [3]. This property may be understood on the basis of the properties of such susy models, and we refer to the discussion of Ref. [3], where a comparison with experimental results is also shown.

INDIRECT DETECTION OF RELIC NEUTRALINO

Indirect detection relies on the possibility to identify signals which are originated by neutralino self–annihilations.

The first type of signal, due to neutralino annihilation taking place in celestial bodies (Earth or Sun) where the neutralinos have been gravitationally captured and accumulated, is a neutrino flux, detected in a neutrino tele-

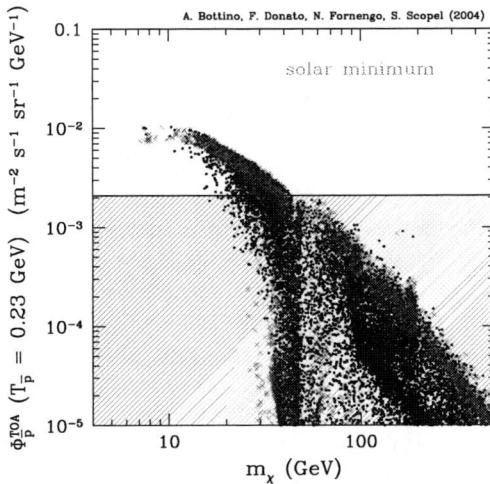

FIGURE 2. Antiproton flux at $T_{\bar{p}} = 0.23$ GeV as a function of the neutralino mass, calculated at solar minimum. The solar modulation is calculated at the phase of solar minimum. Crosses (red) and dots (blue) denote neutralino configurations with $0.095 \leq \Omega_\chi h^2 \leq 0.131$ and $\Omega_\chi h^2 < 0.095$, respectively. The shaded region denotes the amount of primary antiprotons which can be accommodated at $T_{\bar{p}} = 0.23$ GeV without entering in conflict with the experimental BESS data [5, 6] and secondary antiproton calculations [7]. The best fit set for the astrophysical parameters is used [8]

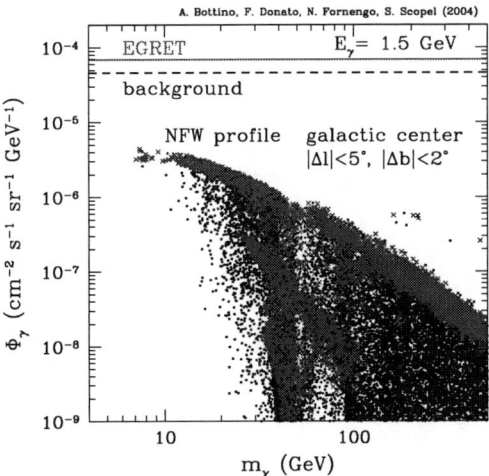

FIGURE 3. Gamma–ray flux from the galactic center inside the angular region $|\Delta l| \leq 5°$, $|\Delta b| \leq 2°$ for a NFW matter density profile and calculated at $E_\gamma = 1.5$ GeV. Crosses and dots denote neutralino configurations with $0.095 \leq \Omega_\chi h^2 \leq 0.131$ and $\Omega_\chi h^2 < 0.095$, respectively. The solid horizontal line shows the gamma–ray flux measured by EGRET [9], the shaded horizontal band denotes the 1σ error bar on the EGRET data and the dashed line is an estimate of the gamma–ray background.

scope as a flux of *up–going muons*. Since the process relevant for accumulation of neutralino is capture, which relies on neutralino scattering off the nuclei of the Earth and the Sun, this detection technique is sensitive only to local properties of the Galaxy, like direct detection. Indirect detection at neutrino telescopes is partially competitive with direct detection, although a large fraction of configurations which are explored by direct detection require more sensitive neutrino telescopes in order to be probed [4].

Neutralino annihilations can take place also in the galactic halo and the signals can consist in: a *diffuse neutrino flux* ; a *diffuse gamma–ray flux* or a *gamma–ray line*; exotic components in cosmic rays: *positrons, antiprotons* and *antideuterium*. In this case, global properties of the halo are relevant, and therefore the matter distribution of neutralinos is an important quantity. In particular, the overdensities which would be present if the halo were clumpy would have the effect of largely enhancing the predicted signals. Fig. 2 shows the theoretical predictions for the antiproton signal, compared to current experimental sensitivities. Theoretical predictions are affected by large uncertainties [8] which can boost or reduce the predicted flux by one order of magnitude [8]. For this reason, from Fig. 2 we can conclude that no firm exclusion on susy models can be driven on a conservative basis, even thouhg the antiproton searches are the most sensitive probe to susy dark matter, especially in the sector of low mass neutralinos [4].

Fig. 3 shows the gamma–ray flux predicted for annihilation of nautralinos in the galactic center, for a dark matter density profile which is relatively steep toward the galactic center, until its very inner regions [4]. We notice that the predicted fluxes are well below experimental sensitivities. The gamma–ray signal is in fact able to explain the excess observed by EGRET [9] only in the case of sizeable clumpiness along the line of sight toward the galactic center [4]. An example is shown in Fig. 4.

CONCLUSIONS

Two are the main issues in particle dark matter studies: *i)* to explain the observed amount of dark matter in the Universe ($0.05 \lesssim \Omega_M h^2 \lesssim 0.3$) by finding suitable particle candidates; *ii)* to detect a relic particle. For both of them there appear to be good prospects of success.

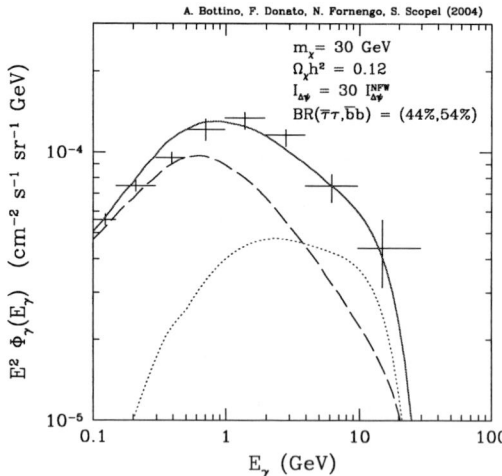

FIGURE 4. Gamma–ray spectru, $\Phi_\gamma(E_\gamma)$, multiplied by E_γ^2, from the galactic center inside the angular region $|\Delta l| \leq 5°$, $|\Delta b| \leq 2°$, as functions of the photon energy. The dotted line is the spectrum for a neutralino with mass $m_\chi = 30$ GeV, calculated for a density profile with a factor 30 of enhancement with respect to the NFW case; the dashed line is the gamma ray background calculated in Ref. [9], reduced by 10%; the solid line is the total flux, sum of the supersymmetric signal and the background; the experimental points are the EGRET data [9]. The numbers quoted in the legend inside parentheses denote the values of the neutralino annihilation branching ratios into $\bar{b}b$ and $\bar{\tau}\tau$.

and quite generally it appears simpler to detect a relic neutralino which is a sub–dominant dark matter component. Nevertheless, there are many susy schemes where relic neutralinos can provide enough cosmological abundance to explain the observed amount of dark matter, and at the same time they can have detection rates large enough to be accessible to direct, and also to some indirect, detection methods. The positive indication of annual modulation in the detection rate of the DAMA/NaI Collaboration [10], which is at the moment, the most compelling indication for a particle dark matter signal, may be interpreted as originated from relic neutralinos and explained in a number of realization of supersymmetry. It is worth noticing that the presence of a signal from dark matter, like the annual modulation effect or signals which could hopefully come in future experiments, can be very important not only for astrophysics and cosmology but also for particle physics, since the need to explain the effect can help in deriving properties of particle physics models and possibly discriminate among different realizations, for instance of supersymmetry.

ACKNOWLEDGMENTS

This work is supported by a joint Research Grant of the Italian Ministero dell'Istruzione, dell'Università e della Ricerca (MIUR) and of the Università di Torino within the *Astroparticle Physics Project* and by a Research Grant from INFN.

As for the candidates, there are many proposed particles which could act as dark matter. Some of these candidates turn out to be quite natural, like *e.g.* the massive neutrino, the axion or the neutralino. Almost all of the proposed candidates can play the role of the dominant dark matter component, although for some of them a non–standard cosmology is required. An important remark is that, from the particle physics point of view, dark matter may naturally be multi–component. A multi–component dark matter scenario offers opportunity for interesting phenomenology not only to the dominant candidate, which would explain the cosmological observation on the Ω_M parameter, but also to the sub–dominant candidates, since usually these are the ones which are easier to detect. The detection of a particle which is a relic from the early Universe would be a very important and exciting result.

As for detection, perspectives are good, both for direct and for indirect detection techniques, especially for the most interesting and studied candidate, the neutralino. The possibility to have detectable rates for neutralinos depends on the specific susy model which is considered,

REFERENCES

1. For a thorough list of references on the subjects discussed in the paper, we refer the reader to references in Refs. [3, 4].
2. A. Bottino, N.Fornengo, S. Scopel, *Phys. Rev. D* **67** (2003) 063519; A. Bottino, F. Donato, N.Fornengo, S. Scopel, *Phys. Rev. D* **68** (2003) 043506
3. A. Bottino, F. Donato, N.Fornengo, S. Scopel, *Phys. Rev. D* **69** (2003) 037302
4. A. Bottino, F. Donato, N.Fornengo, S. Scopel, *Phys. Rev. D* **70** (2004) 015005.
5. S. Orito, *et al* (BESS Collaboration), *Phys. Rev. Lett.* **84**, 1078 (2000).
6. T. Maeno, *et al.* (BESS Collaboration), *Astropart. Phys.* **16**, 121 (2001).
7. F. Donato *et al.*, *Astrophys. J.* **563**, 172 (2001).
8. F. Donato, N. Fornengo, D. Maurin, P. Salati, R. Taillet *Phys. Rev. D* **69** (2003) 063501
9. S.D. Hunter *et al.*, *Astrophys. J.* **481**, 205 (1997).
10. R. Bernabei *et al.* (DAMA/NaI Collaboration), *Phys. Lett.* **B480**, 23 (2000); *Eur. Phys. J.* **C18**, 283 (2000).

Search for the SM and MSSM Higgs Bosons at CDF

Giorgio Cortiana

Dipartimento di Fisica "G.Galilei", Via Marzolo 8, 35131 Padova, Italy

Abstract. The CDF and experiment at the Tevatron has searched for the SM and MSSM Higgs bosons in data collected between 2001 and 2004. Upper limits have been placed on the production cross section times branching ratio to $b\bar{b}$ pairs or W^+W^- pairs as a function of the SM Higgs boson mass. The MSSM cp-odd Higgs boson A has being searched in its decay into τ pairs.

Keywords: Standard Model Higgs, MSSM Higgs, CDF, Tevatron
PACS: 14.80.Bn, 14.80.Cp

INTRODUCTION

In the standard model (SM) and in its minimal supersymmetric extension (MSSM) the Higgs boson sector is crucial for our understanding of the electroweak symmetry breaking dynamic and for gauge bosons and fermions mass generation mechanism[1]. Electroweak precision measurements[2] indicate as $M_H = 126^{+73}_{-48}\,GeV/c^2$ the preferred value of the SM Higgs mass and place at $280\,GeV/c^2$ it upper limit at 95% CL. On the other hand direct searches performed by LEP experiments have set at $114\,GeV/c^2$ the SM Higgs boson mass 95% CL lower limit[3].

In 2003 a joint committee of CDF and DØ members carried out a reassessment of the Tevatron reach in the search for the Higgs boson[4]. By using a more realistic model of the two detectors than the simplified one used in a study performed in 1998 [5], together with real data collected by the experiments, the committee determined that the early claims on the Higgs sensitivity were not off the mark.

In the meantime, the Tevatron has continued to improve its performance, recently surpassing the peak luminosity of $1.2 \times 10^{32} cm^{-2} s^{-1}$. The chances of delivering an integrated statistics of $8 fb^{-1}$ by the end of 2009 appear now sizable. It seems therefore possible to expect that before the CMS and ATLAS collaborations start analyzing their first collisions, CDF and DØ may either discover a $115\,GeV/c^2$ Higgs boson, or exclude it up to $135\,GeV/c^2$ (see Fig. 1).

The search for the SM Higgs boson at CDF is performed by looking for two different final states depending on the particle mass. For masses below $135\,GeV/c^2$, the dominant decay is $H \to b\bar{b}$, and the search channels always include an accompanying W or Z boson signature.

For masses above that threshold, the $H \to W^+W^-$ decay is the most promising one, and both direct production

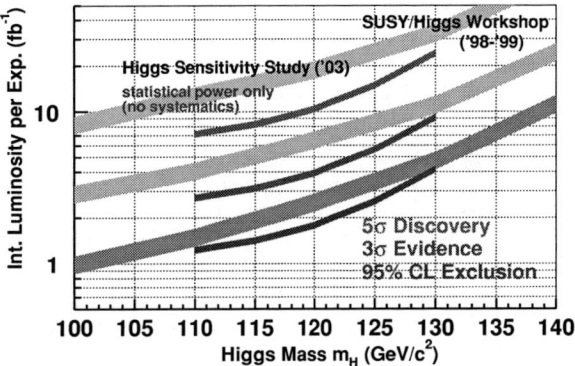

FIGURE 1. The three bands show the integrated luminosity per experiment needed for a 95% exclusion (purple), a 3σ evidence (green), or a 5σ discovery (blue) of the Higgs boson as a function of the particle mass. The 2003 study, which only considers the low mass regime ($M_H < 135$ GeV), shows slightly reduced luminosity thresholds for a given mass, but does not include systematic uncertainties.

and production in association with an additional electroweak boson are considered.

On the other hand, as far as the MSSM Higgs sector is concerned, the search for the CP-odd boson A is performed by looking for its decays into τ pairs.

Results and details of these searches will be given in the following sections.

In general, the smallness of the Higgs boson signal, when compared with competing backgrounds, implies that for a successful search many tools have to be developed:

- a performant lepton identification;
- an efficient b-quark tagging;
- a precise jet energy measurement.

CDF detector provides excellent high-P_T electrons and muons identification that can already be used at trigger

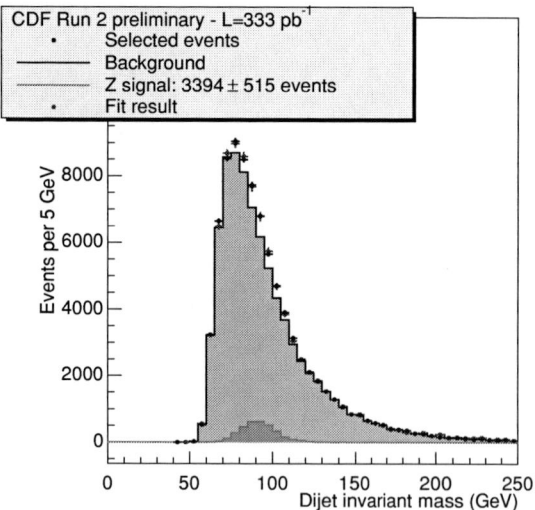

FIGURE 2. The signal of $Z \to b\bar{b}$ decays extracted from 333 pb^{-1} of CDF Run 2 data is visible as an excess of events in the dijet mass region around 90 GeV. The green histogram shows the fitted Z contribution.

level. The analyses described in this paper indeed all start from high-P_T leptons triggered data samples.

The identification and precise measurement of b-quark-originated jets is crucial in particular when searching for light SM Higgs. To successfully identify b-jets CDF detector is equipped with a dedicated silicon detector which can measure the impact parameter of charged tracks with a resolution of few tens of microns, thereby allowing dedicated algorithm to search for secondary vertices originated by B hadrons decays. This secondary vertex tagging algorithm is found to have an efficiency of 45-50% for central b-jet, while the mis-identification rate is kept generally as low as 1%. For completeness, other b-jet identification methods rely on an estimate of the global probability that all tracks in a jet come from the primary interaction vertex, or on the identification of electrons or muons from the semileptonic decay of B particles, but are not used in the analyses here presented.

The goal for the precise measurement of the energy of b-jets was set by the Higgs Sensitivity Working Group. They claimed that by using advanced algorithms a relative resolution $\sigma_{M_{jj}}/M_{jj} \sim 10\%$ was attainable on the dijet invariant mass pair of b-quark jets from $H \to b\bar{b}$ decay [4].

CDF has recently identified a first signal of $Z \to b\bar{b}$ decays from a dedicated dataset triggered which requires jets and displaced tracks identified in the silicon detector. Fig. 2 shows the Z signal, whose size and shape are in good agreement with expectations. Besides providing a precise measurement of the b-jet energy scale, the data will be used in studies aimed at improving the dijet mass resolution, to increase the chances of a Higgs discovery in the $b\bar{b}$ final state.

SEARCHES FOR SM HIGGS BOSON

At Tevatron the dominant Higgs production mechanism is through gluon-gluon fusion. For light Higgs ($M_H < 135\ GeV/c^2$) the dominant decays channel is to $b\bar{b}$, which provides a signature containing two heavy flavor jets. However, this final states is dominated by the large SM QCD multi-jet production. Therefore, the search for light Higgs concentrates on channels where it is produced in association with Z or W bosons. On the other hand, for Higgs masses above $135\ GeV/c^2$ the particle is predicted to decay into a W^+W^- pairs, in which case the $gg \to H$ process can be considered in addition to the associated production with gauge bosons.

Light mass Higgs search

The search for $p\bar{p} \to W^{\pm}H \to l^{\pm}\nu b\bar{b}$ process is performed in 162 pb^{-1} of collected data by looking for a peak in the dijet invariant mass distribution in two jet events. Candidate events are selected by requiring an energetic isolated lepton ($E_T > 20\ GeV$ for electrons, $P_T > 20\ GeV/c$ for muons) in the central detector, two jets with $E_T > 15\ GeV$ and missing E_T greater than $20\ GeV$. Additionally, at least one of the jets is required to be identified as originating from a b-quark by the secondary vertex tagger algorithm. The SM background expectation, from $Wb\bar{b}$, $Wc\bar{c}$, Wc, diboson and top production consists of 66.5 ± 9.0 events, where the expected amount of mis-tagged events is also added[1]. After the all the selection cuts, 62 events are observed in the data sample. Upper limits at 95% CL on the production cross section times branching ratio as a function of the Higgs mass are calculated with a maximum binned likelihood technique on the dijet mass distribution (Fig. 3) and found to be $5 \div 3$ pb for M_H in the range $110 \div 150\ GeV/c^2$.

High Mass Higgs search

For $M_H > 135\ GeV/c^2$ GeV the $H \to WW^{(*)}$ decay mode becomes dominant. When both W bosons decay to an electron-neutrino or muon-neutrino pair the final state is very clean, with residual backgrounds mostly due to Drell-Yan production of lepton pairs. To discriminate direct production of a Higgs boson from non-resonant WW

---

[1] Mis-tags originated from mis-identification of light quarks jet as heavy-flavor jets by the b-tagging algorithm.

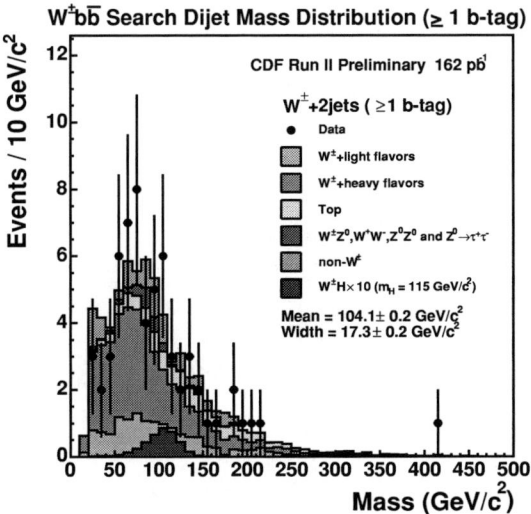

FIGURE 3. The dijet mass distribution of 62 b-tagged W +2 jet candidates (black points) is understood as a sum of several contributing SM backgrounds (left); a $H \to b\bar{b}$ signal 10 times larger than expected is overlaid.

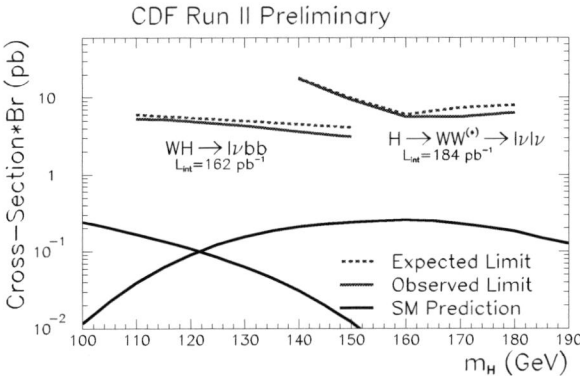

FIGURE 5. Summary of 95% C.L. limits obtained from CDF Run II seaches of the SM Higgs boson.

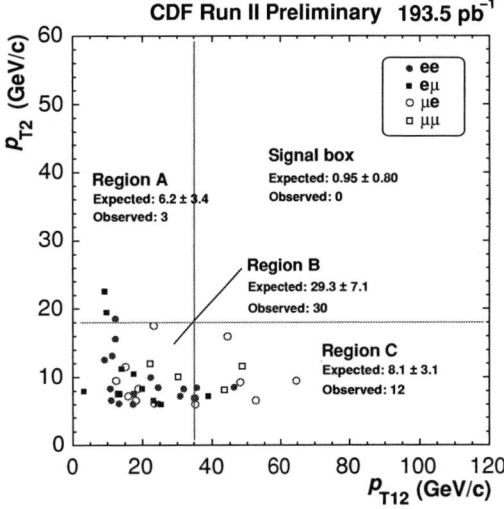

FIGURE 6. Distribution of the vector sum of lepton transverse momenta (P_{T12}) versus subleading lepton transverse momentum (P_{T2}) for same-sign dilepton candidates found by CDF

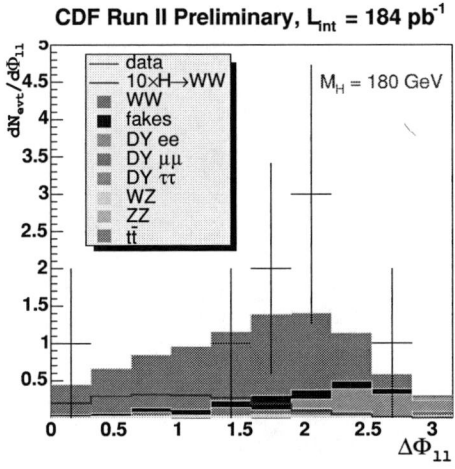

FIGURE 4. Azimuthal angle between charged leptons in the CDF $H \to WW$ analysis

production it is useful to study the azimuthal angle $\Delta\Phi_{ll}$ between the two charged leptons, since the zero spin of the Higgs boson and helicity conservation conspire to produce leptons in the same direction in the transverse plane.

CDF selects W pairs by searching for lepton pairs of opposite charge, $P_T > 20 \ GeV/c$, and then applying a missing E_T cut at 25 GeV and a tight jet veto. A small dilepton mass is then required ($M_{ll} < 55 \ (80) \ GeV/c^2$ for $M_H = 140 \ (180) \ GeV/c^2$). After this selection, 8 events are observed in $184 pb^{-1}$ of data, where 8.9 ± 1.0 are expected from non-Higgs SM sources. A likelihood fit to the $\Delta\Phi_{ll}$ distribution is performed to extract a 95% C.L. limit on the cross section as a function of M_H (see Fig. 4). The result is $\sigma_H \times B(H \to WW) < 5.6 \ pb$ for $M_H = 160 \ GeV/c^2$.

The results of the analyses previously described are summarized in Fig. 5.

CDF also searched in $193.5 pb^{-1}$ of data for the striking signature of associated WH production at high M_H, when the event may yield *three* W bosons. The search starts from a dataset of lepton pairs of same charge, which is understood as a sum of fake lepton backgrounds and SM sources.

Optimized cuts are then applied to the second lepton transverse momentum ($P_{T_2} > 18 \ GeV/c$) and to the vector sum of lepton transverse momenta, $P_{T_{12}} > 35 \ GeV/c$.

Zero events are observed (see Fig. 6), with 0.95 ± 0.80 expected from known sources. A 95% C.L. cross section times branching ratio limit of $8 pb$ was thus set for $M_H = 160\, GeV/c^2$.

SEARCHES FOR MSSM HIGGS BOSONS

The Minimal Supersymmetric extension of the Standard Model predicts the existence of five Higgs particles, three neutral and two charged. The coupling of the neutral pseudo-scalar Higgs boson, A, to the third generation fermions can be enhanced by a factor $\tan\beta$ (the ratio of the vacuum expectation values of the two Higgs fields) relative to the SM. Therefore, the production cross section of A will scale with $\tan^2\beta$. While the branching ratio of $A \to b\bar{b}$ dominates (90%) a search for this channel is very difficult due to the overwhelming multi-jet background. For this reason, the $A \to \tau\tau$ (8%) decay channel can expand the CDF reach.

CDF has searched for the $A \to \tau\tau$ at high $\tan\beta$ in 195 pb^{-1} of data, collected with a set of dedicated τ-triggers requiring a lepton (e,μ) and an isolated track. The signal consists of a τ-pair, in which one of the τ lepton decays leptonically while the other hadronically. After initial selection of events with two reconstructed τ candidates, the multi-jet background was suppressed by requiring large H_T, which is defined as the scalar sum of momenta of τ-decay products and the missing E_T. In addition, the missing transverse energy is required not to point in directions opposite to the τ decay products. After all of the requirements, 236 events are observed, against a background expectation of 263.6 ± 30.1. The dominant source of background come from $Z \to \tau\tau$. To further discriminate signal from background and set a limit on production cross section, a mass-like variable, m_{vis}, is constructed using four momenta of the lepton, of the hadronically decaying tau visible products and missing E_T. Fig. 7 shows the distribution of m_{vis} for the surviving events.

A Maximum binned-likelihood fit is performed to the m_{vis} distribution. Observed limits as a function of M_A on the production cross section times the branching ratio at 95% C.L. for are shown on Fig. 8 along with expected limits from pseudo-experiments.

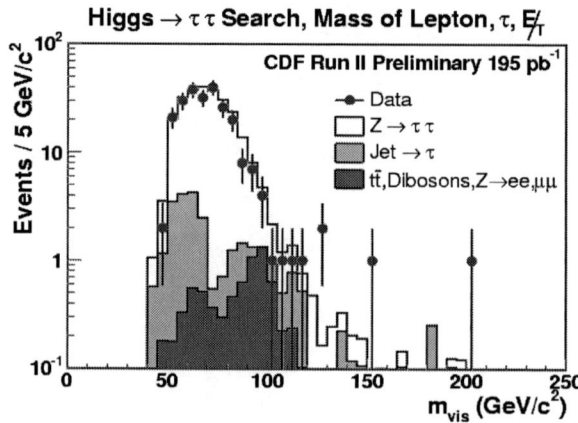

FIGURE 7. Observed partially reconstructed ditau mass and contributions from various SM background processes.

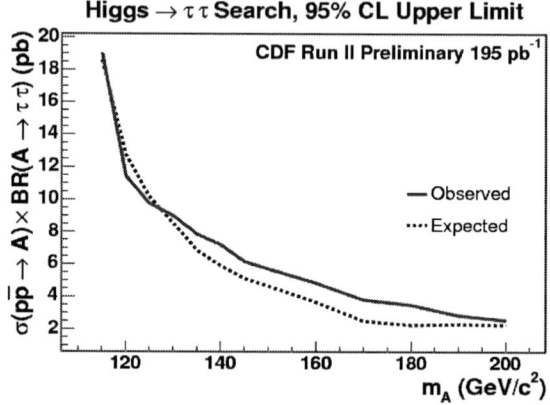

FIGURE 8. Observed and expected upper limits at 95% C.L. on the $\sigma(p\bar{p} \to A) \times BR(A \to \tau\tau)$ as a function of M_A.

CONCLUSION

The SM and MSSM Higgs bosons are being hunted at CDF in most of the promizing channels. No signal evidence are found in the data so far and limits on the production cross section times branching ratio are derived. The sensitivity will improve as CDF continues to take data and new results should be expected soon with larger statistics. The Tevatron remains the energy frontier of collider searches for the Higgs, and provides the best opportunity for its hunting before LHC starts.

REFERENCES

1. J.Gunion *et al*, *The Higgs Hunter's Guide*, Addison-Wesley, New York 1990.
2. The LEP Electroweak Working Group, http://lepewwg.web.cern.ch/LEPWWG/.
3. A.Heister *et. al*, Phys. Lett. B **565**, 61, (2003).
4. CDF and D0 collaborations, *Results of the Tevatron Higgs Sensitivity Study*, FERMILAB-PUB=03/320-E.
5. M.Carena *et. al.*, *Higgs Working Group Report*, hep-ph/0010338.

Searches for New Physics at CDF

Mario Paolo Giordani

Università degli Studi di Udine e INFN Sezione di Trieste (gruppo collegato di Udine)

Abstract. The search for physics beyond the Standard Model is one of the primary motivations of the Run II at the Tevatron. The CDF collaboration has been very active in this field by searching in a variety of final states for signals of new physics processes. The observation of no significant deviation from the Standard Model has led to limits on the cross-sections of the considered processes, which, in turn, have been used to constrain several models that predict such new physics scenarios.

Keywords: Models beyond the standard model
PACS: 12.60-i, 12.10.-g, 04.65+e

INTRODUCTION

Despite all the efforts devoted to precision measurements aimed at probing the current theory of high energy physics, the Standard Model (SM), the question concerning what physics scenario to expect beyond the electroweak scale is still open. Several theoretical arguments support the idea that new physics is likely to be expected above this energy boundary and several models for extending the SM have been suggested.

The CDF collaboration, after extensive detector upgrades, has been involved in searching for new physics signals in a wide range of final states. Details of these searches and preliminary results based on samples collected in the first three years of data-taking are given in the following.

SINGLE-LEPTON FINAL STATES

CDF investigated the production of additional charged vector bosons, W', in events characterized by single isolated leptons in the final state. The existence of such bosons is predicted by models that embed the SM gauge symmetry in a larger group [1]. These models usually specify the strength of the coupling of such bosons to fermions without making any prediction upon their mass. Although not suggested by a specific theoretical model, a W' with SM couplings represents a convenient term of comparison.

CDF has performed a W' search in the assumption that the neutrino emerging from the boson is light and stable. Therefore, events have been selected by requiring a highly energetic isolated electron ($p_T > 25\,\mathrm{GeV}/c$) and large missing E_T ($> 25\,\mathrm{GeV}$), which is expected to emerge due to the neutrino; the dominating QCD background is reduced by balancing the transverse energy associated to the electron candidate with the transverse missing energy recorded for the event.

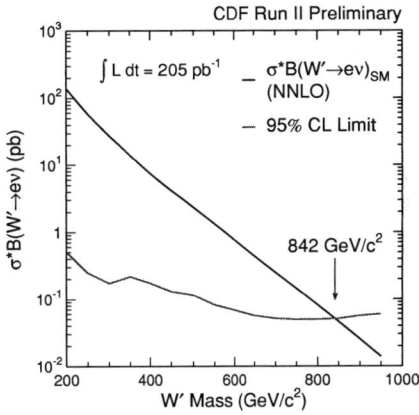

FIGURE 1. 95% C.L. limit on $\sigma(W') \cdot B(W' \to e\nu_e)_{SM}$ as a function of $m(W')$.

The number of events selected from data matches, within experimental errors, the predictions of the SM in all the investigated W' mass range ($200 \div 1000\,\mathrm{GeV}/c^2$), as shown in figure 1; the sensitivity achieved by the study reaches the level of $50 \div 100\,\mathrm{fb}$ and allows to set a 95% C.L. lower limit on $m(W'_{SM})$ at $842\,\mathrm{GeV}/c^2$.

DILEPTON FINAL STATES

A very clean signature of the production of new neutral massive states at hadron colliders is given by their decay into a lepton pair ($\ell = e, \mu, \tau$); examples of such states include tau sneutrinos decaying leptonically in violation of the R-parity [2], additional neutral vector bosons (Z') [3, 4, 5] and Randall-Sundrum (RS) gravitons [6].

The dilepton mass spectrum obtained from events characterized by the presence of two highly energetic, isolated leptons of equal flavour and opposite electric charge can be directly compared to the expectations from different models once the spin-dependent acceptance has been taken into account. Due to the neutrinos generated in the leptonic decay of the tau lepton, the visible mass:

$$m_{vis} \equiv m\left(\ell, \tau_h, E_T^{miss}\right) \quad (1)$$

(where τ_h indicates a tau lepton decaying hadronically) has been used instead of $m_{\ell\ell}$ in the $\tau\tau$ channel.

No high-mass bump in the dilepton spectra has been observeded and data show a good agreement with the SM predictions; the sensitivity achieved by this search ranges at the $50 \div 100$ fb level (see figure 2), allowing to exclude at 95% C.L. masses up to 725 GeV/c^2 for the R-parity violating $\tilde{\nu}_\tau$, 815, 690, 715, 670, 610, 875 GeV/c^2 respectively for Z'_{SM}, Z'_ψ, Z'_η, Z'_χ, Z'_I, Z'_H, 320 GeV/c^2 for straw-man Technicolor ω_T and ρ_T and 700 GeV/c^2 for RS gravitons[1].

Given the fact that, at least from the experimental point of view, photons resemble electrons[2], a search for high-mass diphotons can be carried out in the same way as for electrons. This procedure, which has been applied to the RS graviton, whose rate of decay to diphotons can be sizable, has shown to have a better sensitivity than the dilepton channel in the high mass end; preliminary results show that, for 345 pb^{-1}, the reach in terms of exclusion is approximately that of the combined $ee, \mu\mu$ channels with 200 pb^{-1}.

LEPTONS PLUS JETS SIGNATURES

Typical signatures of leptoquark decays include energetic isolated leptons accompanied by jets. Leptoquarks are hypothetical color triplet bosons of unknown mass that are predicted by several extensions of the SM, such as Grand Unified models; their peculiarity is to carry both lepton and baryon quantum numbers, which makes possible their decay into a lepton-quark pair, with both fermions belonging to the same generation in order not to enhance flavour changing neutral currents beyond the experimental limits. This suggests a natural classification of leptoquarks into first, second and third generation, according to the fermion family they decay into. At the Tevatron, the dominant leptoquark production mechanism is the pair-production through $q\bar{q}$, gg fusion; considering a subsequent decay of each leptoquark into a lepton-quark pair, this leads to events characterized by

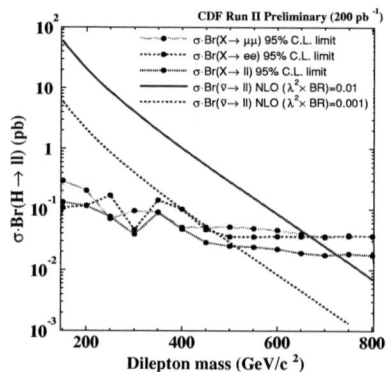

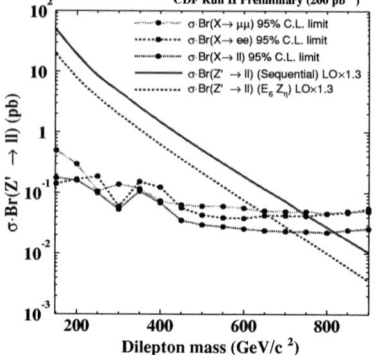

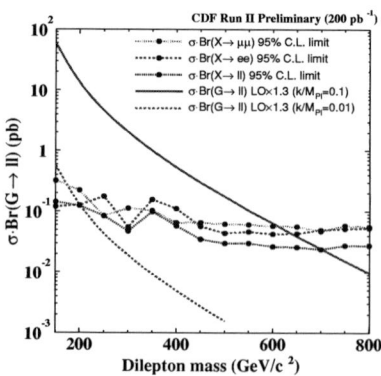

FIGURE 2. 95% C.L. limits on $\sigma(X) \cdot B(X \to \ell\ell)$ ($\ell = e, \mu$) for spin 0 (top), 1 (center) and 2 (bottom) X as functions of $m(X)$.

one or two energetic charged leptons – or, alternatively, by large missing E_T (> 60 GeV) due to the two neutrinos – accompanied by typically two jets. Despite both scalar and vector leptoquarks being supported by theory, the coupling of vector leptoquarks to gluons is model-dependent, which makes scalar leptoquarks the most studied from an experimental point of view.

The major contaminations to the signal come from QCD multijet, Z/W plus jets and $t\bar{t}$ with at least one W decaying leptonically. Background reduction is performed differently according to the signal topology: by

---

[1] All quoted lower mass bound refer to the combined $ee, \mu\mu$ search; a 95% C.L. exclusion at 394 GeV/c^2 has been achieved on $m(Z'_{SM})$ by considering the $\tau\tau$ channel independently.
[2] Photons are in fact identified as electromagnetic clusters reconstructed in the calorimeter that – conversely of electrons – are not associated to any track.

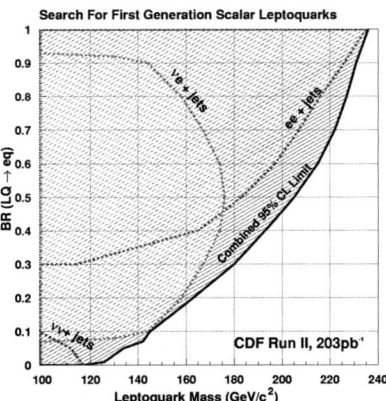

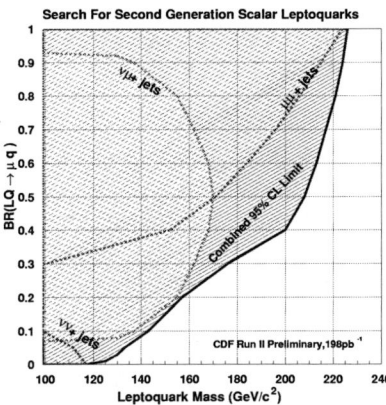

FIGURE 3. 95% C.L. exclusion contours in the $\beta - m(\text{LQ})$ plane (β being the branching fraction of $\text{LQ} \to \ell q$).

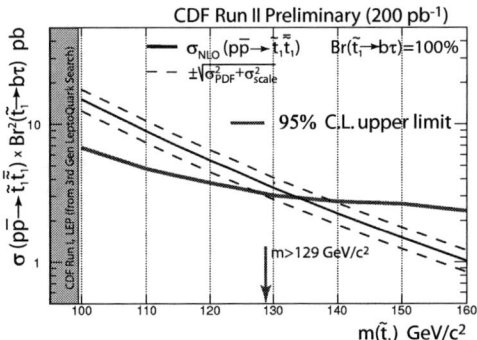

FIGURE 4. 95% C.L. limit on $\sigma(\tilde{t}_1 \bar{\tilde{t}}_1) \cdot B^2(\tilde{t}_1 \to b\tau)$ as a function of $m(\tilde{t}_1)$ (same applies to third generation LQ).

vetoing values of $m(\ell\ell)$ or $m_T(\nu\ell)$ compatible with $m(Z)$ or $m_T(W)$ for $\ell\ell jj$ and $\nu\ell jj$; or by vetoing energetic leptons and by requiring both jets and missing E_T being all well isolated for $\nu\nu jj$. The number of events selected from data is in all cases compatible with the SM prediction; first and second generation leptoquark masses have therefore been excluded at 95% C.L. up to 235 and 224 GeV/c^2 respectively, as shown in figure 3.

Third generation leptoquarks, on the other hand, are characterized by a more challenging experimental signature: due to the decay of the tau leptons, the most powerful final state configuration is that of a lepton accompanied by jets and missing E_T (corresponding to the case in which one tau lepton decays leptonically, the other decaying into hadrons). This same signature can be used to probe the R-parity violating decay of a stop pair ($gg, q\bar{q} \to \tilde{t}_1 \bar{\tilde{t}}_1 \to b\tau^+ \bar{b}\tau^-$). The major background originates from Z plus jets, with $Z \to \tau^+\tau^-$, and is reduced by requiring the visible mass (m_{vis}, see equation (1)) being outside the Z mass window. The number of events selected from data is compatible with the SM prediction, which allows a 95% C.L. exclusion for third generation leptoquarks and R-parity violating stops up to 129 GeV/c^2, as shown by figure 4.

MISSING TRANSVERSE ENERGY SIGNATURE

The presence of missing transverse energy is one of the most characterizing signatures of exotic processes: for instance in SUSY models with R-parity conserved, the lightest supersymmetric particle (LSP) is stable and therefore escape the detector giving rise to missing E_T. However, several factors weaken this signature: first, LSPs are likely to emerge from cascade decays, therefore being expected not to be very energetic; second, there are multiple sources of missing energy in an event, coming not only from new, undetected particles (as the LSPs), but also from standard sources, such as neutrinos. As a consequence, only moderate levels of missing E_T are expected even when exotic processes are considered; this requires a high degree of understanding of beam and detector effects – typically complicated to model – which makes this signature hard to exploit.

Despite its intrinsic difficulty, the missing E_T signature, especially when considered in conjunction with photons, jets or leptons, is expected to achieve the highest sensitivities in the search for new phenomena.

Models relying on gauge-mediated SUSY breaking (GMSB) predict that, if the gravitino is the LSP, then the role of next to lightest supersymmetric particle (NLSP) is played either by the neutralino or by a slepton. If the former is true, then $\tilde{\chi}_1^0 \to \gamma \tilde{G}$ is the only decay mode kinematically accessible by the NLSP; the resulting GMSB signatures are then similar to mSUGRA final states (where the $\tilde{\chi}_1^0$ is the LSP) complemented by a photon pair.

A signature characterized by two energetic photons and large missing E_T (> 40 GeV) has been investigated

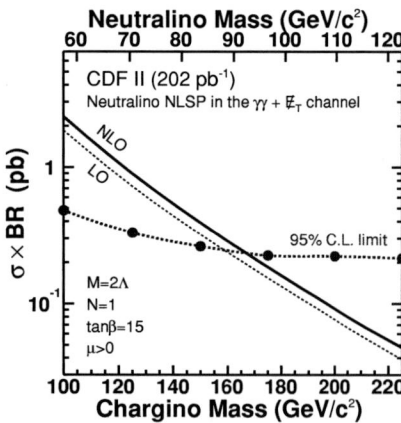

FIGURE 5. 95% C.L. limit on $\sigma \cdot BR$ as a function of $m(\tilde{\chi}_1^\pm)$ and $m(\tilde{\chi}_1^0)$ in the light gravitino scenario.

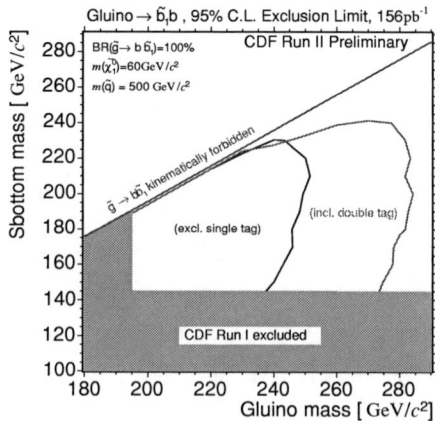

FIGURE 6. 95% C.L. exclusion contours in the $m(\tilde{b}) - m(\tilde{g})$ plane.

by CDF [7]. The missing E_T spectrum observed from data (before the missing E_T cut) is in good agreement with the SM prediction, which is dominated by QCD processes (multijet events with production or misidentification of photons); no excess has been observed from data. As shown in figure 5, a 95% C.L. exclusion for GMSB charginos has been set up to $167\,\text{GeV}/c^2$.

Another process involving the production of missing E_T is the decay of gluino pairs, particles that are again predicted by SUSY models. The production of gluino pairs is of particular interest at the Tevatron, where its cross-section can be fairly large. Several different signatures can characterize the events yielded by this process, depending on the decay mode of the gluino. As an effect of the large mixing affecting the third generation, which can push the $\tilde{b}_1$ and $\tilde{t}_1$ squarks mass eigenstates to low mass values, the $\tilde{g} \to b\bar{b}$ becomes most interesting in conjunction with the decay $\tilde{b} \to b\chi_1^0$. In fact, if both gluinos decay according to this scheme, the final state is characterized by a spectacular four b-jets and missing E_T signature.

CDF has searched for events selected in terms of moderate jet activity (three or more 10 GeV jets within the silicon vertex detector's acceptance), heavy flavour content and missing E_T above 50 GeV. The missing E_T spectra observed in the data samples with one and two jets identified as originated from a b quark before any missing E_T requirement have been checked against the SM predictions and are in good agreement. The number of observed events passing the selection is compatible with the SM predictions in both single and double tag samples; the resulting 95% C.L. limits exclude gluino and sbottom masses up to 280 and $240\,\text{GeV}/c^2$ respectively, as shown in figure 6.

CONCLUSIONS

A collection of results obtained by the CDF collaboration in the searches for new physics has been presented; particular emphasis has been put on the most recent analyses that are based on the first $350\,\text{pb}^{-1}$ collected during the Run II.

CDF is actively searching for new physics signals in a large variety of channels; no evidence has been observed so far. The limits achieved in Run II are rapidly extending those set by previous experiments.

REFERENCES

1. J. C. Pati and A. Salam, *Phys. Rev. D* **10**, 275 (1974); R. N. Mohapatra and J. C. Pati, *Phys. Rev. D* **11**, 566 (1975); ibid., *Phys. Rev. D* **11**, 2558 (1975); G. Senjanovic and R. N. Mohapatra, *Phys. Rev. D* **12**, 1502 (1975).
2. D. Choudhury, S. Majhi and V. Ravindran, *Nucl. Phys. B* **660**, 343 (2003).
3. F. del Aguila, M. Quiros and F. Zwirner, *Nucl. Phys. B* **287**, 457 (1987).
4. T. Han, H. E. Logan, B. McElrath and L. Wang, *Phys. Rev. D* **67**, 095004 (2003).
5. K. D. Lane and S. Mrenna, *Phys. Rev. D* **67**, 115011 (1999).
6. L. Randall and R. Sundrum, *Phys. Rev. Lett.* **83**, 3370 (1999).
7. D. Acosta et al. (The CDF Collaboration), *Phys. Rev. D* **71**, 031104(R) (2005).

Higgs Boson Searches at LHC

D. Giordano

*Università degli studi di Bari and INFN Sezione di Bari,
via Amendola 173, I-70126, Bari, Italy*

Abstract.
The physics potential for Higgs boson discovery with the ATLAS and CMS detectors at the LHC is discussed, in the framework of the Standard Model (SM) and its Minimal Supersymmetric extension (MSSM).

Keywords: Higgs boson, Standard Model, MSSM, LHC, CMS, ATLAS
PACS: 14.80.Bn

INTRODUCTION

The search for the Higgs boson is one of the primary tasks of the two multi-purpose experiments that will take data from June 2007 at the *Large Hadron Collider* (LHC) [1], a proton-proton collider with a centre-of-mass energy of 14 TeV and a bunch crossing rate of 40 MHz that is under construction at the European Organization for Nuclear Research (CERN) in Geneva, Switzerland. The LHC is designed to operate at the "high" luminosity of 10^{34} cm^{-2} s^{-1}, after a first three-year period in which it will collect about 20 fb^{-1}/experiment every year at the "low" luminosity of 2×10^{33} cm^{-2} s^{-1}.

The two multi-purpose detectors that will be placed along the LHC tunnel are ATLAS (*A Toroidal LHC ApparatuS*) [2] and CMS (*Compact Muon Solenoid*) [3]. They have excellent detector performances and outstanding trigger systems, optimized for fulfil a wide physics program ranging from precision measurements in the electroweak sector to search for new particles, as the Higgs boson.

In this report, the Higgs boson discovery potential at LHC with the ATLAS and CMS detectors is discussed in the framework of the Standard Model (SM) and its Minimal SuperSymmetric extension (MSSM) [4, 5].

SEARCH FOR A STANDARD MODEL HIGGS BOSON

In the SM Higgs mechanism, the Higgs boson mass m_H is the only free parameter, bounded from below to $m_H > 114.4$ GeV/c^2 by the LEP measurements [6]. The fits to the electroweak precision measurements favour a light Higgs boson with a central value of $m_H = 129^{+74}_{-49}$ GeV/c^2 and a 95% CL upper limit of 285 GeV/c^2 [7].

Figure 1 shows the production cross sections for the

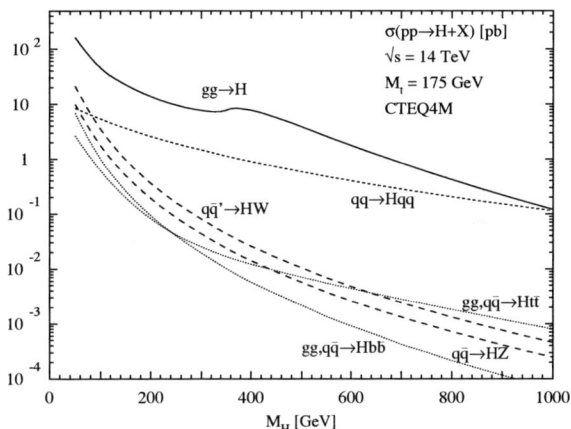

FIGURE 1. Cross sections for the SM Higgs boson production at LHC as a function of m_H.

SM Higgs boson as a function of m_H for different production processes calculated at QCD next-to-leading order [8]. The production is dominated by the gluon-gluon fusion, $gg \to H$, over the mass range 100 GeV/c^2 < m_H < 1 TeV/c^2. The associated production processes, $q\bar{q}' \to$ HW, $q\bar{q} \to$ HZ, $gg/q\bar{q} \to t\bar{t}H$ and $gg/q\bar{q} \to b\bar{b}H$ have cross sections lower by one order of magnitude or more over the full mass range. The production cross section for the vector boson fusion, $qq \to qqH$, is about 10% of the production cross section for gluon-gluon fusion for $m_H < 200$ GeV/c^2, and becomes comparable for $m_H \sim 1$ TeV/c^2. Although characterized by cross sections lower than the gluon-gluon fusion process, these production processes often can be suited to reduce more efficiently the background than in the gluon-gluon fusion process, due to their characteristic event topology.

The branching ratios for the SM Higgs boson are shown in Fig. 2 as a function of Higgs boson mass [9]. For Higgs boson masses below 130 GeV/c^2, the H → $b\bar{b}$ decay channel dominates, followed by the branching

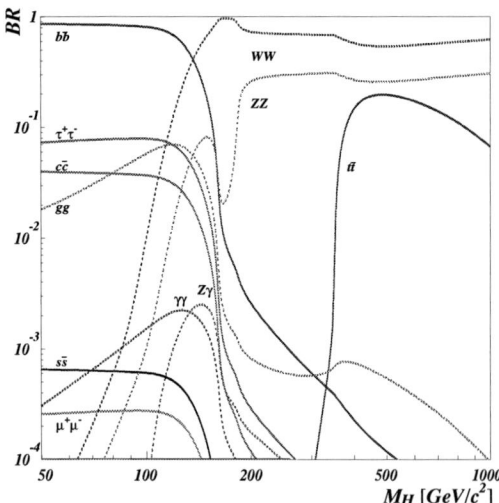

FIGURE 2. Decay branching ratios for the SM Higgs boson as a function of m_H.

ratio for H $\to \tau^+\tau^-$ decay rate that amounts to $\sim 8\%$. The branching ratio for H $\to \gamma\gamma$ is about 1.5×10^{-3} for $m_H \leq 150$ GeV/c^2.

For larger m_H, the decay is almost entirely through the H $\to$ WW*/WW and H $\to$ ZZ*/ZZ channels, although for $m_H > 2m_{top}$ the $t\bar{t}$ branching ratio can be as large as 20%.

A light SM Higgs boson ($m_H < 150$ GeV/c^2), can be searched for in the H $\to \gamma\gamma$, H $\to$ b$\bar{b}$ and H $\to$ ZZ*/WW* decay channels. In this mass range the small natural width, $\Gamma_H \ll 1$ GeV, can be exploited by optimizing the mass resolution in the H $\to \gamma\gamma$ and H $\to$ ZZ*/WW* channels. These channels can be searched for inclusively in the dominant gg $\to$ H production process. A good mass resolution is particulary important for the inclusive search of the H $\to \gamma\gamma$ channel, where the signal can be identified in the di-photon invariant mass distribution as a peak over the large irreducible background from direct pp $\to \gamma\gamma$ + X production. The search for this kind of signature was the main motivation for the choice of LAr (ATLAS) and PbWO$_4$ (CMS) electromagnetic calorimeters with high granularity, uniformity of response, energy resolution less than 1% and excellent γ/π^0 separation. The H $\to \gamma\gamma$ channel can also be searched for in the associated production processes WH and $t\bar{t}$H with an isolated lepton from W $\to \ell\nu_\ell$ and in the H+jet production with a large E_T hadronic jet.

The large H $\to$ b$\bar{b}$ branching ratio for $m_H <$ 130 GeV/c^2 can be exploited only in the associated production channels $t\bar{t}$H and WH due to the huge QCD background and due to the modest ($\sim 11\%$) Higgs boson mass resolution in this decay channel. The most promising channel is the $t\bar{t}$H associated production followed by top quarks decays t $\to$ bW($\to \ell\nu$) ($\ell =$ e, μ) and t $\to$ bW($\to$ q$\bar{q}'$), leading to final states with at least six jets, four of them coming from b quark hadronization and decay, and one isolated lepton used to trigger the event.

The four-lepton final state from H $\to$ ZZ*/ZZ $\to$ $\ell^+\ell^-\ell'^+\ell'^-$ ($\ell, \ell' =$ e, μ) provides a clearly signature over a wide mass range, from $m_H \sim$ 130 GeV/c^2 to about 600 GeV/c^2. Backgrounds from the ZZ, $t\bar{t}$ and Zb$\bar{b}$ can be efficiently suppressed by requiring the four leptons to be isolated in the tracker and coming from a common vertex and by cutting on the di-lepton and four-lepton invariant masses. Above 200 GeV/c^2, the H $\to$ ZZ $\to \ell^+\ell^-\ell'^+\ell'^-$ channel yields the highest sensitivity up to 600 GeV/c^2. If the Higgs boson mass is around 150 GeV/c^2 and above 200 GeV/c^2, the luminosity required to claim the discovery is only 10 fb^{-1}. For the cleanest experimental signature this decay channel is called the "golden" channel at LHC. Above 600 GeV/c^2, this channel is no longer suited for discovery, because the natural width of the Higgs boson is large and the inclusive production cross section of the Higgs boson decreases. The final states with larger branching ratios including jets and E_T^{miss} from the H $\to$ ZZ/WW decays are thus preferred. However a clear mass peak in the Higgs boson reconstructed mass is missing, due to both the large background and the large Higgs boson width.

Around $m_H \sim$ 170 GeV/c^2, where the H $\to$ ZZ$^{(*)}$ branching ratio is smallest, the H $\to$ WW*/WW $\to \ell^+\ell'^-\nu_\ell\bar{\nu}_{\ell'}$ is the main decay channel to be exploited. The backgrounds from the WW, $t\bar{t}$ and Wt can be suppressed by taking advantage of WW spin correlations, which turn into small opening angles between the two charged leptons for the signal. Central-jet vetoing further suppresses the $t\bar{t}$ background. As the Higgs boson mass reconstruction is not possible in this channel, a careful understanding of systematic uncertainties in the background estimation is required.

The Higgs boson production in the vector boson fusion (VBF), qq $\to$ qqH, has been shown to be important for the Higgs boson searches at LHC, in particular in the regions outside the reach of the four-lepton channel for $m_H <$ 130 GeV/c^2 and $m_H >$ 500 GeV/c^2 [10]. In the VBF process the Higgs boson is produced together with two energetic jets in the forward/backward regions of the detector, originating from the final state quarks. In addition, the absence of colour exchange between the scattered quarks leads to a low jet activity in the central rapidity region when the Higgs boson decays into nonhadronic modes of $\gamma\gamma$, WW* and $\tau\tau$. Therefore, the $t\bar{t}$ and the various QCD background processes can be largely reduced with the requirement of a very little jet activity in the central rapidity region and the detection of two energetic jets at large rapidities, with a substantial rapidity gap between the jets. For $m_H >$ 300 GeV/c^2, the

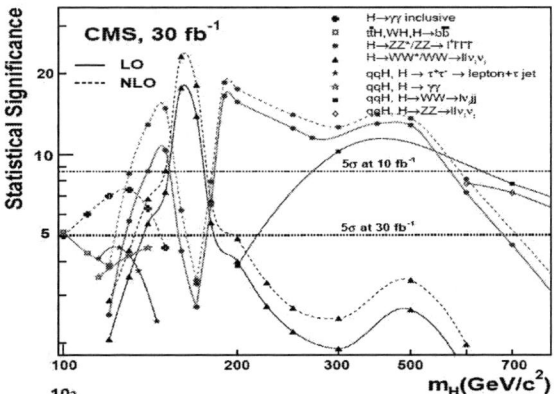

FIGURE 3. Expected statistical significance for the SM Higgs boson as a function of m_H for 30 fb^{-1} with the CMS detector. Results of NLO analysis with k factors for both signal and background are shown for the inclusive $H \to \gamma\gamma$, $H \to ZZ^*/ZZ \to \ell^+\ell^-\ell'^+\ell'^-$, $H \to WW^*/WW \to \ell^+\ell'^-\nu_\ell\bar\nu_{\ell'}$, channels.

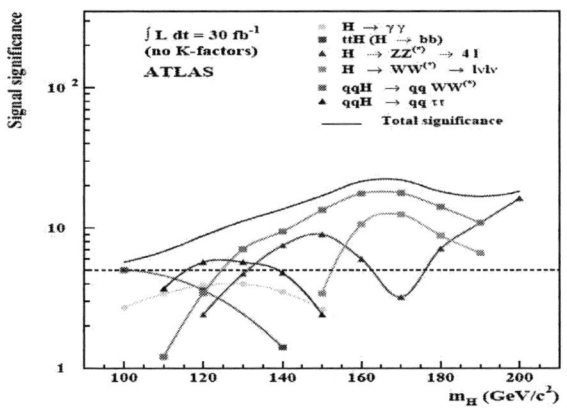

FIGURE 4. Expected statistical significance for the SM Higgs boson as a function of m_H for $m_H <$ 200 GeV/c^2 for 30 fb^{-1} with the ATLAS detector.

Higgs boson discovery potential is improved exploiting the large cross section of the VBF production and the $H \to WW/ZZ$ decay modes in final state topologies containing leptons, jets and E_T^{miss} with large branching ratios.

The statistical significance for the SM Higgs boson search with 30 fb^{-1} is shown both for the CMS experiment, in the mass range of $100 \leq m_H \leq 800$ GeV/c^2 (Fig. 3), and for the ATLAS experiment, for $m_H \leq$ 200 GeV/c^2 (Fig. 4). Combining all the discovery channels, the first 2-3 years of data taking should be enough for claim a discovery over the whole mass spectrum. In the mass range 130 GeV/$c^2 \leq m_H \leq$ 500 GeV/c^2 a discovery is possible in the $H \to WW^*/WW$ and $H \to ZZ^*/ZZ$ decay channels already with an integrated luminosity of 10 fb^{-1} or less.

SEARCH FOR SUPERSYMMETRIC HIGGS BOSONS

The MSSM contains five Higgs bosons: the lighter scalar h, the heavier scalar H, the pseudoscalar A and the two charged bosons H$^\pm$. The MSSM parameter space is in general presented as a function of the pseudoscalar mass m_A and the ratio tanβ of the vacuum expectation values of the two Higgs doublets. In most of the LHC studies, the remaining SUSY parameters are fixed to the values used in the LEP studies [11]. At tree level the h(H) mass is bound to be below(above) the Z boson mass but the radiative corrections, proportional to m_{top}^4, bring the upper (lower) bound to a significantly larger value.

The lighter scalar Higgs boson h behaves like the SM Higgs boson for $m_A > m_h^{max}$ (decoupling region), with production cross sections and decay partial widths close to those of the SM Higgs boson.

At large tanβ, the couplings of the heavy neutral Higgs bosons to the electroweak gauge bosons are strongly suppressed, while those to the down-type fermions are enhanced with tanβ. The production of the heavy neutral MSSM Higgs bosons, H and A, proceeds mainly through $gg \to$ H/A and $gg/q\bar q \to b\bar b$H/A. At large tanβ, the $b\bar b$H/A associated production dominates and is about 90% of the total rate for tan$\beta >$ 10 and $m_A >$ 300 GeV/c^2. The heavy scalar Higgs boson is SM-like near its lower mass bound (i.e. at small m_A and large tanβ) and has in this region a significant cross section in the vector boson fusion process $qq \to qq$H. Light charged Higgs bosons, $m_{H^\pm} < m_{top}$, are predominantly produced in t$\bar{\text{t}}$ events, with a t $\to$ H$^\pm$b decay. Several processes, like $gb \to$ tH$^\pm$, $gg \to$ t$\bar{\text{b}}$H$^\pm$, $q\bar q' \to$ H$^\pm$, $gg \to$ H$^+$H$^-$ and $gg \to$ W$^\pm$H$^\mp$ contribute to the production of heavier ($m_{H^\pm} > m_{top}$) charged Higgs bosons.

The suppression of the couplings to gauge bosons implies different search strategies for H/A bosons. At large tanβ, the coupling enhancement to down-type fermions allows the H/A bosons to be searched for in the H/A $\to \tau^+\tau^-$ and H/A $\to \mu^+\mu^-$ decay channels in the associated production $gg \to b\bar b$H/A. In this production process, the tagging of the associated b jets can be used to suppress the Z, γ^* and QCD multi-jet backgrounds. To take advantage of the hadronic decays in the lepton-plus-τ-jet and two-τ-jet final states from the H/A $\to \tau^+\tau^-$ decay, an efficient hadronic τ trigger and τ-jet identification method is required to suppress further QCD multi-jet and W+jet backgrounds. The Higgs boson mass can be reconstructed in the H/A $\to \tau^+\tau^-$ channels from the E_T^{miss} and the visible τ momenta exploiting the neutrino collinearity with the parent direction. The H/A search

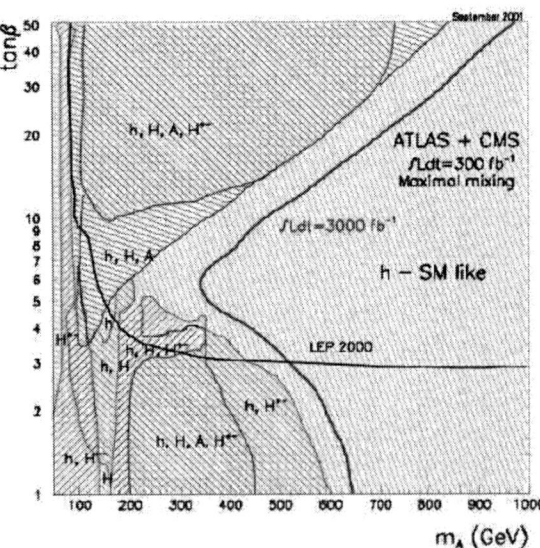

FIGURE 5. Discovery potential for the MSSM Higgs bosons in the (m_A; tanβ) plane, combining the results of ATLAS and CMS experiments for 300 fb^{-1}. The exclusion limit given by LEP and the LHC result with 3000 fb^{-1} are also shown.

through H/A $\rightarrow \mu^+\mu^-$ decay channels proceeds via the identification of a peak in the reconstructed di-muon invariant mass distribution, over the dominant background events $Z/\gamma^* \rightarrow \mu^+\mu^-$. As the two-muon mass resolution of 1% is not good enough to resolve the mass difference between the A and H boson masses, the signal peak results as a superposition of the A and H decays.

The light charged Higgs bosons ($m_{H^\pm} < m_{\text{top}}$) decay to $\tau\nu_\tau$ with an almost 100% branching ratio. For $m_{H^\pm} > 200$ GeV/c^2 the $H^\pm \rightarrow$ tb decay dominates at large tanβ while the $H^\pm \rightarrow \tau\nu_\tau$ branching ratio decreases and is about 10% for $m_{H^\pm} > 400$ GeV/c^2. The branching ratio for the $H^\pm \rightarrow$ Wh decay mode may reach $\sim 10\%$ at small tanβ around $m_{H^\pm} \sim 200$ GeV/c^2. The light charged Higgs bosons ($m_{H^\pm} < m_{\text{top}}$) can be searched for in t$\bar{\text{t}}$ events exploiting the $H^\pm \rightarrow \tau\nu_\tau$ decay channel with hadronic τ decays. The $H^\pm$ signature is characterized by an excess of τ's in t$\bar{\text{t}}$ events relative to electron and muon rates. To search for the heavy charged Higgs bosons ($m_{H^\pm} > m_{\text{top}}$) the $H^\pm \rightarrow \tau\nu_\tau$ and $H^\pm \rightarrow$ tb decay channels can be used in the associated production process gg $\rightarrow$ t$\bar{\text{b}}$H$^\pm$ and also in the direct production $q\bar{q}' \rightarrow H^\pm$. The t$\bar{\text{t}}$ and Wt backgrounds with genuine τ's can be suppressed exploiting the different helicity correlations in the $H^\pm \rightarrow \tau\nu_\tau$ and the $W^\pm \rightarrow \tau\nu_\tau$ decays.

Figure 5 summarizes the discovery potential for the MSSM Higgs bosons as a function of m_A and tanβ, combining the ATLAS and CMS results at the integrated luminosity of 300 fb^{-1}. In most of the parameter space at least two Higgs bosons can be revealed. For high values of m_A the only detectable Higgs boson is the SM-like lighter CP-even h.

CONCLUSIONS

The present understanding of the LHC potential for the SM and MSSM Higgs boson discovery has been presented.

For the SM Higgs boson, when all channels are combined, a discovery is possible for $m_H \geq 114$ GeV/c^2 already with an integrated luminosity of 30 fb^{-1} or less.

The distinctive signatures of VBF significantly enhance the discovery significance for low and medium m_H during the initial low-luminosity phase of LHC.

In the MSSM, the (m_A; tanβ) plane will be almost completely explored with an integrated luminosity of 300 fb^{-1}.

Some modifications to these results can be expected since new systematic studies are ongoing, using more sophisticated analysis methods and more detailed simulation of the detectors.

REFERENCES

1. The LHC Study Group, "The Large Hadron Collider, Conceptual Design", CERN/AC/95-05.
2. ATLAS Collaboration, CERN/LHCC-94-43, LHCC/P2, "ATLAS: A Toroidal LHC ApparatuS Technical Proposal".
3. CMS Collaboration, CERN/LHCC 94-38, LHCC/P1, "The Compact Muon Solenoid Technical Proposal".
4. ATLAS Collaboration, "ATLAS Detector and Physics Performance, Technical Design Report", ATLAS TDR 14, CERN/LHCC 99-14 and ATLAS TDR 15, CERN/LHCC 99-15.
5. A. Abdullin et al., "Summary of the CMS Potential for the Higgs Boson Discovery", CMS NOTE 2003/033 and referencies therein.
6. LEP Higgs Working Group, Phys. Lett. **B565**(2003) 61.
7. The LEP ElectroweakWorking Group, "Status of Summer 2005", http://lepewwg.web.cern.ch/LEPEWWG/.
8. R. Rainwater, M. Spira and D. Zeppenfeld, "Higgs Boson Production at Hadron Colliders", hep-ph/0203187.
9. A. Djouadi, J. Kalinowski and M. Spira, "HDECAY: a program for Higgs Boson Decays in the Standard Model and its Supersymmetric Extension", hep-ph/9704448.
10. D. Zeppenfeld, Int. J. Mod. Phys. **A16** (2001) 831; T. Plehn, D. Rainwater, D. Zeppenfeld, Phys. Rev. Lett. **88** (2002) 051801; Phys. Lett. **B454** (1999) 297; M.L. Mangano et. al., CERN-TH/2002-287.
11. M. Carena, S. Heynemeyer, C.E.M.Wagner and G.Weiglein, "Suggestions for improved benchmark scenarios for Higgs-boson searches at LEP2", CERN-TH/99-374, DESY 99-186, hep-ph/9912223.
12. LEP Higgs Working Group "Searches for the Neutral Higgs Bosons of the MSSM", hep-ex/0107030.

Search for New Physics at ATLAS

Gianluca Comune

Michigan State University, East Lansing, Michigan 48824, USA

Abstract. Several extensions of the Standard Model, other than Supersymmetry, have been proposed. There is a very large variety of such theories and they all aim, more or less successfully, at resolving some of the outstanding theoretical difficulties like the so called 'hierarchy' or the fine tuning problems that affects the Standard Model. We present a brief and by far not exhaustive review of some of the most recent non Supersymmetry studies carried out within the ATLAS collaboration assessing how the ATLAS detector will be able to probe the existence of these theories up to the multi TeV energy scale when the Large Hadron Collider (LHC) will start its operation at CERN.

Keywords: ATLAS Extra Dimensions Heavy Leptons Left Right Symmetric
PACS: 01.30.Cc

INTRODUCTION

A large variety of extensions of the currently accepted formulation of the Standard Model (SM) have been proposed in the last two decades. The most widely accepted is the Supersymmetric extension of the Standard Model that will be covered elsewhere [1]. We shall present a brief selection of investigations recently carried out within the ATLAS collaboration regarding theories contemplating the existence of extra spatial dimensions, leptoquarks, Left-Right symmetry and heavy leptons. All those extension aim at resolving some or all of the theoretical difficulties affecting the current formulation of the SM. All result shown have been obtained using ATL-FAST [2], a fast parameterized simulation of the ATLAS detector response.

EXTRA DIMENSIONS

Large Extra Dimensions

By postulating the existence of n extra spatial dimensions compactified on a circle [3, 4] and allowing the Standard Model fields to propagate only in the usual four dimensional space while leaving the gravity free to propagate in the higher dimensional space we obtain a whole tower of gravitons à la Kaluza-Klein (KK). Depending on the actual number, compactification mechanism and extension of the additional dimensions, the first few KK modes should be accessible to experimental probe at AT-LAS. Under this assumptions a new fundamental mass scale (M_S) appears and is related to the Planck scale by the relation

$$M_{Pl}^2 \sim M_S^{n+2} R^n \qquad (1)$$

where n is the number of extra dimensions and R is their compactification radius. Both direct production [5] and virtual graviton exchange [6] have been investigated. The study has confirmed that ATLAS should be able to see deviations from the SM for masses of the graviton of up to 9.1 (7) TeV with two (three) additional extra dimensions and an integrated luminosity of 100 fb^{-1}.

Direct graviton production ($gg \to gG^{(k)}, gq \to qG^{(k)}, q\bar{q} \to gG^{(k)}$ and $q\bar{q} \to \gamma G^{(k)}$) leads to missing energy and hard jet signature and is expected to be visible as an excess of missing E_T at large values. Discriminating this excess from the SM background implies a throughout understanding and modelling of the underlying backgrounds at very high missing E_T values. An example of direct KK graviton production signal can be found in the diphoton angular distribution (Fig. 1) and it manifests as an excess at small η as compared to the SM predictions

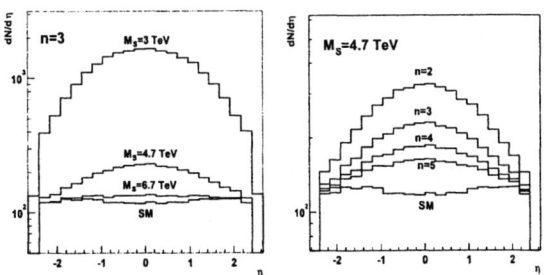

FIGURE 1. Rapidity distribution of signal and background for the diphoton production: left) for $n = 3$ and various values of M_S; right) for $M_S = 4.7$ TeV and various n.

TeV^{-1} Scale Extra Dimensions

The SM fields may not be confined to the space-time dimensions if some of the extra dimensions have the size in the order of TeV^{-1}.

So called M1 [7] and M2 [8] models have been investigated [9] showing that with the integrated luminosity of 100 fb^{-1} it should be possible to see the first KK state peak in the dilepton invariant mass distribution up to values of the compactification scale $M(Z^{(1)}/\gamma^{(1)}) = 5.8\ TeV$. In Fig. 2 it is shown this peak in the dielectron invariant mass distribution. The direct detection of the peak from second Z and γ resonances is impossible but a deviation of the invariant mass distribution at relatively low values would allow to set the discovery reach at ~ 8 TeV for 100 fb^{-1} and at ~ 10.5 TeV for 300 fb^{-1}.

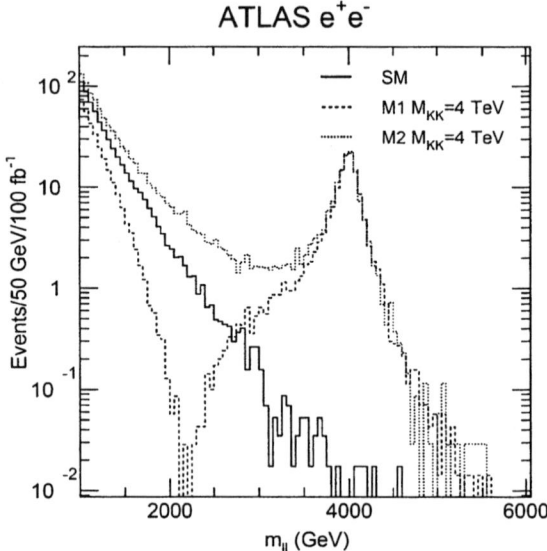

FIGURE 2. Distribution of the lepton-lepton invariant mass for electrons (full line) and muons (dashed line). The distribution assumes 4 TeV for the mass of the lowest lying KK excitation and an integrated luminosity of 100 fb^{-1}

Small warped extra dimensions

In the scenario where gravity propagates in a 5-dimensional space bounded by a 3+1 dimensional space [10, 11] the SM fields are confined on one of the smaller dimensional spaces. This assumption leads to a spectrum of KK gravitons with masses

$$m_n = x_n \Lambda_\pi k / M_{Pl} \qquad (2)$$

where x_n are the roots of the first-order Bessel functions and they lead to the fact that the graviton states are not equally spaced allowing the discrimination of this type of theories. The easiest signature to study at LHC is the dilepton or diphoton final state. The dielectron final state has been studied [12] and the dielectron invariant mass (Fig. 3) clearly shows a peak at the assumed mass of the graviton (1.5 TeV).

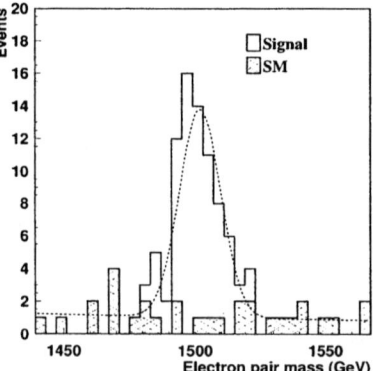

FIGURE 3. Graviton resonance, with $m_G = 1.5\ TeV$ (signal), superimposed on the expected Standard Model background (100 fb^{-1})

LEPTOQUARKS

Leptoquarks (LQs) are an interesting category of particles generally predicted by several theories [14, 15, 16, 17, 18]. Leptoquarks couple to quarks and leptons via a Yukawa potential conserving baryion and lepton numbers. The production mechanisms at hadron colliders include $q\bar{q}$ annihilation and gluon-gluon fusion (Fig. 4). Relaxing the assumption of chiral couplings, the number

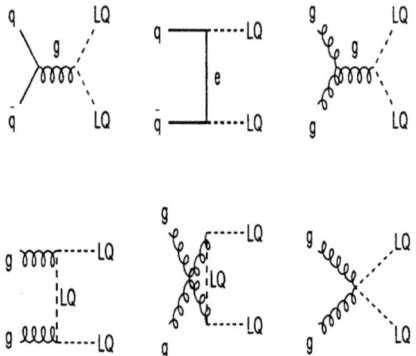

FIGURE 4. Leptoquark production mechanisms at LHC

of LQs stands at ten, the LQs differing by the fractional electric charge ($\pm 5/3, \pm 4/3, \pm 2/3$ and $\pm 1/3$). They decay into a neutral or charged lepton and a quark via the processes

$$LQ \to \ell q; LQ \to \nu q \qquad (3)$$

The characteristic signature is thus [13] the one where the two LQs undergo a decay into charged lepton plus jet leading to the two lepton plus two jets signal used in the analisys. The m_{ej} distribution for two LQ masses is depitced in Fig. 5. The same analysis carried out or muons shows just a slight deterioration but the peak are still clearly visible. The study shows that LQ of mass up to $\sim$ 1.3 TeV should be observed.

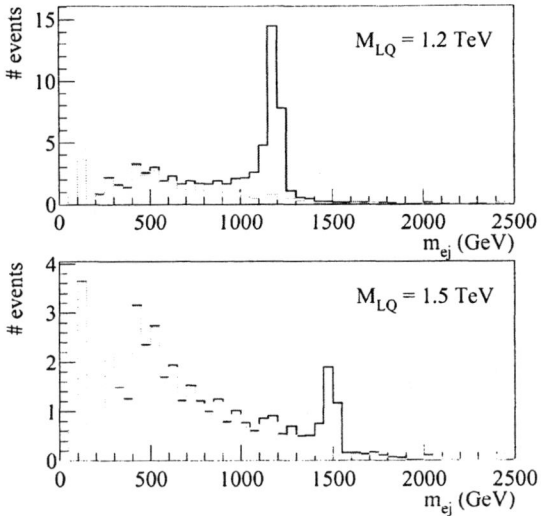

FIGURE 5. m_{ej} distributions for background (light line) and signal plus background (dark line), for various values of LQ mass and for an integrated luminosity of 30 fb^{-1}.

LEFT-RIGTH SYMMETRIC THEORIES

The fact that the weak interactions in the SM are strictly left handed seems rather arbitrary. One can drop this requirement and allow the SM to be invariant under transformations of a larger Left-Right symmetric group. The simplest realization of this idea comes under the name of Left-Right Symmetric Model (LRSM) [19]. The breaking of this symmetry happens at high scale through three complex Higgs fields Δ_R^0, Δ_R^+ and Δ_R^{++}.

The detectability of the Z' [20], of the W_R [21] and of the Z_R [22, 23, 24] has been studied in the past. More recently the complementary study of the doubly charged Higgs decays in the context of LRSM [25] has been carried out. Single production of a doubly charged Higgs at the LHC is possible via vector boson fusion, or via the fusion of a singly charged Higgs with either a W or another singly charged Higgs.

The ATLAS discovery reach for the dilepton decay channel of the Δ_R^{++} is plotted in Fig. 7.

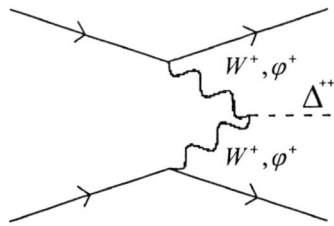

FIGURE 6. Doubly charged Higgs production mechanism at LHC

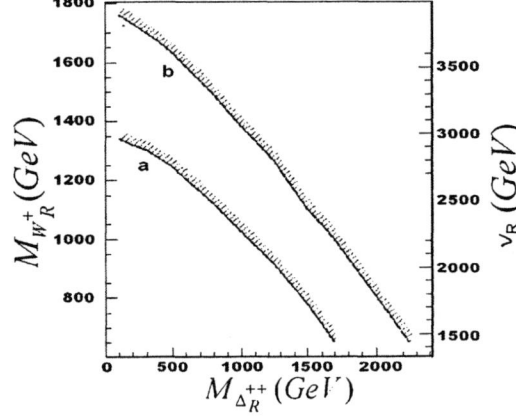

FIGURE 7. Discovery reach for $\Delta_R^{++} \to \ell^+\ell^+$ versus $M_{\Delta_R^{++}}$ (or v_R) for integrated luminosities of 100 fb^{-1}(a) and 300 fb^{-1}(b), and assuming 100% BR to dileptons.

HEAVY LEPTONS

Several extended gauge groups contain the SM $SU_C(3) \times SU_L(2) \times U_Y(1)$ as a subgroup. Those extended groups generally predict the existence of new fermions. LRSM is a good example together with technicolor models [33], composite models [34], grand unified theories [35] and several others. Alongside the new heavy leptons many of those theories include new Z' and W' gauge bosons. Study of the Drell-Yan production of heavy leptons (L) via $q\bar{q}$ annihilation and gluon-gluon fusion with subsequent decay $L \to \ell Z^0$ has been carried out [36] focusing on the 2 leptons + 4 jets and on 6 leptons final states. The mas peak for an assumed $M_L = 0.5\ TeV$ is shown in Fig. 8 while the signal significance for this decay channel is in Fig. 9.

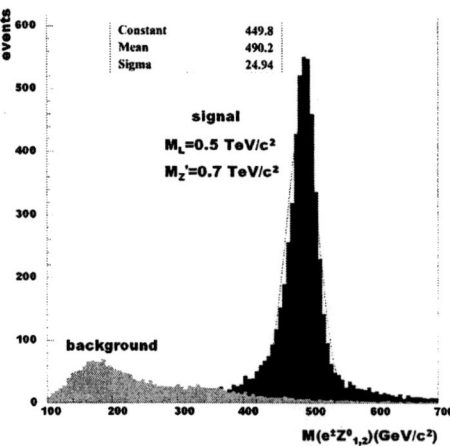

FIGURE 8. Signal to background comparison for $M_L = 0.5\ TeV/c^2$ and $M_{Z'} = 0.7\ TeV/c^2$, for $L \to e + Z^0$

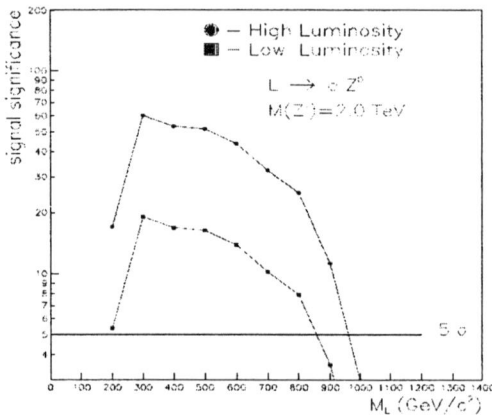

FIGURE 9. Signal significance M_L dependence for high($300 fb^{-1}$) and low ($30 fb^{-1}$) luminosity for $L^{\pm} \to e^{\pm} + Z^0$ decay channel and for $M_{Z'} = 2.0\ TeV$.

CONCLUSIONS

In this report we have shown that if new physics exists, in the selected theoretical frameworks we have considered, it shall be possible to detect it at ATLAS in a large portion of those theories' parameter space.

ACKNOWLEDGMENTS

This paper is a brief summary of the recent activities on searches for new physics at ATLAS. As a result of this the author would like to thank directly the authors of the selected papers.

REFERENCES

1. T. Lari, "SUSY Searches at ATLAS", proceedings of "Incontri di Fisica delle alte Energie", IFAE 2005 Catania.
2. A. Ricther-Was, D. Froidveaux, L. Poggioli, ATLAS internal note PHYS-1996-079.
3. N. Arkani-Hamed, S. Dimopoulos and G. R. Dvali, Phys. Lett. **B429** (1998) 263.
4. I. Antoniadis et al., Phys. Lett. textbf B436 (1998) 257.
5. L. Vacavant and I. Hinchliffe, J. Phys. **G27** (2001) 1839.
6. V. Kabachenko, A. Miagkovy and A. Zenin, (ATL-PHYS-2001-012).
7. T. G. Rizzo, Phys. Rev. **D61** (2000) 055005.
8. N. Arkani-Hamed and M. Schmaltz, Phys. Rev. **D61** (2000) 033005.
9. G. Azuelos, G. Polesello, SN-ATLAS-2003-023.
10. L. Randall and R. Sundrum, Phys. Rev. Lett. **83** (1999) 3370.
11. L. Randall and R. Sundrum, Phys. Rev. Lett. **83** (1999) 4690.
12. B. C. Allanach et al., JHEP 0212 (2002) 039.
13. V. A. Mitsou, N. Ch. Benekos, I. Panagoulias, Th. D. Papadopoulou, hep-ph/0411189
14. J. C. Pati and A. Salam, Phys. Rev. **D10** (1974) 275; H. Georgi and S. L. Glashow, Phys. Rev. Lett. **32** (1974) 438.
15. B. Schrempp and F. Schrempp, Phys. Lett. **B153** (1985) 101; W. Buchmuller, Acta Phys. Austriaca Suppl. **27** (1985) 517; W. Buchmuller and D. Wyler, Phys. Lett. **B177** (1986) 377.
16. S. Dimopoulos and L. Susskind, Nucl. Phys. **B155** (1979) 237; S. Dimopoulos, Nucl. Phys. **B168** (1980) 69; E. Eichten and K. Lane, Phys. Lett. **B90** (1980) 125.
17. V. D. Angelopoulos et al., Nucl. Phys. **B292** (1987) 59; J. L. Hewett and T. G. Rizzo, Phys. Rept. **183** (1989) 193.
18. A. F. Zarnecki, Eur. Phys. J. C **17** (2000) 695 [arXiv:hep-ph/0003271].
19. R.N. Mohapatra and J. C. Pati, Phys. Rev. **D11** (1975) 566
20. ATLAS Detector and Physics Performance Technical Design Report, CERN/LHCC/99-15, May 1999.
21. J. Collot and A. Ferrari, ATL-PHYS-99-018 (also PhD Thesis of A. Ferrari).
22. A. Ferrari and J. Collot, ATL-PHYS-2000-034.
23. D. Benchekroun, C. Driouichi and A. Hoummada, SN-ATLAS-2001-001 published in EPJdirect C3, 1 (2001) DOI 10.1007/s1010501c0003.
24. E. Arik et al., ATL-PHYS-2001-005.
25. G. Azuelos, K. Benslama and J. Ferland, hep-ph/0503096.
26. J. L. Hewett, T. G. Rizzo, Phys. Rep. **183** (1989) 193.
27. L. Djouadi, J. Ng, T. G. Rizzo, in Electroweak Symmetry Breaking and New Physics at the TeV scale, World Scientific, Singapore (1997).
28. M. Davier, Rapporteur talk at the Lepton-Photon Conference, Geneva 1991, Preprint LAL-91-48 (1991).
29. J. Maalampi, M. Roos, Phys. Rep. **186**, 53 (1990).
30. J. Montvay, Phys. Lett. **B205**, 315 (1988).
31. W. Buchmuller, C. Breub, Nucl. Phys. **B363** 345 (1991).
32. W. Robinett, J. L. Rosner, Phys. Rev. **D55**, 5263 (1997).
33. S. Dimopoulos, Nucl. Phys. **B168** (1981) 69.
34. E. J. Eichten, K. D. Lane, M. E. Peskin, Phys. Rev. Lett. **50** (1983) 811.
35. P. Langacker, Phys. Rep. **72** (1981) 185.
36. C. Alexa, ATL-PHYS-CONF-2005-001.

Sudakov Expansions and Top Quark Physics at LHC

Matteo Beccaria

Dipartimento di Fisica and INFN, Via Arnesano Ex Coll. Fiorini, 73100 Lecce, ITALY

Abstract. We review some peculiar features of Sudakov expansions in the calculation of electroweak radiative corrections in the MSSM at high energy. We give specific examples and consider in particular the process $bg \to tW$ of single top quark production relevant for the top quark physics programme at LHC.

Keywords: Minimal Supersymmetric Standard Model, Electroweak Radiative Corrections
PACS: 12.15.Lk,12.60.Jv,13.75.Cs

INTRODUCTION

The analysis of physical processes in the MSSM is complicated by the large number of parameters of the model. Also, radiative correction typically involve a large subset of them making apparently quite difficult to isolate specific physical effects. However, there is a way out to simplify the situation at the price of some controlled approximation. Here we shall be concerned with the electroweak sector of the MSSM. The planned future collider experiments (LHC, ILC) will be characterized by large invariant masses and at very high energy it is known that radiative corrections simplify and become smooth, at least beyond production thresholds. In this regime they can be described by rather simple asymptotic expansions. These are basically power series in the logarithm of the center of mass energy or typical large invariant mass. This logarithmic approximation is called a Sudakov expansion and requires light SUSY scenarios with not too heavy sparticles in order to be accurate [1]. Apart from this basic requirement, it compactly expresses the relevant radiative corrections with a very minimal set of parameters [2]. The deep reason for this important feature is that at high energy soft SUSY breaking operators are suppressed for dimensional reasons and disappear at the leading orders in the expansion.

The origin of large logarithms in the physical amplitudes is nowadays quite clear [3]. There are two different kinds of contributions coming respectively from short and long distance and therefore of ultraviolet or infrared nature. The UV logarithms are governed by standard Renormalization Group equations. They typically arise in gauge boson self energy corrections or in essentially gaugeless sectors of the MSSM, like the important Yukawa one. The IR logarithms are manifestations of mass singularities. They can arise in diagrams with exchange of gauge bosons. In the calculation of exclusive processes the mass of massive gauge bosons plays the role of an infrared regulator and gives rise to energy growing contributions of logarithmic type. For instance, at one loop the leading terms are proportional to $\alpha \log^2(s/M^2)$ where $\alpha = e^2/4\pi$, $\sqrt{s}$ is the center of mass energy and M is a process dependent scale. The unusual counting of two logs per loop is a typical remnant of the IR origin of these contributions.

In the next Section, we shall discuss briefly the classification of the Sudakov logarithms. This will provide the necessary tools to discuss the corrections to single top quark production processes at LHC, as an interesting application.

SOME CLASSIFICATION OF SUDAKOV LOGARITHMS

The electroweak Sudakov logarithms in a $2 \to 2$ process at one loop can be classified according to the following three main categories. We closely follow the notation of [3] and refer to this paper for minor or notational details concerning the next paragraphs.

a) UV Logarithms from gauge boson self energies. If F^{Born} is the amplitude under study, at one loop we get from coupling constants renormalization, the following correction term

$$F^{RG} = -\frac{1}{4\pi^2}\left(g^4 \beta \frac{\partial F^{Born}}{\partial g^2} + g'^4 \beta' \frac{\partial F^{Born}}{\partial g'^2}\right) \log \frac{s}{\mu^2}$$

and of course β, β' depend on the model (SM, MSSM, Split SUSY, etc.)

b) Universal Logarithms. These are terms independent on the scattering angles. They can be computed by analyzing the various external lines associated to initial of final states one after the other. Each line gets a correction factor that we now discuss. The general form of the correction is (p = scalar, fermion or vector)

$$\frac{\alpha}{\pi} c_p^{\text{gauge}} \left(n \log \frac{s}{M_V^2} - \log^2 \frac{s}{M_V^2} \right) + \frac{\alpha}{\pi} c_p^{\text{Yukawa}} \log \frac{s}{M'^2}$$

where $V = \gamma, Z, W^\pm$. The linear logarithm scale is a choice to be fixed at NNLO.

In each model and for each kind of external line (initial of final) we can list the various coefficients (n, c_p, c^{Yukawa}) and write immediately the universal corrections at one loop and next to leading logarithmic order.

c) Angular dependent logarithms. The final type of logarithms is called angular dependent since depends on the scattering angle. These are terms of the form

$$\frac{\alpha}{\pi} \log \frac{s}{M^2} \log \frac{1 \pm \cos \vartheta}{2},$$

where ϑ is the scattering angle. These contributions arise from Standard Model box diagrams and do not receive SUSY additional terms.

An example: correction to fermion or sfermion external lines

To give an explicit example of the above mentioned corrections, we now consider the specific case of external lines associated to a fermion (lepton or quark) or a sfermion (slepton or squark). The universal logarithms are

$$\frac{\alpha}{\pi} c^{\text{gauge}} \left(n \log \frac{s}{M_V^2} - \log^2 \frac{s}{M_V^2} \right) + \frac{\alpha}{\pi} c^{\text{Yukawa}} \log \frac{s}{m_t^2}$$

where m_t is the top quark mass. The gauge part reflects the SU(2) × U(1) structure. With standard notation, it reads

$$c_f^{\text{gauge}} = \frac{1}{8} \left[\frac{I_f(I_f+1)}{s_W^2} + \frac{Y_f^2}{4c_W^2} \right]$$

$$Y_f = 2(Q_f - I_f^3)$$

with $n = 3$ in the Standard Model and $n = 2$ in the MSSM. The Yukawa part is present only for heavy quarks top and bottom and reads (the upper line refers to the Standard Model, the lower to the MSSM)

$$c_{b_L} = c_{t_L} = \begin{cases} -\frac{1}{32 s_W^2} \left(\frac{m_t^2}{M_W^2} + \frac{m_b^2}{M_W^2} \right) \\ -\frac{1}{16 s_W^2} \left(\frac{m_t^2}{M_W^2} \frac{1}{\sin^2 \beta} + \frac{m_b^2}{M_W^2} \frac{1}{\cos^2 \beta} \right) \end{cases}$$

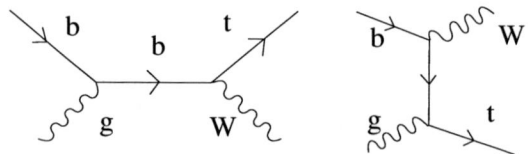

FIGURE 1. $bg \to tW$ at tree level.

$$c_{b_R} = \begin{cases} -\frac{1}{16 s_W^2} \frac{m_b^2}{M_W^2} \\ -\frac{1}{8 s_W^2} \frac{m_b^2}{M_W^2} \frac{1}{\cos^2 \beta} \end{cases} \quad c_{t_R} = \begin{cases} -\frac{1}{16 s_W^2} \frac{m_t^2}{M_W^2} \\ -\frac{1}{8 s_W^2} \frac{m_t^2}{M_W^2} \frac{1}{\sin^2 \beta} \end{cases}$$

The angle β is the common MSSM parameter defined at tree level by $\tan \beta = v_2/v_1$ where $v_{1,2}$ are the vevs of the two Higgs doublets.

For sfermions (sleptons or squarks) the gauge part does not change. Also the Yukawa terms are identical (matching chiralities). This is a non trivial consequence of supersymmetry and of the general fact that breaking tends to be suppressed at high energy.

SINGLE TOP PRODUCTION AT LHC: SUDAKOV CORRECTIONS

Single top quark production at LHC is a very interesting topic for the search of New Physics effects via radiative corrections. It is also remarkable due to its special feature of being a direct measure of $|V_{tb}|^2$. At leading order there are three basic processes that must be considered, i.e. at parton level, the processes $bu \to td$, $u\bar{d} \to t\bar{b}$ and $bg \to tW$. They have quite different features as discussed in details in [4]. Application of the Sudakov technique to the three processes have been discussed in [5]. Here we discuss in some details the associated Wt production $bg \to tW^-$ as a complete application of the previous general discussion.

The two tree level diagrams to be considered are shown in Fig. (1). In the first (a) a bottom quark is exchanged in the s channel; In the second (b) a top quark is exchanged in the $u = (p_b - p_W)^2$ channel. To simplify the discussion let us neglect all masses with the exception of m_t. We neglect the ratio $m_t/\sqrt{s}$, but obviously keep m_t/M_W. Let us denote the helicity amplitude for the process as $F_{\alpha \alpha' \beta \beta'}$ where the four subscripts denote the helicities of the four external states b, g, t, and W. In our approximations we remain with the two non suppressed helicity amplitudes

$$F^{\text{Born a+b}}_{----} \to g_{WL} g_s (\frac{\lambda^l}{2}) \frac{2}{\cos \frac{\vartheta}{2}}$$

$$F^{\text{Born a+b}}_{-+-+} \to g_{WL}\, g_s(\frac{\lambda^l}{2}) 2\cos\frac{\vartheta}{2}$$

$$F^{\text{Born a+b}}_{-++0} \to g_{WL}\, g_s(\frac{\lambda^l}{2})\sqrt{2}\,\frac{m_t}{M_W}\cos\frac{\vartheta}{2}\,\frac{1-\cos\vartheta}{1+\cos\vartheta}$$

The differential cross section is simply (s,t,u are standard Mandelstam variables)

$$\frac{d\sigma^{\text{Born}}}{d\cos\vartheta} \to -\frac{\pi\alpha\alpha_s}{24 s_W^2\, us^2}\left[s^2+u^2+\frac{m_t^2 t^2}{2M_W^2}\right]$$

We can now write in a very simple way the Sudakov electroweak corrections to these two non suppressed amplitudes.

The universal component for producing a transverse W is

$$\frac{F^{\text{Univ}}_{-,\mu,-,\mu}}{F^{\text{Born}}_{-,\mu,-,\mu}} = \frac{1}{2}\left(c^{\text{ew}}(b\bar{b})_L + c^{\text{ew}}(t\bar{t})_L\right) + c^{\text{ew}}(W_T)$$

where $c^{\text{ew}}(f\bar{f})$ is the sum of the gauge and Yukawa correction associated to a f fermion line. We have also introduced the new coefficient

$$c^{\text{ew}}(W_T) = \frac{\alpha}{4\pi s_W^2}\left(-\log^2\frac{s}{M_W^2}\right)$$

In the production of longitudinal W_0^- we have instead

$$\frac{F^{\text{Univ}}_{-,+,+,0}}{F^{\text{Born}}_{-,+,+,0}} = \frac{1}{2}\left(c^{\text{ew}}(b\bar{b})_L + c^{\text{ew}}(t\bar{t})_R\right) + c^{\text{ew}}(W_0)$$

where

$$c^{\text{ew}}(W_0) = \frac{\alpha}{\pi}(\frac{1+2c_W^2}{32 s_W^2 c_W^2})\left(n_G\log\frac{s}{M_W^2}-\log^2\frac{s}{M_W^2}\right)$$

and $n_G = 4, 0$ in the Standard Model or MSSM, respectively.

For reasons of space, we do not write the explicit angular dependent terms. Finally there are SUSY QCD Sudakov logarithms from vertices with gluino exchanges. They read

$$F^{\text{Univ SUSYQCD}}_{-,\mu,-,\mu} = F^{\text{Born}}_{-,\mu,-,\mu}\left(-\frac{\alpha_s}{3\pi}\log\frac{s}{M_{\text{SUSY}}^2}\right)$$

$$F^{\text{Univ SUSYQCD}}_{-,+,+,0} = F^{\text{Born}}_{-,+,+,0}\left(-\frac{\alpha_s}{3\pi}\log\frac{s}{M_{\text{SUSY}}^2}\right)$$

In this specific process there are no RG logarithms.

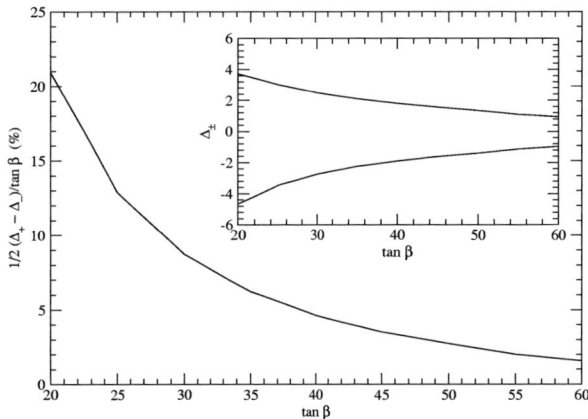

FIGURE 2. Bounds on $\tan\beta$ from the combination of the three single top production processes.

NUMERICAL RESULTS

The same set of corrections can be computed for the other three processes of single top production. The partonic cross sections are then converted in hadronic observables. In particular, one defines the distribution

$$\frac{d\sigma(PP\to tY+...)}{ds} = \frac{1}{S}\int_{-\cos\overline{\vartheta}}^{\cos\overline{\vartheta}}d\cos\vartheta\sum_{ij}L_{ij}(\tau,\cos\vartheta)\frac{d\sigma_{ij\to tY}}{d\cos\vartheta}(s)\,]$$

where $\sqrt{S}$ is the pp energy, L_{ij} is the luminosity from partons i, j and the various kinematical variables are explained in details in [5].

The results for the three single top production processes are summarized in Fig. (3). Large effects can be obtained due to the Yukawa terms. These effects isolate the parameter $\tan\beta$ and can be used to fix bounds on it. To give an example of such a procedure, we have combined the effects in the three production processes assuming measurements in the range $\sqrt{s}=500-1500$ GeV (with 20 GeV spacing) and an overall precision of 10%. Fig. (2) illustrates the results and show the values of $\Delta_\pm$ which appear in the confidence region $(\tan\beta+\Delta_-,\tan\beta+\Delta_+)$ which is determined by a χ^2 analysis of fake data simulated at a certain $\tan\beta$.

In practice, the simple Sudakov approximation displays interesting features and suggest that a full one loop calculation would be certainly worthwhile.

A similar analysis in the case of top - antitop pair production has also been completed and can be found in [6].

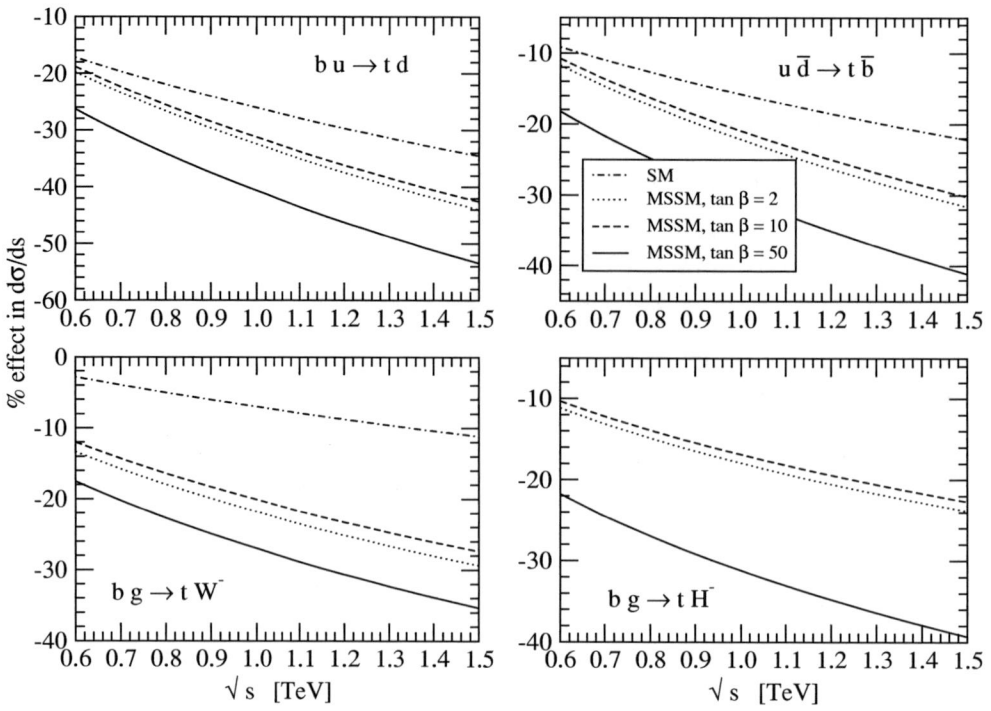

FIGURE 3. Effects in $d\sigma/ds$ for the three single top production processes.

REFERENCES

1. M. Kuroda, G. Moultaka and D. Schildknecht, Nucl. Phys. **B 350**, 25 (1991); A. Denner, S. Dittmaier and R. Schuster, Nucl. Phys. **B 452**, 80 (1995); A. Denner, S. Dittmaier and T. Hahn, Phys. Rev. **D 56**, 117 (1997); A. Denner and T. Hahn, Nucl. Phys. **B 525**, 27 (1998); M. Beccaria, G. Montagna, F. Piccinini, F. M. Renard and C. Verzegnassi, Phys. Rev. **D 58**, 093014 (1998); P. Ciafaloni and D. Comelli, Phys. Lett. **B 446**, 278 (1999).
2. M. Beccaria, S. Prelovsek, F. M. Renard and C. Verzegnassi, Phys. Rev. D **64**, 053016 (2001) [arXiv:hep-ph/0104245]. M. Beccaria, F. M. Renard, S. Trimarchi and C. Verzegnassi, Phys. Rev. D **68**, 035014 (2003) [arXiv:hep-ph/0212167].
3. M. Beccaria, M. Melles, F. M. Renard, S. Trimarchi and C. Verzegnassi, Int. J. Mod. Phys. A **18**, 5069 (2003) [arXiv:hep-ph/0304110].
4. M. Beneke *et al.*, arXiv:hep-ph/0003033.
5. M. Beccaria, F. M. Renard and C. Verzegnassi, Phys. Rev. D **71**, 033005 (2005) [arXiv:hep-ph/0410089].
6. M. Beccaria, S. Bentvelsen, M. Cobal, F. M. Renard and C. Verzegnassi, Phys. Rev. D **71**, 073003 (2005) [arXiv:hep-ph/0412249].

Reconstruction of Sparticle Masses at the LHC

T. Lari

Università di Milano and INFN, Milano, Italy

Abstract. The techniques developed by the ATLAS and CMS collaborations to reconstruct the decays of the supersymmetric particles and measure their masses are presented. The potential of the LHC to reconstruct the Supersymmetry spectrum and the expected resolution on sparticle masses are discussed.

Keywords: Supersymmetry, LHC, mass, focus point
PACS: 11.30.Pb

1. INTRODUCTION

One of the main purposes of the LHC collider is the search for Physics beyond the Standard Model (SM). Supersymmetric (SUSY) extensions of the SM [1] predict the existence of superpartners for all ordinary particles. If Supersymmetry exists at the electroweak scale, it could hardly escape detection at LHC. The center-of-mass energy of 14 TeV, available at the LHC, extends the search for SUSY particles up to masses of 2.5 to 3 TeV/c^2 [2, 3].

In R-parity conserving models, the Lightest SUSY Particle (LSP) is stable and provides a suitable candidate for Dark Matter. The sparticles are produced in pairs and colored states (scalar quarks and gluinos) usually dominates the production cross section at the LHC. Their decay into the LSP, which escapes detection, produce an excess of events with multijets, missing energy, and isolated leptons final states compared to the SM expectations.

In preparation for the LHC startup, simulated data samples are used to develop the analysis strategies. Most studies have been performed in the context of the mSUGRA framework, with five independent parameters: the common gaugino mass $m_{1/2}$, the common scalar mass m_0, the common trilinear couplings A_0, the ratio of the vacuum expectation values of the two Higgs doublets $\tan\beta$ and the sign of the Higgsino mixing parameter μ. These studies [2, 3] show that discovery is going to be relatively easy over most of the parameter space favored by the theory. The determination of the masses of supersymmetric particle is more challenging, because each event contains two LSP, and there are not enough constraints to determine the momenta of these particles. The main goal of this report, is to show the potential of the ATLAS and CMS detectors to reconstruct SUSY particles and the achievable mass resolution.

Detailed studies of this potential have been performed for specific sets of values of the mSUGRA parameters (benchmark points). These were selected taking into account the constraints from experimental data, in particular the LEP experiments, and are also usually located in the (relatively narrow) region of the parameter space where the relic density of the LSP (the lightest neutralino in mSUGRA) matches the cosmological measurements of Dark Matter abundance [4].

In Section 2 the studies performed by ATLAS for the benchmark point SPS1A [5] and those performed by CMS for the point B [6] are reported. They are both located near the "bulk" region of the mSUGRA space with a relic density consistent with Dark Matter abundance. In this region the scalar quarks and gluinos are relatively light and the total SUSY cross at the LHC is relatively large (49.3 pb for point SPS1A and 57.7 pb for point B). In Section 3 some studies performed by ATLAS for a more difficult point in the "focus point" region [7] are reported. The mSUGRA parameters and the total SUSY cross section for the three benchmarks are reported in Table 1.

2. SUSY PARTICLE MEASUREMENTS IN THE BULK REGION OF MSUGRA

2.1. The dilepton kinematic edges

A particularly interesting decay chain, present for both the SPS1A and point B benchmarks, is

$$\tilde{q}_L \to q\chi_2^0 \to q l^\pm \tilde{l}_R \to q l^\pm l^\mp \chi_1^0 \quad (1)$$

The two leptons in the final state provide a natural trigger, and the energy resolution is high. The invariant mass of the two leptons has a kinematical maximum at

$$(M_{ll}^2)\text{edge} = \frac{(m_{\chi_2^0}^2 - m_{\tilde{l}_R}^2)(m_{\tilde{l}_R}^2 - m_{\chi_1^0}^2)}{m_{\tilde{l}_R}^2} \quad (2)$$

TABLE 1. mSUGRA parameters and SUSY production cross section at the LHC, for the three benchmark points discussed in this report.

point	m_0 (GeV)	$m_{1/2}$ (GeV)	A (GeV)	sgn(μ)	tanβ	σ (pb)
SPS1A	100	250	-100	+	10	49.3
B	100	250	0	+	10	57.7
SU2	3550	300	0	+	10	5.0

The analysis performed by ATLAS for the SPS1A point [5] is discussed here. Similar results were obtained by CMS for point B [6].

Events are required to pass selections requiring jets with high transverse energy and missing transverse energy, in order to reject the Standard Model background. Two isolated Opposite-Sign and Same-Flavor (OSSF) leptons (electrons or muons) are also required.

The two leptons, which are the basic signature of the decay chain (1), can also be produced in other processes. If the two leptons are independent of each other, one would expect equal amounts of OSSF leptons and Opposite Sign Opposite Flavor (OSOF) leptons. Their distributions should also be identical. A powerful technique to remove the background from other processes is thus the subtraction of OSOF events from the OSSF distribution.

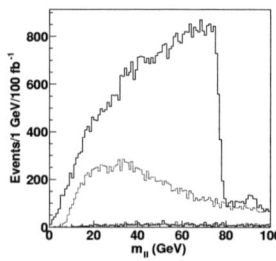
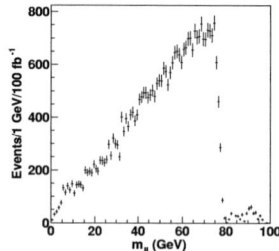

FIGURE 1. Distribution of 2-lepton invariant mass, for an integrated luminosity of 100 fb^{-1} [5]. Left plot: the solid curves are the OS-SF distribution, the dotted curves the OS-OF distribution. The two upper curves are the contribution for the SUSY benchmark SPS1A, the two lower curves the Standard Model contribution. Right plot: flavor subtracted OSSF-OSOF distribution. The two plots were obtained using the fast simulation of the ATLAS detector.

The resulting distributions are shown in Fig. 1 for point SPS1A and an integrated luminosity of 100 fb^{-1}. After the cuts the Standard Model background is negligible and the dominant background is due to leptons from independent decay chains in the SUSY events. After flavor subtractions, the expected triangular distribution with a clear edge is observed.

The position of the edge, which measures the mass combination of Eq. 2, is very accurately determined by fitting it with a triangular shape with gaussian smearing. With an integrated luminosity of 100 fb^{-1} a value of $(77.02 \pm 0.05 \pm 0.08)$ GeV is obtained, where the first error is statistical and the second comes from the uncertainty on the lepton energy scale, assumed to be 0.1%.

2.2. Lepton+jets kinematic edges

Further invariant mass combinations can be measured by combining the leptons with the hardest jets in the event. Four kinematic endpoints, in addition to the dilepton edge, can be obtained [5]. The five endpoints are a function of four unknown masses (the two lightest neutralinos, the slepton, and the scalar quarks), so these can be extracted from the data.

TABLE 2. The RMS of the mass distribution measured by the ATLAS detector with 300 fb^{-1} of data for the benchmark point SPS1A [5].

particle	m (GeV)	Δm (GeV)
χ_1^0	96.1	3.8
χ_2^0	176.8	3.7
$\tilde{l}$	143.0	3.8
q_L	537.2	6.1

In order to determine the precision which can be obtained on the sparticle masses by the ATLAS detector with 300 fb^{-1} of data, a total of 10000 sets of endpoints measurements was simulated, and the equations relating the endpoints to the masses were inverted [5]. Two sets of solutions are in general obtained. One solution give masses close to the nominal values, with a precision reported in Table 2. The other solution has masses lower by 10 to 20 GeV and is generally in poorer agreement with the data; in a fraction of the experiments (of the order of 10%), however, the two solutions cannot be distinguished without additional measurements.

2.3. Gluino and squark mass measurements

In this analysis, the decay chain

$$\tilde{g} \to \tilde{b}b \to bb\chi_2^0 \to bbl^{\pm}\tilde{l}_R \to bbl^{\pm}l^{\mp}\chi_1^0 \quad (3)$$

and the similar chain with bottom (s)quarks substituted by light (s)quarks are used. The analysis performed by CMS for Point B [6] is discussed here; a similar study was done by ATLAS for SPS1A [5].

In addition to some cuts on high-p_T jet and transverse missing energy to reject the SM background [6], two OSSF leptons with $p_T > 15$ GeV and $|\eta| < 2.4$ and an invariant mass within 15 GeV of the edge are also required. This selects events when the two leptons from the decay (1) are emitted back-to-back with the χ_1^0 at rest. In this configuration, the relation:

$$\vec{p}_{\chi_2^0} = \left(1 + \frac{m_{\chi_1^0}}{M_{ll}}\right) \vec{p}_{ll} \qquad (4)$$

allows to reconstruct the four-momentum of the second lightest neutralino, provided that the mass of the χ_1^0 is known.

The χ_2^0 four-momentum is combined with those of the (b-tagged) jets in the event with largest transverse momentum to reconstruct the (b) squark mass. Further kinematic cuts are used to reduce the combinatorial background from wrong jet associations. The gluino mass is reconstructed using the combination of the (bottom) squark with the closest (b) jet in space.

The reconstructed squark, sbottom and gluino mass peaks are shown in Fig. 2. In a scenario similar to that of point B, the reconstruction of squark is feasible already with a luminosity of 1fb^{-1}, while in the case of sbottom and gluino at least 10fb^{-1} are required to achieve a mass resolution better than 10%. Although both $m(\tilde{q})$ and $m(\tilde{g})$ depend on the value assumed for the χ_1^0 mass [6] their difference does not and can be measured with an error of a few percent.

2.4. Other masses

The reconstruction of the mass of other supersymmetric states has been studied [5]. In particular, the mass of the right-handed squark, which decays almost exclusively into $q\chi_1^0$, can be reconstructed selecting events with only two hard jets and missing energy and using an appropriate combination of the jets transverse momenta and missing energy [8]. A similar analysis can be used to observe the direct production of left-handed sleptons, decaying into $l\chi_1^0$.

The right-handed stau mass can be reconstructed using the visible energy of τ pairs to build the dilepton edge; a sharp edge can not be obtained because of the missing energy from neutrinos and the resolution is much poorer, but the edge position can still be determined. This analysis becomes very important for high values of $\tan \beta$, when the decay branching ration of χ_2^0 into third generation leptons becomes much larger than the branching ratio into the first two generations. The reconstruction of the mass of heavier neutralinos and chargino states has also been studied and should be possible for point SPS1a, given enough statistics.

3. SUSY PARTICLE MEASUREMENTS IN THE FOCUS POINT REGION

The benchmark in the "focus-point" region represents a more difficult scenario for the LHC. In this region, scalar particles are very heavy, at or beyond the limit of the LHC reach, and only the gluino and gauginos are accessible. The gaugino direct production is about the 90% of the SUSY cross section for the selected benchmark. However, since the mass of these particles is of the order of the top mass, it is not possible to separate these events from Standard Model processes using the jet and missing transverse signature [9].

The gluino has a mass of 856 GeV and decays into a lighter chargino or neutralino state and two jets (usually third generation jets). The gluino pair production can be efficiently separated from the Standard Model background by requiring four energetic jets and missing transverse energy. However, the cross section is only 0.5 pb, two orders of magnitude lower than that of the SPS1a and point B benchmarks. As a results, a larger integrated luminosity is required to reconstruct the SUSY masses.

Two dilepton decay are available: $\chi_2^0 \to \chi_1^0 l^+ l^-$ and $\chi_3^0 \to \chi_1^0 l^+ l^-$. Since scalars are heavy, there is no intermediate slepton in the decay. The distribution of the invariant mass of the two leptons has an edge corresponding to the difference between the mass of the two neutralinos involved in the decay. With an integrated luminosity of 300 fb^{-1} the ATLAS detector would be able to reconstruct the edges from both decays (Fig. 3).

The gluino mass can be estimated from the production cross section (the number of events which pass the cuts) and the distribution of the effective mass $M_{\text{eff}} = \Sigma_{\text{jets}} p_T + E_T^{\text{miss}}$ which is correlated with the mass of the SUSY particle produced in the pp collision [10]. The difference between the mass of the gluino and the lightest chargino can also be obtained using the decay $\tilde{g} \to \chi_n^{\pm} tb$, reconstructing the top quark hadronic decay and building the tb invariant mass [9].

4. CONCLUSIONS

If SUSY exists at the electroweak scale, it is likely to be discovered in the first months of data taking at the LHC. The reconstruction of the mass spectrum and the decays of the new particles will be more challenging.

In this contribution some of the techniques developed by the ATLAS and CMS collaborations to reconstruct the mass of Supersymmetric particles have been discussed. The physics potential of these techniques was shown for two selected benchmarks of the mSUGRA parameters space. For these models, several masses and mass combinations can be measured at the LHC.

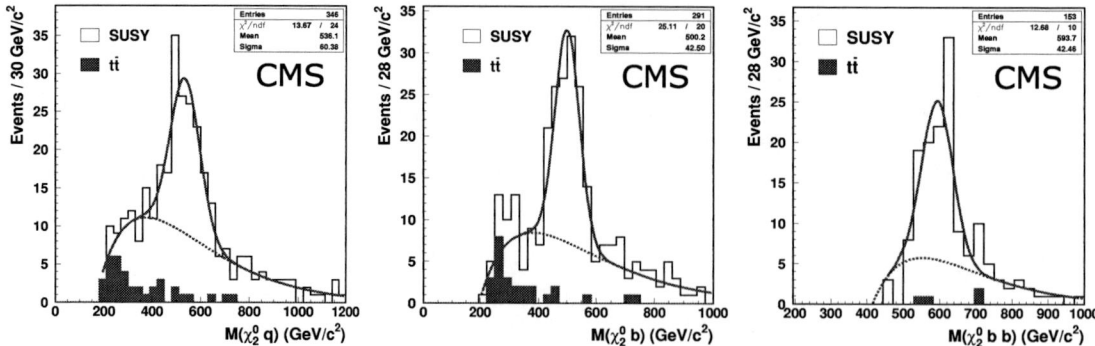

FIGURE 2. Reconstructed invariant mass distribution for squark (left), sbottom (middle) and gluino in the sbottom chain (right) at point B [6]. The integrated luminosity is 1fb^{-1} for the squark peak and 10fb^{-1} for the sbottom and gluino peaks.

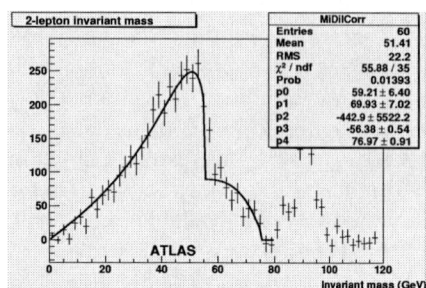

FIGURE 3. Flavor-subtracted invariant mass distribution of electron and muon pairs for the focus-point benchmark. The integrated luminosity is 300 fb^{-1}

The potential of the ATLAS detector for studying the Supersymmetric mass spectrum in a more difficult scenario (a benchmark in the "focus-point" region of mSUGRA parameter space) has also been discussed. For the selected point in the parameter space, the gluino pair production still allows a fast discovery of New Physics. After a few years of data taking at the design LHC luminosity, the mass difference between two pairs of neutralino states can be reconstructed through the two-lepton invariant mass distribution, and the gluino mass can be estimated.

Acknowledgments

The material reported here is the result of a huge amount of work made by the ATLAS and CMS collaborations. The author wishes in particular to express his gratitude to F. Paige, D. Tovey, G. Polesello and C. Troncon from the ATLAS collaboration and M. Maggi and M. Chiorboli from the CMS collaboration for providing material and suggestions. This review would not have been possible without their help.

REFERENCES

1. H.P. Nilles, Phys. Rev. **110** (1984) 1.

2. The ATLAS Collaboration, "ATLAS Detector and Physics Performance Technical Design Report", CERN/LHCC 99-14 (1999).

3. S. Abdullin et al., J. of Phys. **G28** (2002) 469.

4. J.R. Ellis et al., Phys. Lett. **B565** (2003) 176.
   H.W. Baer and C. Balazs, J. Cosmol. Astr, Phys. **0305** (2003) 006.

5. C.K. Gjelsten et al., "A detailed analysis of the measurement of SUSY masses with the ATLAS Detector at LHC", ATLAS Internal note, ATL-PHYS-2004-007 (2004).
   C.K. Gjelsten et al., JHEP **12** (2004) 003.

6. M. Chiorboli, A. Tricomi, *Squark and gluino reconstruction in CMS*, CMS Internal note, CMS-NOTE-2004-029.
   M. Chiorboli, Czech. J. Phys. **54** (2004) A151.

7. J.L. Feng et al., Phys. Lett. **B482** (2000) 388.
   J.L. Feng et al., Phys. Rev. Lett. **84** (2000) 2322.
   J.L. Feng et al., Phys. Rev. **D61** (2001) 075005.

8. C.G. Lester, Phys. Lett. **B463** (1999) 99.

9. T. Lari, *SUSY studies with ATLAS: hadronic signatures and focus-point*, ATL-PHYS-CONF-2005-002, to be published on the Czech. J. Phys.

10. D.R. Tovey, EPJ Direct **4** (2002) 1.

Searches for New Physics in CMS

Mario Galanti

Department of Physics and Astronomy, University of Catania, via Santa Sofia, 64, 95123 Catania, Italy

Abstract. Some of the studies made by the CMS collaboration in order to evaluate the CMS potential in the discovery of physics beyond the Standard Model are presented. In particular, results from studies concerning Minimal Supergravity, non-universal SUSY breaking and Gauge Mediated Supersymmetry Breaking are shown and discussed.

Keywords: CMS, Supersymmetry, mSUGRA, GMSB
PACS: 14.80.Ly

1. INTRODUCTION

The main problem of the searches for new physics is the existence of many alternative models and theories which are consistent with the current experimental data. Data coming from experiments with high energy colliders (especially the LEP and the TeVatron) on one side, and from cosmological observations on the other side, pose costrains on how an extension of the Standard Model should look like.

Nonetheless, within these costrains, several different models and theories can live, each of them predicting new particles with rather different charateristics.

One of the main purposes of the CMS detector [1] is to find signals of new physics beyond the Standard Model. CMS is one of the four detectors that are currently being built at CERN, and will operate at the Large Hadron Collider (LHC) starting from 2007.

Generally speaking, searches for new physics can be divided into two broad categories: inclusive searches try to find an excess of events over the predicted Standard Model background, and exclusive searches study individual channels in detail, trying to reconstruct new particles. In exclusive searches, single decay chains are analyzed, with the aim to measure the mass spectrum and branching ratios of new particles.

In what follows, I will focus on inclusive searches only. The general strategy of looking for excesses of events with respect to the Standard Model background means that, when the experiment takes and analyzes data, significant differences between expected and observed number of events are looked for.

Before data taking, an inclusive search is performed by analyzing available models, in order to find the best experimental signatures for them and to refine the experimental techniques. Good signatures have to cover all the possible event topologies which can stem from the various theories, but yet be distinctive enough to allow discrimination among the different models. Particularly important signatures are also those that can lead to an early discovery, in the first weeks of data taking, even with a possibly miscalibrated or incomplete detector.

2. MINIMAL SUPERGRAVITY

Among the various possible extensions of the Standard Model, Supersymmetry [2] is undoubtedly the most studied and one of the most attractive ones. If SUSY exists, at the Large Hadron Collider the production cross section of strongly interacting supersymmetric particles will be enhanced, so a higher number of squarks and gluinos will be directly produced, with respect to the other sparticles.

The signature to look for in this case would be made of one or more high P_T jets, coming from $\tilde{g}$ and $\tilde{q}$ decays, and one or more high P_T leptons, coming from subsequent decays. If R-parity is conserved, the main signature of Supersymmetry would be the presence of high missing E_T; if not, a high jet multiplicity would be seen. Different strategies have then to be set up in order to deal with the various possible scenarios.

The Minimal Supergravity (mSUGRA) is a supersymmetric model which is based on the minimal supersymmetric extension of the Standard Model (MSSM), with the addition that the free parameters have to unify at the GUT scale. In this way, the 100+ parameters of MSSM reduce to only five: $m_{1/2}$, m_0, $\tan\beta$, sign(μ) and A_0. All parameters except one, sign(μ), are expressed at the GUT scale, and their values are propagated to the electroweak scale using the Renormalization Group Equations (RGE). This unification of parameters is reflected by approximate relations among sparticle masses at the weak scale:

$$M(\tilde{g}) \approx 3M(\tilde{\chi}_1^{\pm}) \approx M(\tilde{\chi}_2^0) \approx 2M(\tilde{\chi}_1^0) \quad (1)$$

$$M(\tilde{g}) \approx M(\tilde{q}) > M(\tilde{\chi}) \quad (2)$$

However, despite these relations, a big indetermination still remains for the values of the masses.

TABLE 1. Some optimized cuts for three of the signatures taken into account.

Signature	mSUGRA point			Cut values					$\frac{S}{\sqrt{S+B}}$
	m_0 (GeV)	$m_{1/2}$ (GeV)	N_{Jets}	E_T^{Miss} (GeV)	E_T^{Jet1} (GeV)	E_T^{Jet2} (GeV)	$\Delta\varphi(P_T^\ell, E_T^{Miss})$	μ isol.	
E_T^{Miss}	500	1200	2	1200	900	600	0	Off	5.77
	1600	1000	7	600	600	300	0	Off	4.2
1ℓ	400	1100	2	900	600	300	20	On	4.66
	1000	1000	4	800	500	300	20	Off	4.68
3ℓ	400	700	2	300	150	80	0	On	6.37
	1400	700	2	300	300	200	0	Off	4.72

TABLE 2. The main backgrounds for the same-sign di-muon channel.

	tb	tqb	$\bar{t}b$	$\bar{t}qb$	ZZ	ZW	WW	$t\bar{t}$	$Zb\bar{b}$	All
$\sigma(pb)$	0.212	5.17	0.129	3.03	18 (NLO)	26.2	70.2	886 (NLO)	232 (NLO)	
N_1	2,120	51,700	1,290	30,300	180,000	262,000	702,000	8,860,000	2,320,000	
N_2	112	1,798	71	1,067	256	727	39.7	142,691	12,924	160,000

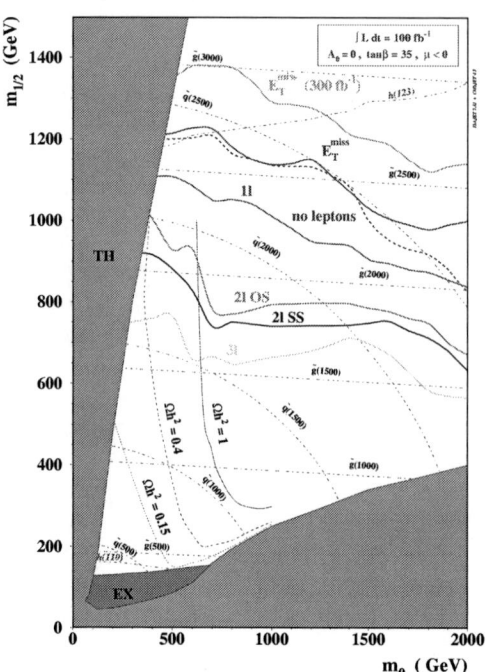

FIGURE 1. The CMS discovery reach for mSUGRA ($\tan\beta = 35$) and 100 fb^{-1} of integrated luminosity.

Several scans of the parameter space have already been carried out in order to determine the discovery reach of CMS. In an old comprehensive inclusive study, Standard Model background events, generated with PYTHIA 5.7 [3], and SUSY signal events, generated with ISAJET 7.32 [4], have been passed through CMSJET 4.51 [5], a fast simulation of the CMS detector. Events passing preselection cuts ($E_T^{Miss} > 200$ GeV and at least 2 jets with $E_T > 40$ GeV) have been classified according to their final-state topology: events with no leptons (0ℓ), at least one lepton (1ℓ), two opposite sign leptons (2ℓOS), two same sign leptons (2ℓSS), three leptons (3ℓ), and so on. Other cuts have then been optimized and applied to these events (see Tab. 1 for an example of the cut values adopted) in order to enhance the signal to background ratio. The scan has given the results shown in Fig. 1. Overall, for most of the mSUGRA parameter space, CMS is expected to see supersymmetry for sparticles with masses not exceeding 2.6-3.0 TeV/c^2 if the inclusive E_T^{Miss} signature is looked for. Other signatures give a narrower discovery reach.

More recently, inclusive scans have been repeated using a more accurate simulation of the detector response, and focusing on specific topologies. As an example, let's consider the same-sign di-muon signature [6]. This channel has been chosen because it is very clean, has a high trigger efficiency and a small background contamination. For this study, SUSY cross sections have been simulated with ISASUGRA 7.69, and the cross sections of the SM backgrounds have been calculated in the Leading Order (LO) with PYTHIA 6.220 and CompHEP 4.2p1 [7]. Next-to-Leading Order (NLO) corrections have also been calculated for some of the background processes using the PROSPINO program [8]. All events were generated with PYTHIA 6.220, and the detector response was simulated using the ORCA [9] package.

The events have been pre-selected requiring at least two same-sign muons with $P_T > 10$ GeV and $|\eta| < 2.5$. The main background processes considered are shown in Tab. 2. The main contribution comes from $t\bar{t}$ and $Zb\bar{b}$ production. Other background processes that have also been taken into account are the production of 3 and 4

TABLE 3. The optimized cuts used for the same-sign di-muon scan.

Set	1	2
E_T^{Miss} (GeV)	>200	>100
E_{T,jet_1} (GeV)	>0	>300
E_{T,jet_3} (GeV)	>70	>100
P_{T,μ_1} (GeV)	>20	>10
P_{T,μ_2} (GeV)	>10	>10

TABLE 4. The various phenomenologies of GMSB models. L is of the order of the linear dimensions of the detector.

	$\tilde{\tau}_1$ is the NLSP	$\tilde{N}_1$ is the NLSP
$c\tau \gg L$	Signature similar to a heavy μ	Same phenomenology as mSUGRA
$c\tau \approx L$	NLSP decays into the detector; possible mean life measurements	
$c\tau \ll L$	NLSP decays to $\tau\tau$	NLSP decays to $\gamma\gamma$

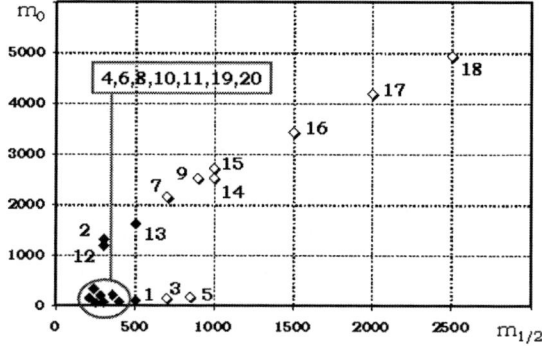

FIGURE 2. The results of the parameter scan for the di-muon channel. The points in black have significance > 5 for an integrated luminosity of 10 fb^{-1}, and the points in white have significance < 5.

vector bosons, and the production of $t\bar{t}$ in association with 1 or 2 vector bosons. Of them, the only two which are not negligible are $t\bar{t}Z$ and $t\bar{t}W$. The SUSY signal has been simulated in a series of 20 benchmark points which cover a wide region of the parameter space.

In order to enhance the signal over background ratio, kinematical cuts have been applied to five variables: missing transverse energy E_T^{Miss}, transverse energy of first and third jet E_{T,jet_1} and E_{T,jet_3}, and transverse momentum of the two muons P_{T,μ_1} and P_{T,μ_2}. Several cuts have been considered for each variable, giving a grand total of > 10^5 possible cut sets. After an optimization process, only the two sets of cuts shown in Tab. 3 have been kept, since they cover all the mSUGRA points with significance greater than five. The results of this study are shown in Fig. 2; it can be seen that the region of the $(m_0, m_{1/2})$ plane in which $m_{1/2} < 650$ GeV will be accessible to the CMS experiment with 10 fb^{-1} of integrated luminosity.

An estimate of the effects of systematic errors has also been done, by reducing the number of signal events by 30% while at the same time increasing the number of background events by 30%. The results show to be quite stable, since only one of the discovery points (#13) goes out of reach.

3. OTHER MODELS

The CMS collaboration has been studying several other models which either extend the Minimal Supergravity by changing some of its postulates, or are different extensions of the Standard Model. Here I will focus on models that drop the condition of universality of SUSY soft breaking parameters at the GUT scale and on Gauge Mediated Supersymmetry Breaking (GMSB).

Generally speaking, omitting the condition of universality of soft SUSY breaking parameters gives more freedom degrees to the mass spectrum, since the approximate relations (1) and (2) do not hold anymore. A scenario that is particularly challenging from the experimental point of view has very heavy ($\sim 5-20$ TeV) first and second generation scalars, and light (< 1 TeV) third generation scalars [10]. In this case only few sparticles could be produced at the LHC. Such a model would have a signature similar to the R-parity conserving mSUGRA, namely $n (\geq 2)$ jets, $m (\geq 0)$ leptons and E_T^{Miss}. The CMS discovery reach has been found to be strongly dependent from the relation between M_{LSP} and min $(M_{\tilde{g}}, M_{\tilde{q}})$. If $M_{LSP} \approx$ min $(M_{\tilde{g}}, M_{\tilde{q}})$ the discovery is hardest because the jets produced have lower energy and the momenta of the two LSP partially cancel out themselves, resulting in a lower E_T^{Miss}. In this situation, CMS should be able to detect Supersymmetry if min $(M_{\tilde{g}}, M_{\tilde{q}}) \leq 1.2 \div 1.5$ TeV or, if the condition $M_{\tilde{q}_{1,2}} \gg M_{\tilde{q}_3}$ holds, if $M_{\tilde{q}_3} \leq 800$ GeV.

In GMSB models [11] the Supersymmetry breaking takes place in a high-scale hidden sector which does not sense the SM interactions. Its effects are propagated to an intermediate-scale "messenger" sector by some new flavor-blind interaction and then to the visible sector by the gauge interactions. In GMSB, the LSP is the gravitino $\tilde{G}$, and the models are classified according to the Next-to-Lightest Super Particle (NLSP). There are two possible candidates for the NLSP, the lighter stau $\tilde{\tau}_1$ and the lightest neutralino $\tilde{N}_1$. They decay to the LSP with a mean life $c\tau$ which is almost uncostrained. The various experimental situations that can arise are shown in Tab. 4.

An example of experimental signature is shown in Fig.

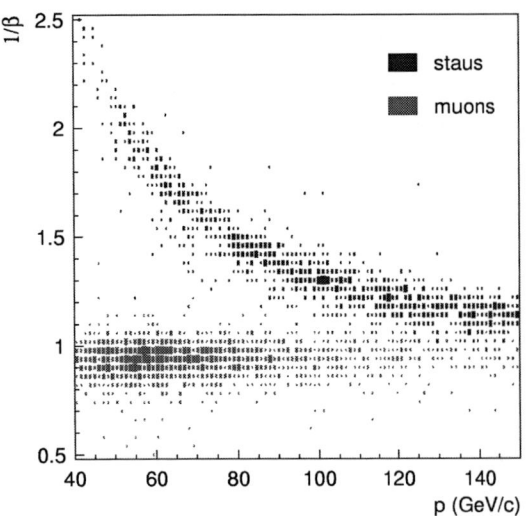

FIGURE 3. Measured $1/\beta$ as a function of the particle momentum.

3 for the case in which $\tilde{\tau}_1$ is the NLSP and $c\tau \gg L$. In this case the staus go out of the detector undecayed, and their passage is detected by the Muon Stations of CMS. Their velocity can then be measured by CMS through a time of flight measurement in the muon chambers; in that way, their signal can be discriminated from the one coming from muons.

4. CONCLUSIONS

The capacity of the CMS detector to discover signals of physics beyond the Standard Model has been shown. Discovery contours in the framework of Minimal Supergravity have been presented, as well as discovery limits in the framework of non-universal Supersymmetry. One of the possible experimental signatures of GMSB has also been illustrated.

ACKNOWLEDGMENTS

I wish to thank the organizers and the conveners for having invited me to speak.

REFERENCES

1. CMS Collaboration, *The Compact Muon Solenoid - Technical Proposal*, **CERN/LHCC 94-38**, CERN 1994.
2. H. P. Nilles, *Phys. Rev.* **110**, 1 (1984).
3. T. Sjöstrand et al., *Comput. Phys. Commun.* **82**, 74 (1994); T. Sjöstrand et al., *Comput. Phys. Commun.* **135**, 238 (2001).
4. H. Baer, F. E. Paige, S. D. Protopopescu, X. Tata, hep-ph/0001086.
5. S. Abdullin, A. Khanov, N. Stepanov, CMS TN/94-180.
6. D. Acosta, A. A. Drozdetski, G. Mitselmakher, CMS CR/2004-041.
7. CompHEP Collaboration, hep-ph/9908288.
8. W. Beenakker et al., hep-ph/9611232.
9. CMS Collaboration, *ORCA (http://cmsdoc.cern.ch/orca/)*.
10. S. I. Bityukov, N. V. Krasnikov, hep-ph/0210269.
11. M. kazana, G. Wrochna, P. Zalewski, CMS CR/1999-019.

The Higgs Boson as a Gauge Field in Extra Dimensions

Marco Serone

SISSA, Via Beirut 2, I-34014 Trieste, Italy

Abstract.
I review, at a general non-technical level, the main properties of models in extra dimensions where the Higgs field is identified with some internal component of a gauge field.

Keywords: Extra Dimensions, Electroweak Symmetry Breaking
PACS: 11.10.Kk; 11.25.Mj; 12.60.-i

The Standard Model of fundamental interactions (SM) represents the best theory at our disposal to describe all high-energy processes we know so far. Most likely, however, it cannot be the final description of Nature. Gravity is excluded from the theory, the origin of many parameters, such as the Yukawa couplings, is unexplained and it is also afflicted by hierarchy problems. The latter are best understood if one considers the SM as an effective field theory valid up to energy scales of order Λ, above which the theory has to be replaced by a more fundamental (and yet unknown) microscopic theory. At the quantum level, one finds that two parameters in the SM heavily depends on the details of the microscopic theory: the cosmological constant and the Higgs mass. For instance, by using a simple cut-off regularization at the scale Λ, one finds that the radiative corrections to the cosmological constant and the Higgs mass are respectively proportional to the quartic and quadratic power of Λ. We do not know the value of Λ, but the phenomenological success of the SM puts a bound on it: $\Lambda_{Exp} \geq$ few Tev (see *e.g.* ref.[1]).[1] Leaving aside the outstanding problem of the cosmological constant hierarchy problem (whose solution might well be due to unknown quantum gravity effects), we have still to face the problem of why and how the electroweak scale (and thus the Higgs mass) is stabilized to a value which is roughly one order of magnitude smaller than the minimal experimentally allowed value for Λ. Sometimes one refers to this problem as the gauge "little hierarchy problem". Although it involves only one order of magnitude, one has to notice that, contrary to the "usual" gauge hierarchy problem in which one takes $\Lambda \sim M_{Planck}$, this is an experimental fact and it does not rely on any assumption about the scale at which new physics should appear.

During the years, many solutions have been proposed to address the gauge hierarchy problem. Independently of the precise nature of the Higgs field that is assumed in each of these proposals, all of them require, in one way or another, the appearance of new physics at $\Lambda \sim$ TeV.

The Minimal Supersymmetric Standard Model (MSSM) is at the moment the best candidate theory of new physics beyond the Standard Model. However, no super particle has been discovered yet and, as far as the little hierarchy problem is concerned, the MSSM needs some unwanted fine tuning. It is thus important to investigate alternative scenarios where radiative corrections to the Higgs mass can be somehow suppressed.

Theories formulated in $D > 4$ space-time dimensions seem to be a promising arena for new ideas along this direction. Being non-renormalizable, these theories must always be seen as effective theories valid up to an UV cut-off scale Λ (not to be confused with the SM cut-off introduced before), above which the extra dimensional theory needs an UV completion. It is in particular important to have an estimate of Λ in order to quantify the relevance of quantum corrections given by higher derivative operators and understand the energy range of validity of the effective theory. A good estimate of Λ (which is typically hard to determine otherwise) is obtained by using Naïve Dimensional Analysis (NDA). The low-energy effective theory is trustable only if $\Lambda \gg 1/L, E$, where L is the typical size of the compact extra dimensions and E is the energy of the process under consideration.

There are several ideas and theoretical frameworks in the context of extra dimensions. We focus here on the idea that the SM Higgs boson arises from the internal component of a higher-dimensional gauge field of a group $G \supset G_{SM} = SU(3)_c \times SU(2)_L \times U(1)_Y$. By choosing suitable gauge groups in the extra dimensions, one can incorporate all SM gauge bosons (γ, $W^\pm$, Z and gluons) and the Higgs field H as arising from different components of the same higher dimensional gauge field A_M, where M runs over all (usual and extra) space-time coor-

---

[1] A remark is in order here. More precisely, the bound is on certain higher derivative operators, suppressed by powers of Λ. The value reported assumes that the coefficients of the operators are of order one. If this is not the case, Λ does not necessarily coincide with the scale at which new physics arises.

dinates.

Due to this common origin of the gauge and the Higgs fields, this idea is sometimes called "gauge-Higgs unification". Its essential point is that the Higgs field, being the component of a gauge field, is protected by radiative quadratic divergencies by the underlying higher-dimensional gauge symmetry.

This idea has been first advocated in refs.[2] but no concrete realization was found. In particular, since the Higgs is a gauge field, the Yukawa couplings are gauge couplings and thus it is not straightforward to get a realistic fermion spectrum. Interestingly enough, it has recently been understood that realistic Yukawa couplings can be obtained in these models and in a manner which provide a natural explanation of the large hierarchy of fermion masses [3, 4, 5]. This has allowed to construct several interesting models of gauge-Higgs unification, both in a supersymmetric [3, 6] and in a non-supersymmetric context [4, 5, 7].

The minimal model that one can consider is a five-dimensional theory compactified on a segment (or $S^1/\mathbf{Z}_2$ orbifold) of length L, with gauge group $G = SU(3)_c \times SU(3)_w$ [5][2]. If one suitably breaks the $SU(3)_w$ gauge group down to $SU(2)_L \times U(1)_Y$ by appropriate orbifold boundary conditions, one ends up with a 5D spectrum of Kaluza-Klein states, in which the only massless fields (zero modes) are the 4D gauge bosons A_μ^a ($a = 1, 2, 3$) and A_μ of $SU(2)_L \times U(1)_Y$ ($\mu = 0, 1, 2, 3$) and a complex scalar doublet H coming from A_5, the Higgs field. Gauge invariance forbids any local potential for A_5 in the interior of the segment (bulk), the only allowed gauge-invariant local operators being built with the field strength F_{MN}. Actually, a remnant of the 5D $SU(3)_w$ gauge symmetry also forbids any local potential for A_5 at the boundaries as well. In fact, at the boundaries there is a symmetry acting non-linearly on the Higgs field [8]:

$$\delta A_5 = \partial_5 \xi, \qquad (1)$$

where ξ are the gauge parameters of $SU(3)_w/[SU(2)_L \times U(1)_Y]$. The only gauge invariant operator that can give rise to a Higgs potential $V(H)$ must then be non-local in the extra dimension and expressed in term of the Wilson line $W = \mathcal{P} \exp(i \int dy A_5) \equiv \exp(i\alpha)$, where $0 \leq \alpha \leq 2\pi$ is the Wilson line phase [9], related to the Higgs vacuum expectation value v by $\alpha \simeq vL$ (notice that α defined here differs by a 2π factor from the α defined in ref.[5]). The crucial and most important property of this construction is that $V(H)$, being a function of W, is necessarily radiatively generated and non-local in the extra dimension. Being a non-local operator, $V(H)$ is finite at all orders in perturbation theory [10]. No dependence on the UV cut-off Λ appears in $V(H)$ and thus the little hierarchy problem is solved. Depending on the field content of the model, one could then have a radiatively induced electroweak symmetry breaking (EWSB), governed by the Wilson line phase α.[3] The EWSB is thus equivalent to a Wilson line symmetry breaking.

As we mentioned, the introduction of matter in this framework is not straightforward. If one assumes that the SM fermions are fields localized at the boundaries, then the symmetry (1) forbids local couplings between them and the Higgs field. On the other hand, if they are 5D fields propagating along the whole segment (bulk fields), their Yukawa couplings will necessarily be all the same and given by the gauge coupling constant. An interesting possibility to overcome this difficulty is obtained by assuming that the SM fermions are localized fields with mixing terms with bulk massive fermions [4, 5]. Since the bulk fermions couple to the Higgs, thanks to the mixing, an effective Yukawa coupling will be induced among the SM fermions. In fact, the effective Yukawa couplings between the Higgs and the SM fermions achieved in this way are roughly given by [5]

$$Y_f \sim \varepsilon_L \varepsilon_R g_4 L M_f e^{-LM_f}. \qquad (2)$$

In eq.(2), M_f is the mass of the bulk fermion coupled with the SM fermion f, g_4 is the 4D gauge coupling constant and $\varepsilon_{L,R}$ are dimensionless couplings which govern the mixing between the bulk fermion and the left- and right-handed SM fermion f. Notice that the couplings $\varepsilon_{L,R}$ are bounded, their values ranging from 0 (no mixing with bulk fermions) to 1 (maximal mixing with bulk fermions). The Yukawa couplings are effective (rather than fundamental) couplings, which depend exponentially on $M_f L$. In this way, we may not only get a realistic pattern of Yukawa couplings, but also have an understanding of their hierarchy in terms of the exponential behaviour appearing in eq.(2).[4]

Given this field content, one can thus compute the one-loop Higgs effective potential. As we have already argued, this is necessarily finite. One finds that an EWSB occurs with a value of the Wilson line phase α at the minimum that is about $\sim 1/2 \div 1$. All the qualitative features of the SM are then nicely reproduced. At the quantitative level, however, there are some problems. Their occurrence can actually be predicted on general

---

[2] In order to get the correct weak-mixing angle, a further 5D $U(1)'$ gauge field has to be introduced, but we can neglect it in the considerations that will follow.

[3] See refs.[11] for studies of the structure of one-loop Wilson line potentials on orbifolds.

[4] An alternative but essentially equivalent way of getting exponentially suppressed Yukawa couplings is obtained by considering massive fermions on the segment. In this case, a relation similar to eq.(2) is found, with $\varepsilon_{L,R} = 1$ (maximal mixing) [3].

grounds and are somehow model independent:

- $V(H)$ is radiatively generated. From a 4D perspective, one thus would expect a small Higgs quartic coupling in general, leading to a too light Higgs mass.
- The effective Yukawa couplings (2) are exponentially suppressed by $M_f L$. This is fine for all the SM fermions, but the top quark. Unless some other mechanism is advocated, one would expect from eq.(2) a too light top mass.
- The compactification scale is determined by the value of the Wilson line phase α at the minimum, since $M_W = \alpha/(2L)$. For $\alpha \sim 1/2 \div 1$, this results in a too low compactification scale, given the current bounds (see *e.g.* [12]).

These problems can be solved, or alleviated, in various ways. One possibility is to increase the value of the 5D gauge coupling constant g_5, which is the microscopic coupling that governs the size of the Yukawa couplings and of the Higgs effective potential. In flat space, g_5 is simply related to the 4D coupling constant g_4 by the simple relation $g_5 = g_4 \sqrt{L}$. Since L and g_4 are fixed by the experimental values of M_W and of the $SU(2)_L$ SM gauge coupling constant, the only way to increase g_5 is to introduce modifications in the model that change the above relation between g_4 and g_5. A simple way to do that is provided by adding kinetic terms for the 4D gauge fields A_μ, localized at the boundaries. If these terms are large enough, their net effect is to increase the Higgs and the top mass to realistic values [5]. Since the relation between M_W and α is also modified in presence of localized gauge kinetic terms, it turns out that one can get phenomenological acceptable values for the compactification scale as well. All the above problems are solved, but unfortunately other potential problems are introduced. They are all related to the fact that these localized gauge kinetic terms introduce mixing among all Kaluza-Klein states. This results in unwanted effects, such as too large deviations to the ρ parameter or to a non-universality of the 4D gauge couplings [5]. An other interesting way (probably closely related to the former) to increase the 5D coupling constant is obtained by considering a warped, rather than flat, space [13]. In this case, the Higgs mass is generally higher than the value obtained in flat space compactifications [7, 14], as well as the Yukawa couplings, which are dynamically generated in a way that is essentially the same as in flat space. The warping, however, produces distortions similar to those given by adding localized gauge kinetic terms in flat space. By suitably imposing a custodial $SU(2)$ symmetry to the Higgs sector, an interesting model of gauge-Higgs unification in warped space has been constructed where the distortions might be under control and small enough to be compatible with the Electroweak Precision Tests (EWPT) [7].[5]

An other possibility to solve the listed three problems is to find some microscopic mechanism to dynamically stabilize the Wilson line phase α to a smaller value, such as 5×10^{-2} or smaller. In this case, the Higgs quartic coupling is effectively enhanced and can give rise to realistic Higgs masses. The compactification scale would also be above the current bounds. The top mass problem is not directly solved in this way, unless this new mechanism also allows for greater Yukawa couplings. Unfortunately, there is no known satisfactory mechanism which allows to get values of $\alpha \sim 5 \times 10^{-2}$. It is interesting to note, however, that massive 5D fermions in very large representations of the gauge group typically tend to give lower values of α and also allows for bigger Yukawa couplings [5, 15]. The representations needed are however very large, and would lead to a breakdown of an effective field theory approach, since they lead to a NDA estimate of the cut-off $\Lambda \sim 1/L$.

So far, we focused on one compact extra dimension, but what happens if one has more extra dimensions? Since the NDA estimate of Λ decreases with the number of extra dimensions and no new interesting features seem to appear in further increasing their number, let us only consider the case of two extra dimensions, namely a 6D theory. In 6D, there are several potentially interesting two-dimensional compact spaces one could consider. The simplest spaces, leading to a 4D chiral spectrum of fermions, are given by orbifolds of tori of the form $T^2/\mathbf{Z}_N$, where $N = 2, 3, 4, 6$. Let us focus on these spaces in the following.

There are two main important features that happen when going to 6D. The first, good feature, is the appearing of a gauge-invariant Higgs quartic coupling at tree-level, simply arising from the non-abelian part of the internal components of the gauge field kinetic term F_{56}^2. A tree-level quartic coupling is welcome, because it can automatically solve the problem of a too light Higgs. The second, bad feature, is the possible appearance of a local, gauge-invariant, operator that contributes to the Higgs mass. This is an operator localized at the fixed-points of the $T^2/\mathbf{Z}_N$ orbifold, with a quadratically divergent coefficient, in general [4, 8, 16, 17]. It is linear in the internal components of the field-strength F. Its abelian term corresponds to a tadpole for certain gauge field components, whereas its non-abelian part represents a mass term for the Higgs field. If there is no symmetry to get rid of this operator, the hierarchy problem is reintroduced. It turns out that a discrete symmetry forbidding this operator can be implemented only for $T^2/\mathbf{Z}_2$ orbifolds, in which case,

---

[5] Interestingly enough, the model of ref.[7] has a purely 4D dual interpretation as a composite Higgs model.

however, one gets two Higgs doublets, rather than one. In this case, the Higgs effective potential has various similarities with the one arising in the Minimal Supersymmetric Standard Model (MSSM). Explicit computations on a given 6D model [18] have shown that the lightest Higgs field turns out to be again too light [19].

Maybe a more interesting possibility is obtained by considering $T^2/\mathbf{Z}_N$ orbifolds, with $N \neq 2$. If $N \neq 2$, one can get 2, 1 or 0 Higgs doublets, depending on the orbifold projection. The most interesting case appears to be given by the 1 Higgs doublet models, for which one finds $M_H = 2M_W$ at tree-level, by geometrical considerations [16]. However, no symmetry forbids the appearance of the localized operator mentioned above, which would spoil the stabilization of the electroweak scale. Even if this operator is put to zero at tree-level, no accidental one-loop cancellation seems to be possible. The best one can do is to advocate a spectrum of 6D fields such that the sum of the one-loop quadratically divergent coefficients over all fixed points vanish (global cancellation). In this case, it actually turns out that the electroweak scale is not destabilized. Contrary to the 5D construction considered before, the quadratic sensitivity to the cut-off would presumably be reintroduced at two-loop level, but a one-loop cancellation might be enough to solve the little hierarchy problem. No concrete model has been yet presented along these lines and thus it is premature to establish whether gauge-Higgs unification in 6D can be a realistic proposal or not.

The idea of a Higgs field as a gauge boson in extra dimensions seems to be a promising candidate to more conventional scenarios of new physics, such as SUSY.

Several aspects of this idea require further study. From a more theoretical side, it is desirable to find some mechanism to increase the Higgs mass without introducing the unwanted distortion effects that appears when one considers warped models or theories in flat space with localized gauge kinetic terms. It has also to be understood whether such theories (as many other theories in extra dimensions) admit a microscopic completion where the orbifold singularities (for 6D models) or the boundaries of the segment (for 5D models) are replaced by an UV model defined on a smooth compact space [20].

From a more phenomenological side, there are several issues which deserves further study: the generic suppression of Flavour Changing Neutral Currents or a systematic classification of all possible CP violating terms would be desirable. The latter study would also shed light on the possibility of having Baryogenesis at the electroweak scale, considering that a moderately strong first-order phase transition can be obtained in these models [21]. It would also be interesting to better understand whether a possible Dark Matter candidate can be found in such theories and under what conditions gauge coupling unification (typically lost in these models) can be recovered (see ref.[22] for a recent proposal).

I thank the hospitality of the Aspen Center for Physics, where this work has been completed.

REFERENCES

1. [LEP Collaboration], arXiv:hep-ex/0312023.
2. N. S. Manton, Nucl. Phys. B **158** (1979) 141; D. B. Fairlie, Phys. Lett. B **82** (1979) 97; J. Phys. G **5** (1979) L55; P. Forgacs, N. S. Manton, Commun. Math. Phys. **72** (1980) 15; S. Randjbar-Daemi, A. Salam, J. Strathdee, Nucl. Phys. B **214** (1983) 491; N. V. Krasnikov, Phys. Lett. B **273** (1991) 246; H. Hatanaka, T. Inami, C. Lim, Mod. Phys. Lett. A **13** (1998) 2601.
3. G. Burdman, Y. Nomura, Nucl. Phys. B **656** (2003) 3;
4. C. Csaki, C. Grojean, H. Murayama, Phys. Rev. D **67** (2003) 085012.
5. C. A. Scrucca, M. Serone, L. Silvestrini, Nucl. Phys. B **669** (2003) 128.
6. L. J. Hall, Y. Nomura, D. R. Smith, Nucl. Phys. B **639** (2002) 307; N. Haba, Y. Shimizu, Phys. Rev. D **67** (2003) 095001; K. w. Choi et al., JHEP **0402** (2004) 037; I. Gogoladze, Y. Mimura, S. Nandi, Phys. Lett. B **560** (2003) 204; ibid. **562** (2003) 307; Phys. Rev. D **69** (2004) 075006.
7. K. Agashe, R. Contino and A. Pomarol, Nucl. Phys. B **719** (2005) 165.
8. G. von Gersdorff, N. Irges, M. Quiros, Phys. Lett. B **551** (2003) 351; G. von Gersdorff, N. Irges, M. Quiros, hep-ph/0206029.
9. Y. Hosotani, Phys. Lett. B **126** (1983) 309; ibid. **129** (1983) 193; Ann. Phys. **190** (1989) 233.
10. N. Arkani-Hamed et al., Nucl. Phys. **B605** (2001) 81; A. Masiero, C. A. Scrucca, M. Serone, L. Silvestrini, Phys. Rev. Lett. **87** (2001) 251601.
11. M. Kubo, C. S. Lim and H. Yamashita, Mod. Phys. Lett. A **17**, 2249 (2002). N. Haba, M. Harada, Y. Hosotani and Y. Kawamura, Nucl. Phys. B **657** (2003) 169 [Erratum-ibid. B **669** (2003) 381]; N. Haba and T. Yamashita, JHEP **0402** (2004) 059.
12. A. Delgado, A. Pomarol, M. Quiros, JHEP **0001** (2000) 030.
13. R. Contino, Y. Nomura and A. Pomarol, Nucl. Phys. B **671** (2003) 148.
14. Y. Hosotani and M. Mabe, Phys. Lett. B **615** (2005) 257.
15. G. Martinelli, M. Salvatori, C. A. Scrucca and L. Silvestrini, arXiv:hep-ph/0503179.
16. C. A. Scrucca, M. Serone, L. Silvestrini and A. Wulzer, JHEP **0402** (2004) 049.
17. C. Biggio and M. Quiros, Nucl. Phys. B **703** (2004) 199.
18. I. Antoniadis, K. Benakli, M. Quiros, New J. Phys. **3** (2001) 20.
19. Y. Hosotani, S. Noda and K. Takenaga, Phys. Lett. B **607** (2005) 276.
20. M. Serone and A. Wulzer, "Orbifold resolutions and fermion localization," arXiv:hep-th/0409229; A. Wulzer, arXiv:hep-th/0506210.
21. G. Panico and M. Serone, JHEP **0505** (2005) 024.
22. K. Agashe, R. Contino and R. Sundrum, arXiv:hep-ph/0502222.

Theories with Extra Dimensions at Finite Temperature[1]

Alessia Gruzza

Dipartimento di Fisica, Università di Parma
and
INFN Gruppo Collegato di Parma,
Parco Area delle Scienze n.7A, 43100 Parma, Italy

Abstract. In 5-dimensional theories on multiply-connected manifolds the fifth component of the gauge fields can be identified with the Higgs field. We consider the Hosotani mechanism on S^1/Z_2 orbifold with an $SU(2)$ gauge group. When A_5 gets a VEV the gauge symmetry is completely broken. The VEV is undetermined at the tree level but a potential is generated at one loop. Finite temperature effects on the effective potential are studied.

Keywords: gauge field theory; dimension, 5; orbifold; Kaluza-Klein model; mass spectrum; effective potential
PACS: 11.15.Ex, 11.10.Kk, 11.10.Wx

INTRODUCTION

We consider a 5-dimensional fermion field gauge theory with boundary fixed points $y = 0$ and $y = \pi R$, where y is the extradimension coordinate and R is the compactification radius [1].
In particular, we consider the gauge group $SU(2)$ on the S^1/Z_2 orbifold.
Let us consider the Lagrangian density:

$$\begin{aligned}\mathscr{L} &= i\bar{\psi}\gamma^M D_M \psi + M\varepsilon\bar{\psi}\psi = \\ &= i\bar{\psi}\gamma^\mu\partial_\mu\psi + i\bar{\psi}\gamma^5\partial_y\psi + i\bar{\psi}\gamma^5(i\frac{\omega\tilde{\sigma}_2}{R})\psi + \\ &+ M\varepsilon\bar{\psi}\psi \;;\end{aligned} \quad (1)$$

where $\psi = (\psi^1, \psi^2)^T$ is a doublet of $SU(2)$, $D_M \equiv \partial_M + i\sigma_2 \frac{\omega}{R}\delta_{M5}$, M is the 5-dimensional bulk mass and

$$\varepsilon(y) = \begin{cases} 1 & if \; 0 < y < \pi R \\ -1 & if \; \pi R < y < 2\pi R \end{cases}.$$

Notice that ψ^1 and ψ^2 must have opposite parity and we choose ψ^1 even and ψ^2 odd. Expressing 5D fermions in terms of 4D chiral fermions as $\psi^i = (\psi^i_L, \psi^i_R)$, the equations of motion can be written as

$$\partial_y \psi_L + i\sigma^\mu \partial_\mu \psi_R + M\varepsilon \psi_L + i\frac{\omega}{R}\tilde{\sigma}_2 \psi_L = 0 \;;$$
$$-\partial_y \psi_R + i\bar{\sigma}^\mu \partial_\mu \psi_L + M\varepsilon \psi_R - i\frac{\omega}{R}\tilde{\sigma}_2 \psi_R = 0 \;. (2)$$

Using Poincare invariance in the $y =$const. slices we can represent the solutions of the equations of motions as $\psi_{L,R}(x,y) = e^{-ip\cdot x}f_{L,R}(y)\chi(x)_{L,R}$, where $\chi(x)_{L,R}$ are 4D chiral fermions satisfying the Dirac equation.
Considering the jump conditions for the $f_{L,R}$, that can be found by integrating over a small interval around $y = 0$ and $y = \pi R$, we arrive to have a discrete Kaluza-Klein (KK) [2] spectrum given by

$$\sin^2(\omega\pi) = \frac{m^2}{\Omega^2}\sin^2(\Omega\pi R) \;; \quad (3)$$

where $\Omega^2 = m^2 - M^2$. For $MR \geq 0.5$ [3], we can find an approximate analytical expression for the "lightest" mode, m_0, setting $\Omega = \sqrt{m^2 - M^2} \approx \pm iM$, from which

$$m_0 = \sin(\pi\omega)\frac{M}{\sinh(M\pi R)} \;. \quad (4)$$

There is a quasi-localized state that corresponds to m_0. Similarly, for the "heavy" modes, we can set $\Omega^2 \gg M^2$, obtaining

$$m_n^2 \approx M^2 + \left(\frac{n \pm \omega}{R}\right)^2 \;. \quad (5)$$

EFFECTIVE POTENTIAL

$T = 0$ contribution

The fermionic contribution to the one-loop effective potential at $T = 0$ is [4]

$$\widehat{V}_0 = -N_f \frac{1}{2}\sum_n \int \frac{d^4p}{(2\pi)^4} \ln(p^2 + m_n^2) \;; \quad (6)$$

where p is the euclidean momentum, N_f is the number of fermionic degrees of freedom, m_n is the 4-dimensional

---

[1] work done in collaboration with L. Pilo

mass of the nth particle, coming from Eq. (3). Using Eq. (6)

$$\widehat{V}_0(\omega) = -N_f \int \frac{d^4p}{(2\pi)^4} \ln\left[\sin^2(\omega\pi) + \frac{p^2}{p^2+M^2}\sinh^2(\sqrt{p^2+M^2}\pi R)\right] \; ; \quad (7)$$

To get rid of ultraviolet divergences we subtract out the $\omega = 0$ contribution form $\widehat{V}_0$ and we get

$$V_0(\omega) = -\frac{N_f}{8\pi^2}\int_0^\infty dp \cdot p^3 \ln\left[\sin^2(\omega\pi) + \frac{p^2}{p^2+M^2}\sinh^2(\sqrt{p^2+M^2}\pi R)\right] + \frac{N_f}{8\pi^2}\int_0^\infty dp \cdot p^3 \cdot \ln\left[\frac{p^2}{p^2+M^2}\sinh^2(\sqrt{p^2+M^2}\pi R)\right]. \quad (8)$$

For $MR = 4$ the potential at $T = 0$ is shown in Fig. 1.

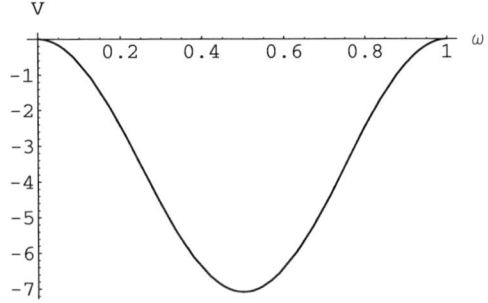

FIGURE 1. The potential multiplied for 10^{12} at $T = 0$ for $MR = 4$ and $N_f = 1$, varying ω

For $2M\pi R \gg 1$ we can find an analytical expression

$$V_0(\omega) \approx \frac{-N_f}{16\pi^6 R^4}\sin^2(\omega\pi)[3 + 6M\pi R + 6(M\pi R)^2 + 4(M\pi R)^3]e^{-2M\pi R} . \quad (9)$$

$T \neq 0$ contribution

The fermionic contribution at finite temperature T is given by

$$V_T(\omega) = -\frac{N_f}{2\beta}\sum_{k=-\infty}^{+\infty}\int\frac{d^3p}{(2\pi)^3}\sum_{n=-\infty}^{+\infty}\ln(v_k^2 + p^2 + m_n^2) \; ; \quad (10)$$

where v_k are the Matsubara frequencies, given by

$$v_k = \begin{cases} \frac{2k\pi}{\beta} & \text{for the gauge bosons} \\ \frac{(2k+1)\pi}{\beta} & \text{for the fermions} \end{cases}.$$

Summing over the Matsubara frequencies, we get

$$V_T(\omega) \equiv V_0 + V_{T\neq 0}$$
$$= V_0 - \frac{N_f}{2\pi^2\beta^4}\sum_{n=-\infty}^{+\infty}\int_0^{+\infty}dx \cdot x^2 \cdot \ln\left[1 + e^{-\sqrt{x^2+\beta^2 m_n^2}}\right] \; ; \quad (11)$$

where m_n are the solutions of (3). As a result, the thermal contribution to the effective potential can be written as

$$V_{T\neq 0}(\omega) = -\frac{N_f}{2\pi^2\beta^4}[J_F(m_0^2\beta^2) + \sum_{n=-\infty,\neq 0}^{+\infty}J_F(m_n^2\beta^2)] \; ; \quad (12)$$

where

$$J_F(m_n^2\beta^2) = \int_0^{+\infty}dx \cdot x^2 \ln\left[1 + e^{-\sqrt{x^2+\beta^2 m_n^2}}\right] . \quad (13)$$

We are mainly interested to the case $MR \geq 0.5$, where a very light first KK state m_0 is present (quasi-zero mode). For a temperature range $T \gg m_n$ and $T \ll m_n$, $n > 0$, we can find an analytically approximate expression for J_F, corresponding to an high temperature expansion for the quasi-zero mode and a low temperature expansion for the remaining part of the KK tower. We get [4][5]

$$V_{T\neq 0}(\omega) \approx N_f\left[\frac{m_0^2}{48\beta^2} - \sum_n \frac{m_n^{3/2}}{\beta^{5/2}(2\pi)^{3/2}}e^{-m_n\beta}\right] \; ; \quad (14)$$

We get a further approximation of the potential considering only the m_0 contribution (for $m_n^2\beta^2 \gg 1$ is a good approximation), that is

$$V_{f_{T\neq 0}}(\omega) \approx N_f\frac{m_0^2}{48\beta^2} = \frac{N_f}{48\beta^2}\left[\frac{M\sin(\pi\omega)}{\sinh(M\pi R)}\right]^2 . \quad (15)$$

TRANSITION TEMPERATURE

The total fermionic contribution is given by $V_f = V_{f_0} + V_{f_{T\neq 0}}$. From the sum of these two contributions we can obtain the temperature of transition. When $\omega \neq 0$ the symmetry $SU(2)$ is completely broken, while $\omega = 0$ corresponds to breaking $SU(2) \to U(1)$. The case $\omega = 1/2$ is special because, for a determined temperature (the temperature of transition), the potential has a minimum. In this case $SU(2) \to U(1)$.

The numerical evaluation is done for $MR = 4$. In this case we have the transition for $1.505 < \beta/R < 1.52$, Fig. 2 and Fig. 3 [6]. A numerical evaluation for larger values of MR is difficult because, for interesting values of β in the transition, the sum of Kaluza-Klein modes must include a large number of excited masses, making the calculation impracticable.

Numerically there is a transition for $MR = 4$ in the range $1.505 < \beta/R < 1.52$, this result is in agreement with the analytical (approximate) computation because $MR > 0.5$, as requested in the approximation of Eq. (3), and we are in the range $m_0^2 \beta^2 \ll 1$ and $m_n^2 \beta^2 \gg 1$. Indeed imposing that the contributions of Eq. (9) and of Eq. (15) are equal, we get

$$\frac{\beta}{R} = \sqrt{\frac{4\pi^6 (MR)^2}{3[3 + 6(M\pi R) + 6(M\pi R)^2 + 4(M\pi R)^3]}} \ ; \quad (16)$$

that is, for $MR = 4$, results $\beta/R = 1.51$, in perfect agreement with our numerical result.

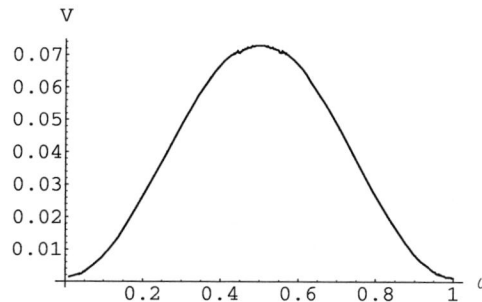

FIGURE 2. Effective potential multiplied for 10^{12} for $MR = 4$, varying ω for $\beta/R = 1.505$

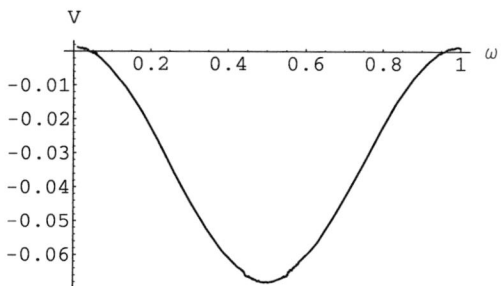

FIGURE 3. Effective potential multiplied for 10^{12} for $MR = 4$, varying ω for $\beta/R = 1.52$

CONCLUSIONS

We considered a 5-dimensional gauge field theory with an $SU(2)$ gauge group, on the S^1/Z_2 orbifold. We analyzed the effects of the quasi-localization of fields at finite temperature.

In the one-loop effective potential there is a minimum for $\omega = 0$ and the $U(1)$ symmetry is restored, resulting in an expression compatible with the Standard Model.

The numerical results for $MR = 4$ are compared to the analytic results in range of validity of the approximations, $T \gg m_n$, $T \ll m_n$ and $2M\pi R \gg 1$, and the transition of the fermionic contribution is in good agreement with the analytical result in the approximate case.

We restricted the analysis at $MR = 4$ but different conclusions, in other ranges, are possible [6].

ACKNOWLEDGMENTS

I would like to thank Luigi Pilo for the collaboration and the extreme availability. I also thank Antonio Riotto for initial useful suggestions. I am also grateful to Antonio Masiero and to Luca Trentadue for useful discussions.

REFERENCES

1. Y. Hosotani, *Phys. Lett. B*, **126** (1983) 309; ibidem, *Phys. Lett. B*, **129**(1983) 193; ibidem, *Annals Phys.* **190** (1989) 233.
2. Th. Kaluza, *Sitzungsber. Preuss. Akad. Wiss. Phys. Math. Klasse* 996 (1921); O. Klein, *Z. F. Physik*, **37** (1926) 895
3. G. von Gersdorff, L. Pilo, M. Quiros, D.A.J. Rayner, A. Riotto, *Phys. Lett. B*, **580** (2004) 93
4. M. Quiros, e-Print Archive: hep-ph/9901312
5. Gian F. Giudice, *Phys. Rev. D* **45** (1992) 3177
6. A. Gruzza, L. Pilo *in preparation*

Search for Extra Dimensions at the LHC

Alessio Ghezzi

Università di Milano Bicocca and INFN Milano

Abstract. In recent years it has been realized that the attempt to solve the problem of the huge difference between the gravity scale (M_{Pl}) and the electroweak scale can lead to extra dimensions scenarios with sizeable effects at the TeV scale. In this paper the prospects of the search at the LHC experiments for some of these models are reviewed.

Keywords: LHC, extra dimensions
PACS: 12.10.-g, 11.10.Kk, 11.25.Mj

INTRODUCTION

The modern string theories which are conceived to provide a common ground to gauge theories and gravitation, presuppose the existence of extra dimensions. Originally it was thought that the extra dimensions produced sizeable effects only at energies as large as the gravity scale $M_{Pl} \sim 1.2 \times 10^{19}$ GeV. Recently it was realized that, in particular models, the fundamental scale of gravity can be lowered to the TeV range. This opened new possibilities to the solution of the the huge difference between the electroweak scale and M_{Pl} (hierarchy problem). A growing interest on this topic has developed a rich spectrum of scenarios whose phenomenology, being the fundamental scale in the TeV range, appears very interesting for the LHC experiments.

This paper presents the prospects of the search, as outlined by the ATLAS [1] and CMS [2] experiments, at the LHC, for three of the main scenarios of extra dimensions at the TeV scale: the ADD model of large extra dimensions, the RS model of warped extra dimensions and the TeV^{-1} sized extra dimensions. Unless differently stated all the results reported refer to an integrated luminosity of 100 fb^{-1}.

THE *ADD* MODEL OF LARGE EXTRA DIMENSION

This model, devised by Arkani-Hamed, Dimopoulos and Dvali [3], foresees the existence of n spatial flat extra dimensions, which are accessible only to the gravity. The other fields of the Standard Model, on the contrary, are localized onto 3D brane. The extra dimensions are compactified on a torus whose size R is related to the fundamental mass scale (M_D) by the equation:

$$M_{Pl}^2 = M_D^{2+n} R^n, \qquad (1)$$

needed to restore at large distances the correct values of the Newton constant. From equation 1 it appears that for M_D values of the order of 1 TeV, R can be, depending on n, as large as tenth of millimeter, motivating the name "large extra dimensions" for this framework. In this model the weakness of the gravity is due to the fact it spreads into the large bulk of the extra dimensions, and the hierarchy is removed.

The compactification of the the extra dimensions gives rise to a tower of massive Kaluza Klein (KK) states with a mass gap $\Delta m \sim 1/R$. The KK gravitons can be produced, at sizeable rates, in high energy particle collisions, and since they are weakly coupled to the matter (coupling of the gravitational interaction) they escape the detector resulting as missing energy. Even if the detailed aspect of the theory above M_D is not known, the graviton emission rates can be calculated in a effective theory which holds for energies below M_D [4]. The search at the LHC for evidences in favour of this model will be performed by detecting an excess of events with a single energetic jet or a single energetic photon from the recoil against the graviton (missing E_T + jet or missing $E_T + \gamma$) with respect to the expected rate of these events from Standard Model processes, mainly from $jet + Z \to \nu\nu$ and $jet + W \to l\nu$.

Figure 1 shows the expected missing E_T distribution, evaluated in a simulation of the reconstruction with the ATLAS detector. It can be seen that the contribution from the gravitons emission becomes relevant at high missing E_T. The signal can be obtained considering the number of events above a threshold in missing E_T after the subtraction of the expected contribution from background processes. Considering a threshold of 1000 GeV the discovery potential for M_D ranges from 7.5 TeV to 5.3 TeV for a number n of extra dimensions between respectively 2 and 4 [5]. This estimate takes also into account a calibration procedure of the $jet + Z \to \nu\nu$ background exploiting the $jet + Z \to ll$ events.

These values indicate the possibility to cover a vast

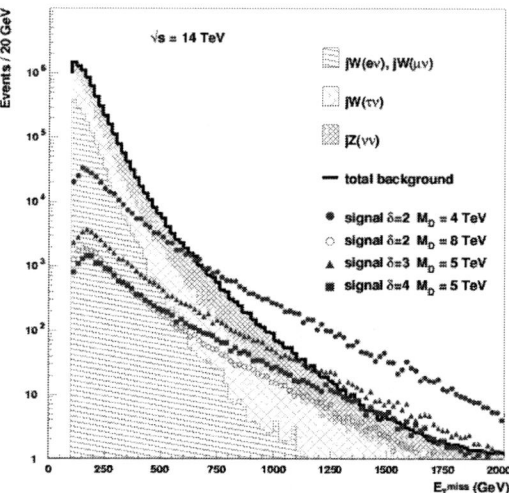

FIGURE 1. Distribution of the missing E_T both for the different Standard Model backgrounds and for the signal, for different values of the number of extra dimensions (δ) and M_D. The number of events refers to an integrated luminosity of 100 fb^{-1}. Figure from [5].

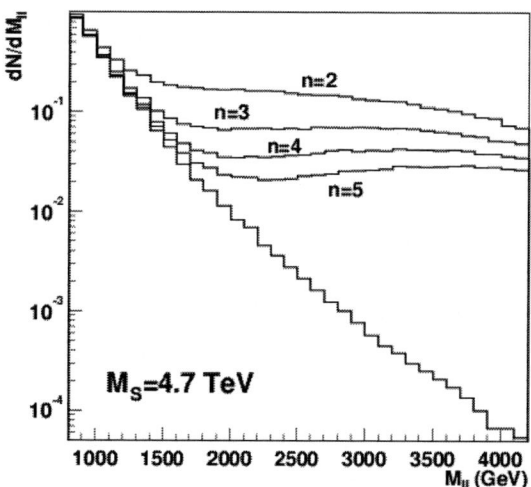

FIGURE 2. Invariant mass distribution for electron pairs both for the Standard Model background and for the signal, for different values of the number of extra dimensions (n) and a mass scale of 4.7 TeV. The number of events refers to an integrated luminosity of 100 fb^{-1}. Figure from [8].

region of the parameters still not excluded by the present limits from collider experiments which set a lower bound on M_D of 1.5 TeV and 0.5 TeV respectively for 2 and 8 extra dimensions. It has to be noticed that strong constraints from astrophysical results [6] exists for $n = 2$ and $n = 3$ ($\sim$ 100 TeV and $\sim$ 8 TeV respectively) but these limits are sensitive, contrarily to the limits from collider experiments, to the low energy region of the KK graviton mass spectra, and they can be avoided by distortions of the compactified space [7].

The presence of ADD extra dimensions can be detected at the LHC also by investigating the contribution of virtual KK graviton exchange to Drell-Yan processes, in particular di-lepton or di-photon production. The graviton virtual exchange produces deviations from the Standard Model expectations in the high invariant mass ($m_{ll} > 2$TeV) region, as shown in figure 2. The reachable sensitivity on the mass scale ranges from 8.1 to 7 TeV for two and five extra dimensions [8]. It has to be pointed out however that the study of these processes properly produce constraints on a cutoff scale (Λ) for the energies in the summation over KK states contributing to the process, and these constraints do not directly translate into limits on the mass scale M_D (for a detailed description of the relation between Λ and M_D see for example [9]).

THE RS MODEL OF WARPED EXTRA DIMENSIONS

Another approach to solve the hierarchy problem within the framework of the extra dimensions was developed by L.Randall and R.Sundrum [10]. The simplest realization of their model foresees the existence of one extra dimension, bounded by two four dimensional branes. The Standard model fields are localized onto one of these branes while gravity lives on both the branes and into the bulk. A distinctive feature of this model is that the metric of the five dimensional space is non-factorizable, with a warping factor of the four dimensional Minkowski metric ($\eta_{\mu\nu}$) which depends on the position along the extra dimension ($|y| \in [0, \pi r_c]$):

$$ds^2 = e^{-k|y|} dx^\mu dy^\nu \eta_{\mu\nu}. \qquad (2)$$

This warp factor eliminates the hierarchy problem: the only fundamental scale is of the order of $\overline{M}_{Pl}$ and the TeV scale (Λ) is generated on the Standard Model brane by the warp factor, i.e.

$$\Lambda = \overline{M}_{Pl} e^{-k\pi r_c}, \qquad (3)$$

with $k\pi r_c \sim 11 - 12$. The value of k is also of the order of $\overline{M}_{Pl}$ [11] therefore the size r_c of the extra dimension in this model is very small. The KK excitations of the graviton have a mass spectrum given by $m_n = k x_n e^{-k\pi r_c}$, where x_n is the nth root of the first Bessel function J_1. Given the value of k and Λ the first excitation have a mass around the TeV and the resonances are well separated.

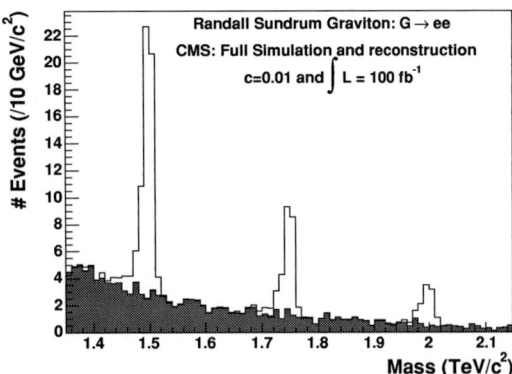
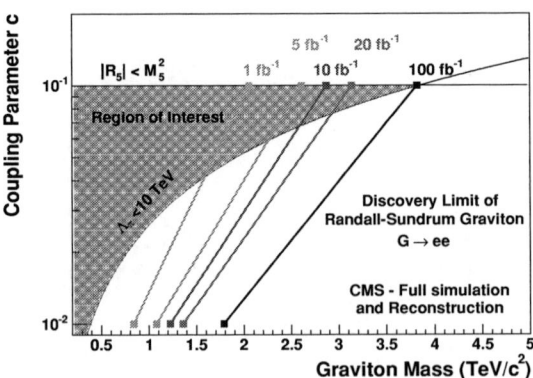

FIGURE 3. Average distribution of the invariant mass for e^+e^- pairs both for the Standard Model background and for the signal, for different mass value hypothesis: $m_G = 1.5$, 1.7 and 2.0 TeV and in the weak coupling scenario (c=0.01). The number of events refers to an integrated luminosity of 100 fb^{-1}. Figure from [12].

FIGURE 4. Reach in the parameters space for the search of RS gravitons in the channel $G \rightarrow e^+e^-$. Different integrated luminosities are reported. Figure from [12].

Contrarily to the ADD model, in this model the gravitons are coupled to the Standard Model fields and they decays into fermions or bosons.

Two parameters determine the model, they can be chosen as:

- the mass m_G of the first excited KK graviton
- the coupling $c = k/M_{Pl}$ which determines the width of the resonances

The appealing signature of the RS gravitons at the LHC is the presence of resonances in the invariant mass spectrum of di-leptons (m_{ll}) or di-photons ($m_{\gamma\gamma}$). The contribution of the standard Drell-Yan processes to the m_{ll} spectrum is a continuous tail and becomes very low at high m_{ll} (~ 2 TeV), as appears from the example of the expected spectrum shown in figure 3. Detailed studies of the $G \rightarrow e^+e^-$ channel have been carried out both by CMS [12] and ATLAS [13]. The analogous search for the $G \rightarrow \mu^+\mu^-$ channel suffers from the poorer resolution on high p_T muons. Figure 4 from [12] summarizes the discovery potential for different integrated luminosities, in the parameters space, concerning the $G \rightarrow e^+e^-$ channel. The region of interest is bounded by the theoretical constraints on the curvature $|R_5|$ and on the scale Λ [14], and, relatively to what is shown in figure 4 it is almost uncovered by the present limits from collider experiments (700 GeV and 200 GeV respectively for the strong $c = 0.1$ and weak $c = 0.01$ coupling). It can also be seen that the LHC experiments will be able to explore the whole allowed region.

The resonance of the second KK state, which would have been a very distinctive signature of this model, is unfortunately out of reach at the LHC for most of the parameters space. Nevertheless the spin-2 nature of the resonance can be determined by looking at the angular distribution of the leptons: it will be detected, at 90% C.L. for m_G values up to 1.70 TeV, with an integrated luminosity of 100 fb^{-1} [13]. Other evidences in favour of the model can be found by comparing the branching ratios in various channels (e.g. $G \rightarrow \gamma\gamma$) since for the graviton it is expected an universal coupling to the Standard Model fields [14].

TEV^{-1} SIZED EXTRA DIMENSIONS

Some variants of the large extra dimensions scenario foresee the existence of m longitudinal extra dimensions of size about 1/TeV, besides to the n large extra dimensions as in the ADD model (see section II). The Standard Model field as well as gravity can populate these "small" extra dimensions. The most attractive of these schemes allows the gauge fields to propagate in the bulk while it confines the fermion fields to a 3-d rigid brane.

The KK tower of the gauge fields would appear at an hadron collider as a set of resonances in the $l^+l^-, l\nu, jj$ channels. This phenomenology has been investigated for a detector at LHC [15] focusing on the $m = 1$ scenario. In this case the model has only one free parameter, a mass scale M_c, with a mass spectrum given by $M_n^2 = (nM_c)^2 + M_0^2$, where M_0 is the mass of the standard boson. The couplings of the KK bosons are proportional to the ones of the standard boson.

The present lower limit on M_c is 4 TeV and it arises from precision electroweak measurements at LEP[1] [16].

---

[1] For models with more than one extra dimensions the derivation of a limit on M_c from these measurements is not straightforward, see

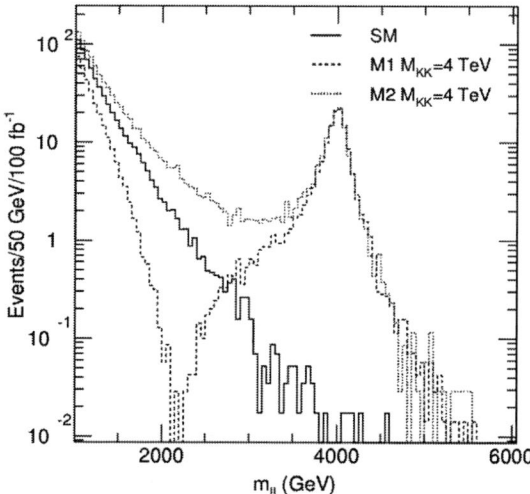

FIGURE 5. Distribution of the invariant mass for di-leptons both for the Standard Model alone and with two possible model of extra dimensions with $M_c = 4$ TeV. The model discussed in the text is labelled $M1$, while $M2$ is a model with quarks and leptons localized onto different branes. The two models present different interferences with the standard bosons. The number of events refers to an integrated luminosity of 100 fb^{-1}. Figure from [15].

An example of the electron pair invariant mass distribution [2] expected for the ATLAS detector is shown in figure 5; it is evident the capability of detecting the resonance.

Focusing on the region of the peak the reach on M_c, at 5σ significance is about 5.8 TeV [15]. This can be improved considering the effect of the interference between the Standard model boson and its KK states, which dominates the region below the peak, as can be seen in figure 5. The achievable reach is about 8 TeV, provided that a control of the background much better than 15% is obtained.

If a resonance is detected the study of the angular distribution of the lepton and of the forward-backward asymmetry as a function of m_{ll} can distinguish between this model and the resonance of a graviton, as described in section III, or the resonance from a Z', on condition that M_c is not too large (below 5 TeV).

CONCLUSIONS

The studies performed by ATLAS and CMS has shown that the phenomenology of the extra dimensions models with a TeV scale is likely to be detected at the LHC, for a large region of the allowed parameters space. In most cases however an unambiguous evidence in favour of the model, probing e.g the spin-2 nature of a resonance, will require high statistics and good control of the systematics.

REFERENCES

1. CERN/LHCC 94-14 e 99-15 ATLAS Collaboration, *ATLAS Detector and Physics Performance Technical Design Report*, (1994)
2. CERN/LHCC 94-14 e 99-15 CMS Collaboration, *CMS Detector and Physics Performance Technical Design Report*, (1994)
3. N. Arkani-Hamed, S. Dimopoulos, G. Dvali, "Phenomenology, Astrophysics and Cosmology of Theories with Sub-Millimeter Dimensions and TeV Scale Quantum Gravity", Phys.Rev. D59 (1999) 086004, arXiv:hep-ph/9807344
4. G.F. Giudice, R. Rattazzi, J.D. Wells,"Quantum Gravity and Extra Dimensions at High-Energy Colliders", Nucl.Phys. B544 (1999) 3-38,arXiv:hep-ph/9811291
5. L. Vacavant and I. Hinchliffe, "Signal of models with large extra dimensions in ATLAS", J. Phys. G: Nuvl. Pat. Phys. 27 (2001) 1839-1850
6. J. Hewett, M. Spiropulu, "Particle Physics Probes Of Extra Spacetime Dimensions", Ann.Rev.Nucl.Part.Sci. 52 (2002) 397-424 arXiv:hep-ph/0205106; Y. Uehara, "A Mini-Review of Constraints on Extra Dimensions", Mod.Phys.Lett. A17 (2002) 1551-1558, arXiv:hep-ph/0203244, and reference therein
7. G.F. Giudice, T. Plehn, A. Strumia, "Graviton collider effects in one and more large extra dimensions",Nucl.Phys. B 706 (2005) 455-483
8. V. Kabachenko, A. Miagkov and A. Zenin, "Sensitivity of the ATLAS detector to extra dimensions in di-photon and id-lepton production processes", ATL-PHYS-2001-012
9. G. F. Giudice, A. Strumia, "Constraints on extra-dimensional theories from virtual-graviton exchange", Nucl.Phys. B663 (2003) 377-393, arXiv:hep-ph/0301232
10. L. Randall and R. Sundrum, "A large mass hierarchy from a small extra dimension", Phys. Rev. Lett. 83 (1999) 3370 arXiv:hep-ph/9905221
11. H.Davoudiasl, J.L. Hewett and T.G. Rizzo, "Warped Phenomenology", hep-ph/9909255
12. C.Collard, M.-C. Lemaire, "Search with the CMS detector for Randall-Sundrum excitations of gravitons decaying into electron pairs", CMS NOTE 2004/024
13. B.C. Allanach, K. Odagiri, Ma. Parker and R. Webber, "Searching for narrow graviton resonances with the ATLAS detector at the Large Hadron Collider", JHEP 9 (2000) 19
14. H. Davoudiasl, J.L. Hewett, T.G. Rizzo, "Experimental Probes of Localized Gravity: On and Off the Wall", Phys.Rev. D63 (2001) 075004
15. G. Azuelos and G. Polesello, "Prospects for the detection of Kaluza-Klein excitations of gauge bosons in the ATLAS detector at the LHC", SN-ATLAS-2003-023
16. T.G. Rizzo, "Testing the nature of Kaluza-Klein excitations at future lepton collider", Phys Rec D61, 055005

---

e.g. [16].

[2] When considering the muons pair channel the peak in the distribution broadens due to the poorer momentum resolution for high p_T muons.

Split Supersymmetry

G.F. Giudice

CERN, Theory Division, 1211 Geneva 23, Switzerland

Abstract. The naturalness criterion applied to the cosmological constant implies a new-physics threshold at 10^{-3} eV. Either the naturalness criterion fails, or this threshold does not influence particle dynamics at higher energies. It has been suggested that the Higgs naturalness problem may follow the same fate. We investigate this possibility and, abandoning the hierarchy problem, we use unification and dark matter as the only guiding principles. The structure of Split Supersymmetry and its phenomenological consequences are briefly discussed.

Keywords: Split Supersymmetry

INTRODUCTION

For decades the naturalness (or hierarchy) problem of the Higgs mass term has been the guiding principle to construct theories beyond the Standard Model (SM). The criterion of naturalness has the exciting implication that the SM should stop to be valid at a scale around the TeV, and new dynamics should occur at energies reachable by present or near-future colliders. Although no clear indications for any SM failure at electroweak energies has emerged so far, a conclusive resolution of this issue has to wait for the LHC.

From a field-theoretical point of view, the cosmological constant problem appears to be very similar to the naturalness problem of the Higgs mass, since both of them are related to ultraviolet power divergences. The same naturalness criterion, applied to the cosmological costant, leads to the existence of a threshold of new dynamics at 10^{-3} eV. We do not know if some hidden dynamics actually takes place at that scale, or if the resolution of the problem comes without any modification of the dynamics. What we know is that present particle physics calculations, valid at energies much larger than 10^{-3} eV, can be safely performed by setting the cosmological constant to zero and ignore any effect caused by the mechanism ultimately responsible for the solution of the problem. This fact has been justified by invoking the anthropic principle [1], which could be operating in presence of a large number of meta-stable vacua, as in string theory [2].

It is conceivable to ponder whether such an explanation could also apply to the hierarchy problem, imagining a mechanism (not necessarily based on the anthropic principle) which allows to extrapolate SM calculations to energies much larger than the TeV, without the need of introducing new dynamics, besides the Higgs.

At first sight, this sounds like a devastating proposal. But, if we are willing to abandon the hierarchy problem, we can try to use other clues to drive the search for the theory beyond the SM. Gauge coupling unification could be one such clue: it is motivated by a theory that addresses questions related to the fundamental structure of the SM particle content.

The failure of exact unification of gauge couplings in the SM suggests the existence of new particles, belonging to incomplete GUT irreps, which mend the mismatch. It is well known that low-energy supersymmetry provides precisely the necessary particles with the appropriate quantum numbers. Recently, the proposal of Split Supersymmetry was put forward [3, 4, 5]. Setting aside the hierarchy problem, gauge-coupling unification can be achieved in a supersymmetric model where all scalars, but one Higgs doublet, are much heavier than the electroweak scale. Most of the unpleasant aspects of supersymmetry (excessive flavour and CP violation, fast dimension-5 proton decay, tight constraints on the Higgs mass) are eliminated, but the unification is retained.

If supersymmetry plays no rôle in solving the hierarchy problem, there is no reason to insist that the spectrum is (partially) supersymmetric. Why should the new states appear at the weak scale, if the hierarchy problem is not the guiding principle of the analysis? Here we can use a second observational clue: the evidence for dark matter and the observation that a particle with weak cross section and mass around the Fermi scale is a natural candidate for it. We stress that the unification and dark-matter arguments are not in general sufficient to insure that new physics be within the LHC discovery reach, contrary to the naturalness criterion. Nevertheless, in some cases there are interesting experimental consequences to be investigated.

THE STRUCTURE OF SPLIT SUPERSYMMETRY

The spectrum of Split Supersymmetry contains the higgsino components $\tilde{H}_{u,d}$, the gluino ($\tilde{g}$), the W-ino ($\tilde{W}$) and B-ino ($\tilde{B}$), and the SM particles with one Higgs doublet. The most general renormalizable Lagrangian with a matter parity, besides gauge-invariant kinetic terms, is given by

$$\begin{aligned}\mathscr{L} &= m^2 H^\dagger H - \frac{\lambda}{2}\left(H^\dagger H\right)^2 - \left[h^u_{ij}\bar{q}_j u_i \varepsilon H^* + h^d_{ij}\bar{q}_j d_i H \right.\\ &\quad + h^e_{ij}\bar{\ell}_j e_i H + \frac{M_3}{2}\tilde{g}^A\tilde{g}^A + \frac{M_2}{2}\tilde{W}^a\tilde{W}^a + \frac{M_1}{2}\tilde{B}\tilde{B}\\ &\quad + \mu\tilde{H}_u^T \varepsilon \tilde{H}_d + \frac{H^\dagger}{\sqrt{2}}\left(\tilde{g}_u \sigma^a \tilde{W}^a + \tilde{g}'_u \tilde{B}\right)\tilde{H}_u\\ &\quad \left.+ \frac{H^T \varepsilon}{\sqrt{2}}\left(-\tilde{g}_d \sigma^a \tilde{W}^a + \tilde{g}'_d \tilde{B}\right)\tilde{H}_d + \text{h.c.}\right], \end{aligned} \quad (1)$$

where $\varepsilon = i\sigma_2$.

The Lagrangian in eq. (1) describes the effective theory obtained by removing squarks, sleptons, charged and pseudoscalar Higgs from the supersymmetric SM. These states are assumed to be heavy and, for simplicity, we will assume them to be all degenerate with mass $\tilde{m}$. The coupling constants of the effective theory at the scale $\tilde{m}$ are obtained by matching the Lagrangian in eq. (1) with the interaction terms of the supersymmetric Higgs doublets H_u and H_d,

$$\begin{aligned}\mathscr{L}_{\text{susy}} &= -\frac{g^2}{8}\left(H^\dagger_u \sigma^a H_u + H^\dagger_d \sigma^a H_d\right)^2\\ &\quad -\frac{g'^2}{8}\left(H^\dagger_u H_u - H^\dagger_d H_d\right)^2\\ &\quad + \lambda^u_{ij} H_u^T \varepsilon \bar{u}_i q_j - \lambda^d_{ij} H_d^T \varepsilon \bar{d}_i q_j - \lambda^e_{ij} H_e^T \varepsilon \bar{e}_i \ell_j\\ &\quad - \frac{H^\dagger_u}{\sqrt{2}}\left(g\sigma^a\tilde{W}^a + g'\tilde{B}\right)\tilde{H}_u\\ &\quad - \frac{H^\dagger_d}{\sqrt{2}}\left(g\sigma^a\tilde{W}^a - g'\tilde{B}\right)\tilde{H}_d + \text{h.c.}\end{aligned} \quad (2)$$

Once the Higgs doublet $H = -\cos\beta\varepsilon H_d^* + \sin\beta H_u$ is fine-tuned to have small mass term, the matching conditions of the coupling constants in eq. (1) at the scale $\tilde{m}$ are obtained by replacing $H_u \to \sin\beta H$, $H_d \to \cos\beta\varepsilon H^*$ in eq. (2):

$$\lambda(\tilde{m}) = \left[g^2(\tilde{m}) + g'^2(\tilde{m})\right]/4 \times \cos^2 2\beta, \quad (3)$$
$$h^u_{ij}(\tilde{m}) = \lambda^{u*}_{ij}(\tilde{m})\sin\beta, \quad h^{d,e}_{ij}(\tilde{m}) = \lambda^{d,e*}_{ij}(\tilde{m})\cos\beta, \quad (4)$$
$$\tilde{g}_u(\tilde{m}) = g(\tilde{m})\sin\beta, \quad \tilde{g}_d(\tilde{m}) = g(\tilde{m})\cos\beta, \quad (5)$$
$$\tilde{g}'_u(\tilde{m}) = g'(\tilde{m})\sin\beta, \tilde{g}'_d(\tilde{m}) = g'(\tilde{m})\cos\beta. \quad (6)$$

In the context of a supersymmetric theory it is possible to argue that the gaugino masses M_i or the higgsino mass μ are much smaller than the typical scale because they are protected by an R-symmetry and a PQ symmetry, respectively. However, the symmetries of the effective Lagrangian in eq. (1) are not enhanced if we set $M_i = 0$ or $\mu = 0$, separately. This is because supersymmetry has been explicitly broken by eliminating the scalar fields. Nevertheless, if we simultaneously set $M_i = \mu = 0$, the effective theory remains invariant under the product of an R-symmetry (with R-charges $R[H_u] = 2$, $R[H_d] = 0$) and a supersymmetric PQ symmetry (with charges $PQ[H_u] = PQ[H_d] = -1$). In other words, eq. (1) is invariant under a global $U(1)$ symmetry with charges $S[\tilde{B}] = S[\tilde{W}] = S[H] = -S[\tilde{H}_d]/2, S[\tilde{H}_u] = 0$, and with quarks and leptons with the appropriate charges. This symmetry is spontaneously broken by the Higgs vev and explicitly broken by M_i or μ.

This result shows that the choice of keeping only light gauginos or light higgsinos, but not both of them simultaneously, is not radiatively stable. It is interesting that the only consistent choice (forgetting the tuning of the Higgs mass) of splitting the supersymmetric spectrum is the one that successfully reproduces gauge-coupling unification. Another consequence of this result is that the μ parameter mixes with $M_{1,2}$ under renormalization effects.

A particular situation occurs when $\tan\beta \to \infty$, since h^d, h^e, $\tilde{g}_d$, $\tilde{g}'_d$ vanish in this limit. Then eq. (1), for $\mu = 0$, has an additional global symmetry (with charges $([\tilde{B}] = [\tilde{W}] = 0, [H] = [\tilde{H}_u] = [\tilde{H}_d])$. When we set $M_i = 0$ and keep μ non-vanishing, we find a different global symmetry (with charges $([\tilde{B}] = [\tilde{W}] = [H]/2 = [\tilde{H}_u] = -[\tilde{H}_d])$. Therefore, we expect no renormalization mixing between $M_{1,2}$ and μ in the limit of large $\tan\beta$.

UNIFICATION AND HIGGS MASS

To make a precise assessment on gauge-coupling constant unification in Split Supersymmetry, it is necessary to study the 2-loop renormalization group evolution, including one-loop threshold effects. Between the unified scale M_{GUT} and the scale of heavy scalars $\tilde{m}$, the theory is exactly supersymmetric. Below the scale $\tilde{m}$, we use the spectrum of Split Supersymmetry, with gauginos and higgsinos included in the 2-loop evolution. Then we include separate thresholds for the gluino M_3 and for weak gaugino and higgsino masses, and properly evolve between the two scales. Thresholds of the top quark and SM gauge bosons are taken into account in the usual way [6].

The prediction for $\alpha_s(M_Z)$, shown in ref. [4], is consistent with present experimental data. A value of $\tilde{m}$ larger than in conventional supersymmetric models improves the agreement between the theoretical prediction and experimental data. However, because of the theoretical uncertainty due to unknown GUT thresholds and supersymmetric thresholds at the scale $\tilde{m}$, it is not possible to extract firm bounds on the parameters. A dependence of $\alpha_s(M_Z)$ on $\tan\beta$ arises from two-loop effects proportional to the Yukawa couplings and the gaugino couplings $\tilde{g}$, but the numerical contribution is marginal. On the other hand, the effect of the gaugino and higgsino threshold is important.

The present lower bound on the SM Higgs mass, $m_H > 114.4$ GeV at 95% CL [7], provides a strong constraint on the parameters of low-energy supersymmetry. This constraint is relaxed in Split Supersymmetry, because the Higgs boson mass receives large radiative correction in the evolution from $\tilde{m}$ to the weak scale. The result for the prediction of the Higgs mass in Split Supersymmetry is shown in ref. [4]. The Higgs-mass dependence on $\tan\beta$ comes primarily from the boundary condition in eq. (3), and to a lesser extent from the renormalization-group evolution. Because of the large renormalization effect, values of $\tan\beta$ equal to 1 are allowed, but they require large values of $\tilde{m}$ and low values of gaugino and higgsino masses.

DARK MATTER

As discussed in the introduction, in the absence of the naturalness criterion, dark matter can provide the link between new physics and the electroweak scale. It is therefore crucial to study what are the implications of the request that the lightest neutralino is the dark matter particle. Differently than in ordinary low-energy supersymmetry, the parameter μ is not determined by electroweak symmetry breaking, but uniquely by the relic abundance calculation. In this section, we study this relation.

Let us first consider the case in which the lightest neutralino is mostly B-ino. Since squarks and sleptons are decoupled and the B-ino is a gauge singlet, its only interaction is through its coupling $\tilde{g}'_{u,d}$ with Higgs and higgsinos, given in eq. (1). Therefore, if $\mu \gg M_1$, the B-ino is nearly decoupled and it annihilates too weakly in the early universe. This means that we need to consider values of μ comparable with M_1, and thus the lightest neutralino χ is always a mixture of gaugino and higgsino and never a pure state. Through the mixing, this state annihilates efficiently into Higgs and gauge bosons. The dominant contribution, when μ is not much larger than M_1 comes from p-wave annihilation into longitudinal gauge bosons which, for $M_1 \gg M_Z$, gives a χ relic abundance

$$\Omega_\chi h^2 \simeq 0.1 \frac{\mu^2(M_1^2+\mu^2)^2}{m_\chi^4\,\text{TeV}^2}. \qquad (7)$$

Now we turn to the case in which the lightest neutralino is mainly a higgsino. The higgsino has gauge interactions which survive in the limit $M_{1,2} \gg \mu$, and therefore it can be the dark matter, even in a pure state. Actually the coupling of the lightest higgsino with the Z vanishes when $\mu \gg M_Z$. However, in this limit the other neutral and charged higgsinos are nearly degenerate in mass (and the lightest state is neutral) and off-diagonal couplings of the gauge bosons to higgsinos are allowed. When computing the relic abundance, it is therefore important to include the coannihilation of the various channels [8]. When this is done, we find that the relic abundance of a heavy higgsino in Split Supersymmetry is[1]

$$\Omega_{\tilde{H}}h^2 = 0.09 \left(\frac{\mu}{\text{TeV}}\right)^2. \qquad (8)$$

Using the 2-σ range of dark matter density preferred after WMAP data [10]

$$0.094 < \Omega_{DM}h^2 < 0.129, \qquad (9)$$

we find that a higgsino dark matter should have a mass in the range 1.0 to 1.2 TeV.

Next we consider a W-ino lightest neutralino. The W-ino has also a neutral component that could play the rôle of the dark matter. It is not usually considered a standard candidate, because in ordinary low-energy supersymmetry with gaugino unification condition, it can never be the lightest state. However, the W-ino can become the LSP in anomaly mediation [11]. Both tree-level and one-loop effects contribute to make the neutral state belonging to the $SU(2)$ triplet lighter than the charged one [12]. In the limit of pure W-ino, the relic abundance in Split Supersymmetry is

$$\Omega_{\tilde{W}}h^2 = 0.02 \left(\frac{M_2}{\text{TeV}}\right)^2. \qquad (10)$$

Using eq. (9), we find that the mass range of a W-ino dark matter is 2.0 to 2.5 TeV.

Let us now study the dependence of the neutralino relic abundance as we vary the parameters of Split Supersymmetry. For small values of M_2, the lightest neutralino is mainly B-ino, but as explained before, the higgsino component is non-negligible. When M_2 is larger than the gauge-boson masses, eq. (7) is a good approximation and therefore the correlation between M_2 and μ is roughly $\mu \propto M_2^{2/3}$. As M_2 grows, the gap between μ and M_1 gets reduces until we have the transition to the higgsino region, where the value of μ is uniquely determined by eq. (8).

Dark matter particles are actively searched for in underground experiments through their scatterings with nuclei. In the case of Split-Supersymmetry, since squarks

---

[1] The numerical calculations of the dark-matter relic abundances and detection rates presented in this paper have been performed using the fortran package DarkSUSY [9] adapted to the case of Split Supersymmetry.

and sleptons are very heavy, and the Z–χ^0–χ^0 coupling gives only a spin-dependent neutralino interaction with nuclei, the only contribution to the spin-independent cross section comes from the exchange of the Higgs boson [13]. To compare different experiments, it is customary to consider the neutralino scattering cross-section off a proton, which is given by

$$\sigma_p = (N_{11}\tan\theta_W - N_{12})^2 (N_{13}\cos\beta - N_{14}\sin\beta)^2 \times \left(\frac{115 \text{ GeV}}{m_H}\right)^4 4 \times 10^{-43} \text{ cm}^2. \quad (11)$$

Here m_H is the Higgs mass and $N_{1,i}$ are the gaugino and higgsino components of the lightest neutralino, in standard notations. As shown by eq. (11), the H–χ^0–χ^0 coupling vanishes when the neutralino is a pure gaugino ($N_{13} = N_{14} = 0$) or pure higgsino ($N_{13} = N_{14} = 0$). Interestingly, Split Supersymmetry predicts that χ^0 is a mixed state, as long as M_2 is not too large. However, the scattering rate is rather small and it drops with the Higgs mass as M_H^{-4}. The detection rate is negligible when χ^0 becomes higgsino ($M_2 \gtrsim$ TeV). Nevertheless, in the case of a mixed state and in a large range of the allowed Higgs mass, the signal is within the reach of future experiments, which will have a sensitivity up to 10^{-44}–10^{-45} cm^2 for $m_\chi = 1$ TeV.

ELECTRIC DIPOLE MOMENTS

Split Supersymmetry resolves the difficulties with flavour and CP violation, encountered in generic supersymmetric models, because squarks and sleptons are taken to be very heavy, with masses of the order of $\tilde{m}$. Nevertheless, the effective theory below $\tilde{m}$ still contains physical CP-violating phases.

Since the new phases are in the chargino–neutralino sector, which does not couple at tree level to quarks and leptons, CP violation in processes involving SM fermions occurs only at two loops. The most interesting effect appears in the fermion electric dipole moments (EDM). These are induced by the effective operator $\bar{f}\sigma_{\mu\nu}\gamma_5 f F^{\mu\nu}$, where f is a generic fermion ($f = e$ for the electron EDM, and $f = u,d,s$ for the neutron EDM). As gluinos carry no phases, there are no contributions from either chromomagnetic or 3-gluon operators at the leading order. The fermion EDM is generated by two-loop diagrams. The leading contribution in the limit $|M_2|, |\mu| > m_H$ is

$$\frac{d_f}{e} = -\frac{\alpha Q_f m_f K_{\text{QED}} \tilde{g}_u \tilde{g}_d \sin\Phi}{16\pi^3 M_2 \mu} \left(\ln\frac{M_2\mu}{m_H^2} + 2 + \frac{M_2^2 + \mu^2}{M_2^2 - \mu^2} \ln\frac{M_2}{\mu}\right) + \mathcal{O}\left(\frac{M_W^2}{M_2\mu}, \frac{m_H^2}{M_2\mu}\right). \quad (12)$$

Here all variables indicate the absolute value of the corresponding quantity and $\Phi = \arg(\tilde{g}_u^* \tilde{g}_d^* M_2 \mu)$.

The prediction for the electron EDM in Split Supersymmetry is shown in ref. [5]. For weak-scale chargino masses and maximal CP-violating phase, the result is very close to the present experimental limit $d_e < 1.9 \times 10^{-27}$ e cm at 95% CL. In ordinary low-energy supersymmetry, EDMs are generated at one loop and therefore suppressed phases ($\lesssim 10^{-2}$) are necessary to reconcile theory with experiments. Because of the two-loop suppression, Split Supersymmetry makes the exciting prediction that EDM are on the verge of experimental testability, if phases have the most natural value of order unity.

ACKNOWLEDGMENTS

The work presented here is based on collaborations with A. Romanino and with N. Arkani-Hamed, S. Dimopoulos, and A. Romanino.

REFERENCES

1. A. Vilenkin, Phys. Rev. Lett. **74** (1995) 846 [arXiv:gr-qc/9406010]; S. Weinberg, Phys. Rev. Lett. **59** (1987) 2607.
2. R. Bousso and J. Polchinski, JHEP **0006**, 006 (2000) [arXiv:hep-th/0004134]; S. Kachru, R. Kallosh, A. Linde and S. P. Trivedi, Phys. Rev. D **68**, 046005 (2003) [arXiv:hep-th/0301240]; L. Susskind, arXiv:hep-th/0302219; F. Denef and M. R. Douglas, arXiv:hep-th/0404116.
3. N. Arkani-Hamed and S. Dimopoulos, arXiv:hep-th/0405159.
4. G. F. Giudice and A. Romanino, Nucl. Phys. B **699** (2004) 65 [arXiv:hep-ph/0406088].
5. N. Arkani-Hamed, S. Dimopoulos, G. F. Giudice and A. Romanino, Nucl. Phys. B **709** (2005) 3 [arXiv:hep-ph/0409232].
6. L. J. Hall, Nucl. Phys. B **178**, 75 (1981).
7. R. Barate et al., Phys. Lett. B **565** (2003) 61 [arXiv:hep-ex/0306033].
8. J. Edsjo and P. Gondolo, Phys. Rev. D **56** (1997) 1879 [arXiv:hep-ph/9704361].
9. P. Gondolo, J. Edsjo, P. Ullio, L. Bergstrom, M. Schelke and E. A. Baltz, arXiv:astro-ph/0211238.
10. C. L. Bennett et al., Astrophys. J. Suppl. **148** (2003) 1 [arXiv:astro-ph/0302207]; D. N. Spergel et al., Astrophys. J. Suppl. **148** (2003) 175 [arXiv:astro-ph/0302209].
11. L. Randall and R. Sundrum, Nucl. Phys. B **557**, 79 (1999) [arXiv:hep-th/9810155]; G. F. Giudice, M. A. Luty, H. Murayama and R. Rattazzi, JHEP **9812**, 027 (1998) [arXiv:hep-ph/9810442].
12. T. Gherghetta, G. F. Giudice and J. D. Wells, Nucl. Phys. B **559**, 27 (1999) [arXiv:hep-ph/9904378].
13. R. Barbieri, M. Frigeni and G. F. Giudice, Nucl. Phys. B **313**, 725 (1989).

Parallel Session III:

Heavy Flavour Physics

Conveners:

M. Bona
University of Torino and INFN, Italy

M. Ciuchini
INFN Roma III, Italy

Recent Theoretical Progress in Kaon Physics

Federico Mescia

Dip. di Fisica, Univ. degli Studi "Roma TRE", via della Vasca Navale 84, I-00146 Roma, Italy
INFN, Laboratori Nazionali di Frascati, Via E. Fermi 40, I-00044 Frascati, Italy

Abstract. Recent progress and future prospects in Kaon physics are summarized. I will mainly concentrate on the developments concerning the following four observables: ε'/ε, ε_K, the Cabibbo angle and the rare Kaon decays.

Keywords: Kaon, Rare decays, CP violation
PACS: 11.15.-q,12.15.Hh,12.39.Fe,13.20.Eb,13.25.Es

INTRODUCTION

Kaon decays have contributed significantly to the development of the Standard Model (SM). The first observation of the decay $K_L \to \pi^+\pi^-$ by Christenson, Cronin, Fitch and Turlay 40 years ago [1] has shown for the first time that the CP symmetry is violated in nature and guided Kobayashi and Maskawa, a decade later, to introduce a third quark family into their quark mixing scheme [2]. Low decay rates observed in the decay $K_L \to \mu^+\mu^-$, lead to the discovery of the GIM mechanism [3] and to the prediction of the charm quark.

In this write-up, the subject is divided into three parts. We first sketch the present scenario for direct and indirect CP violation. The second part is devoted to semileptonic Kaon, the best probe for the measurement of the CKM matrix element $|V_{us}|$. Finally, we discuss the potentiality of rare decays, as next frontier in exploring fundamental physics by Kaon physics.

CP VIOLATION

ε_K and ε'/ε are the two famous quantities parametrizing CP violation respectively in $K^0 - \bar{K}^0$ mixing and neutral Kaon to two pion decays.

A few years ago, the direct violation of the CP symmetry in the decay amplitude (ε'/ε), predicted by the Standard Model and first observed by NA31 with 3σ evidence [4], has been firmly established by NA48 [5, 6] and KTeV [7]. However, despite this experimental effort, present theoretical uncertainties are still large and make impossible to convert the $Re(\varepsilon'/\varepsilon)$ into a meaningful CKM constraint. The situation is instead different for ε_K, which already allows to over-constrain the CKM $\bar{\rho} - \bar{\eta}$ plane. Namely, $\varepsilon_K \propto \hat{B}_K \bar{\eta} \left[(1-\bar{\rho}) + \text{const}\right]$., where $\hat{B}_K$ controls the non-perturbative QCD effects, through the matrix element $\langle \overline{K^0} | \bar{s}\gamma_\mu(1-\gamma_5)d\ \bar{s}\gamma^\mu(1-\gamma_5)d | K^0\rangle = \frac{8}{3} B_K(\mu) f_K^2 m_K^2$. In lattice (quenched) simulations, the B_K determinations have reached over the years a high level of accuracy (see [8] for a recent review). We

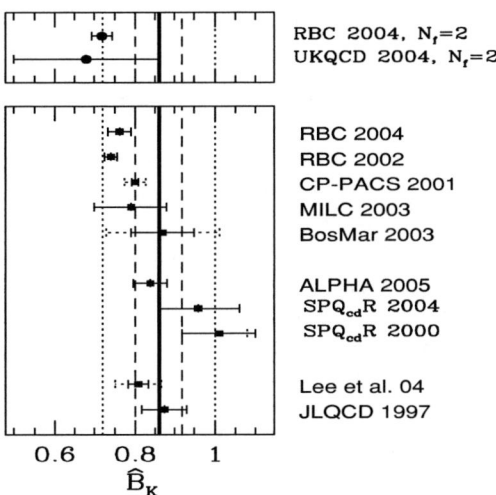

FIGURE 1. Recent lattice calculations of B_K. Only the first two numbers are unquenched.

summarize for completeness in fig. 1 recent lattice estimates of B_K. The present average corresponds to a renormalization group invariant (RGI) value $\hat{B}_K = 0.79(6)(9)$. The second error is an estimate of quenching errors obtained combining the two recent exploratory unquenched calculations [9, 10]. Present uncertainty in ε_K is dominated by the theoretical error in B_K and improvement should come from lattice simulations thanks to next generation of machines.

THE CABIBBO ANGLE AND CKM UNITARITY

Recent experimental and theoretical results on Kaon semileptonic decays have significantly improved the determination of the CKM matrix element V_{us}.

The determination of the Cabibbo angle provides at present the most stringent unitarity test of the CKM matrix, $|V_{ud}|^2 + |V_{us}|^2 + |V_{ub}|^2 = 1$. The value of $|V_{ud}|$ is accurately determined from nuclear super-allowed $0^+ \to 0^+$ beta decays [16]

$$|V_{ud}| = 0.9740 \pm 0.0005.\,^{1} \quad (1)$$

Using this value and by imposing the CKM unitarity, the Cabibbo angle ($|V_{us}|$) amounts to

$$|V_{us}|^{\text{uni.}} = 0.2265 \pm 0.0022. \quad (2)$$

Testing the unitarity of the 1st-row of the CKM matrix means a comparison of this value with $|V_{us}|$ deduced directly from the processes governed by the $s \to u$ transition. Although theoretical constraints on $|V_{us}|$ from the semileptonic hyperon decays [11, 12, 13], $\tau \to K\nu_\tau$ [14] and leptonic Kaon decays [15, 16] ($K_{\ell 2}$) recently became promising too, the best determination of $|V_{us}|$ is still obtained from $K \to \pi \ell \nu$ decay modes ($K_{\ell 3}$).

$K_{\ell 3}$ decays: old vs new experimental results

From the experimental $K_{\ell 3}$ rates, we can directly access to $|V_{us}| \cdot f_+(0)$, where $f_+(0)$ is the vector form factor at zero four-momentum transfer. By averaging old experimental results for $K_{\ell 3}$ decays with the recent measurement by E865 at BNL [21], and using the estimate $f_+^{K^0\pi^-}(0) = 0.961 \pm 0.008$, the PDG quotes $|V_{us}| = 0.2200 \pm 0.0026$ [20]. This value, once combined with eq. (1), implies about 2σ deviation with $|V_{us}|^{\text{unit.}}$ of eq. (2).

With respect to the PDG 2004 analysis, however, a significant novelty is represented by several new experimental results, for both charged and neutral $K_{\ell 3}$ decays, which have been recently presented by KTeV[22], NA48[23] and KLOE[24]. Expressed in terms of $|V_{us}| \cdot f_+(0)$, these determinations are shown in fig. 2. Remarkably, these new results [25] are consistent among themselves, but they disagree with the old ones. In particular, the new results implies that CKM unitarity is recovered (see next section).

The experimental average from new results reads

$$(|V_{us}|f_+(0))^{exp} = 0.2160 \pm 0.0005, \quad (3)$$

which is represented in fig. 2 by the dark-shaded band.

$K_{\ell 3}$ decays: Theory status

The remaining ingredient to extract $|V_{us}|$ from eq. (3) is $f_+(0)$. Its expansion in chiral perturbation theory (ChPT) reads,

$$f_+(0) = 1 + f_2 + f_4 + \ldots, \quad (4)$$

where $f_+(0) = 1$ reflects the CVC in the SU(3) limit, while f_2 and f_4 stand for the leading and next-to-leading

---

[1] The estimate of $|V_{ud}|$ from nuclear super-allowed decays has been updated by Marciano at the CKM 2005 Workshop on the Unitarity Triangle. The new estimate, whose uncertainty is further reduced, reads $|V_{ud}| = 0.9739 \pm 0.0003$ [17].

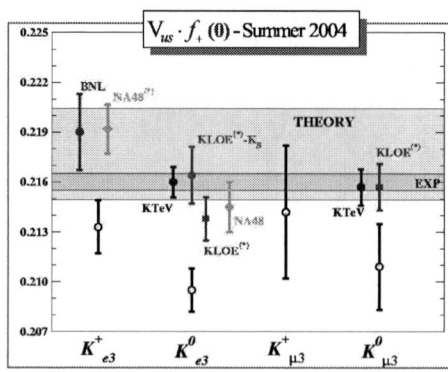

FIGURE 2. Experimental results for $|V_{us}| \cdot f_+(0)$. The "EXP" and "THEORY" bands indicate respectively the average of the new experimental results and the unitarity prediction combined with the LR and lattice determination of the vector form factor.

chiral corrections. AGT ensures that the SU(3) breaking corrections are quadratic in $(m_s - m_u)$ and $f_2 = -0.023$ is a clean prediction by ChPT, i.e. no unknown couplings enter at $\mathcal{O}(p^4)$. The calculation of the chiral loop contribution, $\Delta(\mu)$ in f_4 has been recently completed [26, 27]. The estimate of f_4, however, still suffers from the uncertainty due to the lack of knowledge of the low energy constants. The PDG quotes the value obtained in the quark-model calculation by Leutwyler-Roos [19] (LR),

$$f_4 = -0.016 \pm 0.008 \to f_+(0) = 0.961(8), \quad (5)$$

based on parameterization of the asymmetry between kaon and pion wave functions. If the estimate of $f_+(0)$ in eq. (5) is used along with the experimental average of $|V_{us}|f_+(0)$, eq. (3), one gets

$$|V_{us}| = 0.2248 \pm 0.0018_{f_+(0)} \pm 0.0005_{exp}, \quad (6)$$

in good agreement with the value obtained by imposing the CKM unitarity [cfr eq. (2)]. This compatibility is also observed in fig. 2 where the light-shaded band. The LR estimate has been corroborated this year by a (quenched) lattice QCD study that gave [28]

$$f_4 = -0.017 \pm 0.009 \to f_+(0) = 0.960(9). \quad (7)$$

In addition to the lattice estimates, three more calculations [26, 29, 30] appeared this year, yielding respectively:

$$f_4 = -0.001 \pm 0.010 \to f_+(0) = 0.976(10), \quad (8)$$
$$f_4 = -0.003 \pm 0.011 \to f_+(0) = 0.974(11) \quad (9)$$
$$f_4 = +0.007 \pm 0.012 \to f_+(0) = 0.984(12). \quad (10)$$

However they both contain model-dependent assumptions and disagree with each other. Notice also that by using the values $f_+(0)$ (8)-(10), unitarity is violated by about $1.5 - 2.0\sigma$.

RARE K DECAYS

In the quest to find New Physics, rare Kaon decays are very promising. Whereas in the SM they are GIM-suppressed processes, in the presence of new physics they can get enhancement factors. Moreover, these decays offer a possibility of fully determining the CKM unitarity triangle using exclusively rare kaon decays.

The leading parts of the effective Hamiltonians relevant to the study of these modes are

$$H_{eff}(\bar{s}d \to \nu\bar{\nu}) \sim \frac{G_F}{\sqrt{2}}(y_\nu(\bar{s}d)_{V-A}(\nu\bar{\nu})_{V-A}) \quad (11)$$

$$H_{eff}(\bar{s}d \to \ell^+\ell^-) \sim \frac{G_F}{\sqrt{2}}(y_{7V}(\bar{s}d)_{V-A}(\ell^+\ell^-)_V \quad (12)$$

$$+ y_{7A}(\bar{s}d)_{V-A}(\ell^+\ell^-)_A) \quad (13)$$

$y_\nu = \frac{\alpha}{2\pi}\sum \lambda_q \frac{X_0(x_q)}{\sin^2\theta_W}$, $y_{7A} = \frac{\alpha}{2\pi}\sum \lambda_q \frac{-Y_0(x_q)}{\sin^2\theta_W}$, $y_{7V} = \frac{\alpha}{2\pi}\sum \lambda_q \left[\frac{Y_0(x_q)}{\sin^2\theta_W} - 4Z_0(x_q)\right]$ with $\lambda_q = V_{qs}^*V_{qd}$, $x_q = m_q^2/M_W^2$, summation over $q = u,c,t$ is implicit.

According to the relative strength of the top contribution, these decays can be divided into two classes:

Pure FCNC decays: $K_L \to \pi^0 \nu\bar{\nu}$ and $K^+ \to \pi^+ \nu\bar{\nu}$. These decays are dominated by short-distance dynamics which can be calculated perturbatively and the hadronic matrix element, $\langle K|(\bar{s}d)_{V-A}|\pi\rangle$, can be obtained from the well measured $K^+ \to \pi^0 e^+\nu$ decay. This leads to an extremely precise (to few %) relation between decay rates and the combination of CKM matrix elements $\lambda_t = V_{ts}^*V_{td}$.

Decays with both short-distance and long-distance contribution: $K_L \to \pi^0 l^+ l^-$ ($l = e, \mu$). Here, the long-distance amplitudes must be determined to a sufficient precision in order to access the short-distance physics. The determination of the long-distance contributions is done in the framework of Chiral Perturbation Theory (see later). The SM theoretical predictions are

$$\begin{aligned}BR(K_L \to \pi^0 \nu\bar{\nu}) &= (3.0 \pm 0.6) \times 10^{-11} \\ BR(K^+ \to \pi^+ \nu\bar{\nu}) &= (7.8 \pm 1.2) \times 10^{-11} \\ BR(K_L \to \pi^0 e^+ e^-) &= 3.7^{+1.1}_{-0.9} \times 10^{-11} \\ BR(K_L \to \pi^0 \mu^+ \mu^-) &= (1.5 \pm 0.4) \times 10^{-11}\end{aligned} \quad (14)$$

$K \to \pi\nu\bar{\nu}$

The decay amplitude of $K_L \to \pi^0 \nu\bar{\nu}$, entirely due to direct CP violation, is proportional to $Im\lambda_t$ and determines the height of the CKM unitarity triangle. The present experimental limit [31] is many orders of magnitude above the SM expectation [33, 32], but progress is expected from new experiments.

The decay $K^+ \to \pi^+ \nu\bar{\nu}$ amplitude depends instead on both $Re\lambda_t$ and $Im\lambda_t$. For the real part, the CKM structure compensates the GIM suppression, and the t and c contributions are similar in size (68% vs. 32%). In detail, the top quark effects [33] are known at NLO within 3%, $X(x_t) = 1.529 \pm 0.042$. For the c-quark effects, $X(x_c) = \lambda^4 (0.49 \pm 0.07)$ and corresponds at NLO to a 18% error.

In ref. [34], two sub-leading effects, needed to get down to a few % precision at the BR level, have been analysed. First there are the c-quark dimension eight operators [35] and then, residual u-quark effects [36], which are purely LD.

For dim. 8 operators $Q_c^{(8)}$ like $(\bar{s}d)_{V-A}\partial^2(\nu\bar{\nu})_{V-A}$, we have confirmed by an OPE at the charm scale the basis found in [35]. For u-quark effects, the amplitude is computed at one-loop in ChPT. An important improvement over previous analyzes [36] was to include in the $\mathcal{O}(G_F^2 p^2)$ operator basis all the effective interactions arising from the integration of heavy modes, in particular the local $Z^\mu(\bar{s}d)_{V-A}$ non-gauge invariant one.

Numerically, the combined effect of c-quark dim. 8 operators and of non-local long-distance u-quark loops amount to an enhancement of about 6% for the total rate.

Combining E787 [37] and E789 [38] results, the experimental branching ratio is

$$BR(K^+ \to \pi^+ \nu\bar{\nu}) = (14.7^{+13.0}_{-8.9}) \times 10^{-11} \quad (15)$$

This result is compatible with SM prediction. However, the fact that the central value is about a factor of two above the expectation and the uncertainties are large, leaves an exciting window for NP models [32]. The experiment E949 may take more data in the future and improve the present uncertainty by a factor two or three.

$K_L \to \pi^0 l^+ l^-$

While the experimental state of the art is slowly reaching the level of being capable to seriously search for the $K \to \pi\nu\bar{\nu}$ decays, appealing decays are also the $K_L \to \pi^0 l^+ l^-$. The advantage of this decay is that all the decay products can be detected.

The short-distance direct CP-violating (DCPV) contribution has been calculated in [33]. Unlike $K_L \to \pi^0 \nu\bar{\nu}$, these decays receive also two long-distance contributions. The CP-conserving (CPC) contribution proceeds through two virtual photons and has been recently determined by a precise study of the decay $K_L \to \pi^0 \gamma\gamma$ [41] in the low invariant $\gamma\gamma$ mass region.

The CP-conserving decay $K_1 \to \pi^0 l^+ l^-$, through $K^0 - \bar{K}^0$ mixing, provides an extra long-distance indirect CP-violating (ICPV) contribution to the $K_L \to \pi^0 l^+ l^-$. In addition, DCPV and ICPV components interfere with each other. Clearly, in order to determine this long-distance component the knowledge of $BR(K_S \to \pi^0 l^+ l^-)$ is necessary. Both $K_S \to \pi^0 e^+ e^-$ and $K_S \to \pi^0 \mu^+ \mu^-$ decays were successfully found by NA48/1. In principle, having measured both decays, the form factor, parametrised as

$W_S \sim a_S + b_S(m_{ll}/m_K)^2$, should be fully determined, at least up to a sign ambiguity. Due to low statistics, however, only $|a_S|$ can be measured. Averaging both decay modes one obtains $|a_S|_{ll} = 1.21^{+0.22}_{-0.18}$.

With these measurements, the BR's [39, 40] are :
$BR(K_L \to \pi^0 e^+ e^-) \approx (17_{ICPV} \pm 9_{INTF} + 4_{DCPV}) \times 10^{-12}$
$BR(K_L \to \pi^0 \mu^+ \mu^-) \approx (8_{ICPV} \pm 3_{INTF} + 2_{DCPV} + 5_{CPC}) \times 10^{-12}$.

The sign of the interference term (INTF) depends on the unknown sign of the a_S parameter. A constructive interference is favored now by two theoretical groups [39, 44].

The best present limits on the BR's of $K_L \to \pi^0 l^+ l^-$ decays are provided by the KTeV collaboration:

$$BR(K_L \to \pi^0 e^+ e^-) \leq 2.8 \times 10^{-10} \quad (16)$$
$$BR(K_L \to \pi^0 \mu^+ \mu^-)) \leq 3.8 \times 10^{-10} \quad (17)$$

These limits are still far from SM predictions, however, a combined observation of the two modes increases the sensitivity to New Physics and in addition, information on its nature can be extracted. In particular, a rather recent model of enhanced electroweak penguins predicts an order of magnitude enhancement of the direct CP-violating component in both decays [47]. Therefore, even slight improvement of these limits can have selective power on models beyond SM.

SUMMARY AND PROSPECTS

Most recent developments in the experimental and theoretical kaon physics have been reviewed. Theoretical efforts in rare K decays have shed more light on FCNC processes which seem to be one of the best windows to the new physics in the future. Their measure can then be considered complementary to the search for new particles (LHC) or precision measurements (Super B-Factories).

ACKNOWLEDGMENTS

We thank M. Antonelli, P. Franzini, G Isidori, A. Sibidanov, C. Smith and S. Trine for useful discussions. The work of F.M. is partially supported by IHP-RTN, EC contract No. HPRN-CT-2002-00311 (EURIDICE).

REFERENCES

1. J.H. Christenson et al., Phys. Rev. Lett. **13** (1964) 138.
2. M. Kobayashi and K. Maskawa, Prog. Theor. Phys **49** (1973) 652.
3. S.L. Glashow, J. Iliopoulos and L. Maiani, Phys. Rev. D **2** (1970) 1285.
4. G. Barr et al. (NA31), Phys. Lett. B **317** (1993) 233.
5. A. Lai et al. (NA48), Eur. Phys. J. C **22** (2001) 231.
6. J.R. Batley et al. (NA48), Phys. Lett. B **544** (2002) 97.
7. A. Alavi-Harati et al. (KTeV), Phys. Rev. D **67** (2003) 012005.
8. S. Hashimoto, hep-ph/0411126.
9. J. M. Flynn, F. Mescia and A. S. B. Tariq [hep-lat/0406013].
10. Y. Aoki et al., hep-lat/0411006.
11. N. Cabibbo, E. C. Swallow and R. Winston, hep-ph/0307214; hep-ph/0307298.
12. R. Flores-Mendieta, hep-ph/0410171.
13. D. Guadagnoli et al., hep-lat/0409048.
14. E. Gamiz et al., hep-ph/0408044.
15. C. Aubin et al. [MILC Collaboration], hep-lat/0407028.
16. A. Czarnecki, W. J. Marciano and A. Sirlin, hep-ph/0406324.
17. W. J. Marciano, presented at the CKM 2005 Workshop, San Diego, California, 15-18 March 2005.
18. M. Ademollo and R. Gatto, Phys. Rev. Lett. **13** (1964) 264.
19. H. Leutwyler and M. Roos, Z. Phys. C **25** (1984) 91.
20. S. Eidelman et al. [Particle Data Group Collaboration], Phys. Lett. B **592** (2004) 1.
21. A. Sher et al. [E865 Coll.], Phys. Rev. Lett. **91**, 261802 (2003).
22. T. Alexopoulos et al. [KTeV Collaboration], Phys. Rev. Lett. **93** (2004) 181802.
23. A. Lai et al. [NA48 Collaboration], Phys. Lett. B **602** (2004) 41; Phys. Lett. B **604** (2004) 1.
24. P. Franzini [Kloe Coll.], hep-ex/0408150.
25. F. Mescia, hep-ph/0411097.
26. J. Bijnens and P. Talavera, hep-ph/0303103.
27. P. Post and K. Schilcher, hep-ph/0112352.
28. D. Becirevic et al., hep-ph/0403217.
29. M. Jamin, J. A. Oller and A. Pich, hep-ph/0401080.
30. V. Cirigliano et al, hep-ph/0503108.
31. A. Alavi-Harati et al. (KTeV), Phys. Rev. D **61** (2000) 072006.
32. G. Isidori, hep-ph/0307014.
33. A.J. Buras, M.E. Lautenbacher, M. Misiak and M. Münz, Nucl. Phys. **B423** (1994) 349; G. Buchalla, A.J. Buras and M.E. Lautenbacher, Rev. Mod. Phys. **68** (1996) 1125; A. J. Buras, F. Schwab and S. Uhlig, [hep-ph/0405132].
34. G. Isidori, F. Mescia and C. Smith, hep-ph/0503107. C. Smith, arXiv:hep-ph/0505163.
35. A. F. Falk, A. Lewandowski and A. A. Petrov, Phys. Lett. **B505** (2001) 107.
36. M. Lu and M. B. Wise, Phys. Lett. **B324** (1994) 461.
37. S. Adler et al. (E787), Phys. Rev. Lett. **88** (2002) 041803.
38. V.V. Anisimovski et al. (E949), hep-ex/0403036.
39. G. Buchalla, G. D'Ambrosio and G. Isidori, Nucl. Phys. B **672** (2003) 387.
40. G. Isidori, C. Smith and R. Unterdorfer, hep-ph/0404127.
41. A. Lai et al. (NA48), Phys. Lett. B **536** (2002) 229, A. Alavi-Harati et al. (KTeV), Phys. Rev. Lett. **83** (1999) 917.
42. J.R. Batley et al. (NA48), Phys. Lett. B **576** (2003) 43.
43. J.R. Batley et al. (NA48), Phys. Lett. B **599** (2004) 197.
44. S. Friot, D. Greynat and E. de Rafael, Phys. Lett. B **595** (2004) 301.
45. A. Alavi-Harati et al. (KTeV), hep-ex/0309072.
46. A. Alavi-Harati et al. (KTeV), Phys. Rev. Lett. **84** (2000) 5279.
47. A.J. Buras et al., hep-ph/0402112.

Recent KLOE Results on Kaon Physics

M. Martini[1]

Laboratori Nazionali di Frascati, INFN

Abstract. The KLOE experiment at DAΦNE has collected ~ 450 pb^{-1} of e^+e^- collisions at center of mass energy $W \sim 1.02$ GeV. Final results on the BR($K_S \to 3\pi^0$), K_L dominant Branching ratio, K_L lifetime and the extraction of the CKM parameter V_{us} are presented.

Keywords: e^+e^- collisions, DAΦNE, KLOE, CP, V_{us}
PACS: 11.30.Er,11.30.Qc

INTRODUCTION

DAΦNE, the Frascati ϕ factory, is an e^+e^- collider working at $W \sim m_\phi \sim 1.02$ GeV with a design luminosity of $5 \cdot 10^{32}$ cm^{-2} s^{-1}. ϕ mesons are produced, essentially at rest, with a visible cross section of ~ 3.2 μb and decay into K^+K^- ($K_S K_L$) pairs with BR of $\sim 49\%$ ($\sim 34\%$). These pairs are produced in a pure $J^{PC} = 1^{--}$ quantum state, so that observation of a K_S (K^+) in an event signals (tags) the presence of a K_L (K^-) and viceversa; highly pure and nearly monochromatic K_S, K_L, K^+ and K^- beams can be obtained. Neutral kaons get a momentum of ~ 110 MeV/c which translates in a slow speed, $\beta_K \sim 0.22$. K_S and K_L can therefore be distinguished by their mean decay lengths: $\lambda_S \sim 0.6$ cm and $\lambda_L \sim 340$ cm.

The KLOE detector consists essentially of a drift chamber, DCH, surrounded by an electromagnetic calorimeter, EMC. The DCH [1] is a cylinder of 4 m diameter and 3.3 m in length which constitutes a large fiducial volume for K_L decays (1/2 of λ_L). The momentum resolution for tracks at large polar angle is $\sigma_p/p \leq 0.4\%$. The EMC is a lead-scintillating fiber calorimeter [2] consisting of a barrel and two end-caps which cover 98% of the solid angle. The energy resolution is $\sigma_E/E \sim 5.7\%/\sqrt{E(\text{GeV})}$. The intrinsic time resolution is $\sigma_T = 54$ ps/$\sqrt{E(\text{GeV})} \oplus 50$ ps. A superconducting coil surrounding the barrel provides a 0.52 T magnetic field.

During 2002 data taking, the maximum luminosity reached by DAΦNE was $7.5 \cdot 10^{31}$ cm^{-2} s^{-1}. Although this is lower than the design value, the performance of the machine was improving during the years and, at the end of 2002, we collected ~ 4.5 pb^{-1}/day. The whole sample (2001-2002) amounts to 450 pb^{-1}, equivalent to 1.4 billion ϕ decays. Recently, the machine has been upgraded and KLOE has resumed its data taking in spring 2004.

DIRECT SEARCH OF $K_S \to 3\pi^0$

The decay $K_S \to 3\pi^0$ is a pure CP violating process. The related CP violation parameter η_{000} is defined as the ratio of K_S to K_L decay amplitudes: $\eta_{000} = A(K_S \to 3\pi^0)/A(K_L \to 3\pi^0) = \varepsilon + \varepsilon'_{000}$ where ε describes the CP violation in the mixing matrix and ε'_{000} is a direct CP violating term. In the Standard Model we expect η_{000} to be similar to η_{00}. The expected branching ratio of this decay is therefore $\sim 2 \cdot 10^{-9}$, making its direct observation really challenging. The best upper limit on the BR (i.e. on $|\eta_{000}|^2$) has been set to $1.5 \cdot 10^{-5}$ by SND [3] where, similar to KLOE, it is possible to tag a K_S beam. The other existing technique is to detect the interference term between $K_S K_L$ in the same final state which is proportional to η_{000}. The best published result using this method comes from the NA48 collaboration [4], and they set an upper limit at BR($K_S \to 3\pi^0$) $\leq 7.4 \times 10^{-7}$ at 90% C.L.

---

[1] The KLOE collaboration: F. Ambrosino, A. Antonelli, M. Antonelli, C. Bacci, P. Beltrame, G. Bencivenni, S. Bertolucci, C. Bini, C. Bloise, V. Bocci, F. Bossi, D. Bowring, P. Branchini, R. Caloi, P. Campana, G. Capon, T. Capussella, F. Ceradini, S. Chi, G. Chiefari, P. Ciambrone, S. Conetti, E. De Lucia, A. De Santis, P. De Simone, G. De Zorzi, S. Dell'Agnello, A. Denig, A. Di Domenico, C. Di Donato, S. Di Falco, B. Di Micco, A. Doria, M. Dreucci, G. Felici, A. Ferrari, M. L. Ferrer, G. Finocchiaro, C. Forti, P. Franzini, C. Gatti, P. Gauzzi, S. Giovannella, E. Gorini, E. Graziani, M. Incagli, W. Kluge, V. Kulikov, F. Lacava, G. Lanfranchi, J. Lee-Franzini, D. Leone, M. Martini, P. Massarotti, W. Mei, S. Meola, S. Miscetti, M. Moulson, S. Müller, F. Murtas, M. Napolitano, F. Nguyen, M. Palutan, E. Pasqualucci, A. Passeri, V. Patera, F. Perfetto, L. Pontecorvo, M. Primavera, P. Santangelo, E. Santovetti, G. Saracino, B. Sciascia, A. Sciubba, F. Scuri, I. Sfiligoi, T. Spadaro, M. Testa, L. Tortora, P. Valente, B. Valeriani, G. Venanzoni, S. Veneziano, R. Versaci, G. Xu.

K_S tagging procedure and analysis method

The tagging of K_L and K_S is the basis of each KLOE analysis for neutral kaons. K_S's are tagged by identifying a K_L interaction, K_L-crash, in the calorimeter, which has a very distinctive signature given by a late ($\beta_K = 0.2$) high-energy cluster not associated to any track. Reconstruction of one kaon establishes the trajectory of the other one with an angular resolution of $1°$ and a momentum resolution of ~ 2 MeV.

Our selection starts by requiring a K_L-crash tag and six neutral clusters coming from the interaction point, IP. A tight constraint on β and moderate requirements on energy and angular acceptance are applied in order to have a large control sample for the background, while retaining large selection efficiency for the signal. On 450 pb^{-1} we have an initial sample of 39 k events dominated by $K_S \to 2\pi^0 +$ 2 false γ. To reduce the sample, a kinematic fit which imposes K_S mass, K_L 4-momentum conservation and $\beta = 1$ for each γ is applied. Only the events with $\chi^2_{\text{fit}}/\text{ndf} < 3$ are retained for further analysis. However, this cut improves the rejection power only of a factor ~ 3 and, to better discriminate $2\pi^0$ vs $3\pi^0$ final state, we build two pseudo-χ^2 variables: $\chi^2_{3\pi}$, which is based on the 3 best π^0 mass estimates and $\chi^2_{2\pi}$, which selects 4 out of the 6 photons providing the best kinematic agreement with the background decay. A signal box region in the $\chi^2_{2\pi}$ vs $\chi^2_{3\pi}$ plane has been defined by optimizing the upper limit in the MC sample.

At the end of analysis we find 2 events in the signal box with an estimate background of $N_b = 3.13 \pm 0.82_{stat} \pm 0.37_{sys}$. To derive the upper limit on the number of signal counts, we build the background probability distribution function, taking into account the uncertainties on the MC calibration factors. This function is folded with a Gaussian of width equivalent to the entire systematic uncertainty on the background. Using the Neyman construction [5] we limit the number of $K_S \to 3\pi^0$ decays observed to 3.45 at 90% C.L. with a total reconstruction efficiency of $(24.36 \pm 0.11_{stat} \pm 0.57_{sys})\%$. In the same tagged sample, we count 3.78×10^7 $K_S \to 2\pi^0$ events used as normalization. Finally, using the value BR($K_S \to 2\pi^0) = 0.3105 \pm 0.0014$ [6] we obtain BR($K_S \to 3\pi^0) \leq 1.2 \times 10^{-7}$ at 90% C.L. [7] which represents an improvement by a factor ~ 6 with respect to the previous limit.

This result can also be translated into a limit $|\eta_{000}| < 0.018$ at 90% C.L. which confirms that the uncertainty for this decay is now negligible in the calculation of $\Im(\delta)$.

MEASUREMENT OF THE DOMINANT K_L BRANCHING RATIOS

The K_L absolute branching fractions are determined on a tagged K_L events sample by counting the number of K_L decays for each channel, N_i, in the used fiducial volume and correcting for acceptance, reconstruction efficiency and tagging efficiency. The $BR(K_L \to i)$ is then:

$$\frac{N_i}{N_{tag}} \times \frac{1}{\varepsilon_{rec}(i) \times \varepsilon_{FV}(\tau_L) \times \varepsilon_{tag}(i)/\varepsilon_{tag}(all)} \quad (1)$$

where $i = 3\pi^0$, $\pi^{\pm}e^{\mp}\nu$, $\pi^{\pm}\mu^{\mp}\nu$, $\pi^+\pi^-\pi^0$, N_{tag} is the number of tagging events, ε_{rec} is the reconstruction efficiency ($\sim 45\%$ for $\pi^+\pi^-\pi^0$ events, $\sim 60\%$ for $\pi^{\pm}e^{\mp}\nu$, $\pi^{\pm}\mu^{\mp}\nu$ events and $\sim 100\%$ for $3\pi^0$ events) and $\varepsilon_{FV}(\tau_L)$ is the fiducial volume geometrical acceptance which depends on the K_L lifetime. Finally, the ratio $\varepsilon_{tag}(i)/\varepsilon_{tag}(all)$ represents the fractional variation of the tagging efficiency when the K_L decays in a given channel with respect to the average over all channels. We define this ratio as *tag bias*.

K_L tagging procedure

The tag is provided by $K_S \to \pi^+\pi^-$ events selected requiring the presence of a vertex with two opposite curvature tracks within a cylinder around the IP.

The main source of tag bias is due to the dependence of the trigger efficiency on the K_L behavior. The calorimeter trigger which requires two local energy deposits above some thresholds is used for the present analysis. The trigger efficiency is essentially 100% for $3\pi^0$, while ranging between $95 - 85\%$ for charged K_L decays. To reduce the tag bias due to trigger efficiency we require the trigger conditions to be satisfied by the pions from K_S.

The *FV* used for the analysis is defined inside the drift chamber by 35 cm $< \sqrt{x^2+y^2} < 150$ cm and $|z| < 120$ cm, where (x,y,z) are the K_L decay vertex position coordinates. The defined FV contains $\sim 26.1\%$ of the K_L decays. This choice minimizes the difference in tag bias among the decay modes. The average tag bias is 0.985, 0.99 and 1.02 for $\pi^{\pm}e^{\mp}\nu$ or $\pi^{\pm}\mu^{\mp}\nu$, $\pi^+\pi^-\pi^0$ and $3\pi^0$ decays respectively.

Charged K_L decays

The K_L in charged decay modes are selected by requiring the presence of two good tracks forming a vertex in the FV and not belonging to the K_S decay tree. A track is associated to the K_L if the point of closest approach to

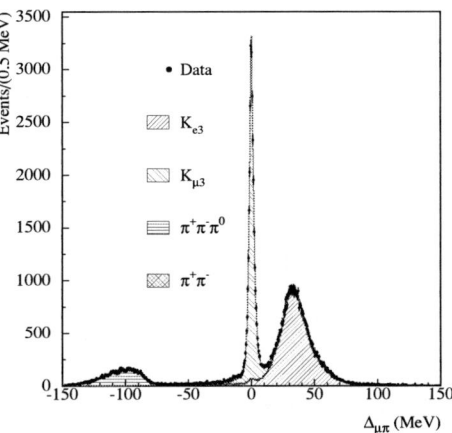

FIGURE 1. $\Delta_{\pi\mu}$ distribution for an event subsample. Dots are data, solid histograms are Monte Carlo.

the K_L line of flight has a distance with respect to the K_L line of flight $d_c < a\sqrt{x^2+y^2}+b$, with $a = 0.03$ and $b = 5$ cm.

The tracking efficiency has been determined by counting the number of events with at least one found K_L track and the number of events in which there are two opposite sign decay tracks. The tracking efficiency is also evaluated from Monte Carlo simulation and it is 60.5% for K_{e3}, 58.5% for $K_{\mu3}$ and 43.0% for $\pi^+\pi^-\pi^0$. A correction to the tracking efficiency has been applied by comparing the results for data and Monte Carlo simulation using independently selected samples of $K_L \to \pi^+\pi^-\pi^0$ and $K_L \to \pi^\pm e^\mp \nu$. The correction is evaluated as a function of the track momentum and ranges between 1.03 and 0.99 depending on the decay channel.

The variable to discriminate among the different charged decay modes is the smallest of the two values of $\Delta_{\pi\mu} = |\vec{p}_{mis}| - E_{miss}$, where $\vec{p}_{mis}$ is the missing momentum and E_{miss} is the missing energy evaluated for the two mass assignments $\pi^+\mu^-$ or $\pi^-\mu^+$. An example of this distribution is shown in Fig. 1 where the different components are shown.

The $\Delta_{\pi\mu}$ distribution obtained with data is fitted with a linear combination of three Monte Carlo distributions (for $\pi^\pm e^\mp \nu$, $\pi^\pm \mu^\mp \nu$ and $\pi^+\pi^-\pi^0$ events) by leaving free their relative weights. The contribution from the CP violating decay $K_L \to \pi^+\pi^-$ and the $K_L \to \pi^0\pi^0\pi^0$ with Dalitz conversion is kept fixed in the fit and amounts to 0.3%. The radiative corrections which affect mainly the K_{e3} decays have been properly taken into account in the Monte Carlo simulation [8].

Neutral K_L decays

$K_L \to \pi^0\pi^0\pi^0$ decays are selected counting neutral clusters and estimating the position of the K_L vertex by using the photon arrival times on the EMC. Each photon defines a time of flight triangle. The three sides are the K_L decay length, L_K; the distance from the decay vertex to the calorimeter cluster centroid, L_γ; and the distance from the cluster to the ϕ vertex, L.

The accuracy of this method is checked with $K_L \to \pi^+\pi^-\pi^0$ decays, comparing the position of the K_L decay vertex from tracking using the $\pi^+\pi^-$ pair with the one from timing with the two photons from π^0. The vertex reconstructed by the calorimeter has shows an offset of 2 mm from the charge one which is almost uniform in the fiducial volume.

To select the $K_L \to \pi^0\pi^0\pi^0$ events we require at least three photons with energy greater than 20 MeV originating from the same vertex.

The main sources of inefficiencies are: 1) geometrical acceptance; 2) cluster energy threshold; 3) merging of clusters; 4) accidental association to a charged track; 5) Dalitz decay of one or more π^0's. The effect of these inefficiencies is to modify the relative population for events with 3, 4, 5, 6, 7 and ≥ 8, clusters with a loss of efficiency of $\sim 0.8\%$.

Background contamination affects only events with three or four clusters. The main source of background comes from $K_L \to \pi^+\pi^-\pi^0$ decays where one or two charged pions produce a cluster not associated to a track and neither track is associated to the K_L vertex. This background is rejected by requiring at least one cluster with at least 50 MeV energy and a polar angle satisfying $|\cos\theta| < 0.88$. Other sources of background are $K_L \to \pi^0\pi^0$ decays, possibly in coincidence with machine background particles ($e^\pm$ or γ) that shower in the QCAL and generate soft neutral particles and $K_S \to \pi^0\pi^0$ following the $K_L \to K_S$ regeneration in the drift chamber material.

Results

A total of ~ 13 millions of tagged K_L are used to compute the branching fractions, almost ~ 40 millions to evaluate systematic uncertainties. Using the published result of the K_L lifetime ($\tau_L = 51.54 \pm 0.44$ ns)[9] we obtain the following results:

$$BR(K_L \to \pi^\pm e^\mp \nu) = (40.49 \pm 0.10_{stat} \pm 0.31_{syst})\%$$
$$BR(K_L \to \pi^\pm \mu^\mp \nu) = (27.26 \pm 0.08_{stat} \pm 0.22_{syst})\%$$
$$BR(K_L \to \pi^0\pi^0\pi^0) = (20.18 \pm 0.04_{stat} \pm 0.26_{syst})\%$$
$$BR(K_L \to \pi^+\pi^-\pi^0) = (12.76 \pm 0.06_{stat} \pm 0.16_{syst})\%$$

TABLE 1. Summary of systematic uncertainties on the absolute branching fractions measurements.

	$\pi^\pm e^\mp \nu$	$\pi^\pm \mu^\mp \nu$	$\pi^+\pi^-\pi^0$	$\pi^0\pi^0\pi^0$
Selection	0.0011	0.0007	0.0004	0.0020
Sig. Shape	0.0006	0.0009	0.0010	-
Tag Bias	0.0013	0.0008	0.0007	0.0005
lifetime	0.0023	0.0017	0.0007	0.0012

where the sources of systematic uncertainties are shown in Table 1. The sum of all measured branching fraction above, plus the PDG value for rare decays, 0.0036, is $\sum BR_i = 1.0104 \pm 0.0018_{\text{cor.}} \pm 0.0074_{\text{uncor.}}$ where the correlated error includes all contributions to the uncertainties on the branching ratios that are 100% correlated between channels, such as the uncertainty in the value of the K_L lifetime. This result depends on the value of the K_L lifetime through the acceptance (Eq. 1). Turning the argument around, by normalizing the sum to 1 we obtain an indirect estimate of the K_L lifetime: $\tau(K_L) = (50.72 \pm 0.14_{\text{stat}} \pm 0.36_{\text{syst}})$ns and a new set of values for the branching fractions:

$$BR(K_L \to \pi^\pm e^\mp \nu) = (40.07 \pm 0.06_{stat} \pm 0.14_{syst})\%$$
$$BR(K_L \to \pi^\pm \mu^\mp \nu) = (26.98 \pm 0.06_{stat} \pm 0.14_{syst})\%$$
$$BR(K_L \to \pi^0\pi^0\pi^0) = (19.97 \pm 0.05_{stat} \pm 0.19_{syst})\%$$
$$BR(K_L \to \pi^+\pi^-\pi^0) = (12.63 \pm 0.05_{stat} \pm 0.11_{syst})\%$$

DIRECT MEASUREMENT OF THE K_L LIFETIME

We have measured the K_L lifetime using $\sim 15 \times 10^6$ events of the fully neutral decay $K_L \to \pi^0\pi^0\pi^0$ tagged by $K_S \to \pi^+\pi^-$ events.

The K_L lifetime is measured using events with a vertex reconstructed in the region 40 cm $< L_K <$ 165 cm and with a flight direction defined by a polar angle θ with respect to the beam axis between $40° < \theta < 140°$. These two conditions define the fiducial volume.

The K_L proper time, t^*, is obtained event by event dividing the decay length L_K by $\beta\gamma$ of the K_L in the laboratory, $t^* = L_K/(\beta\gamma c)$. The residual background is subtracted bin by bin using Monte Carlo predictions.

The statistical uncertainty of the efficiency values ($\sim$ 0.1%) has been taken into account by adding it in quadrature to the statistical fluctuation of the entries in each bin of the t^* distribution after the background subtraction.

The distribution is fitted with an exponential function inside the fiducial volume. The major sources of systematic uncertainties come from the background evaluation, tagging and selection efficiency and from the estimate of the K_L nuclear interactions in the drift chamber material. The total systematic error is $\sim 0.5\%$. With $\sim 8.5 \cdot 10^6$ events inside the fit region we obtain

$$\tau(K_L) = (50.87 \pm 0.17_{\text{stat}} \pm 0.25_{\text{syst}})\text{ns}$$

with a $\chi^2 = 58$ for 62 degrees of freedom. It is compatible at 1.3σ level with the other measurement[9] and only at 1.7σ level with the PDG 2004 fit [6].

DETERMINATION OF $|V_{US}|$

$|V_{us}|$ is proportional to the square root of the semileptonic BR of K mesons. For K_L^{e3} decays we can write [10]:

$$|V_{us}| \times f_+^{K^0\pi}(0) = \sqrt{\frac{128\pi^3 BR(K_L \to \pi e \nu)}{\tau_L G_\mu^2 M^5 S_{ew} I_K^e(\lambda_+, \lambda'_+)}} \times \frac{1}{1+\delta_{\text{em}}^{Ke}} \quad (2)$$

In the extraction of V_{us} we use the values of λ_+ and λ'_+ obtained by KTeV experiment from a quadratic fit [13]. We evaluate the value of $|V_{us}| \times f_+^{K^0\pi}(0)$ using the KLOE preliminary measurement [14] of $BR(K_S \to \pi^\pm e^\mp \nu) = (7.09 \pm 0.07_{stat} \pm 0.08_{syst}) \cdot 10^{-4}$ and the $\pm 1\sigma$ band from the $|V_{ud}|$ and unitarity where we use $f_+^{K^0\pi}(0) = 0.961 \pm 0.008$ following Ref. [15].

REFERENCES

1. M. Adinolfi et al., KLOE Collaboration, Nucl. Instrum. Methods A **488**, 51 (2002).
2. M. Adinolfi et al., KLOE Collaboration, Nucl. Instrum. Methods A **482**, 364 (2002).
3. M.N. Achasov et al., SND Collaboration, Phys. Lett. B **459**, 674 (1999).
4. A. Lai et al, NA48 Collaboration, Phys. Lett. B 610 (2005) 165.
5. G. J. Feldman and R. Cousins, Phys. Rev. D57 (1998), 57.
6. S. Eidelman et al., Phys. Lett. B592 (2004) 1
7. F. Ambrosino et al., KLOE collaboration, arXiv: hep-ex/0505012 (2005).
8. C. Gatti, Kloe Note 194 (2004) http://www.lnf.infn.it/kloe/pub/knote/kn194.ps.
9. K. G. Vosburgh et al., Phys. Rev. Lett. 26 (1971), 866.
10. M. Battaglia et al., arXiv: hep-ph/0304132.
11. V. Cirigliano et al., Eur. Phys. J. C23,121 (2002).
12. V. Cirigliano et al., Eur. Phys. J. C**35**,53 (2004).
13. T. Alexopoulos et al., KTeV Collaboration],
14. T. Spadaro, *Recent results from KLOE at DAΦNE*, Proceedings of Les Rencontres de Physique de la Vallee d'Aoste, La Thuile 2004, ed. M. Greco.
15. H. Leutwyler, M. Roos, Z. Phys. C **25**, 91 (1984).

Kaon Physics at CERN: Recent Results by NA48/2 Experiment

G. Lamanna

Università & INFN di Pisa
Largo Pontecorvo 3, 56100, Pisa (Italy)

Abstract. NA48/2 is devoted to study direct CP violation in three charged pions kaon decay. A novel design for the beam line and an upgrade of the NA48 detector are adopted to allow the simultaneous detection of decays from K+/K- unseparated beams. The first preliminary measurement of the direct CP violating parameter A_g is presented. A value for the charge asymmetry in the $K^\pm \to \pi^\pm \pi^+ \pi^-$ of $A_g = (0.5 \pm 3.8) \cdot 10^{-4}$ has been measured using using 1.6 billions K^+/K^- decays. The analysis of the charge asymmetry in the "neutral" channel $K^\pm \to \pi^\pm \pi^0 \pi^0$ and a new method to extract the $\pi\pi$ scattering length studying accurately the Dalitz Plot in the "neutral" mode are also shortly discussed. The future program on kaon physics at CERN is briefly outlined.

Keywords: kaons, CP violation, NA48, charge asymmetry
PACS: 11.50.Er,14.40.Aq,13.25.Es,12.15.Hh

INTRODUCTION

The kaon physics, from the early times of the particle physics, represents an useful tool to understand the intimate nature of the fundamental interactions and in particular the interesting phenomena of CP violation is an important part in this topic. It took more than three decades since the discover of CP violation in the neutral kaons by Christenson, Cronin, Fitch and Turlay [1] until that a new kind of CP violation [1] was definitively discovered. After the unconfirmed indication by NA31 [2], KTev [3] and NA48 [4] have demonstrated, with high significance, that direct CP violation exists in the decays of neutral kaons into two pions. Only recently, the B-factory experiments have found the direct CP violation also in the B sector [5].

However the deep understanding of the CP violation still plays an important role both in the test of Standard Model (SM) and in the discovery of the physics beyond the SM. Indeed a possible non-SM structure could appear in the heavy quarks loops which are responsible of the direct CP violating processes. Besides the already measured quantity ε', the most promising observables are the charge asymmetry in the $K \to 3\pi$ decay and the decay rates of the GIM suppressed decays. The main goal of the NA48/2 experiment is to measure the charge asymmetry in the 3π mode of the charged kaons.

The three body decay has low Q value, which allows to expand the amplitude as:

$$|A(K \to 3\pi)| \sim a + b \cdot u + O(u^2, v^2)$$

where the *Dalitz variables* u and v are defined as [2]:

$$u = \frac{(s_3 - s_0)}{m_\pi^2}, \quad v = \frac{(s_2 - s_1)}{m_\pi^2},$$

the direct CP violation might rise from the difference between the weak and strong phases for the two charge conjugate modes. In principle a difference in the partial decay widths $|M|^2$ could be signal of CP violation too, but this difference is highly suppressed, therefore is very challenging from an experimental point of view. Anyway the direct CP violation is also visible in the *shape* of the Dalitz Plot. This can be formalized expanding the matrix element as:

$$|M|^2 \propto 1 + gu + hu^2 + kv^2 + \ldots,$$

where the coefficient g is called *linear slope*[3]. NA48/2 wants to measure the slope asymmetry:

$$A_g = \frac{g_+ - g_-}{g_+ + g_-},$$

with a precision of $2.2 \cdot 10^{-4}$ for the $\pi^\pm \pi^+ \pi^-$ (A_g^c, "charge") decay mode and $3.5 \cdot 10^{-4}$ for the $\pi^\pm \pi^0 \pi^0$

---

[1] the so called *Direct CP violation* that happens directly in the decay.

[2] where $s_i = (p_K - p_i)^2$ with index 3 for the odd pion, and $s_0 = (s_1 + s_2 + s_3)/3$.
[3] the *quadratic slopes* h and k are small with respect to g.

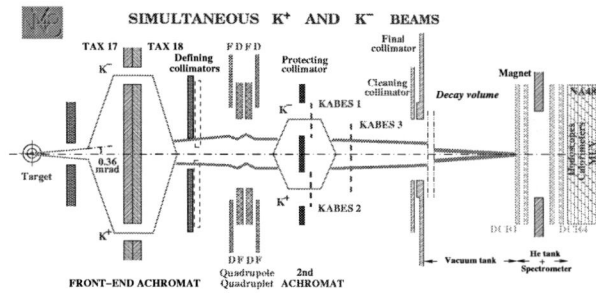

FIGURE 1. The NA48/2 beam line and detector. Not in scale.

(A_g^n, "neutral") decay mode, increasing by a factor 10 the present experimental sensitivity. The best present measurement are [6] $A_g^c=(22\pm15\pm37)\cdot10^{-4}$ (prelim.) and [7] $A_g^n=(2\pm19)\cdot10^{-4}$ based on $54\cdot10^6$ and $620\cdot10^3$ events respectively.

The theoretical prevision for this quantity depends from the calculation of the strong loops giving results from 10^{-6} to 10^{-5} but all the authors concord that a value of A_g greater then $5\cdot10^{-5}$ could be signal of new physics.

To match this goal NA48/2 took data in two periods (50 days in 2003 and 60 days in 2004) using two high intensity, simultaneous and focused K+/K- beams, collecting 4 billions of $K^\pm\to\pi^\pm\pi^+\pi^-$ and 200 millions of $K^\pm\to\pi^\pm\pi^0\pi^0$. Due to this high statistics in the neutral mode has been also possible to study accurately the Dalitz Plot to extract precise information about the $\pi\pi$ scattering in a complete new way.

THE NA48/2 EXPERIMENT

The NA48/2 detector is based essentially on the existing NA48 setup and infrastructure, with a novel beam line design to transport simultaneously K+ and K- beams that overlap in the decay region. The hadron beams are produced using primary protons from the SPS impinging a beryllium target rod 40 cm long and 2 mm in diameter at zero degrees. The particles of opposite charge are split, selected in a narrow momentum band ($p_k=60\pm3GeV$) and recombined by passing trough an achromatic device. A quadruplet of quadrupoles focuses the two beams so they overlap with accuracy better than 1mm in the fiducial region. A second "achromat" houses a collimator for absorbing neutrons and 3 stations of a MICROMEGAS based detector (KABES, KAon Beam Spectrometer) to measure the kaon momentum with 1% resolution. The particles rate at the KABES is $\sim40MHz$ and the $K+/K-=1.8$. The fiducial volume is located in a $\sim114m$ vacuum chamber to minimize scattering of secondary particles. The beam pipe is large enough to contain all the decay products of the pions present in the beams (the beam contains 10 times pions that kaons).

The central NA48 detector is described elsewhere [8]. The spectrometer magnet was operated in order to give $Pt_{kick}=120MeV/c$ and the resolution in momentum (GeV/c) is $\sigma_p/p=1.0\%\oplus0.044$. To detect the photons, for the neutral mode, the LKR calorimeter provides a very good energy (GeV) resolution of $\sigma_E/E=3.2\%/\sqrt{E}\oplus9\%/E\oplus0.42\%$. Thanks to this good detectors the resolution on the recostructed kaon mass is good enough (1.7 MeV in the charge mode, 1.2 MeV in the neutral mode) which allow for precise calibration and monitoring of the detector instabilities itself.

A two level trigger is employed to reduce the rate of the data collection. An hardware level trigger (L1) using fast information from a hodoscope counter is followed by a L2 selection based on the number of online recon-structed tracks (for the charge mode) and the kinematic rejection[4] of the $\pi^+\pi^0$ for the events with one track only. The final trigger rate is 10KHz.

ASYMMETRY ANALYSIS

The method to extract A_g is based on the comparison between the U distribution of the K^+ and K^- decays. In case of $\pi^\pm\pi^+\pi^-$ decay the value [9] of $g=(-0.2154\pm0.0035)$ authorizes some approximations to obtain:

$$R(u)=\frac{N_{K^+}}{N_{K^-}}\propto(1+\Delta g\cdot u) \quad,$$

where $\Delta g=g_+-g_-$. A linear fit is used to extract Δg connected to A_g by the relation $A_g=\Delta g/2g$.

However the presence of magnets both in the beam sector (achromats, focusing magnets, etc.) and in the detector (magnetic spectrometer) introduces a intrinsic charge asymmetry of the apparatus. In order to equalize this charge asymmetry the mains magnetic fields are frequently reversed during the data taking (the beams magnets are inverted on a weekly basis, the spectrometer magnet on a daily basis). In this way, during the 2003 data taking, are defined 4 periods (Super Samples) in which all the combination of magnetic fields are present. For each achromat polarities it's possible to define ratios in which the same part of the spectrometer is involved for opposite charge of the kaons. In particular the product of the two ratios[5]:

$$R_J(u)=\frac{N_{K^+}^{B^-}}{N_{K^-}^{B^+}}\quad,\quad R_S(u)=\frac{N_{K^+}^{B^+}}{N_{K^-}^{B^-}}\quad,$$

---

[4] In the one track event the missing mass (assuming the kaon direction of flight) is required to be different from the π^0.
[5] The letters J (Jura) and S (Saleve) represent the two mountain on the left and right with respect to the kaons directions.

assures that the acceptance differences, due to the spectrometer magnet, are cancelled automatically. The full cancellation of all the acceptance asymmetries is obtained in the *quadruple ratio*[6]:

$$R(u) = R_{US}R_{UJ}R_{DS}R_{DJ} \sim \bar{R}(1 + 4\Delta g u) \quad .$$

The presence of stray magnetic fields (earth field, etc.) is taking into account by measuring directly and applying corrections in the reconstruction. Due to the fact that the quantities in the numerator and in the denominator of each ratios are collected in different periods, this method is sensitive only from the time instabilities of the detector which have a characteristic time smaller than the corresponding field alternation period. Due to the superposition of the two beams and the cancellation of the acceptance asymmetries, the measurement does not need a Monte Carlo. Nevertheless a detailed GEANT-based Monte Carlo simulation has been developed for systematic studies.

The only detectors involved in the "charged" analysis are the spectrometer and the hodoscope for trigger purpose. The event selection is quite simple due to the fact that the channel $K^\pm \to \pi^\pm\pi^+\pi^-$, with a $BR = 5.576 \pm 0.031)\%$ [9], is essentially background free. Three tracks are required asking for quality cuts on the reconstruction in symmetric way for K^+ and K^-. The good resolution on the kaon mass ($\sim 1.7 MeV$) allows to cut $9MeV/c^2$ around the nominal mass to decrease the contribution of the secondary pions decay. As mentioned above particular care is necessary to check the stability in time of the detector and beams. The two mains effects are:

- Time variation of the beams geometry: the beams aren't exactly the same in every sub period.
- Time variation of the spectrometer alignment: a small relative drift of the chambers has been observed.

The beams movements can introduce difference in the acceptance cuts, mainly due to the inner cut (due to the physical presence of the beam pipe) in the chambers. To avoid this, the position of the center for both beams is recorded by measuring the center of gravity in every run and a cut around the actual center is performed separately for K^+ and K^-. The movement of the beam during the spill (observed directly using a beam monitor in the last part of the beam pipe) does not introduce additional corrections. Also the dependence of the beam position from the kaon momentum (non perfectly monochromatic) is taken into account. In this way the K^+ and the K^- acceptance cancel entirely. A conservative limit on resid-

TABLE 1. Limits on systematic uncertainties on Δg in units of 10^{-4}

Acceptance and beam geometry	0.5
Spectrometer alignment	0.1
Analysing magnet	0.1
Pion decay	0.4
Calculation of u and fitting	0.5
Pile-up	0.3
Trigger	0.9

ual systematic uncertainty, $\delta(\Delta g) = 0.5 \cdot 10^{-4}$ was determined studying the sensitivity of the result to various acceptance cuts.

The variations of the characteristics of the spectrometer are observed by looking for the difference between the K^+ and K^- masses in the same run. A residual horizontal shift[7] can introduce this splitting in the measured masses and a charge asymmetric mis-measurement of the momenta. Also a non perfect inversion of the spectrometer magnet can introduce a variation in the measured kaon mass (coherent between opposite charge in the same run). These two effects are corrected forcing the kaon masses to be equal for opposite charge and also equal to the nominal PDG value.

Other important source of systematic errors is the trigger. The level 1 trigger is very efficient ($> 99\%$) and no correction is applied on the data. Small correction is applied for the L2, by measuring the trigger efficiency that varies with time ($> 98\%$) as a function of u. However the systematic bias due to the trigger is fully dominated from the statistic in the control samples. Tab 1 summarises the mains systematic uncertainties attributed for this preliminary result.

The preliminary result presented in this paper on the "charge" mode is based on the whole sample collect in 2003, which includes $1.6 \cdot 10^9$ $K^\pm \to \pi^\pm\pi^+\pi^-$. The results calculated independently in each super samples are statistically consistent. The stability of the result was checked with respect to several variables (kaon energy, vertex position, etc.) without finding any significant dependence. Others ratios, using the same kaon charge in numerator and in denominator, can be constructed to check the detector asymmetries. Such effects are at the 10^{-4} level and are fully reproduced by the Monte Carlo that include time variation of the detector efficiency and of the beam optics. The valuation of the systematic is conservative at this level and the trigger component of the uncertainty could be reduced. The preliminary result (average of three independent analysis that give consis-

---

[6] The letters U (Up) and D (Down) represent the path (upper or lower) in which the K^+ runs in the achromat.

[7] the chambers are aligned using special parallel muons runs.

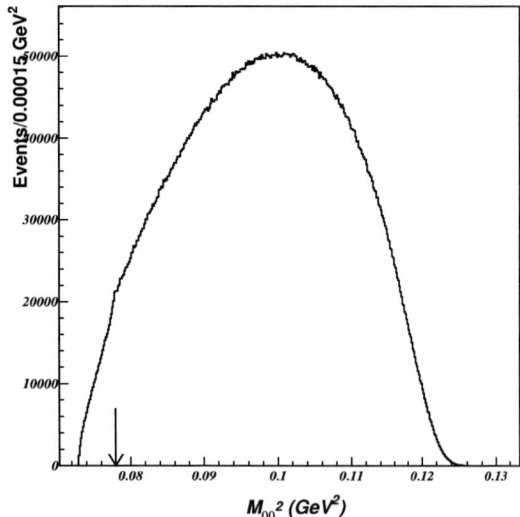

FIGURE 2. The $M^2_{\pi^0\pi^0}$ distribution. The arrow is in the threshold $(2\cdot m_{\pi^+})^2$ position.

tent results) is[8]

$$A_g = (0.5 \pm 2.4_{stat} \pm 2.1_{stat(trig)} \pm 2.1_{syst}) \cdot 10^{-4}$$
$$= (0.5 \pm 3.8) \cdot 10^{-4} \quad .$$

This result is consistent with no CP violation with a precision one order of magnitude better than earlier experiments. The 2004 data sample, which is being analyzed, contains more data than in 2003 and the systematic error could be lower thanks to the more frequent change of the magnets polarities.

THE NEUTRAL MODE

NA48/2 collected the largest samples of $K^\pm \to \pi^\pm \pi^0 \pi^0$. The analysis of the asymmetries is quite advanced and the final result could be competitive with the charge mode. In spite of the lower statistics ($\sim 60 \cdot 10^6$ events in 2003) the most favourable Dalitz Plot distribution and the greater value of $g = 0.638 \pm 0.020$ give a similar statistical error on A'_g. The systematics biases are completely different with respect to the charge mode, due to the fact that the detector involved in the studies of this decay is, mainly, the LKr calorimeter. The U distribution can be constructed in two complementary way, using the LKr only or using KABES and the spectrometer. The two results can cross check each other.

---

[8] using the PDG value for the linear slope g.

An interesting effect is observed for the first time in the Dalitz Plot distribution in the neutral mode. In the left part of the $M^2_{\pi^0\pi^0}$ distribution an evident distortion in the $2 \cdot m_{\pi^+}$ position can be seen (fig.2). Following the suggestion of Cabibbo [10] and Isidori [11] this effect is explained by the strong rescattering $\pi^+\pi^- \to \pi^0\pi^0$ that gives a real (and negative) contribution to the matrix element of $K^\pm \to \pi^\pm \pi^0 \pi^0$ below the threshold and an imaginary contribution above the threshold. The size of this effect is directly dependent from the scattering length $(a_0 - a_2)$. Using the model proposed by Cabibbo and Isidori and fixing the number of possible electromagnetic $\pi\pi$ bound state (pionic atoms) the fit gives a $\chi^2 = 149/147$ with a preliminary result of

$$(a_0 - a_2) = 0.281 \pm 0.007_{stat} \pm 0.014_{syst} \pm 0.014_{th}$$

where the systematic error is computed checking the stability at different cuts definitions. This result is based on one half of the 2003 statistic. A total error less than 5% can be easily achieved by improuving the theoretical description of the effect.

THE FUTURE

Members of the NA48 collaboration is looking for the possibility to measure the BR of the very rare decay $K^+ \to \pi^+ \nu \bar{\nu}$ with 80 events [9], using a high energy unseparated K^+ beam. The V_{td} CKM matrix element, could be measured, from this BR, at 10% level, with small theoretical uncertainties. The proposal P326 has been submitted to the CERN committee. Description, motivation and status can be found in [12] and [13].

REFERENCES

1. J. H. Christenson et al. *Phys. Rev. Lett.* **13**, 138 (1964)
2. G. Barr et al. (NA31), *Phys. Lett. B* **317**, 233 (1993)
3. A. Alavi-Harati et al. (KTeV) *Phys. Rev. D* **67**, 012005 (2003), Erratum: *Phys. Rev. D* **70**, 079904 (2004)
4. A. Lai et al (NA48), *Eur, Phys. J. C* **22**, 231 (2001) J. R. Batley et al. (NA48), *Phys. Lett. B* 544, 97 (2002)
5. K. Abe et al. (Belle), *Phys. Rev. Lett.* **93**, 021601 (2004) B. Aubert et al. (Babar), *Phys. Rev. Lett.* **93**, 131801 (2004)
6. W. S. Choong, Ph.D. thesis, Berkeley (2000) LBNL-47014
7. G. A. Akopdzhanov et al, *Eur. Phys. J. C* **40**, 343 (2005)
8. V. Fanti et al. (NA48) *Phys. Lett. B* **465**, 335 (1999)
9. S. Eidelman et al. (PDG) *Phys. Lett. B* **592**,1 (2004)
10. N. Cabibbo *Phys.Rev.Lett.* **93**, 121801 (2004)
11. N. Cabibbo and G. Isidori, *JHEP* **0503**,021 (2005)
12. CERN-SPSC-2005-013, SPSC-P-326
13. http://na48.web.cern.ch/NA48/NA48-3

---

[9] assuming the SM prevision of BR$\sim 10^{-10}$.

Charmonium Spectroscopy and Production: New Results

Enrico Robutti

INFN Genova: Enrico.Robutti@ge.infn.it

Abstract. Recent results in charmonium spectroscopy are briefly reviewed, including new evidences for the $^1P_1$ state (h_c) and the study of the properties of the $X(3872)$ state. Also, new results on charmonium production in B decays and $e^+e^- \to (c\bar{c})(c\bar{c})$ are presented.

Keywords: charmonium, spectroscopy, production, h_c, X(3872)
PACS: 13.25.Hw, 13.66.Bc, 14.40.Gx

INTRODUCTION

The observation, in 2002, of the $\eta_c(2S)$, whose first evidence had been lacking firmer confirmation for many years, and the more recent discovery of the $X(3872)$ state decaying to $J/\psi \pi^+ \pi^-$, have greatly revived the interest in the field of charmonium spectroscopy. At the same time, new measurements of charmonium states production in B decays and in e^+e^- interactions have shown behaviours which are difficult to explain in the framework of current theoretical models.

Several experiments are now contributing to studies aimed at a better understanding of these observations, and in the meanwhile new surprising effects and yet other new, unexpected states are being found. A very brief review of the most recent results is reported here.

THE $^1P_1$ STATE (h_c)

Observation of the $^1P_1$ state of charmonium, named h_c, was reported by the E760 experiment at Fermilab in 1992 [1] at a mass close to the "center of gravity" of the triplet vector states of charmonium ($\chi_{c0,1,2}$). Since then, no other search has reported positive evidence: only very recently two new measurements have lead to the re-discovery of the h_c.

Results from CLEO

Moving from E760 result in $h_c \to J/\psi \pi^0$ decays, CLEO has studied h_c production in the $\psi(2S) \to h_c \pi^0$ process, making use of the large (about 3 million) $\psi(2S)$ sample collected in e^+e^- formation. The h_c is then looked for in radiative decays $h_c \to \eta_c \gamma$: this mode is expected to account for about 50% of the h_c decay width.

CLEO-c has conducted two different analyses: an inclusive one and an exclusive one, where "inclusive" or

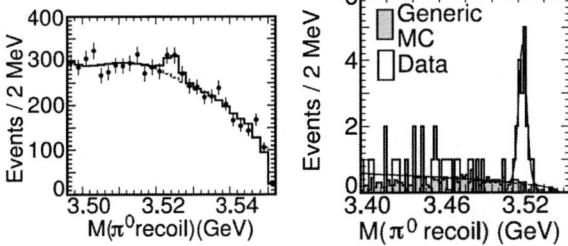

FIGURE 1. Search for $\psi(2S) \to h_c \gamma$ at CLEO: invariant mass recoiling against a reconstructed π^0 for (left) all selected events with photon energy consistent with $h_c \to \eta_c \gamma$ decay and (right) events with an η_c exclusively reconstructed in one of several decay modes.

"exclusive" refers to the η_c reconstruction. In both cases, π^0 candidates are formed from photon pairs, and in addition the presence of a photon and at least two charged tracks is required.

In the inclusive analysis, the photon is required to have an energy consistent with that expected in the h_c radiative decay. The spectrum of the mass recoiling against the π^0 candidate, shown in Fig. 1 (left), is then fit to a smooth background plus a resonant shape: the fit yields a signal contribution of 150 ± 40 events, with a statistical significance of 3.8σ.

In the exclusive analysis, η_c candidates are reconstructed in seven known hadronic decay modes. Again, the π^0 recoil mass spectrum (Fig. 1) is fit to a background plus resonance shape: the fit yields 17.5 ± 4.5 events, with a statistical significance of 6.1σ.

Combining the results of the two analyses, the following measurements are obtained for the mass value and the product of branching ratios: $m(h_c) = (3524.4 \pm 0.6 \pm 0.4)\,\mathrm{MeV}/c^2$, $\mathscr{B}(\psi(2S) \to \pi^0 h_c, h_c \to \eta_c \gamma) = (4.0 \pm 0.8 \pm 0.7) \times 10^{-4}$ [2].

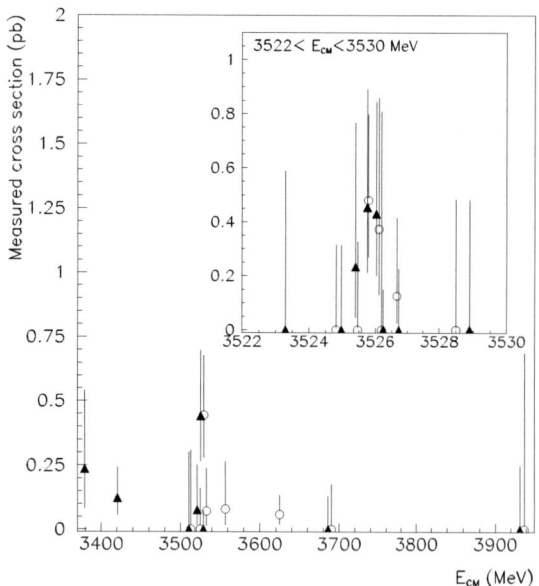

FIGURE 2. Measured cross section for selected $p\bar{p} \to \eta_c \gamma$ events at E835, as a function of the $p\bar{p}$ center-of-mass energy. The inset shows an enlarged view of the region around the $\chi_{c0,1,2}$ center of gravity.

Results from E835

The E835 experiment at Fermilab studies charmonium formation in $p\bar{p}$ interaction, with subsequent decays to electromagnetic final states. Searches are conducted by counting the number of signal candidates at each given value of the $p\bar{p}$ center-of-mass energy and studying the resulting excitation shape. A dedicated scan has been performed around the χ_c center of gravity, with about $90\,\text{pb}^{-1}$ integrated luminosity and an average separation of about $250\,\text{keV}/c^2$ between nearby points.

A search for $h_c \to J/\psi \pi^0$ decays, where J/ψ mesons are reconstructed in e^+e^- has yielded an upper limit for the product of branching fraction of $\mathcal{B}(h_c \to p\bar{p}, h_c \to J/\psi \pi^0) < 0.6 \times 10^{-7}$ at 90% C.L. (for $m(h_c) = 3526.2\,\text{MeV}/c^2$, $\Gamma(h_c) = 750\,\text{keV}$), thus not confirming E760 result.

A different search has been conducted in $h_c \to \eta_c \gamma$, where η_c mesons are reconstructed in the electromagnetic mode $\eta_c \to \gamma\gamma$. Kinematic fits, angular distributions and a π^0 veto have been exploited for the selection of events: 30 candidates are found in $215\,\text{pb}^{-1}$, 13 of which in $30\,\text{pb}^{-1}$ collected around the χ_c center of gravity (Fig. 2). Interpreting this excess as a resonance yields a statistical significance for the signal of about 3σ, and $m(h_c) = (3525.8 \pm 0.2 \pm 0.2)\,\text{MeV}/c^2$, $\Gamma(h_c \to p\bar{p}) \times \mathcal{B}(h_c \to \eta_c \gamma) = (11 \pm 5 \pm 3)\,\text{eV}$ [3].

THE $X(3872)$ STATE

Since its first observation by Belle in B decays [4], the existence of the $X(3872)$ state has received several confirmations, from CDF [5], D0 [6] and BaBar [7]. Nevertheless, the true nature of this state is still not understood, and the "natural" assignment to a charmonium state is put in doubt by several observations disfavouring all specific quantum number assignments [8]. Several other interpretations have been proposed, including a D^*D bound state [9], a hybrid charmonium state [10], a diquark-antidiquark state [11].

New results at B-factories

Recently, Belle has reported an update to its $B \to X(3872)K$ analysis, with $X(3872) \to J/\psi \pi^+ \pi^-$, based on a sample of 275 million $B\bar{B}$ pairs [12]. The analysis yields a new result for the product of branching fractions of $\mathcal{B}(B \to X(3872)K, X(3872) \to J/\psi \pi^+ \pi^-) = (1.31 \pm 0.24 \pm 0.13) \times 10^{-5}$. More importantly, a detailed analysis of the invariant mass distribution of the dipion system and of angular distributions leads to a preference for a $J^{PC} = 1^{++}$ assignment for the $X(3872)$ (although the 2^{++} possibility is not ruled out), while strongly disfavouring the 1^{-+} and 2^{-+} assignments.

In an analogous update on a sample of 232 million $B\bar{B}$ pairs [13], BaBar has specifically studied the neutral mode $B^0 \to X(3872)K_S^0$. Some DD^* molecule models predict a suppression for this production mode of an order of magnitude with respect to the charged one. Moreover, the diquark-antidiquark model predicts a different state, with a mass differing by about $7\,\text{MeV}/c^2$, to be produced in this reaction. The analysis yields a 2.5 σ significant signal (Fig. 3), with 90% C.L. interval for the product of branching fractions of $1.34 \times 10^{-6} < \mathcal{B}(B^0 \to X(3872)K^0, X(3872)^- \to J/\psi \pi^- \pi^0) < 10.3 \times 10^{-6}$, and a mass difference of $(2.7 \pm 1.3 \pm 0.2)\,\text{MeV}/c^2$ for the states produced in the two (charged and neutral) decay modes.

BaBar has also looked for a charged partner of the $X(3872)$, foreseen by some of the models where $X(3872)$ is an isovector state [14]. The search has been conducted on a sample of $212\,\text{fb}^{-1}$ in $B^- \to X(3872)^- K_S^0$ and $B^0 \to X(3872)^- K^+$ decays, with $X(3872)^- \to J/\psi \pi^- \pi^0$, and has yielded negative results, with 90% C.L. upper limits of $\mathcal{B}(B^0 \to X(3872)^- K^+, X(3872)^- \to J/\psi \pi^- \pi^0) < 5.4 \times 10^6$, $\mathcal{B}(B^- \to X(3872)^- K^0, X(3872)^- \to J/\psi \pi^- \pi^0) < 22 \times 10^6$.

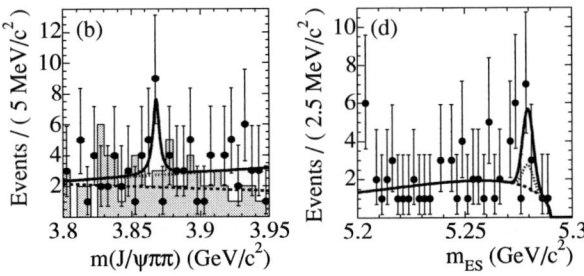

FIGURE 3. Distribution of (left) the $J/\psi\pi^+\pi^-$ invariant mass and (right) the m_{ES} variable for selected $B^0 \to X(3872)K_S^0$ events at BaBar. The superimposed lines show projections of the 2-dimensional fit result for the signal (solid), combinatorial background (dashed) and B background (dotted) contributions.

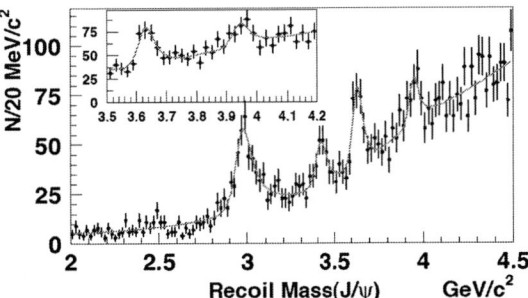

FIGURE 4. Invariant mass recoiling against a J/ψ candidate at Belle. The superimposed line shows the result of the fit. The inset shows a detail of the high mass region.

Production in $p\bar{p}$ interactions at Tevatron

CDF and D0 have studied the production properties of $X(3872)$ in $p\bar{p}$ interactions on a sample of $230\,\mathrm{pb}^{-1}$.

CDF has performed a fit to the time-of-flight distribution of $X(3872)$ candidates, with the purpose of separating the contributions from prompt production and from B decays [15]. The fraction from B decays turns out to be $f = (16,1 \pm 4,9 \pm 2,0)\%$, not far from the value of about 28% measured for $\psi(2S)$ candidates.

D0 has looked at a wide set of variables, including production vertex, transverse momentum, angular distributions, candidate isolation, and compared results for $X(3872)$ and $\psi(2S)$ candidates [6]. In all cases the measured distributions are very similar in the two cases.

DOUBLE CHARMONIUM PRODUCTION IN e^+e^- INTERACTIONS.

The surprisingly large cross section measured by Belle [16] for the $e^+e^- \to J/\psi c\bar{c}$ process (where the $c\bar{c}$ pair hadronizes either in a charmonium state or in a final state with charmed mesons) poses a challenge for the currently adopted non-relativistic QCD models (with or without the inclusion of color-octet contributions) [17]. In particular, the simultaneous high production rate for $J/\psi\eta_c$ and non-observation of $J/\psi J/\psi$ still remains unexplained.

Recently, BaBar has confirmed Belle's result on double charmonium production on a sample corresponding to an integrated luminosity of $124\,\mathrm{fb}^{-1}$ [18]. Clear peaks for η_c, χ_{c0}, $\eta_c(2S)$ are seen in the spectrum of mass recoiling against J/ψ candidates reconstructed in $J/\psi \to e^+e^-, J/\psi \to \mu^+\mu^-$. Measured cross sections are generally in good agreement with Belle's, with a slightly lower value for $e^+e^- \to J/\psi\eta_c$ $((17.7 \pm 2.8 \pm 2.1)\mathrm{fb})$.

In a recent update to its analysis on $155\,\mathrm{fb}^{-1}$ [19], Belle has extended the analysis to the recoil of $\psi(2S)$ candidates reconstructed in the $J/\psi\pi^+\pi^-$ decay mode, where significant (between 3 and 4 σ) signals are found for the same η_c, χ_{c0}, $\eta_c(2S)$ states seen on the recoil of J/ψ. No evidence for other known charmonium states is found yet, in both the J/ψ and the $\psi(2S)$ recoil. Also, the helicity distributions for J/ψ are measured.

By extending the study of the recoil mass region beyond the $\eta_c(2S)$ mass, a new structure appears at mass slightly below $4\,\mathrm{GeV}/c^2$ (Fig. 4). A fit to the distribution with a resonant shape yields 149 ± 33 events (4.5σ significance) and a mass $m(X) = (3940 \pm 12)\,\mathrm{MeV}/c^2$. A search for decays of this state to $D\bar{D}$ and $D\bar{D}^*$ yields a significant signal for the latter mode, while no evidence is found for the former.

Very recently Belle has published a further update on $387\,\mathrm{fb}^{-1}$ [20], where the parameters of the new state are measured with improved accuracy.

CHARMONIUM PRODUCTION IN $B \to (c\bar{c})K$ DECAYS

Production of charmonia in B decays, besides representing an abundant source of these states in B-factories, is also a very useful tool to study predictions of theoretical models based on factorisation. Despite the large samples of B mesons accumulated, and their often sizeable branching fractions to final states containing charmonium, in some cases difficulties arise from the small rates, or the poor knowledge, of decays of charmonium states to given exclusive final states.

As the size of B-factories data samples grows, a new interesting possibility arises from the study of the system recoiling against completely reconstructed B-mesons. BaBar has performed such a study on a sample of events where one B has been completely reconstructed in one of an extremely large set of exclusive hadronic decay

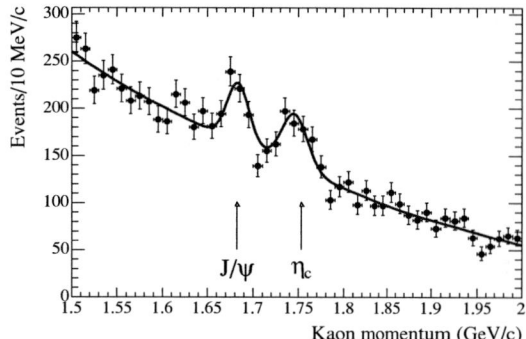

FIGURE 5. Momentum spectrum for charged kaons recoiling against a reconstructed $B^{\pm}$ at BaBar, in the region corresponding to $B \to \eta_c K$, $B \to J/\psi K$ decays. The superimposed line shows the result of the fit.

modes containing charmed mesons. By selecting charged kaons in the rest of the event, and looking at their momentum distribution, one is able to observe the spectrum of X states produced in two-body decays of the type $B \to XK^{\pm}$, independently of the X decay mode. In particular, for reconstructed B^0, the spectrum of charmonium states produced in such processes can be studied.

The analysis makes use of additional selections based on "topological" variables in order to improve the purity of the sample, and to enrich it with kaons from prompt production. The resulting momentum spectrum in the region between 1.5 and 2 GeV/c, corresponding to X masses in the low region of the charmonium spectrum, is shown in Fig. 5: clear signals corresponding to the J/ψ and η_c are seen, both with significance around 7σ. In particular, the η_c signal allows to obtain an improved, absolute measurement of the $B^{\pm} \to \eta_c K^{\pm}$ branching fraction, $\mathscr{B}(B^{\pm} \to \eta_c K^{\pm}) = (8.7 \pm 1.5) \times 10^{-4}$, and from this, in turn, an improved measurement of the "secondary" $\eta_c \to K_S^0 K^{\pm} \pi^{\mp}$ branching fraction.

In the low part of the momentum spectrum (high charmonium masses), significant signals are seen for $\chi_{c1} + h_c$ (6σ) and $\psi(2S)$ (3.2σ); upper limits are calculated for the other states. In particular, the upper limit on $X(3872)$ production allows to set a lower limit on its decay mode $X(3872) \to J/\psi \pi^+ \pi^-$, by combining it with the measured rate for the product branching fraction: $\mathscr{B}(X(3872) \to J/\psi \pi^+ \pi^-) > 4.2\%$ at 90% C.L.

REFERENCES

1. T. A. Armstrong et al. (E760 Collaboration), *Phys. Rev. Lett.* **69**, 2337 (1992).
2. J. L. Rosner et al. (CLEO Collaboration), hep-ex/0505073 (2005).
3. C. Patrignani et al. (FNAL-E835 Collaboration), *Nucl. Phys. Proc. Suppl.* **142**, 98 (2005).
4. S. K. Choi et al. (Belle Collaboration), *Phys. Rev. Lett.* **91**, 262001 (2003).
5. D. Acosta et al. (CDF Collaboration), *Phys. Rev. Lett.* **93**, 072001 (2004).
6. V. M. Abazov et al. (D0 Collaboration), *Phys. Rev. Lett.* **93**, 162002 (2004).
7. B. Aubert et al. (BaBar Collaboration), *Phys. Rev.* **D71**, 071103 (2005).
8. K. Abe et al. (Belle Collaboration), hep-ex/0408116 (2004); S. Dobbs et al. (CLEO Collaboration), *Phys. Rev. Lett.* **94**, 032004 (2005).
9. N. Tornqvist, *Phys. Lett.* **B 590**, 209 (2004); M. B. Voloshin, *Phys. Lett.* **B 579**, 316 (2004); F. Close and P. Page, *Phys. Lett.* **B 578**, 119 (2004); C. Y. Wong, *Phys. Rev. C* **69**, 055202 (2004); E. Braaten and M. Kusunoki, *Phys. Rev.* **D 69**, 074005 (2004); E. Swanson, *Phys. Lett.* **B 588**, 189 (2004).
10. F. Close and S. Godfrey, *Phys. Lett.* **B 574**, 210 (2003).
11. L. Maiani, F. Piccinini, A. D. Polosa, and V. Riquer, *Phys. Rev.* **D71**, 014028 (2005).
12. K. Abe et al. (Belle Collaboration), hep-ex/0505038 (2005).
13. B. Aubert et al., BaBar Collaboration, hep-ex/0507090 (2005).
14. B. Aubert et al. (BaBar Collaboration), *Phys. Rev.* **D71**, 031501 (2005).
15. G. Gomez-Ceballos (CDF Collaboration), *J. Phys. Conf. Ser.* **9**, 72 (2005).
16. K. Abe et al. (Belle Collaboration), *Phys. Rev. Lett.* **89**, 142001 (2002).
17. E. Braaten and J. Lee, *Phys. Rev. D* **67**, 054007 (2003); Y. J. Zhang, Y. J. Gao and K. T. Chao, hep-ph/0506076; A. E. Bondar and V. L. Chernyak, *Phys. Lett.* B **612**, 215 (2005); G. T. Bodwin, J. Lee and E. Braaten, *Phys. Rev. D* **67**, 054023 (2003); G. T. Bodwin, J. Lee and E. Braaten, *Phys. Rev. Lett.* **90**, 162001 (2003); S. J. Brodsky, A. S. Goldhaber and J. Lee, *Phys. Rev. Lett.* **91**, 112001 (2003).
18. B. Aubert et al. (BaBar Collaboration), hep-ex/0506062 (2005).
19. P. Pakhlov (Belle Collaboration), hep-ex/0412041 (2004).
20. K. Abe et al. (Belle Collaboration), hep-ex/0507019 (2005).

Semi-Inclusive Charmless B Decays

Giulia Ricciardi

Dipartimento di Scienze Fisiche,
Università di Napoli "Federico II"
and I.N.F.N., Sezione di Napoli, Italy.

Abstract. We give a brief review on semi-inclusive charmless B decays, with focus on $B \to X_u l \nu$, $B \to X_s \gamma$ and $B \to X_s l^+ l^-$.

Keywords: QCD, bottom mesons, semi-leptonic and radiative decays
PACS: 13.20.He, 12.38.-t

SEMILEPTONIC DECAYS

One reason of interest towards inclusive semileptonic decays of heavy mesons is the possibility of using, in order to calculate decays rates, quark-hadron duality, meaning that a quantity can be computed directly at the parton level if it is inclusive enough and the energy involved is large enough. However, in fact, the differential spectra, in the lepton energy and invariant mass of the leptonic system, for the $B \to X_q l \nu$ and $b \to q l \nu$ decays ($q = u$ or c), are generally not the same. We can relate them after smearing over the electron energy with an appropriate smooth function f (for smearing of a function g we mean $\int dE f(E) g(E)$) [1]. Such result is obtained by using the short distance operator product expansion (OPE) and the heavy quark mass expansion of a Green's function, the physical rate given by its imaginary part (discontinuity across the cut). One therefore gets, in addition, a way to improve the expansion systematically, order by order in $1/m_b$. Moreover, by choosing the function of smearing appropriately, one can find new sum rules, which allow, for example, experimental determination of the non perturbative parameters of the expansion, by computing moments of the spectrum [2].

For the total rate we have

$$\Gamma(B \to X_q l \nu) = \frac{G_F^2 m_b^5}{192\pi^3}|V_{qb}|^2 [c_3 <O_3> +$$
$$+ c_5 \frac{<O_5>}{m_b^2} + c_6 \frac{<O_6>}{m_b^3} + ...] \quad (1)$$

The coefficients c_d are calculated from short-distance physics and O_d are local operators of dimension d. It is basically the same cast of operators, albeit with different weights, that appears in semi-leptonic, radiative and non-leptonic rates as well as distributions. The matrix elements

$$<O_d> \propto <B|O_d|B>$$

take into account non perturbative effects. While we can identify the operators and their dimensions, which then determine the power of $1/m_b$, in general we cannot (yet) compute their hadronic expectation values from first principles. Note the absence of contributions at order $1/m_b$ [3]. This is due to the absence of independent gauge invariant operators of dimension 4. In principle, it is possible to write the operator $i\bar{b} \gamma_\mu D^\mu b$; however, the equations of motion imply that between physical states this operator can be replaced by $m_b \bar{b} b$. For the hadronic expectation value of the leading operator $O_3 = \bar{b}b$ one has $<O_3> = 1 + O\left(\frac{1}{m_b^2}\right)$. The fact that non perturbative, bound state effects in inclusive decays are strongly suppressed (at least two powers of the heavy quark mass) explains a posteriori the success of parton model in describing such processes.

$\bar{B} \to X_u l \bar{\nu}$ decay

The $\bar{B} \to X_u l \bar{\nu}$ decay is used in the determination of the magnitude of the Cabibbo-Kobayashi-Maskawa (CKM) matrix parameter $|V_{ub}|$, the coupling strength of the b quark to the u quark. If it were not for the huge background from decays to charm, such determination would be straightforward. Inclusive B decays rates can be computed using OPE and the results are expressed as series in α_s and Λ_{QCD}/m_b. At leading order, the B meson decay rate is equal to the b quark decay rate. Higher orders start at order Λ_{QCD}^2/m_b^2 and leading non perturbative corrections are characterized by heavy quark effective theory (HQET) matrix elements, which may be determined by other physical quantities. Unfortunately, such direct model independent approach breaks down for not sufficiently inclusive observables, which is the case here, since the $\bar{B} \to X_u l \bar{\nu}$ rate can only be measured imposing cuts on the phase space, to eliminate the huge $\bar{B} \to X_c l \bar{\nu}$ background. Experiments are forced to measure partial branching fractions and the extrapolation to the full phase space needs to rely on QCD based theoretical calculations. The results are sensitive to different theoretical calculations, and therefore to different sources of systematic uncertainties.

Event selection is generally based on three kinematical variables, the lepton energy E_l, the hadronic invariant mass m_X and the leptonic invariant mass squared q^2, each having its own advantages and disadvantages. The first measurements [4] used the lepton energy E_l and were restricted to the high end of the lepton spectrum; at

$$E_l > \frac{m_B^2 - m_D^2}{2 m_B} \quad (2)$$

charm production is suppressed. One drawback is that only a small fraction of events are above the endpoint of the $\bar{B} \to X_c l \bar{\nu}$ background, f. i. about 10% of $\bar{B} \to X_u l \bar{\nu}$ decays have a lepton momentum larger than 2.3 GeV/c.

Another kinematical region where charm background is absent, is the region of small hadronic invariant mass, $m_X < m_D$ [5]. This region has a large usable fraction of events, of the order of 70%.

In both the previous kinematical regions, the rate receives large contributions from long distance effects related to soft interactions between the heavy quark and the light degrees of freedom. In fact, they both include a subregion where

$$E_X \sim m_b, \quad m_X^2 \sim \Lambda_{QCD} m_b \ll m_b. \quad (3)$$

This subregion has sufficient phase space for many different resonances to be produced in the final state, so an inclusive description of the decays may still be considered appropriate. However, in this subregion, the differential rate is very sensitive to non perturbative effects, mainly due to the so-called Fermi motion.

The Fermi motion can be roughly described as the motion resulting from the momentum exchange between the heavy quark and the valence quark inside the meson. The rate becomes affected by Fermi motion since in the subregion (3) an amplification effect occurs, according to which a small virtuality of the heavy flavor in the initial state produces relatively large variations of the fragmentation mass in the final state. A fluctuation in the heavy quark momentum of order Λ_{QCD} in the initial state produces a variation of the final invariant mass of the hard subprocess of order

$$\delta m_X^2 \sim O(\Lambda_{QCD} E_X). \quad (4)$$

An amplification by a factor E_X has occurred, as anticipated. We expect, therefore, the differential rate to be sensitive at leading order to a non perturbative distribution function called shape function, $f(k_+)$, which describes the distribution of the light cone component of the residual momentum k_+ of the b quark inside the meson and factorizes long distance effects. The shape function cannot be calculated analytically, so the rate is model dependent even at leading order. However, since such long distance effects are common to any B decay into a light quark, the shape function may be related to the analogous distribution function in the $B \to X_s \gamma$ spectrum.

Long distance effects manifest themselves also in perturbation theory, under the form of series of large infrared logarithms, which need to be resummed to validate the perturbative result [6].

A recent determination of $|V_{ub}|$ from a measurement of the inclusive charmless semi-leptonic branching fraction of $\bar{B} \to X_u l \bar{\nu}$ has been presented by BaBar with a kinematical cut on $m_X < 1.55 \,\text{GeV}/c^2$ [7]. The measurement was dominated by theoretical uncertainty on the underlying kinematic variable distributions, and therefore on the extrapolation to the full phase space (about 17% on the branching ratio, i.e. 8.5% on $|V_{ub}|$).

For the dilepton invariant mass spectrum, the charm background is suppressed when

$$q^2 \equiv (p_l + p_{\bar{\nu}})^2 > (m_B - m_D)^2. \quad (5)$$

Such a cut forbids the hadronic final state from moving fast in the B rest frame, setting simultaneously $E_X < m_D$ and $m_X < m_D$. It suppresses the subregion (3), where the differential rate is more affected by the details of the shape function. Therefore the q^2 distribution results less sensitive to the shape function, and less dependent on the calculation [8]. As a drawback, it is quite sensitive to the m_b mass and only a small fraction of events (about 20%) is usable with a pure q^2 selection.

There are two more recent proposals to eliminate the charm background and reduce the theoretical error on the extrapolation to the total rate. One is to apply simultaneous cuts on m_X and q^2 [9, 10]; recent measurements of $|V_{ub}|$ using a combination of cuts on m_X and q^2 have already been presented [11, 12]. The idea here is to cut on m_X to remove the charm background, and to cut on q^2 to keep the sensitivity to the shape function small; f.i. it has been estimated a fraction of surviving events of $(45 \pm 5)\%$ by using the combined cuts $m_X < m_D$ and $q^2 > 6\,\text{GeV}^2$ [9].

Another recent proposal is to separate $b \to u$ from $b \to c$ using E_X and m_X together with $p_+ \equiv E_X - |p_X|$, where $|p_X|$ is the magnitude of the three-momentum of the hadronic system. The product of the light cone variables $p_+ \equiv E_X - |p_X|$ and $p_- \equiv E_X + |p_X|$ is the squared invariant hadronic mass $p_+ p_- = m_X^2$ and it is obvious that the boundary of phase space is given by

$$m_\pi^2/p_- < p_+ < p_- < m_B. \quad (6)$$

In order to avoid large background from $b \to c$ transitions, all measurements of $|V_{ub}|$ are in one way or another restricted to the region of phase space where $p_+ p_- <$

m_D^2. If p_- is allowed to reach its maximum value near m_B, it follows that

$$p_+ < \frac{m_D^2}{m_B} \sim 0.66\,\text{GeV}/c^2, \quad (7)$$

which is the proposed cut [13]. This cut includes slightly less phase space space than the hadronic cut, and the results still bear a dependence on the shape function and subleading corrections. However, it has been argued that this cut offers relations to the differential rate of $B \to X_s \gamma$ which are simpler and less plagued by theoretical errors [13]. Indeed, the choice of kinematical variables is crucial for a correct relation between semileptonic and radiative decays [14]. The first analysis using the variable p_+ has been made by the Belle collaboration [15]; values of $|V_{ub}|$ are given for three kinematical regions, $p_+ < 0.66\,\text{GeV}/c^2$, $m_X < 1.7\,\text{GeV}/c^2$ and $m_X < 1.7\,\text{GeV}/c^2$ combined with $q^2 > 8\,\text{GeV}^2/c^2$. The precision of the $|V_{ub}|$ determination is better than in previous measurements [16], due to the use of both a large data sample and improved theoretical predictions.

$B \to X_S$ RARE DECAYS

Flavor Changing Neutral Current (FCNC) processes

$$q_i \to q_j + (\gamma, l^+ l^-, \bar{\nu}\nu)$$

are within the Standard Model (SM) forbidden at the tree level; they can only proceed via higher order loop diagrams.

The branching ratio of $\bar{B} \to X_s \gamma$ decay is one of the most important constraints for many new physics models, due to a rather clean theoretical determination and an accurate experimental measure. At present, high quality data is available [17, 18, 19]. All measurements are performed with a lower cut on the photon energy $E_\gamma < E_0$, which until recently was around $E_0 = 2\,\text{GeV}$. Last year the cut was lowered by Belle to $E_0 = 1.815\,\text{GeV}$ [18], and this year BaBar has presented measurements of partial branching fractions and truncated first and second moments of the true photon energy distribution in the B rest frame, above the minimum photon energy values of $E_0 > 1.9, 2.0, 2.1, 2.2\,\text{GeV}$ [19].

The present world average [20]

$$\text{Br}(\bar{B} \to X_s \gamma) = (3.52 \pm 0.30) \times 10^{-4} \quad (8)$$

is in good agreement with the SM prediction [21]

$$\text{Br}(\bar{B} \to X_s \gamma) = (3.70 \pm 0.30) \times 10^{-4}. \quad (9)$$

The theoretical total error (about 8%) is dominated by the error on the charm mass (about 6%). The charm mass enters the calculation through the normalization, with the ratio $|V_{ub}/V_{cb}|^2$, and through the two loops NLO matrix elements involving one charmed loop. As the related LO diagrams vanish, the definition of m_c is mostly a NNLO issue and will be partially resolved when the complete NNLO calculation will become available. The effective lagrangian for radiative decay is usually written as

$$L_{eff} = \frac{4G_F}{\sqrt{2}} V_{ts}^\star V_{tb} \sum_{i=1}^{10} C_i(\mu) O_i. \quad (10)$$

The operators O_i are:

$$\begin{array}{ll}
(\bar{s}\Gamma_i c)(\bar{c}\Gamma_i' b) & i = 1,2 \\
(\bar{s}\Gamma_i b)\Sigma_q(\bar{q}\Gamma_i' q) & i = 3,4,5,6 \\
(em_b/16\pi^2)\,\bar{s}_L \sigma^{\mu\nu} b_R F_{\mu\nu} & i = 7 \\
(gm_b/16\pi^2)\,\bar{s}_L \sigma^{\mu\nu} T^a b_R G_{\mu\nu}^a & i = 8 \\
(e^2/16\pi^2)(\bar{s}_L \gamma_\mu b_L)(\bar{l}\gamma^\mu l) & i = 9 \\
(e^2/16\pi^2)(\bar{s}_L \gamma_\mu b_L)(\bar{l}\gamma^\mu \gamma_5 l) & i = 10
\end{array} \quad (11)$$

where Γ_i, Γ_i' are combinations of gamma and color matrices, and $q = u, d, s, c, b$.

In order to perform the calculation, three major steps have to be completed:

- the matching of the effective amplitude with the SM amplitude at high scale ($\mu_0 \sim m_W, m_t$), which results in the determination of the operator coefficients at high scale;

- the operator mixing, in order to evolve the coefficients to the lower dynamical scale ($\mu \sim m_b$) and resum the large ultraviolet logarithms $\alpha_s^n(\mu) \log(\mu/\mu_0)^m$ with $m \leq n$. It requires the calculus of the anomalous dimension matrix (ADM) for the operators;

- the calculation of the on-shell matrix elements. It includes bremsstrahlung contributions $b \to s\gamma g$.

Recently, the following parts of NNLO calculation have been completed

- NNLO matching of C_7 and C_8 [22].
- three loop NNLO elements of ADM for the dimension-five dipole operators O_7 and O_8 [23]
- parts of the three loop NNLO matrix elements [24]
- the two loop matrix element of Q_7 [25]
- the dominant part of NNLO spectrum [26]

Let us observe that the first out of the three step (NNLO matching) outlined above is now completed. The effect of the NNLO matching alone is scheme and scale dependent. In the $\overline{\text{MS}}$ scheme with $m_W < \mu_0 < m_t$, it stays within 2% of the decay width [22], i. e. it is significantly smaller than the present theoretical uncertainty. The present theoretical uncertainty is expected to reduce significantly, after the completion of the remaining two steps of the NNLO calculation.

As far as the second step is concerned, the three-loop $O(\alpha_s^3)$ ADM is completed as well.

Still missing are the four loop elements of the ADM, the three loop matrix elements with charm and the subdominant two loop matrix elements.

The NNLO calculation of the decay $B \to X_s l^+ l^-$ corresponds to the NLO calculation of $\bar{B} \to X_s \gamma$ as far as the number of loops in the diagrams is concerned. Including the leading power corrections in $1/m_b$ and $1/m_c$, the branching ratios for the decays $B \to X_s l^+ l^-$ at NNLO (complete) are [27]

$$\text{Br}(B \to X_s e^+ e^-)_{SM} \simeq \text{Br}(B \to X_s \mu^+ \mu^-)_{SM} = (4.2 \pm 0.7) \times 10^{-6}, \quad (12)$$

where a dilepton invariant mass cut, $m_{ll} > 0.2\,\text{GeV}/c^2$ has been assumed. The theoretical calculations have to be compared with the current world average of the branching ratio, at the same cut $m_{ll} > 0.2\,\text{GeV}/c^2$ [28]

$$\text{Br}(B \to X_s l^+ l^-) = \left(4.46^{+0.98}_{-0.96}\right) \times 10^{-6}. \quad (13)$$

Thus, with the current experimental accuracy, which is around 25%, data and the SM are in agreement with each other. The measurement of $b \to s$ transitions provide valuable constraints on beyond-the-SM scenarios. For example, one possibility of new physics, in principle, would be an effective coefficient of the operator O_7 similar in magnitude, but opposite in sign, to the SM effective coefficient (as given in some SUSY models with large $\tan \beta$). However, it has been recently argued that this case has been ruled out by the analysis of data on $b \to s$ rare decays [29].

REFERENCES

1. J. Chay, H. Georgi and B. Grinstein, *Phys. Lett.*, **B247**, 399–405 (1990).
2. A. F. Falk, M. E. Luke and M. J. Savage *Phys. Rev.*, **D53**, 2491–2505 (1996); A. V. Manohar and M. B. Wise *Phys. Rev.*, **D49**, 1310–1329 (1994); B. Blok et al., *Phys. Rev.*, **D49**, 3356–3366 (1994) [erratum-ibidem **D50**, 3572 (1994)]; T. Mannel, *Nucl. Phys.*, **B413**, 396–412(1994).
3. M. Voloshin and M. Shifman, *Sov. J. Nucl. Phys.*, **47**, 511 (1988) [*Yad. Fiz.*, **47**, 801–806 (1988)]; M. Luke *Phys. Lett.*, **B252**, 447–455 (1990).
4. CLEO Collaboration, R. Fulton et al., *Phys. Rev. Lett.*, **64**, 16–20 (1990); ARGUS Collaboration, H. Albrecht et al., *Phys. Lett.*, **B234**, 409–416 (1990); CLEO Collaboration, A. Bornheim et al., *Phys. Rev. Lett.*, **88**, 231803 (2002).
5. V. Barger, C. S. Kim, and R. J. Phillips, *Phys. Lett.*, **B251**, 629–633 (1990); A. F. Falk, Z. Ligeti, and M. B. Wise, *Phys. Lett.*, **B406**, 225–231 (1997); I. Bigi, R. D. Dikeman and N. Uraltsev, *Eur. Phys. J.*, **C4**, 453–461 (1998).
6. U. Aglietti and G. Ricciardi, *Phys. Rev.*, **D66**, 074003 (2002); U. Aglietti and G. Ricciardi, *Phys. Rev.*, **D70**, 114008 (2004); J. R. Andersen, E. Gardi, *JHEP*, 0506:030 (2005); and references within.
7. BaBar Collaboration, B. Aubert at al., *Phys. Rev. Lett.*, **92**, 071802 (2004).
8. C. W. Bauer, Z. Ligeti and M. E. Luke, *Phys. Lett.*, **B479**, 395–401 (2000).
9. C. W. Bauer, Z. Ligeti and M. E. Luke, *Phys. Rev.*, **D64**, 113004 (2001).
10. B. O. Lange, M. Neubert and G. Paz, hep-ph/0504071.
11. BaBar Collaboration, B. Aubert at al., hep-ex/0507017.
12. Belle Collaboration, H. Kakuno et al., *Phys. Rev. Lett.*, **92**, 101801 (2004).
13. S. W. Bosch, B. O. Lange, M. Neubert and G. Paz, *Nucl. Phys.*, **B699**, 335–386 (2004); S. W. Bosch, B. O. Lange, M. Neubert and G. Paz, *Phys. Rev. Lett.*, **93**, 221801 (2004).
14. U. Aglietti, G. Ricciardi and G. Ferrera, preprint ROME1/1407/05, DSFNA1/25/2005.
15. Belle Collaboration, I. Bizjak et al., hep-ex/0505088.
16. ALEPH Collaboration, R. Barate et al., *Eur. Phys. J.*, **C6**, 555–574 (1999); L3 Collaboration, M. Acciarri et al., *Phys. Lett.*, **B436**, 174–186 (1998); Delphi Collaboration, P. Abreu. et. al, *Phys. Lett.*, **B478**, 14–30 (2000); OPAL Collaboration, G. Abbiendi et. al, *Eur. Phys. J.*, **C21**, 399–410 (2001); CLEO Collaboration, A. Bornheim, et. al, contributed to 31st International Conference on High Energy Physics (ICHEP 2002), Amsterdam, 24-31 Jul 2002, hep-ex/0207064.
17. CLEO Collaboration, S. Chen et al., *Phys. Rev. Lett.*, **87**, 251807 (2001).
18. Belle Collaboration, P. Koppenburg et al., *Phys. Rev. Lett.*, **93**, 061803 (2004).
19. BaBar Collaboration, F. Bucci et al., hep-ex/0507001.
20. Heavy Flavor Averaging Group, J. Alexander et al, hep-ex/0412073.
21. P. Gambino and M. Misiak, *Nucl. Phys.*, **B611**, 338–366 (2001); A. J. Buras, A. Czarnecki, M. Misiak and J. Urban, *Nucl. Phys.*, **B631**, 219–238 (2002).
22. M. Misiak and M. Steinhauser, *Nucl. Phys.*, **B683**, 277–305 (2004).
23. M. Gorbahn, U. Haisch, M. Misiak, hep-ph/0504194.
24. K. Bieri, C. Greub, M. Steinhauser, *Phys. Rev.*, **D67**, 114019 (2003); H. M. Asatrian, C. Greub, A. Hovhannisyan, T. Hurth, V. Poghosyan, hep-ph/0505068.
25. I. Blokland, A. Czarnecki, M. Misiak, M. Slusarczyk, F. Tkachov, hep-ph/0506055.
26. K. Melnikov and A. Mitov, hep-ph/0505097.
27. A. Ali, E. Lunghi, C. Greub, G. Hiller, *Phys. Rev.*, **D66**, 034002 (2002); H. H. Asatryan, H. M. Asatryan, C. Greub, M. Walker, *Phys. Rev.*, **D65**, 074004 (2002); A. Grinculov, T. Hurth, G. Isidori and Y.P. Yao, *Nucl. Phys.*, **B685**, 351 (2004); C. Bobeth, P. Gambino, M. Gorbhan and U. Haisch *JHEP*, 0404:071 (2004).
28. Heavy Flavor Averaging Group, http://www.slac.stanford.edu/xorg/hfag, April 2005.
29. P. Gambino, U. Haisch, M. Misiak, hep-ph/0410155.

$b \to s\gamma$ Decays at BABAR

F. Bucci

Dipartimento di Fisica dell'Università degli Studi di Pisa

Abstract.
Recent BABAR results on $b \to s\gamma$ decays are reviewed. Both the branching fraction and the CP violating asymmetry in radiative penguin decays are sensitive to physics beyond the Standard Model. Furthermore, the measurement of the photon energy spectrum allows the determination of the Heavy Quark Effective Theory parameters related to the b-quark mass and its motion inside the B meson. New results on the $B \to X_s\gamma$ branching fraction and CP violating asymmetry are presented, as well as the measurement of the photon energy spectrum. Also reported is the preliminary measurement of the time-dependent CP asymmetry in the $B^0 \to K^{*0}\gamma$ ($K^{*0} \to K^0_S\pi^0$) decay.

PACS: 13.20.-v,13.20.He,14.40.Nd

1. INTRODUCTION

In the Standard Model (SM) the radiative decays involving the flavor-changing neutral current transition $b \to s$ are dominated by a one-loop penguin diagram containing a top quark and a W boson. Therefore, they are promising places to look for physics beyond the SM: new particles could show up in the loop modifying the SM predictions for branching ratios and CP asimmetries.

Next to the leading order (NLO) calculations for the $b \to s\gamma$ rate give $\mathcal{B}(b \to s\gamma) = (3.57 \pm 0.30) \times 10^{-4}$ for $E_\gamma > 1.6$ GeV [1]. The SM prediction for the CP asymmetry $A_{CP} = \frac{\Gamma(b \to s\gamma) - \Gamma(\bar{b} \to \bar{s}\gamma)}{\Gamma(b \to s\gamma) + \Gamma(\bar{b} \to \bar{s}\gamma)}$ is $0.0044^{+0.0024}_{-0.0014}$, but this could be increased to ≈ 0.10 by new physics contributions [2]. If $b \to s\gamma$ and $b \to d\gamma$ decays are not distinguished the CP asymmetry is exactly zero in the U-spin symmetry limit, $m_d = m_s$. New physics models with additional flavour violation can enhance it to a few percent [2].

The photon energy spectrum in $B \to X_s\gamma$ provides access to the distribution function of the b-quark inside the B meson [3]. The knowledge of this shape function is a crucial input in the extraction of $|V_{ub}|$ from the inclusive semileptonic $B \to X_u l\nu$ measurements. The heavy quark parameters m_b and μ^2_π, which describe the effective b-quark mass and kinetic energy inside the B meson, can be determined either from the fits to the photon energy spectrum or from the fits to its moments [4] [5] [6].

The SM also predicts a strongly suppressed time-dependent CP asymmetry in $B^0 \to K^{*0}\gamma$ ($K^{*0} \to K^0_S\pi^0$). The final state $K^0_S\pi^0\gamma$ is accessible both to the B^0 and the $\bar{B}^0$ through $K^0 - \bar{K}^0$ mixing. The interference between the decays with and without mixing can produce a time dependent CP asymmetry:

$$A_{CP}(\Delta t) = S\sin(\Delta m \Delta t) - C\sin(\Delta m \Delta t)$$

In the Standard Model the photon polarization in $b \to s\gamma$ ($\bar{b} \to \bar{s}\gamma$) is predominantly left handed (right handed). Therefore, the final states in the decays $B^0(t) \to (K^0_S\pi^0)\gamma$ and $\bar{B}^0(t) \to (K^0_S\pi^0)\gamma$ are dominantly orthogonal and the time dependent CP asymmetry is strongly suppressed [7]:

$$S \approx -2m_s/m_b \sin 2\beta \approx -0.04, \quad |C| < 1\%$$

In some new physics scenarios the photon polarization is mixed. Thus, any significant sizable asymmetry would be a clear signal of physics beyond the SM.

In this report, preliminary results from the BABAR experiment [8] on the inclusive $B \to X_s\gamma$ decay and the exclusive $B^0 \to K^{*0}\gamma$ decay with $K^{*0} \to K^0_S\pi^0$ are presented.

2. $B \to X_S\gamma$ DECAYS

2.1. Analysis

To calculate the $B \to X_s\gamma$ signal efficiency and optimize the selection cuts, the m_{X_s} spectrum computed by Kagan and Neubert (KN) is used [3]. This spectrum is a smooth function of m_{X_s}: strange resonances are not explicitly modeled in the mass spectrum. It is known experimentally that the mass region $m_{X_s} < 1.1$ GeV/c^2 is dominated by the $K^*(892)$ resonance. Resonances higher than $K^*(892)$ are broad and spaced closely enough, such that the smooth approximation of the KN treatment is justified, but the spectrum needs to be modified to correctly describe the decays $B \to K^*(892)\gamma$. In practice, a relativistic Breit-Wigner distribution modeling the $B \to K^*(892)\gamma$ decays is used for $m_{X_s} < 1.1$ GeV/c^2 and the KN spectrum is used above the cutoff. The $K^*\gamma$ part is normalized to the fraction of the KN spectrum below that m_{X_s} cutoff. The KN calculation has two empirical parameters, m_b and μ_π^2, which describe the effective b-quark mass and its kinetic energy inside the B meson, respectively. As default parameters in the model, $m_b = 4.80$ GeV/c^2 and $\mu_\pi^2 = 0.3$ GeV^2 are used. m_b is varied between 4.55 and 4.80 GeV/c^2 when evaluating systematics.

Backgrounds arise from $B\bar{B}$ events, mainly due to high energy photons from π^0 and η decays, and from continuum events ($e^+e^- \to q\bar{q}$, $q = u,d,s,c$ and $\tau^+\tau^-$) where, in addition to π^0 and η decays, a high energy photon can be radiated from one of the colliding $e^\pm$ beam (Initial State Radiation, ISR).

BABAR has performed two analysis of the $B \to X_s\gamma$ channel: a semi-inclusive analysis, which aims to reconstruct a large part of the total $B \to X_s\gamma$ rate by summing many exclusive reconstructed modes and a fully inclusive measurement, where no requirements are made on the hadronic state (X_s). Both of these analysis are based on approximatively 89 million $B\bar{B}$ pairs.

The semi-inclusive analysis sums 38 exclusive decay modes (and their charge coniugates), which represent about 55% of the total inclusive rate in the region $M_{X_s} = 1.1 - 2.8$ GeV/c^2. This analysis uses event shape information to remove continuum background (in the CM frame continuum background exhibits a jet-like structure while the B decay is more spherical) and the kinematic constraint provided by the $\Upsilon(4S)$ initial state. The energy of the B candidate in the $\Upsilon(4S)$ frame must infact be equal to the energy of the single beam in the center of mass system, E^*_{beam}. We define the energy-substituted mass $m_{ES} = \sqrt{E^{*2}_{beam} - p^{*2}_B}$ and the energy difference $\Delta(E) = E^*_B - E^*_{beam}$, where p^*_B and E^*_B are the momentum and the energy of the reconstructed B candidate in the $\Upsilon(4S)$ frame.

The signal yield in bins of m_{X_s} is extracted by performing a fit to the m_{ES} distribution. The m_{X_s} spectrum is then converted into the photon energy spectrum by using the kinematic relationship for the decay of a B meson of mass m_B:

$$E_\gamma = \frac{m_B^2 - m_{X_s}^2}{2m_B}$$

This gives a measurement of the E_γ spectrum in the B rest frame with a resolution of 1 to 5 MeV. This analysis has a significant systematic uncertainty due to the missing fraction, the part of the $B \to X_s\gamma$ rate which is not reconstructed.

In the fully inclusive analysis, the distinctive feature of the signal decays is an isolated photon with CM energy in the range 1.6 – 2.8 GeV. Photons that are consistent with originating from π^0 or η decay are explicitly vetoed. Event shape variables are used to remove continuum events. Further suppression of continuum background is obtained by requiring a high momentum lepton in the event. About 20% of B mesons decay semileptonically to either an electron or a muon or and the large mass of the B meson tends to impart large momentum to its lepton daughter. Leptons from fragmentation in continuum events tend to be at lower momentum. Off-resonance data are used to subtract the residual continuum background. The remaining $B\bar{B}$ background is subtracted by using data control sample for each of the $B\bar{B}$ sources. The main systematics uncertainty comes to $B\bar{B}$ background subtraction and it is mostly due to the statistical uncertainties on the correction factors derived from the π^0 and η control samples. The photon energy spectrum is measured in the $\Upsilon(4S)$ frame.

2.2. Results

Figures 1 and 2 show the resulting photon energy spectra for the semi inclusive and the fully inclusive analysis, respectively. In the semi inclusive analysis the photon energy is actually inferred from the hadronic invariant mass, which has resolution of ≈ 10 MeV/c^2. The latter converts into an E_γ resolution of 1-5 MeV, so the K^* peak can be clearly seen.

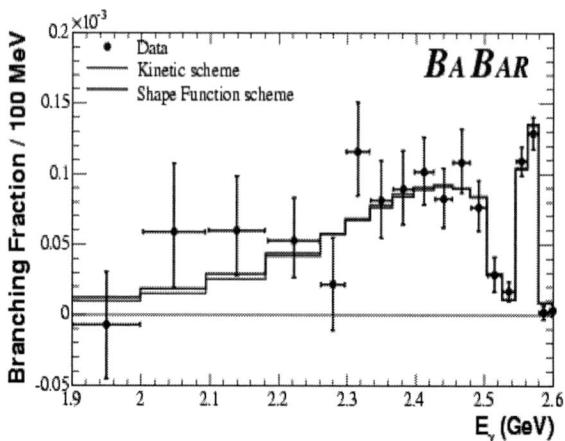

FIGURE 1. The photon energy spectrum, background subtracted and efficiency corrected from the semi inclusive analysis. The spectrum is measured in the B-meson rest frame. The peak due to $B^0 \to K^{*0}\gamma$ decays at high photon energy is visible due to the good resolution.

In Table 1 the moments of the photon energy spectrum as a function of the lower threshold on the photon energy are presented. The moments are given in the B-meson rest frame. Since in the fully inclusive analysis the photon energy spectrum is measured in the $\Upsilon(4S)$ frame, a

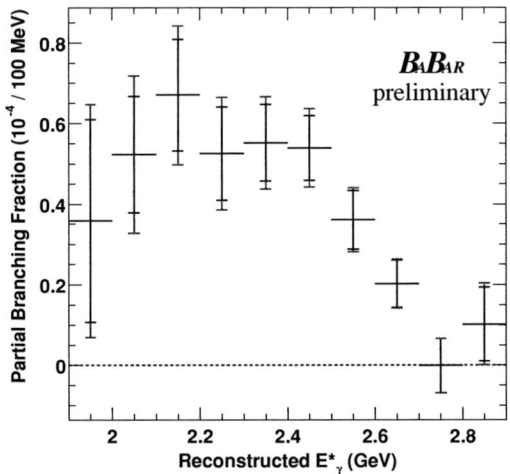

FIGURE 2. The photon energy spectrum, background subtracted and efficiency corrected from the fully inclusive analysis. The spectrum is measured in the $\Upsilon(4S)$ frame.

correction is applied to bring the observed moments into the B-meson rest frame. An additional small correction accounts fro cutting on the reconstructed photon energy in the $\Upsilon(4S)$ frame instead of on the true photon energy in the B-meson rest frame. These moments can be directly compared to theoretical calculations to give information on the parameters of the Heavy Quark Effective Theory (HQET).

The semi inclusive analysis also performed a fit to the spectrum to extract the HQET parameters $\bar{\Lambda} = m_B - m_b$ and μ_π^2. To perform the fit two theoretical schemes were used: the kinetic scheme [4] and the shape function scheme [5]. Results are shown in Table 2. We note that the parameter μ_π^2 is not defined the same way in the two schemes.

The spectrum fit also allows to extrapolate the branching fraction down to $E_\gamma > 1.6$ GeV. The average of the results from the two theoretical schemes is:

$$BF(b \to \gamma) = (3.38 \pm 0.19^{+0.64+0.07}_{-0.41-0.08}) \times 10^{-4}$$

which is consistent with the Standard Model calculation [1] and with the world experimental average [9].

In the fully inclusive analysis the use of lepton tagging on the non-signal side B for continuum suppression allows a straightforward determination of the CP asimmetry:

$$A_{CP} = \frac{\Gamma(b \to s\gamma + b \to d\gamma) - \Gamma(\bar{b} \to \bar{s}\gamma + \bar{b} \to \bar{d}\gamma)}{\Gamma(b \to s\gamma + b \to d\gamma) + \Gamma(\bar{b} \to \bar{s}\gamma + \bar{b} \to \bar{d}\gamma)}$$

The measured asymmetry is $A_{CP}(b \to s\gamma + b \to d\gamma) = -0.110 \pm 0.115 \pm 0.017$, where the first uncertainty is statistical and the second systematic. A measurement of the direct CP asymmetry in the inclusive $b \to s\gamma$ decays was previously performed at BABAR with the semi inclusive technique by reconstructing a set of 12 exclusive $b \to s\gamma$ final states [10]. This measurement led to $A_{CP}(b \to s\gamma) = 0.025 \pm 0.050 \pm 0.015$. Within the large statistical uncertainties both the A_{CP} measurements are consistent with the SM predictions.

3. $B^0 \to K^{*0}\gamma$

3.1. Analysis

The time-dependent CP violating asymmetry in the $B^0 \to K_S^0 \pi^0 \gamma$ decays is extracted from the distribution of the difference of the proper decay times, $\Delta t \equiv t_{CP} - t_{tag}$, where t_{CP} refers to the decay time of the signal B (B_{CP}) and t_{tag} to that of the other B (B_{tag}).

The analysis is performed on 232 million $B\bar{B}$ pairs. $B^0 \to K_S^0 \pi^0 \gamma$ decays have a rich resonance structure. In this analysis $K^{*0} \to K_S^0 \pi^0$ decays from $K_S^0 \pi^0$ combinations with an invariant mass $0.8 < m_{K_S^0 \pi^0} < 1.1$ GeV/c^2 are selected. To identify $B^0 \to K_S^0 \pi^0 \gamma$ decays the energy-substituted mass m_{ES} and the energy difference ΔE are used. Event topology is exploited to further suppress the background from continuum events. The signal efficiency is 16%.

The proper time difference is extracted from the separation of the B_{CP} and B_{tag} decay vertices. The B_{tag} vertex and flavor are determined inclusively from the remaining charged particles in the event. The B_{CP} decay point is obtained intersecting the K_S^0 flight direction with the beams line. The Δt resolution strongly depends on the K_S^0 flight direction and on the number of the SVT layers traversed by the K_S^0 decay daughters. In about 70% of the events both pion tracks from K_S^0 decay are reconstructed with at least 4 SVT hits. The average Δt resolution in these events is about 1.0 ps.

3.2. Results

The signal yield and the CP violating asymmetry are extracted with an unbinned maximum-likelihood fit to m_{ES}, ΔE and Δt. The fit gives $N_{K^*\gamma} = 157 \pm 16$ signal events, with

$$S = -0.21 \pm 0.40 \pm 0.05$$
$$C = -0.40 \pm 0.23 \pm 0.03$$

Figure 3 shows the background subtracted distribution of Δt for B^0- and $\bar{B}^0$-tagged events and the asymmetry

TABLE 1. Partial branching fractions (PBF) and moments of the photon energy spectrum from $B \to X_s\gamma$ decays as a function of the lower threshold on the photon energy. The first four rows are from the fully inclusive analysis, the remaining rows are from the semi inclusive analysis. Values are given for the B-meson rest frame. The uncertainties are statistical, systematic and model-dependent (for the fully inclusive analysis), respectively.

Min E_γ (GeV)	PBF (10^{-4})	1st Moment (GeV)	2nd Moment (GeV^2)	3rd Moment (GeV^3)
1.9	$3.67 \pm 0.29 \pm 0.34 \pm 0.29$	$2.288 \pm 0.025 \pm 0.017 \pm 0.012$	-	-
2.0	$3.41 \pm 0.27 \pm 0.29 \pm 0.23$	$2.316 \pm 0.016 \pm 0.010 \pm 0.012$	-	-
2.1	$2.97 \pm 0.24 \pm 0.25 \pm 0.17$	$2.355 \pm 0.014 \pm 0.007 \pm 0.010$	-	-
2.2	$2.42 \pm 0.21 \pm 0.20 \pm 0.13$	$2.407 \pm 0.012 \pm 0.005 \pm 0.008$	-	-
1.897	-	$2.321 \pm 0.044^{+0.037}_{-0.026}$	$0.0253 \pm 0.0116^{+0.0049}_{-0.0042}$	$0.0006 \pm 0.0085^{+0.0041}_{-0.0032}$
1.999	-	$2.314 \pm 0.025^{+0.027}_{-0.014}$	$0.0273 \pm 0.0039^{+0.0037}_{-0.0022}$	$0.0009 \pm 0.0036^{+0.0036}_{-0.0022}$
2.094	-	$2.357 \pm 0.018^{+0.016}_{-0.007}$	$0.0183 \pm 0.0023^{+0.0021}_{-0.0012}$	$0.0005 \pm 0.0017^{+0.0016}_{-0.0009}$
2.181	-	$2.396 \pm 0.013^{+0.008}_{-0.004}$	$0.0115 \pm 0.0014^{+0.0010}_{-0.0006}$	$0.0001 \pm 0.0008^{+0.0006}_{-0.0003}$
2.261	-	$2.425 \pm 0.009^{+0.004}_{-0.002}$	$0.0075 \pm 0.0007^{+0.0003}_{-0.0002}$	$-0.0001 \pm 0.0003^{+0.0002}_{-0.0001}$

TABLE 2. Heavy quark parameters $\bar{\Lambda}$ and μ_π^2 as obtained by fitting the photon energy spectrum in the kinetic [4] and in the shape function scheme [5]. Errors are the sum of statistical and systematic uncertainties and do not include the theoretical uncertainties.

	$\bar{\Lambda}$ (GeV)	μ_π^2 (GeV^2)
kinetic scheme	$0.59^{+0.05}_{-0.04}$	$0.30^{+0.07}_{-0.05}$
shape function scheme	$0.63^{+0.04}_{-0.04}$	$0.19^{+0.06}_{-0.05}$

$A_{K^*\gamma}(\Delta t)$. Within the large statistical uncertainties the measurement is consistent with the SM expectation.

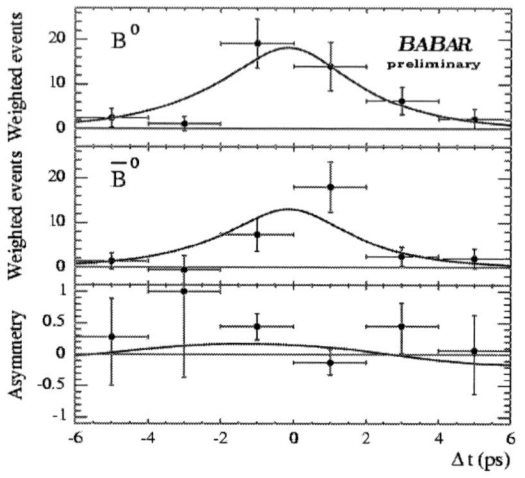

FIGURE 3. Background subtracted distribution of Δt for B^0- (top) and $\bar{B}^0$-tagged events (center), and the asymmetry $A_{K^*\gamma}$ (bottom).

4. CONCLUSIONS

$b \to s\gamma$ decays are directly sensitive to physics beyond the Standard Model, but no deviation has been found yet. The precision on both the A_{CP} and the branching fraction measurement can be improved with higher statistics. The measurement of the photon energy spectrum, especially of its moments, is becoming useful for the determination of the HQET parameters.

REFERENCES

1. P. Gambino, and M. Misiak, *Nucl. Phys. B*, **611**, 338 (2001); A. J. Buras, A. Czarnecki, M. Misiak, and J. Urban, *Nucl. Phys. B*, **631**, 219 (2002).
2. T. Hurth, E. Lunghi, and W. Porod, *Nucl. Phys. B*, **704**, 56–74 (2005).
3. A. L. Kagan, and M. Neubert, *Eur. Phys. J. C*, **7**, 5 (1999);
4. D. Benson, I. I. Bigi, and N. Uraltsev, *Nucl. Phys. B* **710** 371–401 (2005);
5. M. Neubert, *Eur. Phys. J. C*, **40**, 165–186 (2005);
6. B. O. Lange, M. Neubert, and G. Paz, hep-ph/0504071.
7. D. Atwood, M. Gronau, and A. Soni, *Phys. Rev. Lett.* **79**, 185–188 (1997).
8. *BABAR* Collaboration, B. Aubert et al., *Nucl. Instrum. Meth. A* **479** 1 (2002).
9. Heavy Flavor Averaging Group (HFAG), hep-ex/0412073.
10. *BABAR* Collaboration, B. Aubert et al., *Phys. Rev. Lett.* **93**, 021804 (2004)

Measurements of $|V_{cb}|$ and $|V_{ub}|$ at BaBar

M. Rotondo

Dipartimento di Fisica Galileo Galilei
Via Marzolo 8,
Padova 35131, Italy
E-mail: rotondo@pd.infn.it

Abstract. We report results from the BABAR Collaboration on the semileptonic B decays, highlighting the measurements of the magnitude of the Cabibbo-Kobayashi-Maskawa matrix elements V_{ub} and V_{cb}. We describe the techniques used to obtain the matrix element $|V_{cb}|$ using the measurement of the inclusive $B \to X_c \ell \nu$ process and a large sample of exclusive $B \to D^* \ell \nu$ decays. The $|V_{ub}|$ matrix elements has been measured studying different kinematic variables of the $B \to X_u \ell \nu$ process, and also with the exclusive reconstruction of $B \to \pi(\rho)\ell \nu$ decays.

Keywords: Semileptonic B decays, $|V_{ub}|$, $|V_{cb}|$, CKM matrix
PACS: 13.20.He, 12.15.Hh, 14.40.Nd

INTRODUCTION

The measurement of the parameters $|V_{cb}|$ and $|V_{ub}|$ provides important inputs to test the unitarity of the CKM matrix. To determine these parameters we have to measure the decay rates of the $b \to c$ and $b \to u$ transitions. The weak parameters are obscured by the hadronization effects and by the interaction between initial and final state. The common approach is to use the semileptonic decays $B \to X_{u,c} \ell \nu$ which reduce strongly the non calculable hadronic effects compared to the fully hadronic B decays. The parameters $|V_{cb}|$ and $|V_{ub}|$ are extracted in many different way to get a better control of the remaining non-perturbative effects, which still limit the precision of their measurements.

MEASUREMENTS OF $|V_{CB}|$

The measurement of the partial decay width for the process $B \to X_c \ell \nu$ can be used to determine $|V_{cb}|$ throw the relation $|V_{cb}|^2 = \Gamma(B \to X_c \ell \nu)/\gamma_{th}$, where $\Gamma(B \to X_c \ell \nu)$ is the semileptonic Cabibbo favored partial width. The Operator Product Expansion (OPE) allow to write the γ_{th} as a double series in α_s and $1/m_b$, where m_b is the b quark mass, which is the key parameter of these kind of expansions. The leading non-perturbative corrections appears at the $1/m_b^2$ order and are parameterized by the quantity μ_π^2, related to the Fermi motion of the b quark inside the B meson, and μ_G^2, related to the expectation value of the chromomagnetic operator [1]. At $\mathcal{O}(1/m_b^3)$ two additional parameters enter, ρ_D^3 and ρ_{LS}^3, the expectation values of the Darwin and spin-orbit dimension-six operators. All these parameters are a function of the scale μ that separates the short-distance effects from the long-distance QCD contributions. Other experimental quantities, like the moments of the invariant mass m_X of the hadronic system recoiling against the lepton-ν pair and the moment of the lepton energy spectra, can be written as a function of the same non-perturbative parameters. Therefore we have the opportunity to extract these HQE parameters from the data itself.

BABAR exploit the high statistics data sample collected by performing the m_X study on the recoil of fully reconstructed B decays [2]. We use a sample of $B \to D^{(*)}Y$ (B_{reco}) where Y denoted a collection of hadrons with total charge ± 1. On the signal side a lepton with momentum $p_\ell^* > 0.9$ GeV/c is required. All remaining tracks and neutral showers are combined into the hadronic system X. The extracted moments are extracted as a function of the minimum lepton momentum cut. The results are consistent with the recent CLEO, Belle and CDF measurements of the hadronic moments as a function of the lepton momentum cut, and also with a measurement performed by DELPHI without any lepton cut [3]. A fit to all BABAR hadronic moments and lepton spectra moments is performed in the kinetic-scheme [4]. The results, at the scale $\mu = 1$ GeV, up to $\mathcal{O}(1/m_b^3)$, are $m_b = 4.61 \pm 0.05_{exp} \pm 0.04_{HQE}$ GeV, $m_c = 4.61 \pm 0.05_{exp} \pm 0.04_{HQE}$ GeV, $\mu_\pi^2 = 0.45 \pm 0.04_{exp} \pm 0.04_{HQE}$ GeV^2, and $\mu_G^2 = 0.27 \pm 0.06_{exp} \pm 0.04_{HQE}$ GeV^2. The fit extract the semileptonic branching ratio $\mathcal{B}_{c e \nu} = (10.61 \pm 0.16_{exp} \pm 0.06_{HQE})\%$, and $|V_{cb}| = (41.4 \pm 0.4_{exp} \pm 0.7_{HQE}) \times 10^{-3}$. The main source of theoretical errors are due to the limited knowledge of the expression for the decay rate which includes various perturbative corrections and higher order non-perturbative corrections. The choice of the scale μ instead, has a very small impact on $|V_{cb}|$. The lepton spectra and hadronic-mass moments have different sensitivity to the fit parameters, but the results of the separate fits are compatible between them and with the global fit.

Bauer et al. [5] have performed an extensive global fit using measurements from BABAR, Belle, CLEO, CDF and DELPHI. Using the 1S scheme [6], they find $|V_{cb}| = (41.4 \pm 0.6 \pm 0.1) \times^{-3}$, the first error include both theoretical and experimental errors, and the second errors is due to the B meson lifetime. The results in different mass schemes are in good agreement. These results allow to determine $|V_{cb}|$ with an error close to 2% and demostrate the reliability of the OPE framework to predict the inclusive decay rates.

The study of the decay $B^0 \to D^{*+}\ell^-\bar{\nu}$ can also be used to extract $|V_{cb}|$. The differential decay rate for the process $B^0 \to D^{*+}\ell^-\bar{\nu}$ is determined as a function of the four-velocity product, $w = v_B \cdot v_D^* = (m_B^2 + m_{D^*}^2 - q^2)/2m_B m_D^*$ where q^2 is the square of the 4-momentum transfered to the two leptons. The theoretical tool to extract the hadronic matrix element is the Heavy Quark Effective Theory (HQET) [7]. The relation $d\Gamma/dw \propto |V_{cb}|^2 \mathscr{F}^2(w) \mathscr{K}(w)$ expresses the decay width as the product of the phase space factor $\mathscr{K}(w)$ times the hadronic form factor $\mathscr{F}(w)$ square, which corresponds to the Isgur-Wise function in the limit of infinite b and c masses. This form factor can be expressed as $\mathscr{F}(w) = \mathscr{F}(1)g(w)$, where the shape function $g(w)$ is an almost linear function. Using some general analytical constraints by QCD [8], the shape $g(w)$ can written as a function depending on just one single parameter (ρ^2), which is usually determined from data. In the limit $w \to 1$ the HQET predict $\mathscr{F}(1) = 1$, some recent quenched lattice QCD calculations gives $\mathscr{F} \approx 0.913^{+0.030}_{-0.035}$ [9], which is compatible with other calculations. The product $\mathscr{F}(1)|V_{cb}|$ is measured by extrapolating $d\Gamma/dw$ at the point of zero recoil $w \to 1$, where the phase space $\mathscr{K}(w)$ is null. The measurement of $|V_{cb}|$ using this technique has been performed both at the B-Factories and LEP experiments. In BABAR we select events with a lepton with $p^* > 1.2$ GeV and a reconstructed $D^{*+0} \to D^0\pi$, where the D^0 is reconstructed in the $D^0 \to K^-\pi^+$, $D^0 \to K^-\pi^+\pi^-\pi+$ and $D^0 \to K^-\pi^+\pi^0$ modes. A sample of ~ 57.000 signal events are selected. Kinematics constraints are used to extract from the data the amount of the decay $B \to D^*X\ell\nu$, which is the most dangerous physics background, (often called simply D^{**} background). The result of the fit to the w distribution, combined with the value of $\mathscr{F}(1)$ showed above, is $|V_{cb}| = (38.7 \pm 0.3_{stat} \pm 1.7_{syst} {}^{+1.5}_{-1.3theo}) \times 10^{-3}$. The branching fraction $\mathscr{B}(B \to D^{*+}\ell^-\bar{\nu}) = (4.90 \pm 0.07 \pm 0.36)\%$ is determined by integrating the differential w distribution. This measurement is compatible with the result of the inclusive measurement. The total error of 5% of the ex-clusive measurements is dominated by the uncertainty on $\mathscr{F}(1)$, which is expected to decrease with the future lattice calculations.

MEASUREMENTS OF $|V_{UB}|$

The measurement of $B \to X_u\ell\nu$ decays is a difficult experimental task due to the high $B \to X_c\ell\nu$ background. This kind of background have to be reduced by restricting the phase space in the analysis. There are different kinematic variables of the $X\ell\nu$ final state that can be used to reduce the $B \to X_c\ell\nu$ contamination: the lepton energy E_ℓ, the hadronic mass m_X and the lepton-neutrino invariant mass squared q^2. Usually we cut on one or more of these three variables, but we need to know well the fraction of the $B \to X_u\ell\nu$ events that survive the experimental cuts. The OPE framework can reliably predict the inclusive $B \to X_u\ell\nu$ decay rate, but the experimental cuts introduce large uncertainties due to the non-perturbative correction of the order $\mathscr{O}(\Lambda_{QCD}/m_b)$, which can be described by the distribution function of the momentum of the b quark inside the B meson [10]. This distribution function, the *shape function*, have to be extracted experimentally. The parameters of the shape function can be related to parameters m_b and μ_π^2 determined from the OPE fit to the $B \to X_c\ell\nu$ moments described before. But the shape function can also be determined directly from the data using the photon spectrum in the $b \to s\gamma$ decays, which, in the leading order of Λ_{QCD}/m_b is described by the same shape function. The $b \to s\gamma$ measurement by CLEO, Belle and BABAR, are used to constrain the parameters of the shape function. The agreement between the shape function extraction with $b \to c\ell\nu$ and $b \to s\gamma$ spectra shows clearly that the theoretical uncertainties are under control, as can be inspected from Fig.1(left). There are some theoretical issues that have to be solved to reduce the error on the extraction of $|V_{ub}|$: the shape functions that governs $B \to X_u\ell\nu$ and $b \to s\gamma$ are the same only at the $\mathscr{O}(\Lambda_{QCD}/m_b)$, the subleading shape functions can introduce further corrections. The weak-annihilation diagrams, in which the b and the $\bar{d}$ quarks annihilate into a W^- boson, gives contributions in some relevant regions of the phase space, and may introduce other theoretical uncertainties.

Historically the first approach to extract $B \to X_u\ell\nu$ was to measure the lepton spectrum beyond the kinematic cutoff for $B \to X_c\ell\nu$ decays, which is at $p_\ell^* > 2.3$ GeV (CLEO). The disadvantage is that only about 10% of all charmless semileptonic decays are detected, therefore the extrapolation to the full phase space is significant with corresponding uncertainties. Recent measurements use the E_ℓ spectrum also in the region below the $B \to X_c\ell\nu$ endpoint. BABAR measured the electron mo-

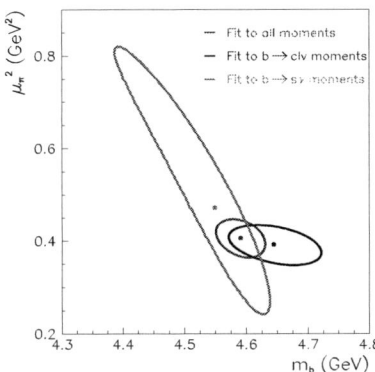 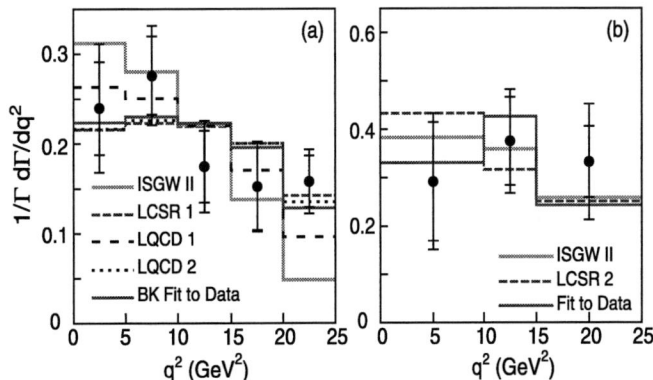

FIGURE 1. Left: contours in the m_b-vs-μ_π^2 plane separated from the $b \to c\ell\nu$ moments fit and the $b \to s\gamma$ moments fit, the results of the global fit is also reported. Right: comparison of the differential decay rates as a function of q^2 for $B \to \pi\ell\nu$ (a) and $B \to \rho\ell\nu$ (b) with various form-factor predictions.

TABLE 1. $|V_{ub}|$ results of different experiments. The numbers below use input from the HQ parameters m_b and μ_π^2 from a global fit to $B \to X_c\ell\nu$ and $b \to s\gamma$ moments in the kinetic scheme, which gives $m_b = (4.60 \pm 0.04)$ GeV and $\mu_\pi^2 = (0.20 \pm 0.04)$ GeV^2 (Fig.1).

	$|V_{ub}| \times 10^{-3}$	Remarks	Ref.
CLEO (endpoint)	$3.93 \pm 0.46 \pm 0.33$	$2.1 < p_\ell < 2.6$ GeV	[14]
Belle (endpoint)	$4.82 \pm 0.45 \pm 0.31$	$1.9 < p_\ell < 2.6$ GeV	[15]
BABAR (endpoint)	$4.20 \pm 0.34 \pm 0.30$	$2.0 < p_\ell < 2.6$ GeV	[11]
BABAR (E_ℓ, q^2)	$4.00 \pm 0.34 \pm 0.30$		[12]
Belle (m_X)	$4.03 \pm 0.27 \pm 0.25$	Breco tag	[16]
Belle (m_X, q^2)	$4.32 \pm 0.46 \pm 0.29$	tag with annealing method	[17]
BABAR (m_X, q^2)	$4.70 \pm 0.34 \pm 0.31$	Breco tag	[18]
HFAG Average	$4.39 \pm 0.20 \pm 0.27$	$\chi^2/\text{d.o.f} = 6.4/6$ (CL=0.38)	[19]

mentum in the interval $2.0 - 2.6$ GeV, which access a signal fraction f_u as large as 28% [11]. This was possible thanks to the better knowledge of the $B \to X_c\ell\nu$ backgrounds.

A different approach uses the leptons and the neutrino reconstructed from the missing momentum of the event, to calculate the q^2 of the lepton-neutrino system [12]. A suitable cut on the set (E_e, q^2) removes a large fraction of $B \to X_c\ell\nu$ events. With this techniques the signal-to-background is close to $1/2$, which have to be compared with a signal-to-noise ratio of $1/20$ using only the cut on the lepton momentum. An other approach uses the invariant mass m_X of the hadron system recoiling against the lepton-ν pair to identify the charmless component which typically involve lighter m_X. The cuts on the m_X retain $50 - 80\%$ of all $B \to X_u\ell\nu$ decays, depending which cut is used. BABAR has exploited the high statistics using the B_{reco} sample described before. A lepton with minimum momentum in the B rest frame $p_\ell^* > 1$ GeV/c is required in the signal side. The m_X, however, is affected by a strong shape-function dependence near the D threshold. A simple cut on q^2 at $(m_B - m_D)^2$ provides only a moderate efficiency, $f_u \approx 20\%$. To avoid this kind of limitations Bauer $et.al$ [13] proposed a combined cut on both m_X and q^2 that maximize f_u and minimize the shape function dependence.

The partially branching fractions measured by experiments can be translated into $|V_{ub}|$ by $|V_{ub}| = \sqrt{(\Delta\mathcal{B}/\overline{\Gamma}_{th}\tau_B)}$ where the reduced decay rate $\overline{\Gamma}_{th}$ is related to the partial width of the $B \to X_u\ell\nu$ decay into the phase space of interest. The calculations of $\overline{\Gamma}_{th}$ needs the shape function. In the most recent results reported in Tab.?? the shape functions are obtained by a global fit to the $B \to X_c\ell\nu$ and $b \to s\gamma$. The agreement between the different measurements agree well with each other, and the error on the average is close to the 10%. Future improvements on the shape function determination (in particular better knowledge of m_b) may further reduce the error on the .

The measurement of can be performed also using the exclusive reconstruction of $B \to \pi\ell\nu$ and $B \to \rho\ell\nu$, etc.

TABLE 2. Branching fractions of $B \to \pi\ell\nu$. The HFAG average assumes equality of the semileptonic decays widths for the B^0 and B^+ decays, and is quoted for the B^0.

	$\mathcal{B} \times 10^{-4}$	$\Delta\mathcal{B}$ ($q^2 > 16$ GeV^2)	Ref.
BABAR $B^+ \to \pi^0 \ell^+ \nu$ SL tag	$3.31 \pm 0.68 \pm 0.42$	-	[23]
BABAR $B^+ \to \pi^0 \ell^+ \nu$ Breco tag	$1.68 \pm 0.52 \pm 0.24$	-	[24]
BABAR $B^0 \to \pi^- \ell^+ \nu$ SL tag	$1.02 \pm 0.25 \pm 0.13$	$0.21 \pm 0.14 \pm 0.06$	[25]
BABAR $B^0 \to \pi^- \ell^+ \nu$ Breco tag	$0.89 \pm 0.34 \pm 0.12$	-	[24]
Belle $B^0 \to \pi^- \ell^+ \nu$ SL tag	$1.79 \pm 0.28 \pm 0.20$	0.46 ± 0.17	[26]
CLEO $B \to \pi\ell\nu$ untagged	$1.32 \pm 0.18 \pm 0.11$	$0.25 \pm 0.09 \pm 0.05$	[27]
BABAR $B \to \pi\ell\nu$ untagged	$1.38 \pm 0.10 \pm 0.18$	$0.49 \pm 0.05 \pm 0.07$	[28]
HFAG average	1.36 ± 0.11	0.40 ± 0.05	[19]

The differential decay rate of the $B \to \pi\ell\nu$ decay is given by $d\Gamma/dq^2 \propto |V_{ub}|^2 |f_+(q^2)|^2 p_\pi^3$, where $f_+(q^2)$ is the form factor, which is harder to calculate than the $D \to D^{(*)}$ form factors. There are different techniques which include quenched lattice QCD, light-cone sum rules (LCSR) and quark model. The most recent calculations from Ball and Zwichy [20] quote an uncertainty of 13% in the small q^2 region, with a LCSR calculations. Some results with the LQCD calculations [21, 22] quote uncertainty at large q^2 about 13%. To fully exploit such kind of theoretical calculations, the experimental measurements of the differential decay rate as a function of q^2 is required. The techniques with allow to obtain the largest sample of $B \to \pi\ell\nu$ is the untagged method, with the neutrino-reconstruction. In this techniques the neutrino is inferred from the missing momentum of the event, which is combined with the lepton and a pion to reconstruct a $B \to \pi\ell\nu$. In Fig.1(right) the distribution of the differential decay rates as a function of q^2 is compared with different theoretical calculations. An other techniques use a sample of events with a B meson tagged via the semileptonic $B \to D(*)\ell\nu$ decays. With the semileptonic tagged sample the signal-to-noise is better compared to the signal-to-noise of the untagged sample, but the statistics is limited. The results of the branching fraction for the $B \to \pi\ell\nu$ using different methods are reported in Tab.2. Using the partial branching fractions ($q^2 > 16$ GeV^2) and the FNAL04 lattice calculations of the form factor, gives $|V_{ub}| = (3.75 \pm 0.27^{+0.64}_{-0.42}) \times 10^{-3}$ where the first error is experimental and the second is due to the normalization uncertainty in the form factor calculation. Calculations using the region at low q^2 along with LCSR calculations are in progress.

SUMMARY AND OUTLOOK

Thanks to the developments of new experimental techniques and to the rapid progress in the theoretical knowledge of the semileptonic B decays, allow to measure and with errors of 2% and 10% respectively. The extraction with the inclusive approaches, is limited by the uncertainties on the shape function, which can be improved with the future measurement of the $B \to X_c \ell\nu$ and $b \to s\gamma$ spectra. The extracted using the exclusive approach is limited by the lattice QCD calculations of the form-factors. Further development are expected in the next future which allow, together with the inclusive measurement, to extract the ratio $|V_{ub}/V_{cb}|$ with an error close to 5% which will give a very stringent test of the Standard Model.

REFERENCES

1. A. H. Hoang, Phys. Rev. D**59**, 014039 (1999)
   N. Uraltsev, J. Phys, G **27**, 1081 (2001)
2. B. Aubert *et al.*, Phys.Rev.Lett.**93**, 011803 (2004)
3. DELPHI Collaboration, hep-ex/0210046
4. P.Gambino and N.Uraltsev, Eur. Phys. J **C34**, 181 (2004); I.Bigi *el al.* Phys. Rev. D**56**, 074017 (1999)
5. C.W. Bauer *el al.* Phys. Rev. D**70**, 094017 (2004)
6. A.H. Hoang *el al.*, Phys. Rev. D**59**, 074017 (1999)
7. N. Isgur and M. B. Wise, Adv. Ser. Direct. High Energy Phys. **10**, 546 (1992)
8. I. Caprini and M. Neubert, Phys.Lett.**B380** (1996) 376
9. S. Hashimoto *et.al* Phys.Rev.D **66**, 014503 (2003)
10. M. Neubert, Phys. Rev. D**49**, 4623 (1994)
11. BABAR Collaboration, hep-ex/0408075
12. BABAR Collaboration, hep-ex/0506036
13. C.W. Bauer *el al.* Phys. Rev. D**64**, 113004 (2001)
14. A. Bornheim *et al.*, Phys. Rev. Lett. **88**, 231803 (2002)
15. Belle Collaboration, hep-ex/0504046
16. Belle Collaboration, hep-ex/0505088
17. Belle Collaboration, Phys. Rev. Lett. **92**, 101801 (2004)
18. BABAR Collaboration, hep-ex/0507017
19. HFAG group, *http://www.slac.stanford.edu/xorg/hfag/semi*
20. P. Ball and Z. Zwicky, Phys. Rev. D**71**, 014015 (2005)
21. J.Shigemitsu *et al.*, hep-lat/0408019
22. M. Okamoto *et al.* hep-lat/0409116
23. BABAR Collaboration, hep-ex/0506065
24. BABAR Collaboration, ICHEP conference paper
25. BABAR Collaboration, hep-ex/0506064
26. BABAR Collaboration, hep-ex/0408145
27. CLEO Collaboration, Phys.Rev.D **68**, 072003 (2003)
28. BABAR Collaboration, hep-ex/0507003

γ Measurement at the B-factories

Luigi Li Gioi

Università di Roma "La Sapienza" and INFN Roma

Abstract.
The integrated luminosity recorded from the B-factories allow to obtain a measurement of the angle γ of the Unitarity Triangle. The different methods using the $B \to D^0 K$ will be discussed and the results from the *BaBar* and *Belle* collaborations will be summarized.

Keywords: gamma B-factory
PACS: 13.25.Hw, 14.40.Nd

INTRODUCTION

After 40 years from its discovery in K^0 system[1], the violation of the *CP* asymmetry is one of the open problem in particle physics. In recent years the *CP* violation in the *B* meson system has been clearly established [2], and although there is good agreement with the expectations of the Standard Model, further measurements of *CP* violation in *B* decays are needed to overconstrain the Unitarity Triangle and look for New Physics effects. Among those a crucial test will be represented by the measurement of γ, which is the complex phase of the CKM-element V_{ub} [3]. Various methods using $B \to D^0 K$ decays have been proposed to measure the Unitarity Triangle angle γ, basically all exploiting the fact that a charged B^- can decay into a $D^0 K^-$ final state[1] via V_{cb} or into $\bar{D}^0 K^-$ via V_{ub} mediated processes. CP violation can be detected if the D^0 and $\bar{D}^0$ decay into the same final state. The measurement of the direct *CP* violation is sensitive to the phase difference between V_{ub} and V_{cb} and thus to the angle γ. Fig. 1 shows the Feynman diagrams for both V_{ub} and V_{cb} processes.

THE GLW METHOD

The first method, originally proposed by Gronau, Wyler and London[4], is based on the interference between $B^- \to D^0 K^-$ and $B^- \to \bar{D}^0 K^-$ when the D^0 and $\bar{D}^0$ decay to *CP* eigenstates. Defining the ratios R and $R_{CP\pm}$ of Cabibbo-suppressed to Cabibbo-favored branching fractions

$$R_{(CP\pm)} \equiv \frac{Br(B^- \to D^0_{(CP\pm)} K^-) + Br(B^+ \to \bar{D}^0_{(CP\pm)} K^+)}{Br(B^- \to D^0_{(CP\pm)} \pi^-) + Br(B^+ \to \bar{D}^0_{(CP\pm)} \pi^+)} \quad (1)$$

with the neutral *D* meson reconstructed in non-*CP* (D^0) or *CP*-even/odd eigenstates ($D^0_{CP\pm}$) channels, and the direct *CP* asymmetry

$$A_{CP\pm} \equiv \frac{Br(B^- \to D^0_{CP\pm} K^-) - Br(B^+ \to \bar{D}^0_{CP\pm} K^+)}{Br(B^- \to D^0_{CP\pm} K^-) + Br(B^+ \to \bar{D}^0_{CP\pm} K^+)} \quad (2)$$

Neglecting the $D^0 - \bar{D}^0$ mixing and the ratio $r_\pi = A(B^- \to \bar{D}^0 \pi^-)/A(B^- \to D^0 \pi^-)$ of the amplitudes of the $B^- \to \bar{D}^0 \pi^-$ and $B^- \to D^0 \pi^-$ processes ($|r_\pi| < 0.02$), it is $R_\pm \equiv R_{CP\pm}/R = 1 + r^2 \pm 2r \cos\delta \cos\gamma$ and $A_{CP\pm} = \pm 2r \sin\delta \sin\gamma/(1 + r^2 \pm 2r \cos\delta \cos\gamma)$. Here $r = |A(B^- \to \bar{D}^0 K^-)/A(B^- \to D^0 K^-)|$ is the magnitude of the ratio of the amplitudes for the processes $B^- \to \bar{D}^0 K^-$ and $B^- \to D^0 K^-$, expected from theory to be about 0.1 – 0.2, and δ is the relative strong phase between these two amplitudes [4]. The measurement of $R_\pm$ and $A_{CP\pm}$ allows to constrain the three unknowns r, δ and the CKM angle γ.

Both *BaBar* and *Belle* use the decay modes $B^\mp \to DK^\mp$, $B^\mp \to D^*K^\mp$ and $B^\mp \to DK^{*\mp}$. Both experiments use the *CP* even D^0 decay final states K^+K^- and $\pi^+\pi^-$ in all three modes and only the $D^{*+} \to D^0 \pi^0$ decay, which gives $CP(D^*) = CP(D)$. For *CP* odd D^0 decay final states, *BaBar* uses $K_S \pi^0$ only for $DK^\mp$ analysis. For $DK^{*\mp}$ analysis *BaBar* also uses $K_S \phi$ and $K_S \omega$ (an asymmetric systematic error due to *CP* even pollution in these *CP* odd channels [5] is assigned). *Belle* uses $K_S \pi^0$, $K_S \eta$ and $K_S \phi$ in all modes.

The results of $R_{CP\pm}$ and $A_{CP\pm}$ from *BaBar* [5] and *Belle* [6] collaboration are summarized in Tab.1. All the measurements are dominated by the statistic uncertainty. The current measurement precision doesn't allow to obtain a constraint on the CKM angle γ.

---

[1] Reference to the charge-conjugate state is implied here and throughout the text unless otherwise stated.

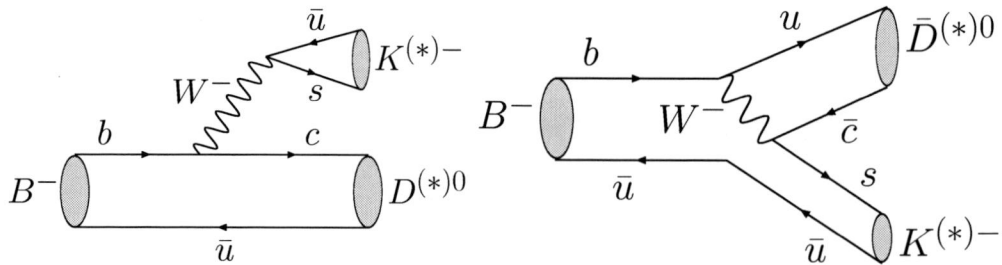

FIGURE 1. Feynman diagrams for $B^- \to D^{(*)0}K^{(*)-}$ and $B^- \to \bar{D}^{(*)0}K^{(*)-}$. The latter is color-suppressed respect to the former.

TABLE 1. Results of $R_{CP\pm}$ and $A_{CP\pm}$ from *BaBar* [5] and *Belle* [6] collaboration

Experiment	A_{CP+}	A_{CP-}	R_{CP+}	R_{CP-}
$D_{CP}K^-$				
BaBar	$0.40\pm0.15\pm0.08$	$0.21\pm0.17\pm0.07$	$0.87\pm0.14\pm0.06$	$0.80\pm0.14\pm0.08$
Belle	$0.07\pm0.14\pm0.06$	$-0.11\pm0.14\pm0.05$	$0.98\pm0.18\pm0.10$	$1.29\pm0.16\pm0.08$
$D_{CP}^*K^-$				
BaBar	$-0.10\pm0.23^{+0.03}_{-0.04}$		$1.06\pm0.26^{+0.10}_{-0.09}$	
Belle	$-0.27\pm0.25\pm0.04$	$0.26\pm0.26\pm0.03$	$1.43\pm0.28\pm0.06$	$0.94\pm0.28\pm0.06$
$D_{CP}K^{*-}$				
BaBar	$-0.09\pm0.20\pm0.06$	$-0.33\pm0.34\pm0.10^{+0.00}_{-0.06}$	$1.77\pm0.37\pm0.12$	$0.76\pm0.29\pm0.06^{+0.04}_{-0.14}$
Belle	$-0.02\pm0.33\pm0.07$	$0.19\pm0.50\pm0.04$		

THE ADS METHOD

The ADS method, proposed by Atwood Dunietz and Soni [7], considers the interference between $B^- \to D^0 K^-$ and $B^- \to \bar{D}^0 K^-$ when the D^0 and $\bar{D}^0$ decay to $K^+\pi^-$. In these processes, the favored B decay followed by the doubly CKM-suppressed D decay interferes with the suppressed B decay followed by the CKM-favored D decay. In the decays of interest, the sign of the kaon from B decay is opposite to that of the kaon from D decay. It is convenient to define ratios of rates between these decays and the similar decays where the two kaons have the same sign. The decays with same-sign kaons have much higher rate and proceed almost exclusively through the CKM-favored and color favored B transition, followed by the Cabibbo-favored D-decay. The advantage in taking ratios is that most theoretical and experimental uncertainties cancel. Thus, ignoring the possible effects of D mixing and taking into account the effective strong phase difference of π between the D^{*0} decays in $D^0\gamma$ and $D^0\pi^0$[8], it is possible to define the charge-integrated ratios:

$$R_{K\pi} \equiv \frac{\Gamma(B^-\to[K^+\pi^-]_D K^-)+\Gamma(B^+\to[K^-\pi^+]_D K^+)}{\Gamma(B^-\to[K^-\pi^+]_D K^-)+\Gamma(B^+\to[K^+\pi^-]_D K^+)}$$
$$= r_B^2 + r_D^2 + 2r_B r_D \cos\gamma\cos\delta \quad (3)$$

$$R^*_{K\pi, D\pi^0(D\gamma)} \equiv$$
$$\frac{\Gamma(B^-\to[K^+\pi^-]_{D^*\to D\pi^0(D\gamma)} K^-)+\Gamma(B^+\to[K^-\pi^+]_{D^*\to D\pi^0(D\gamma)} K^+)}{\Gamma(B^-\to[K^-\pi^+]_{D^*\to D\pi^0(D\gamma)} K^-)+\Gamma(B^+\to[K^+\pi^-]_{D^*\to D\pi^0(D\gamma)} K^+)}$$

$$= r_B^{*2} + r_D^2 + (-)2r_B^* r_D \cos\gamma\cos\delta^* \quad (4)$$

where

$$r_B \equiv \left|\frac{A(B^-\to\bar{D}^0 K^-)}{A(B^-\to D^0 K^-)}\right| \quad (5)$$

$$r_B^* \equiv \left|\frac{A(B^-\to\bar{D}^{*0} K^-)}{A(B^-\to D^{*0} K^-)}\right| \quad (6)$$

and $\delta_B^{(*)}$ and δ_D are strong phase differences between the two B and D decay amplitudes, respectively. The results of the yields, $R_{ADS}^{(*)}$ and $A_{ADS}^{(*)}$ from *BaBar* [9] and *Belle* [10] collaborations are summarized in Tab.2. *BaBar* founds an upper limit on $R_{ADS} < 0.030$ at 90% C.L. that gives an upper limit on $r_B < 0.223$ at 90% C.L. and an upper limit on $R_{ADS}^* < 0.16$ at 90% C.L.. *Belle* founds un upper limit on $r_B < 0.27$ at 90% C.L..

THE DALITZ METHOD

This method based on the analysis of the Dalitz distribution of the three-body decay $D^0 \to K_s\pi^+\pi^-$ [11]. The primary advantage of this method is that it involves the entire resonant structure of the three-body decay, with the interference of doubly Cabibbo-suppressed (DCS), Cabibbo-allowed (CA) and CP eigenstate amplitudes providing the sensitivity to γ. Assuming no CP asymmetry in D decays [12], the $B^\mp \to D^{(*)}K^\mp$,

TABLE 2. Yields, $R_{ADS}^{(*)}$ and $A_{ADS}^{(*)}$ from *BaBar* [9] and *Belle* [10] collaborations.

Experiment	Yields	A_{ADS}	R_{ADS}
$DK^-, D \to K^+\pi^-$			
BaBar	$5.0^{+4.0}_{-3.0}$	$0.013^{+0.011}_{-0.009}$	
Belle	$8.5^{+6.0}_{-5.3}$	$0.88^{+0.77}_{-0.62} \pm 0.06$	$0.023^{+0.016}_{-0.014} \pm 0.001$
$D^*K^-, D^* \to D\pi^0, D \to K^+\pi^-$			
BaBar	$-0.2^{+1.3}_{-0.7}$	$-0.001^{+0.010}_{-0.006}$	
$D^*K^-, D^* \to D\gamma, D \to K^+\pi^-$			
BaBar	$1.2^{+2.1}_{-1.4}$	$0.011^{+0.019}_{-0.013}$	

TABLE 3. *CP*-violating parameters $(x_\pm^{(*)}, y_\pm^{(*)})$ obtained from *BaBar* [13]. The first error is statistical, the second is the experimental systematic uncertainty and the third reflects the Dalitz model uncertainty.

CP parameter	Result
$x_- \equiv Re\{r_B e^{i(\delta_B - \gamma)}\}$	$0.077 \pm 0.069 \pm 0.026 \pm 0.019$
$y_- \equiv Im\{r_B e^{i(\delta_B - \gamma)}\}$	$0.064 \pm 0.092 \pm 0.037 \pm 0.042$
$x_+ \equiv Re\{r_B e^{i(\delta_B + \gamma)}\}$	$-0.129 \pm 0.070 \pm 0.030 \pm 0.032$
$y_+ \equiv Im\{r_B e^{i(\delta_B + \gamma)}\}$	$0.019 \pm 0.079 \pm 0.023 \pm 0.021$
$x_-^* \equiv Re\{r_B^* e^{i(\delta_B^* - \gamma)}\}$	$-0.131 \pm 0.093 \pm 0.028 \pm 0.021$
$y_-^* \equiv Im\{r_B^* e^{i(\delta_B^* - \gamma)}\}$	$-0.143 \pm 0.105 \pm 0.022 \pm 0.025$
$x_+^* \equiv Re\{r_B^* e^{i(\delta_B^* + \gamma)}\}$	$0.140 \pm 0.093 \pm 0.028 \pm 0.025$
$y_+^* \equiv Im\{r_B^* e^{i(\delta_B^* + \gamma)}\}$	$0.013 \pm 0.120 \pm 0.037 \pm 0.056$

$D^{*0} \to D^0\pi^0, D^0\gamma, D^0 \to K_s\pi^+\pi^-$ decay chain amplitude $A_\mp^{(*)}(m_-^2, m_+^2)$ can be written as

$$A_D(m_\mp^2, m_\pm^2) + \kappa r_B^{(*)} e^{i(\delta_B^{(*)} \mp \gamma)} A_D(m_\pm^2, m_\mp^2) \quad (7)$$

where m_-^2 and m_+^2 are the squared invariant masses of the $K_S\pi^-$ and $K_S\pi^+$ combinations, respectively, and $A_D(m_-^2, m_+^2)$ is the $D^0 \to K_s\pi^+\pi^-$ decay amplitude. Here, $r_B^{(*)}$ and $\delta_B^{(*)}$ are the amplitude ratios and relative strong phases between the amplitudes $A(B^- \to D^{(*)0}K^-)$ and $A(B^- \to \bar{D}^{(*)0}K^-)$. As a consequence of parity and angular momentum conservation in the D^{*0} decay, the factor κ takes the value $+1$ for $B^- \to D^0K^-$ and $B^- \to D^{*0}(D^0\pi^0)K^-$, and -1 for $B^- \to D^{*0}(D^0\gamma)K^-$ [8]. The amplitude $A_D(m_-^2, m_+^2)$ is determined through a Dalitz analysis of a high-statistics sample of tagged D^0 mesons from inclusive $D^{*+} \to D^0\pi^+$ decays reconstructed in data. Then a simultaneous fit to the $|A_-^{(*)}(m_-^2, m_+^2)|^2$ and $|A_+^{(*)}(m_-^2, m_+^2)|^2$ distribution is performed in order to determine the *CP* parameters $r_B^{(*)}, \delta_B^{(*)}$, and γ.

Both *BaBar* [13] and *Belle* [14] use the isobar formalism described in Ref. [15] to express A_D as a sum of two-body decay-matrix elements (subscript r) and a non-resonant (subscript NR) contribution

$$A_D(m_-^2, m_+^2) = \Sigma_r a_r e^{i\phi_r} A_r(m_-^2, m_+^2) + a_{NR} e^{i\phi_{NR}} \quad (8)$$

where each term is parameterized with an amplitude a_r and a phase ϕ_r. The function $A_r(m_-^2, m_+^2)$ is the Lorentz-invariant expression for the matrix element of a D^0 meson decaying into $K_S\pi^-\pi^+$ through an intermediate resonance r, parameterized as a function of the position in the Dalitz plane. It is parametrized by a Breit-Wigner with a functional form dependent on the spin of the resonance. Detail of the Dalitz parametrization of *BaBar* and *Belle* can be find in [13] and [14]. Both consider resonances of of type $(\pi\pi)$ and $(K_s\pi^-)$ plus most of the corresponding DCS $(K_s\pi^+)$. All the considered resonances are well established except the two scalars π, π resonances, σ_1 and σ_2, whose masses and widths are obtained from data.

BaBar choices as *CP*-violating parameters $(x_\pm^{(*)}, y_\pm^{(*)})$ defined as the real and imaginary parts of the complex amplitude ratios $r_B^{(*)} e^{i(\delta_B^{(*)} \pm \gamma)}$. These variables are more suitable fit parameters than $r_B^{(*)}, \delta_B^{(*)}$ and γ because they are better behaved near the origin, especially in low-statistics samples. Tab.3 shows the results of *CP*-violating parameters $(x_\pm^{(*)}, y_\pm^{(*)})$ obtained from *BaBar* [13]. Figures 2 show the confidence-level contours (*BaBar* and *Belle*) in the $(x_\pm^{(*)}, y_\pm^{(*)})$ planes for $D^0 K^-$ and (b) D^{*0} and separately for B^- and B^+. The separation between the B^- and B^+ regions in these planes is an indication of direct *CP* violation. *BaBar* found a evidence of direct *CP* violation at a level of 2.7σ in the combined DK and D^*K sample.

Since the measured values of $r_B^{(*)}$ are positive definite, and the error on γ depends on the value of r_B, some statistical treatment is necessary to correct for bias. Both *BaBar* and *Belle* use frequentist treatments. Tab.4 shows the results of $r_B^{(*)}, \delta_B^{(*)}$ and γ from *BaBar* and *Belle*.

REFERENCES

1. J. H. Christenson, J. W. Cronin, V. L. Fitch and R. Turlay, Phys. Rev. Lett. **13**, 138 (1964).
2. BABAR Collaboration, B. Aubert *et al.*, Phys. Rev. Lett. **89**, 201802 (2002); BELLE Collaboration, K. Abe *et al.* Phys. Rev. D **66**, 071102 (2002).

TABLE 4. $r_B^{(*)}$, $\delta_B^{(*)}$ and γ from *BaBar* and *Belle* collaborations. The first error is statistic, the second systematic and the third from the uncertainty on the Dalitz model.

Experiment	γ (°)	δ_B (°)	r_B
$DK^-, D \to K_S\pi^+\pi^-$			
BaBar		$104 \pm 45 ^{+17}_{-21} ^{+16}_{-24}$	$0.12 \pm 0.08 \pm 0.03 \pm 0.04$
Belle	$64 \pm 19 \pm 13 \pm 11$	$157 \pm 19 \pm 11 \pm 21$	$0.21 \pm 0.08 \pm 0.03 \pm 0.04$
$D^*K^-, D^* \to D\pi^0$ or $D\gamma$			
BaBar		$296 \pm 41 ^{+14}_{-12} \pm 15$	$0.17 \pm 0.10 \pm 0.03 \pm 0.03$
Belle	$75 \pm 57 \pm 11 \pm 11$	$321 \pm 57 \pm 11 \pm 21$	$0.12 ^{+0.16}_{-0.11} \pm 0.02 \pm 0.04$
DK^- and D^*K^- combined			
BaBar	$70 \pm 31 ^{+12}_{-10} ^{+14}_{-11}$		
Belle	$68 ^{+14}_{-15} \pm 13 \pm 11$		
$DK^{*-}, D \to K_S\pi^+\pi^-$			
Belle	$112 \pm 35 \pm 9 \pm 11 \pm 8$	$353 \pm 35 \pm 8 \pm 21 \pm 49$	$0.25 \pm 0.18 \pm 0.09 \pm 0.04 \pm 0.08$

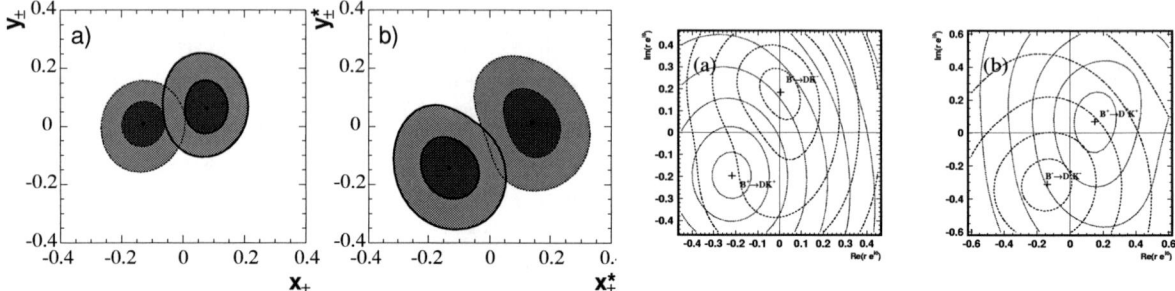

FIGURE 2. Left: *BaBar* Contours at 68.3% (dark) and 95% (light) confidence level (statistical only) in the $(x_\pm^{(*)}, y_\pm^{(*)})$ planes for (a) D^0K^- and (b) $D^{*0}K^-$, separately for B^- (thick and solid) and B^+ (thin and dotted). **Right:** Same results from *Belle* for (a) D^0K^- and (b) D^{*0}. Contours indicate integer multiples of the standard deviation. Dotted contours are from B^+ data, dashed contours are from B^- data.

3. N. Cabibbo, Phys. Rev. Lett. **10**, 531 (1963); M. Kobayashi and T. Maskawa, Prog. Theor. Phys **49**, 652 (1973).
4. M. Gronau and D. Wyler, Phys. Lett. **B265**, 172 (1991); M. Gronau and D. London, Phys. Lett. **B253**, 483 (1991).
5. B. Aubert *et al.* [BABAR Collaboration], arXiv:hep-ex/0408082; BABAR Collaboration, B. Aubert *et al.*, Phys. Rev. D **71**, 031102 (2005); B. Aubert *et al.* [BABAR Collaboration], arXiv:hep-ex/0408069.
6. K. Abe *et al.* [Belle Collaboration], BELLE-CONF-0443 K. Abe *et al.* [BELLE Collaboration], arXiv:hep-ex/0307074.
7. D. Atwood, I. Dunietz, and A. Soni, Phys. Rev. Lett. **78**, 3257 (1997); Phys. Rev. **D63**, 036005 (2001).
8. A. Bondar and T. Gershon, Phys. Rev. D**70**, 091503 (2004).
9. B. Aubert *et al.* [BABAR Collaboration], rXiv:hep-ex/0408028.
10. K. Abe *et al.* [BELLE Collaboration], Phys. Rev. Lett. **94**, 091601 (2005).
11. A. Giri, Yu. Grossman, A. Soffer and J. Zupan, Phys. Rev. D**68**, 054018 (2003).
12. BaBar Collaboration, B. Aubert *et al.*, Phys. Rev. Lett. **91**, 121801 (2003).
13. B. Aubert *et al.* (BABAR Collaboration), hep-ex/0504039, submitted to Phys. Rev. Lett.
14. K. Abe *et al.* (BELLE Collaboration), BELLE-CONF-0476, hep-ex/0411049; K. Abe *et al.* (BELLE Collaboration), BELLE-CONF-0502, hep-ex/0504013.
15. CLEO Collaboration, S. Kopp *et al.*, Phys. Rev. D **63**, 092001 (2001); CLEO Collaboration, H. Muramatsu *et al.*, Phys. Rev. Lett. **89**, 251802 (2002); Erratum-ibid: **90** 059901 (2003).

Status and Future Perspectives on the Measurement of the Unitarity Triangle Angle α

Angelo Carbone

INFN Bologna, via Irnerio 46, 40126 Bologna

Abstract. A summary on the status of the measurements of the Unitarity Triangle angle α using the decays $B \to \pi\pi$, $B \to \rho\pi$ and $B \to \rho\rho$ and on the future perspectives at the Large Hadron Collider experiment LHCb is presented. The UTfit collaboration, combining the BABAR measurements from $B \to \rho\pi$ and $B \to \rho\rho$, give as Standard Model solution the value $\alpha = (106 \pm 8)°$. The LHCb Collaboration has estimated the sensitivity on α from the $B^0 \to \rho\pi$ decay obtaining a relative error $\frac{\sigma_\alpha}{\alpha} \le 0.1$, in one nominal year of data taking.

Keywords: α angle, CKM matrix, Unitarity Triangle, CP violation
PACS: 12.15.Hh

INTRODUCTION

In the Standard Model (SM) the Cabibbo-Kobayashi-Maskawa (CKM) complex matrix [1] describes charged current weak couplings of the quarks. The Unitary Triangle (UT), defined by one of the unitarity relations of the CKM matrix, $V_{ub}^\star V_{ud} + V_{cb}^\star V_{cd} + V_{tb}^\star V_{td} = 0$, is a useful graphical representation, and measurements of its sides and angles give us the possibility to test the validity of the SM picture. After the striking success in measuring the β angle [2, 3], the B-factories are now providing measurements of the other angles as well, in particular the α angle. The α angle defined by

$$\alpha = -\arg\left|\frac{V_{td}V_{tb}^\star}{V_{ud}V_{ub}^\star}\right|, \quad (1)$$

can be measured using the decays $B \to \pi\pi$, $B \to \rho\pi$ and $B \to \rho\rho$. In the next sections a summary on the status of the measurements of the α angle and what are the future perspectives of the LHCb experiment will be given.

STATUS OF MEASUREMENTS OF THE ANGLE α

The BABAR and BELLE experiments are working on the measurements of the α angle using the modes $B \to \pi\pi$, $B \to \rho\pi$ and $B \to \rho\rho$. These decays are difficult to study because of the small branching ratios and the need to reconstruct neutral final states.

Extracting α from $B_d^0 \to hh$

The decays $B_d^0 \to hh$ ($h = \pi$ or ρ) proceed through tree-level processes (T) and penguin processes (P). The CP-violating asymmetry coefficients S_{hh} and C_{hh} can be written as:

$$S_{hh} = \frac{2Im\lambda}{1+|\lambda|^2}, \quad C_{hh} = \frac{1-|\lambda|^2}{1+|\lambda|^2}, \quad (2)$$

where the parameter S_{hh} reflects the CP-violation induced by the interference between mixing and decay, while C_{hh} accounts for the direct CP-violation in the decay process.

The parameter λ is related to the α angle by

$$\lambda = e^{2i\alpha}\frac{1 - \frac{|V_{td}^\star V_{tb}|}{|V_{ud}^\star V_{ub}|}P/Te^{-i\alpha}}{1 - \frac{|V_{td}^\star V_{tb}|}{|V_{ud}^\star V_{ub}|}P/Te^{i\alpha}}, \quad (3)$$

where V_{ij} are the coefficients of the CKM matrix. α can then be extracted from the decay rates:

$$\begin{aligned}f_{Qtag}^{hh}(\Delta t) &= \frac{e^{-|\Delta t|/\tau}}{4\tau} \\ &\quad [1 + Q_{tag}S_{hh}sin(\Delta m_d \Delta t) \\ &\quad - Q_{tag}C_{hh}cos(\Delta m_d \Delta t)],\end{aligned} \quad (4)$$

where Δt is the decay time difference between the B decaying to the hh final state and the decay of the other B in the event, denoted by B_{tag}. τ is the neutral B-meson lifetime and Δm_d is the mass difference between the mass eigenstates of the $B - \overline{B^0}$ system. Q_{taq} is set to 1($-$1) if the B_{tag} is the $B(\overline{B^0})$.

$B^0 \to \pi\pi$

The decay $B^0 \to \pi^+\pi^-$ was originally thought to be a simple path to α, but as it was later understood, the presence of a sizeable pollution from penguin diagrams complicates considerably this task. However, by combining the isospin-related decays $B^0 \to \pi^0\pi^0$, $B^+ \to \pi^+\pi^0$ and $B^- \to \pi^-\pi^0$, one can extract α up to discrete ambiguities [5].

The decay amplitudes of the isospin-related final states obey the relations

$$A_{\pi\pi}^{+0} = \frac{1}{\sqrt{2}} A_{\pi\pi}^{+-} + A_{\pi\pi}^{00}, \quad (5)$$

$$\overline{A}_{\pi\pi}^{+0} = \frac{1}{\sqrt{2}} \overline{A}_{\pi\pi}^{+-} + \overline{A}_{\pi\pi}^{00}, \quad (6)$$

$$\left|A_{\pi\pi}^{+0}\right| = \left|\overline{A}_{\pi\pi}^{+0}\right|, \quad (7)$$

where $A_{\pi\pi}^{+-}$, $A_{\pi\pi}^{+0}$ and $A_{\pi\pi}^{00}$ are the amplitudes for $B \to \pi^+\pi^-$, $B \to \pi^0\pi^+$ and $B \to \pi^0\pi^0$ respectively. The amplitudes $A_{\pi\pi}^{+-}$ and $A_{\pi\pi}^{00}$ and their conjugates are related to the observables $S_{\pi\pi}$, $C_{\pi\pi}$, $S_{\pi\pi}^{00}$, and $C_{\pi\pi}^{00}$, while the amplitude $A_{\pi\pi}^{+0}$ is related to the charge asymmetry

$$A_{CP} = \frac{N(B^+ \to \pi^+\pi^0) - N(B^- \to \pi^-\pi^0)}{N(B^+ \to \pi^+\pi^0) + N(B^- \to \pi^-\pi^0)}. \quad (8)$$

By considering that each amplitude depends on T and P contributions:

$$A_{\pi\pi}^{+0} = A_{\pi\pi}^{+0}(T^{+0}, P^{+0}), \quad (9)$$
$$A_{\pi\pi}^{+-} = A_{\pi\pi}^{+-}(T^{+-}, P^{+-}), \quad (10)$$
$$A_{\pi\pi}^{00} = A_{\pi\pi}^{00}(T^{00}, P^{00}), \quad (11)$$

all the above relations lead to 13 unknowns and 8 observables: $\mathcal{B}(B^0 \to \pi^+\pi^-)$, $\mathcal{B}(B^+ \to \pi^+\pi^0)$, $\mathcal{B}(B^0 \to \pi^0\pi^0)$, $S_{\pi\pi}$, $C_{\pi\pi}$, $S_{\pi\pi}^{00}$, $C_{\pi\pi}^{00}$ and A_{CP}. The observable $S_{\pi\pi}^{00}$ is hard or impossible to measure and the system is reduced to 13 unknowns and 7 observables. Thanks to the isospin symmetry and to the presence of one global phase the number of unknowns is reduced to 7 and the system can be solved to extract α with 8-fold ambiguity within $[0, \pi]$.

Table 2 summarizes the measurements from BABAR and BELLE. The existing measurements give a weak constraint on α: BABAR gives $0° < \alpha < 29°$ and $61° < \alpha < 180°$ within 90% confidence level [6], while the BELLE results are $0° < \alpha < 19°$ and $71° < \alpha < 180°$ within 95.4% confidence level [7].

$B^0 \to \rho\rho$

The decay $B^0 \to \rho^+\rho^-$ can provide a better determination of the angle α since it has the advantage of a larger decay rate (about 5 times more then $\pi^+\pi^-$) and a smaller pollution from penguin contributions. Since there are two vector particles in the final state, it is required to properly

TABLE 1. BABAR results from the *CP* analysis of the decay $B^0 \to \rho\rho$ [8]. The meaning of the parameters is described in the text.

	BABAR
$S_L^{\rho\rho}$	$-0.33 \pm 0.014^{+0.021}_{-0.029}$
$C_L^{\rho\rho}$	$-0.03 \pm 0.18 \pm 0.09$
A_{CP}	$-0.19 \pm 0.023 \pm 0.03$
$\mathcal{B}(B^0 \to \rho^0\rho^0) \times 10^6$	< 1.1 at 90% CL
$\mathcal{B}(B^+ \to \rho^+\rho^0) \times 10^6$	$26.4^{+6.1}_{-6.4}$
$\mathcal{B}(B^0 \to \rho^+\rho^-) \times 10^6$	24.0 ± 2.5

account for *CP*-even and *CP*-odd components. The two particles $\rho^+\rho^-$ can be present in three angular momentum states, $L = 0, 1, 2$, corresponding to different *CP* eigenstates, $CP(L = 0, 2) = +1$ and $CP(L = 1) = -1$. BABAR has measured the longitudinal polarization f_L finding it is dominant: $f_L = (0.978 \pm 0.014^{+0.020}_{-0.028})$ [9]. The analysis to extract α with this method was performed by BABAR [9]. They used the isospin symmetry and the measurements reported in Table 1. With the new measurements of the branching ratio $\mathcal{B}(B^0 \to \rho^0\rho^0)$ the constraint on α due to the penguin contribution was improved and a 68% (90%) CL limit $\Delta\alpha_{\rho\rho} = \alpha - \alpha_{eff}$ of $\pm 11°$ ($\pm 14°$) was so obtained. Since the central value of the fit of the constraint is $\Delta\alpha_{\rho\rho} = 0$, the central value of α obtained from the isospin analysis is the same of α_{eff}, which can be calculated from $sin(\alpha_{eff}) = \frac{S_L^{\rho\rho}}{(1-C_L^{\rho\rho})^{1/2}}$ and is measured with the $B^0 \to \rho^+\rho^-$ decay to be $\alpha = (100 \pm 13)°$ [9].

Extracting α from $B^0 \to \rho\pi$

In the case of the $(\overline{B^0})B^0 \to \rho^\pm\pi^\mp$ decay, the time-dependent decay rate reads

$$f_{Qtag}^{\rho^\pm\pi^\mp}\Delta t = (1 \pm A_{\rho\pi})\frac{e^{-|\Delta t|/\tau}}{4\tau} \quad (12)$$
$$[1 + Q_{tag}(S_{\rho\pi} \pm \Delta S_{\rho\pi})sin(\Delta m_d \Delta t)$$
$$- Q_{tag}(C_{\rho\pi} \pm \Delta C_{\rho\pi})cos(\Delta m_d \Delta t)],$$

where in addition to the case of $B^0 \to \pi\pi$, here we have also $\Delta S_{\rho\pi}$ and $\Delta C_{\rho\pi}$, which arise from the fact that two production modes of the ρ are possible. They are dilution terms and have no *CP* content. Using the isospin symmetry [4] the decay amplitudes of the isospin-related final states obey the relations

$$\sqrt{2}(A_{\rho\pi}^{+0} + A_{\rho\pi}^{0+}) = 2A_{\rho\pi}^{00} + A_{\rho\pi}^{+-} + A_{\rho\pi}^{-+}, \quad (13)$$
$$\sqrt{2}(\overline{A}_{\rho\pi}^{+0} + \overline{A}_{\rho\pi}^{0+}) = 2\overline{A}_{\rho\pi}^{00} + \overline{A}_{\rho\pi}^{+-} + \overline{A}_{\rho\pi}^{-+}. \quad (14)$$

TABLE 2. BABAR and BELLE results from the CP analyses of the decay $B^0 \to \pi\pi$ [8]. The meaning of the parameters is described in the text.

	BABAR	BELLE	Average HFAG
$S_{\pi\pi}$	$-0.30 \pm 0.17 \pm 0.03$	$-0.67 \pm 0.16 \pm 0.06$	-0.50 ± 0.12
$C_{\pi\pi}$	$-0.09 \pm 0.15 \pm 0.04$	$-0.56 \pm 0.12 \pm 0.06$	-0.37 ± 0.10
$C_{\pi\pi}^{00}$	$-0.12 \pm 0.56 \pm 0.06$	$-0.43 \pm 0.51^{+0.16}_{-0.17}$	-0.28 ± 0.39
A_{CP}	$-0.01 \pm 0.10 \pm 0.02$	$-0.02 \pm 0.10 \pm 0.01$	-0.02 ± 0.07
$\mathcal{B}(B^0 \to \pi^0\pi^0) \times 10^6$	$1.17 \pm 0.32 \pm 0.10$	$2.32^{+0.44+0.22}_{-0.48-0.18}$	1.51 ± 0.28
$\mathcal{B}(B^+ \to \pi^+\pi^0) \times 10^6$	$5.8 \pm 0.6 \pm 0.4$	$5.0 \pm 1.2 \pm 0.5$	5.5 ± 0.6
$\mathcal{B}(B^0 \to \pi^+\pi^-) \times 10^6$	$4.7 \pm 0.6 \pm 0.2$	$4.4 \pm 0.6 \pm 0.3$	4.6 ± 0.4

where $A_{\rho\pi}^{ij}$ are the amplitudes for the respective decays. As in the case of $B^0 \to \pi\pi$ the amplitudes are related to the tree and penguin contributions

$$A_{\rho\pi}^{+0} = A_{\rho\pi}^{+0}(T^{+0}, P^{+0}), \quad (15)$$
$$A_{\rho\pi}^{0+} = A_{\rho\pi}^{0+}(T^{0+}, P^{0+}), \quad (16)$$
$$A_{\rho\pi}^{+-} = A_{\rho\pi}^{+-}(T^{+-}, P^{+-}), \quad (17)$$
$$A_{\rho\pi}^{-+} = A_{\rho\pi}^{-+}(T^{-+}, P^{-+}), \quad (18)$$
$$A_{\rho\pi}^{00} = A_{\rho\pi}^{00}(T^{00}, P^{00}), \quad (19)$$

plus the conjugate relations.

With the use of these relations, 21 unknowns are to be determined while 13 observables are available. Using the isospin symmetry and one global phase the system can be solved. In Table 3 the actual measurements by BABAR and BELLE for these observables are reported: the overall CP violating asymmetry coefficients in the $\rho^{\pm}\pi^{\mp}$ system $A_{CP}^{\rho\pi}$, $S_{\rho\pi}$, $\Delta S_{\rho\pi}$, $C_{\rho\pi}$, $\Delta C_{\rho\pi}$, the branching fractions $\mathcal{B}(B^0/\overline{B^0} \to \rho^{\pm}\pi^{\mp})$, $\mathcal{B}(B^0/\overline{B^0} \to \rho^0\pi^0)$, $\mathcal{B}(B^+ \to \rho^+\pi^0)$, $\mathcal{B}(B^+ \to \rho^0\pi^+)$, the CP violating asymmetry coefficients in the $B^0/\overline{B^0} \to \rho^0\pi^0$ decay, $S_{\rho\pi}^{00}$ and $C_{\rho\pi}^{00}$, and the two direct CP asymmetries in $B^+ \to \rho^+\pi^0$ and $B^+ \to \rho^0\pi^+$ A_{CP}^{+0}, A_{CP}^{0+}.

Combining all the results on α

BABAR and BELLE have published results on the asymmetry on all the three modes, $\pi\pi$, $\rho\rho$, $\rho\pi$. The best constraints now come from the $B \to \rho^+\rho^-$ mode, which is more abundant, has smaller penguin contribution and, as we now know from the angular analysis, is practically a pure CP eigenstate. The UTfit Collaboration [11] has combined the $\rho\rho$, $\rho\pi$ results from BABAR ignoring the results from $\pi\pi$ because of the experimental confusing situation. The fit gives for the SM solution $\alpha = (106 \pm 8)°$.

α AT LHCB

The LHCb experiment will start operating at the Large Hadron Collider (LHC) by the 2007, with the aim of performing precise measurements of CP-violating effects arising in the beauty-quark flavor-dynamics [12]. Since in the 14 *TeV* primary collisions of LHC, large statistics of B_s^0 and B_c mesons, as well as other b-baryons, can be produced, LHCb provides a good opportunity to probe SM predictions in the b-quark sector beyond the $B^0 - \overline{B}^0$ system. The LHCb detector is a single-arm forward spectrometer, extremely specialized in selecting B-meson decays. A preliminary study with the purpose to investigate the impact of various experimental effects from a time-dependent Dalitz plot analysis of the $B^0 \to \rho\pi$ decay, in order to extract a value of α was performed. In the hadronic environment the measurement is complicated by the presence of large particle multiplicity and background. Monte Carlo studies on the π^0 reconstruction efficiency show that a single π^0 can be reconstructed with about a 50% of efficiency. LHCb will be able to collect about 10^4 $B^0 \to \rho\pi$ decays assuming an integrated luminosity of $2fb^{-1}/year$, with a background-to-signal ratio less than 3 at 90%CL. The effective tagging power is estimated to be about 6%. Using the isospin symmetry hypothesis a simulation was performed by means of a toy Monte Carlo in order to estimate the sensitivity on α. A simulation of the physics of the $B^0 \to \rho\pi$ and CP-violation was realized by generating several finite size samples corresponding to the expected annual yields, with the background events simulated as well. In the toy Monte Carlo the effects of the resolutions, correlations, acceptance and tagging dilution were included, as from a full MC simulation. Then an event-by-event fit on toy data was performed in order to extract α. Considering two scenarios for α, $\alpha^{gen} = 77.35°$ and $\alpha^{gen} = 106.0°$, the expected standard deviation on α is estimated to be $\frac{\sigma_\alpha}{\alpha} \leq 0.1$ with one year of data. Assuming a background-to-signal ratio equal to 0.8, one has $\sigma_\alpha \simeq 6°$ with $\alpha^{gen} = 77.35°$ and $\sigma_\alpha \simeq 8.5°$ with $\alpha^{gen} = 106.0°$. More details are reported in the LHCb

TABLE 3. BABAR and BELLE results from the CP analysis of the decay $B^0 \to \rho\pi$ [8]. The meaning of the parameters is described in the text.

	BABAR	BELLE	Average HFAG
$S_{\rho\pi}$	$-0.10 \pm 0.14 \pm 0.04$	$-0.28 \pm 0.23^{+0.10}_{-0.08}$	-0.13 ± 0.13
$C_{\rho\pi}$	$0.34 \pm 0.11 \pm 0.05$	$0.25 \pm 0.17^{+0.02}_{-0.06}$	0.31 ± 0.10
$\Delta S_{\rho\pi}$	$0.22 \pm 0.15 \pm 0.03$	$-0.30 \pm 0.24 \pm 0.09$	0.09 ± 0.13
$\Delta C_{\rho\pi}$	$0.15 \pm 0.11 \pm 0.03$	$0.38 \pm 0.18^{+0.02}_{-0.04}$	0.22 ± 0.10
$C^{00}_{\rho\pi}$	$-0.12 \pm 0.56 \pm 0.06$	$-0.43 \pm 0.51^{+0.16}_{-0.17}$	-0.28 ± 0.39
$S^{00}_{\rho\pi}$	$-0.12 \pm 0.56 \pm 0.06$	$-0.43 \pm 0.51^{+0.16}_{-0.17}$	-0.28 ± 0.39
$A_{CP}^{\rho\pi}$	$-0.088 \pm 0.049 \pm 0.013$	$-0.16 \pm 0.10 \pm 0.02$	-0.102 ± 0.045
A_{CP}^{0+}	$-0.19 \pm 0.11 \pm 0.02$		-0.19 ± 0.11
A_{CP}^{+0}	$0.24 \pm 0.16 \pm 0.06$	$0.06 \pm 0.19^{+0.04}_{-0.06}$	0.16 ± 0.13
$\mathcal{B}(B^0 \to \rho^+\pi^-) \times 10^6$	$22.6 \pm 1.8 \pm 2.2$	$29.1^{+5.0}_{-4.9} \pm 4.0$	24.0 ± 2.5
$\mathcal{B}(B^+ \to \rho^0\pi^+) \times 10^6$	$9.4 \pm 1.3 \pm 1.2$	$8.0^{+2.3}_{-2.0} \pm 0.7$	$9.1^{+1.4}_{-1.3}$
$\mathcal{B}(B^+ \to \rho^+\pi^0) \times 10^6$	$10.9 \pm 1.9 \pm 1.9$	$13.2 \pm 2.3^{+1.4}_{-1.9}$	12.0 ± 2.0
$\mathcal{B}(B^0 \to \rho^0\pi^0) \times 10^6$	$1.4 \pm 0.6 \pm 0.3$	$5.1 \pm 1.6 \pm 0.9$	1.8 ± 0.6

note [13].

REFERENCES

1. N. Cabibbo, *Phys. Rev. Lett.*, **10**, 531 (1963); M. Kobayashi and T. Maskawa, *Prog. Theor. Phys.*, **49**, 652 (1973).
2. B. Aubert *et al.*, [BaBar Collaboration], *Phys. Rev. Lett.*, **89**, 201802 (2002).
3. K. Abe *et al.*, [Belle Collaboration], *Phys. Rev. D.*, **66**, 071102 (2002).
4. M. Gronau and D. London, *Phys. Rev. Lett.*, **65**, 3381 (1990).
5. H.J. Lipkin, Y. Nir, H.R. Quinn and A. Snyder, *Phys. Rev. Lett.*, D **44**, 1454 (1991).
6. B. Aubert *et al.*, [BaBar Collaboration], *Phys. Rev. Lett.*, **94**, 181802 (2005).
7. J. Charles *et al.*, [CKMfitter Group], *Eur. Phys. J. C.*, C **41**, 1 (2005).
8. Heavy Flavor Averaging Group [HFAG]. URL http://www.slac.stanford.edu/xorg/hfag/index.html.
9. B. Aubert *et al.*, [BaBar Collaboration], hep-ex/0503049 (2005).
10. B. Aubert *et al.*, [BaBar Collaboration], hep-ex/0408099 (2004).
11. Unitarity Triangle fit Collaboration [UTFit]. URL http://utfit.roma1.infn.it/.
12. Technical Design Report [LHCb Collaboration] CERN-LHCC 2003-030.
13. O. Deschamps. P. Perret and A. Robert [LHCb note] LHCb 2005-024.

Unitarity Triangle Analysis beyond the Standard Model

M. Pierini

Dep. Of Physics, Univ. Of Wisconsin, 53706 Madison WI, USA

Abstract. Starting from a (New Physics free) tree level determination of $\bar{\rho}$ and $\bar{\eta}$, we perform the Unitarity Triangle analysis in a general extension of the Standard Model with arbitrary new physics contributions to loop-mediated processes. Using a simple parametrization, we determine the allowed ranges of non-standard contributions to $\Delta F = 2$ processes. The updated results of the analysis are available at the URL http://www.utfit.org.

Keywords: CKM matrix, New Physics
PACS: 43.35.Ei, 78.60.Mq

INTRODUCTION

The Unitarity Triangle (UT) analysis shows the impressive success of the CKM picture in describing CP violation in the Standard Model (SM). UT parameters have been consistently determined using both CP-conserving ($\left|\frac{V_{ub}}{V_{cb}}\right|$, Δm_d and $\frac{\Delta m_d}{\Delta m_s}$) and CP-violating ($\varepsilon_K$ and $\sin 2\beta$) processes [1]. Additional measurements of several unitarity angle combination, especially γ from DK and α from charmless B decays, confirm this picture [1].

This success becomes a puzzle once the SM is considered as an effective theory valid up to energies not too higher than the electroweak scale, as required by the gauge hierarchy problem. Indeed, even in the favourable case in which the theory above the cutoff is weakly coupled, such as the Minimal Supersymmetric Standard Model, huge contributions to Flavour Changing Neutral Current (FCNC) and CP-violating processes are expected to arise[2], clashing with the large amount of accurate experimental data now available on these transitions. This is due to the presence of additional sources of flavour and CP violation beyond the CKM matrix (see for example Ref. [3] for the SUSY case).

Using the large amount of experimental bouns now available, we generalize the Unitarity Triangle analysis to constraint these additional sources of Flavour Violation to be consistent with observations, parameterizing their effect in a general way and extracting the values of the parameters together with the $\bar{\rho}$ and $\bar{\eta}$ terms of the CKM matrix.

Making the hypothesis that NP enters observables in the flavour sector only at the loop level, it is possible to determine two regions in the $\bar{\rho} - \bar{\eta}$ plane independently of NP contributions, using tree-level B decays. The CKM elements V_{ub} and V_{cb} are determined using semileptonic inclusive and exclusive B decays. The angle γ is obtained by measuring the phase of V_{ub} appearing in the interference between $b \to c$ and $b \to u$ transitions to DK final states.[1] The combination of the two is shown Fig. 1.[2]

TABLE 1. *Results for several UT parameters, obtained using the constraints from $\left|\frac{V_{ub}}{V_{cb}}\right|$ and γ*

UT Fit - using only $\left|\frac{V_{ub}}{V_{cb}}\right|$ and γ		
	SM Solution	2nd Solution
$\bar{\rho}$	0.21 ± 0.10	-0.21 ± 0.10
$\bar{\eta}$	0.36 ± 0.06	-0.36 ± 0.06
$\sin 2\beta$	0.724 ± 0.074	-0.556 ± 0.089
γ	$(59.4 \pm 16.8)°$	$(-120.3 \pm 17.2)°$
α	$(95 \pm 15)°$	$(-43 \pm 15)°$
$\sin(2\beta + \gamma)$	0.921 ± 0.080	-0.45 ± 0.12

The results reported in Tab. 1 can be used as reference for model-building and phenomenology in any extension of the SM with loop-mediated contributions to FCNC processes.

A further step in the analysis is done using the available experimental information on loop-mediated processes to constrain the NP contributions in $|\Delta F|=2$ transitions. NP models introduce in general a large number of new parameters: flavour changing couplings, short distance coefficients and matrix elements of new local operators. The specific list and the actual values of these parameters can only be determined within a given model. Nevertheless, each of the mixing processes listed in Tab. 2, being described by a single amplitude, can be effectively parameterized in a completely general way in terms of two parameters, which quantify the difference of the complex amplitude with respect to the SM one [7].

---

[1] We neglect possible NP contributions to D^0-$\bar{D}^0$ mixing, since their contribution is expected to be well below the present experimental accuracy [4]. In the future, it might become necessary to take them into account following Ref. [5].
[2] Fig. 1 first appeared in [1]. Similar results were recently obtained in [6].

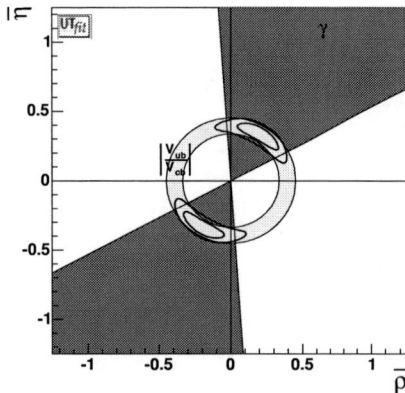

FIGURE 1. *The selected region on $\bar{\rho}$-$\bar{\eta}$ plane obtained from the determination of $\left|\frac{V_{ub}}{V_{cb}}\right|$ and γ (using DK final states). Selected regions corresponding to 68% and 95% probability are shown, together with 95% probability regions for γ and $\left|\frac{V_{ub}}{V_{cb}}\right|$.*

Thus, for instance, in the case of $B_q^0 - \bar{B}_q^0$ mixing we define

$$C_{B_q} e^{2i\phi_{B_q}} = \frac{\langle B_q^0 | H_{eff}^{full} | \bar{B}_q^0 \rangle}{\langle B_q^0 | H_{eff}^{SM} | \bar{B}_q^0 \rangle} \qquad (q=d,s) \qquad (1)$$

where H_{eff}^{SM} includes only the SM box diagrams, while H_{eff}^{full} includes also the NP contributions. By definition, in the absence of NP effects, $C_{B_q} = 1$ and $\phi_{B_q} = 0$. The experimental quantities determined from the $B_q^0 - \bar{B}_q^0$ mixings and listed in Tab. 2 are related to their SM counterparts and the NP parameters by the following relations:

$$\Delta m_d^{exp} = C_{B_d} \Delta m_d^{SM} \qquad (2)$$

$$\sin 2\beta^{exp} = \sin(2\beta^{SM} + 2\phi_{B_d}) \; ; \; \alpha^{exp} = \alpha^{SM} - \phi_{B_d} \qquad (3)$$

in a self-explanatory notation. One can also add the constraint coming from the semileptonic CP asymmetry $A_{CP}(B^0 \to Xl\nu)$, which depends on both C_{B_d} and ϕ_{B_d}. [8]

As far as the $K^0 - \bar{K}^0$ mixing is concerned, we find it convenient to introduce a single parameter which relates the imaginary part of the amplitude to the SM one:

$$C_{\varepsilon_K} = \frac{\text{Im}[\langle K^0 | H_{eff}^{full} | \bar{K}^0 \rangle]}{\text{Im}[\langle K^0 | H_{eff}^{SM} | \bar{K}^0 \rangle]}. \qquad (4)$$

This definition implies in fact a simple relation for the measured value of ε_K,

$$\varepsilon_K^{exp} = C_{\varepsilon_K} \varepsilon_K^{SM} \qquad (K^0 - \bar{K}^0 \text{ mixing}) \qquad (5)$$

Δm_K is not considered due to the fact that the long distance effects are not well under control. Therefore, all NP effects which enter the present analysis are parameterized in terms of three real coefficients, C_{B_d}, ϕ_{B_d}, and C_{ε_K}.

C_{B_d}, ϕ_{B_d} and C_{ε_K} are fitted together with $\bar{\rho}$ and $\bar{\eta}$, using the experimental information on $\left|\frac{V_{ub}}{V_{cb}}\right|$, $B \to DK$ decays (γ), ε_K, $B \to \rho\rho$, $\rho\pi$, and $\pi\pi$ decays, $B \to J/\Psi K^{(*)}$ decays, and $A_{CP}(B^0 \to Xl\nu)$. In this way, we are using the determination of $\bar{\rho}$ and $\bar{\eta}$ from tree level processes (see Tab. 1) and the knowledge of the relevant hadronic parameters ($f_{B_d}\sqrt{B_{B_d}}$ for C_{B_d} and $\hat{B}_K$ for C_{ε_K}) to obtain the SM determination of a given quantity (Δm_d, ε_K,...) and we can translate the experimental bound on the various quantities into a determination of the corresponding NP parameters.

The output p.d.f.'s for C_{B_d}, C_{ε_K}, C_{B_d} vs. ϕ_{B_d}, and γ vs. ϕ_{B_d} are shown in Fig. 2. The two plots of Fig. 3 give the most clear representation of our result: the Standard Model solution is favoured even in this generalized analysis, but the present uncertainties still allow the presence of NP as a correction to the Standard Model contributions. On the other side, $\mathcal{O}(1)$ effects from NP are still possible only in a region corresponding to $\phi_{B_d} \sim 40°$, which is anyhow already excluded if we consider anly the 68% probability region. The uncertainty related to NP parameters, which is what adetermines the spread of the selected region in the 2D plot, is dominated by the knowledge of the tree level determination of $\bar{\rho}$ and $\bar{\eta}$. In particular, the allowed range of C_{B_d} for $\phi_{B_d} \sim 0$ is much larger than what would be given by the error on $f_{B_d}\sqrt{B_{B_d}}$), it is interesting to see that it starts being comparable to the first one, and will eventually become smaller in the near future.

Anyhow, it is important to remark that the overconstraining coming from the experimental observables allows to increase the precision on $\bar{\rho}$ and $\bar{\eta}$, respect to the the output of the tree-level determination. This is clear comparing Fig. 1 to Fig. 4.

In Tab. 3 we give the numerical results related to the NP parameters and to some of the relevant UT quantities. A special comment is needed for the case of Δm_s: The output distribution is obtained generalizing the constraints related to $K-\bar{K}$ and $B_d - \bar{B}_d$ beyond the Standard Model. On the other side, the B_s sector is complitely unconstrained in this analysis, so that the numerical result cannot be taken as a prediction of Δm_s in a general NP scenario.

REFERENCES

1. M. Bona, M. Ciuchini, E. Franco, V. Lubicz, G. Martinelli, F. Parosi, M. Pierini, P. Roudeau, C. Schiavi, L. Silvestrini, and A. Stocchi, *JHEP* **07** 028 (2005).

TABLE 2. Different processes and corresponding measurements contributing to the determination of $\bar{\rho}$, $\bar{\eta}$, C_{B_d}, ϕ_{B_d}, C_{B_s}, ϕ_{B_s} and C_{ε_K}. Δm_K is not considered due to the fact that the long distance effects are not well under control.

Tree-Level	B_d^0 mixing	K^0 mixing	B_s^0 mixing
$\left\|\frac{V_{ub}}{V_{cb}}\right\|$	Δm_d	ε_K	Δm_s
$\gamma(DK)$	$A_{CP}(B \to J/\psi K)$		$A_{CP}(B_s^0 \to J/\psi \phi)$
	$A_{CP}(B \to \rho/\pi\rho/\pi)$		
	$A_{CP}(B^0 \to Xl\nu)$		

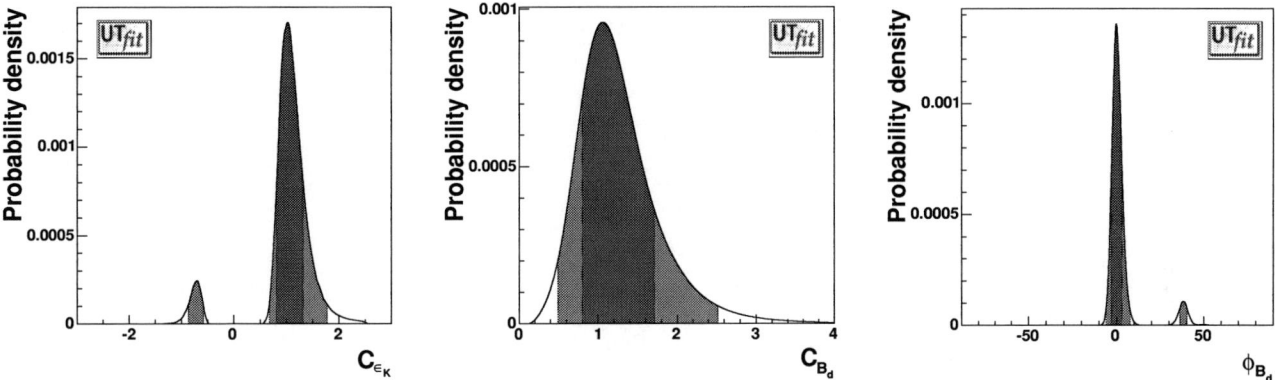

FIGURE 2. Output P.d.f.'s for C_{ε_K} (left), C_{B_d} (center), ϕ_{B_d} (right). Dark (light) areas correspond to the 68% (95%) probability region.

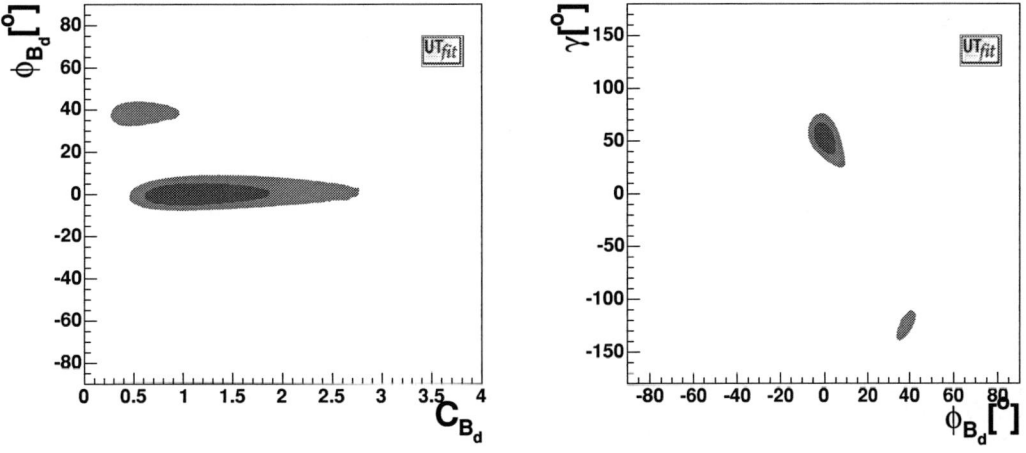

FIGURE 3. 2D distributions of ϕ_{B_d} vs. C_{B_d} (left) and γ vs. ϕ_{B_d} (right). Dark (light) areas correspond to the 68% (95%) probability region.

TABLE 3. Results of the NP generalized analysis on UT parameters. The values for C_{B_d}, ϕ_{B_d} and C_{ε_K} are reported. The results related to Standard Model solutions have 68% probability. Since the second solution is exluded at this probability level, we quote 95% ranges.

	Generalized UTfit analysis in the presence of NP	
	Standard Solution ($\gamma > 0$)	Non-Standard Solution ($\gamma < 0$)
	UT parameters	
$\bar{\rho}$	0.251 ± 0.059	[-0.268, -0.195] @95%
$\bar{\eta}$	0.324 ± 0.049	[-0.367, -0.303] @95%
$\sin 2\beta$	0.727 ± 0.058	[-0.559, -0.465] @95%
γ	$(53 \pm 10)°$	[-131 -118]° @95%
α	$(104 \pm 11)°$	[-46 -34]
$2\beta + \gamma$	$(100 \pm 13)°$	[-0.67, -0.393] @95%
Δm_s	18.8 ± 5.5 ps^{-1}	
	NP related parameters	
C_{B_d}	1.25 ± 0.45	
ϕ_{B_d}	$(0.1 \pm 3.1)°$	[-6.0, -7.7]^o @95%
C_{ε_K}	1.07 ± 0.25	[-0.87, -0.59]^o @95%

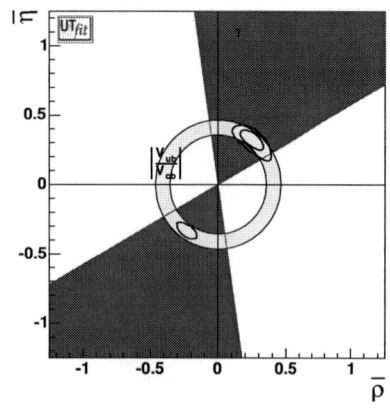

FIGURE 4. The selected region on $\bar{\rho}$-$\bar{\eta}$ plane obtained from the NP generalized analysis. The regions corresponding to the tree level constraints $\left|\frac{V_{ub}}{V_{cb}}\right|$ and γ (from DK final states) are also drawn as a refrence. Selected regions corresponding to 68% and 95% probability are shown, together with 95% probability regions for γ and $\left|\frac{V_{ub}}{V_{cb}}\right|$.

8. S. Laplace, Z. Ligeti, Y. Nir, and G. Perez, *Phys. Rev. D* **65** *094040 [arXiv:hep-ph/0202010]* (2002).

2. F. Gabbiani, E. Gabrielli, A. Masiero, and L. Silvestrini, *Nucl. Phys. B* **477** *321[arXiv:hep-ph/9604387]* (1996).
3. L. J. Hall, V. A. Kostelecky, and S. Raby, *Nucl. Phys. B* **267** *415* (1986).
4. A. Giri, Y. Grossman, A. Soffer, and J. Zupan, *Phys. Rev. D* **68** *054018[arXiv:hep-ph/0303187]* (2003).
5. A. Amorim, M. G. Santos, and J. P. Silva, *Phys. Rev. D* **59** *056001[arXiv:hep-ph/9807364]* (1999).
6. F. J. Botella, G. C. Branco, M. Nebot, M. Rebelo, and N., *arXiv:hep-ph/0502133* (2005).
7. J. M. Soares, and L. Wolfenstein, *Phys. Rev. D* **47** *1021*. (1993).

Search for B_s^0 Flavour Oscillations with the CDF Experiment

Sandro De Cecco

INFN Roma, piazzale Aldo Moro 5, 00185 Roma, Italy
dececco@fnal.gov

Abstract. We present here the search for B_s^0 flavour oscillations using semileptonic $B_s^0 \to D_s^- l^+ \nu$ and fully reconstructed hadronic $B_s^0 \to D_s^- \pi^+$ decays recorded with the CDF detector. Analysis is based on data collected in run II of the Fermilab Tevatron collider, corresponding to an integrated luminosity of 355 pb^{-1}. The D_s meson is reconstructed in three decay modes: $D_s^- \to \phi \pi^-$, $D_s^- \to K^{0*} K^-$ and $D_s^- \to \pi^+ \pi^- \pi^-$. Opposite side lepton and jet charge tags provide information about the B_s^0 production flavour. Time dependent flavour asymmetry on these events is then probed with the amplitude scan method. We obtain a 95% confidence level limit on the oscillation frequency $\Delta m_s > 7.9 ps^{-1}$ and a sensitivity of $8.4 ps^{-1}$ combining the two data samples.

Keywords: B, mixing, CDF
PACS: 12.15.Hh

INTRODUCTION

The neutral $B_s - \bar{B}_s$ meson system eigenstates undergo flavour oscillation through second order weak interaction processes with $|\Delta S| = 2$. The mass eigenstates B_s^H and B_s^L are then mixtures of flavour states; mass and lifetime differences can be defined as: $\Delta m_s = m_H - m_L$ and $\Delta \Gamma_s = \Gamma_H - \Gamma_L$. The probability P_{mix} for a B_s^0 meson produced at time $t = 0$ to decay as a $\bar{B}_s^0$ at proper time $t > 0$ is given by:

$$P_{mix} = P(B_s^0 \to \bar{B}_s^0) = \frac{1}{2}\Gamma e^{-\Gamma t}[1 - cos(\Delta m_s t)] \quad (1)$$

if CP violation is negligible in the mixing and lifetime difference is small. The mass difference Δm_s is then the B_s oscillation frequency. Oscillations have been observed in the B_d^0 system and Δm_d precisely measured to be $(0.502 \pm 0.007) ps^{-1}$. oscillation frequencies Δm_q (where $q = d, s$) in the neutral B systems are proportional to CKM quark mixing matrix elements $|V_{tq}|^2$. The determination of mixing frequencies further add information to SM description of CP violation. This is represented by the unitarity triangle whose sides and angles are proportional to CKM matrix elements and CP violating phases. In the SM the ratio $\frac{\Delta m_s}{\Delta m_d}$ can be calculated accurately with lattice QCD since part of hadronic uncertainties cancels. Global CKM fits, considering all present experimental and theoretical information on unitarity triangle elements in the S.M. predicts $\Delta m_s = (20.4 \pm 2.8) ps^{-1}$ [1].

A direct determination of Δm_s will then be a crucial confirmation for the SM picture of quark mixing. However since several New Physics models can enhance its value, the experimental exclusion of S.M. preferred range could lead to a strong hint for N.P.

Considering that $|V_{ts}| >> |V_{td}|$, Δm in the B_s system is much higher than in the B_d case. Since mixing frequency is measured through time dependent analysis an excellent proper time resolution is needed to resolve fast oscillations which constitutes an experimental challenge. The current limit on Δm_s from LEP, SLD and CDF I analysis combined is $\Delta m_s > 14.5 ps^{-1}$ at 95% C.L.[2]

In the following is reported the first search for Δm_s performed in the run II of the Tevatron $p\bar{p}$ collider by the CDF II experiment. The detector is described in detail elsewhere [3]. It is worth notice that major CDF II upgrades in the trigger system like the Secondary Vertex Trigger [4] based on Silicon detector and in the PID capabilities with the introduction of a Time of Flight detector [5], were primarily designed in order to enhance CDF sensitivity to Δm_s.

ANALYSIS STRATEGY

From eq. 1 we have the time dependent probability $P_{mix}(t)$ for a B_s^0 meson to decay as a $\bar{B}_s^0$ at proper time t. If the B_s decays in a flavour specific state the oscillation frequency Δm_s can actually be extracted from the observable time dependent mixing asymmetry:

$$A_{mix}(t) = \frac{N_{mix}(t) - N_{unmix}(t)}{N_{mix}(t) + N_{unmix}(t)} = -cos\Delta m_s t \quad (2)$$

Analysis will proceed as follows. B_s signals are reconstructed in flavour specific final states. Proper time $t = L\frac{m_B}{p_B}$ is measured through the decay length L with respect to the primary vertex and the reconstructed momentum. Then in order to decide if the event has mixed or not, initial flavour at production must be tagged. In practice the initial state flavour will be correctly tagged with a probability P_{tag} smaller than one. The measured mixing asymmetry will be reduced:

$$A_{mix}^{meas} = A_{mix}D = -Dcos\Delta m_s t \quad (3)$$

by a Dilution factor $D = 2P_{tag} - 1$. It is important to note that while setting a limit on oscillation frequency Δm_s the Dilution D of flavour tagging algorithms must be known a priori.

Finally on measured mixing asymmetry the Amplitude scan method [6] is used to search for Δm_s. An amplitude A is introduced in the mixing (unmixing) probabilities $P_{mix,unmix} = \frac{1}{2}\Gamma e^{-\Gamma t}[1 \pm A \cdot D \cdot cos(\Delta m_s t)]$. The B_s oscillation amplitude A and its error σ_A are extracted as a function of fixed test values of Δm_s using a Likelihood fit method. The statistical Significance of a B_s oscillation signal can be approximated by[6]:

$$Significance \simeq \sqrt{\frac{\varepsilon D^2}{2}} \frac{S}{\sqrt{S+B}} e^{-\frac{1}{2}(\Delta m_s \sigma_t)^2} \quad (4)$$

where S and B are Signal and background events, ε and D are efficiency and Dilution of Flavour tagging algorithms and σ_t is the proper time resolution. Because of proper time resolution effect (from eq. 4) the uncertainty σ_A on the amplitude grows with Δm_s.

If the tested Δm_s is the true value an amplitude statistically compatible with $A = 1$ is expected as a result of the fit to the data. If Δm_s is tested away from its real value an amplitude consistent with $A = 0$ should be measured. Region in Δm_s that exhibits A statistically incompatible with 1 can then be excluded. A method to set a 95% C.L. limit is to start amplitude scan at $\Delta m_s = 0$ and find the lowest value of Δm_s at which the condition $A + 1.645\sigma_A < 1$ is violated. Therefore the sensitivity of the amplitude scan is defined as the Δm_s value at which the condition $1.645 \cdot \sigma_A < 1$ is violated.

Because of proper time resolution effect (from eq. 4) the uncertainty σ_A on the amplitude grows with Δm_s.

B_s Signal Selection

This analysis is based on an integrated luminosity of 355 pb^{-1} collected by CDF up to august 2004. Heavy flavour enriched data sample is selected with the SVT trigger that at Level 2 stage is able to measure the impact parameter of charged tracks with respect to the average $p\bar{p}$ interaction point. The presence of displaced tracks in an event (i.e. non zero impact parameter) is likely a sign that they are the decay products of long lived particles as B and D mesons. Several trigger strategies including SVT information are available at CDF and are used for the extraction of B_s signals in this analysis.

B_s signals used come in two different signatures. The first are the semileptonic $B_s^0 \rightarrow D_s^- l^+ \nu$ decays selected by specific trigger requiring one SVT track with impact parameter greater than 120 μm and less than 1 mm and one identified lepton (e,mu) with $p_T > 4 GeV/c$. The second are hadronic $B_s^0 \rightarrow D_s^- \pi^+$ decays selected by a trigger requiring two displaced SVT tracks.

In both cases D_s meson is reconstructed in three decay modes: $D_s^- \rightarrow \phi\pi^-$, $D_s^- \rightarrow K^{0*}K^-$ and $D_s^- \rightarrow \pi^+\pi^-\pi^-$. Semileptonic B_s decays have high yield due to high BR but presents the disadvantage to have a missing momentum carried by the neutrino in the final state. This turns in an extra proper time resolution smearing from the incomplete B_s momentum measurement, causing a rapid degradation of oscillation amplitude sensitivity (eq. 4) with growing Δm_s. Since complete B_s mass cannot be reconstructed, in Fig. 1 the reconstructed D_s signal is shown when the lepton is coming from the same secondary vertex and having the right charge correlation (in red).

Hadronic B_s decay have a lower BR ratio turning in a lower yield but can be fully reconstructed, resulting in a excellent S/B due to mass resolution (see Fig 2). Proper time resolution is independent of Δm_s giving to these decay modes high potential to explore high Δm_s region.

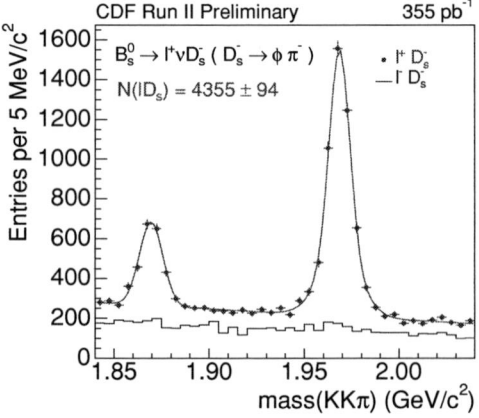

FIGURE 1. Semileptonic $B_s \rightarrow D_s l\nu$ decays with $D_s \rightarrow \phi\pi$

In the semileptonic decays main sources of backgrounds come from double charm B decays like $B_{d,u} \rightarrow D_s D X$ where the second D meson decays semileptonically and residual contribution from prompt D_s paired with a track from primary vertex faking a lepton. Both cases contributes to right charge correlation lepton-D_s pairs as in the B_s signal. The first background is evaluated from Montecarlo simulation and accounts for ~20% of the signal. The second is derived from wrong sign charge correlation lepton-D_s pairs and is around 4% of the sample. In the fully reconstructed modes reflection backgrounds from partially reconstructed B mesons are modest and taken into account with montecarlo simula-

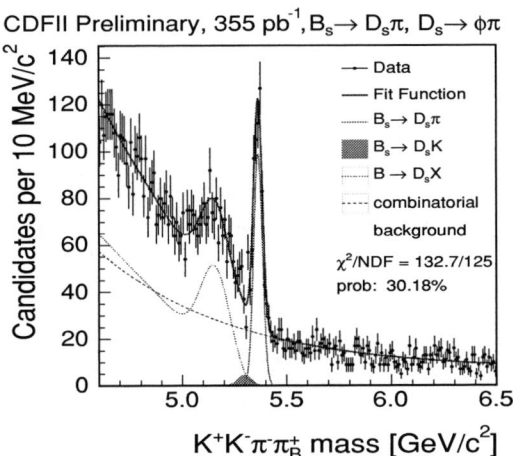

FIGURE 2. Hadronic $B_s \to D_s\pi$ decays with $D_s \to \phi\pi$

tion in the B_s mass shape (Fig. 2).

In summary B_s signal yields are: 4355 ± 94, 1750 ± 83, 1573 ± 88 (total ~ 7700) for semileptonic and 526 ± 33, 254 ± 21, 116 ± 18 (total ~ 900) for hadronic decays, given in both signatures correspondingly to $D_s \to \phi\pi$, $D_s \to K^*K$ and $D_s \to \pi\pi\pi$ modes.

Proper Time Reconstruction

A proper time analysis has to be established with a consistent B_s lifetime measurement before proceed with time oscillation study. In both sample proper time $ct = \frac{m_{B_s} L_{B_s}}{p(B_s)}$ is reconstructed via transverse plane L_{xy} decay distance and momentum p_T. The ct distributions obtained do not follow pure exponential behaviour because of the SVT trigger biases on the impact parameter of one or two tracks of the B_s candidate but have instead a more complicated shape of the form $e^{-ct/c\tau}\varepsilon(ct)$. The trigger and reconstruction acceptance $\varepsilon(ct)$ is obtained from detailed Montecarlo simulation.

In the semileptonic sample there is a further complication coming from the missing momentum in the reconstruction of the $p_T(B_s)$. The observable time is then a pseudo proper decay time $ct^* = \frac{m_{B_s} L_{B_s}}{p(l^+D_s^-)}$. A correction factor $K = p(l^+D_s^-)/p(B_s)$ is then introduced and its distribution taken from simulation model of semileptonic $B_s \to D_s l\nu$ decay including contributions from D_s^* and D_s^{**}. The K factor distribution contributes to the proper time resolution and its width is smaller the higher the lD_s mass. For this reason the lD_s is included in the fit.

In both samples ct templates for combinatorial backgrounds are taken from mass sidebands and those from long lived b-hadron reflections from Montecarlo simula-

tion.

Simultaneous mass and proper time un-binned Likelihood fits are performed and B_s lifetime is found to be $(442.7 \pm 9.5)\mu m$ in the semileptonic sample where error is statistical only and the result is obtained averaging all three D_s decay modes. Lifetime compares well with P.D.G 04 world average value of $(438 \pm 17)\mu m$. In the hadronic modes statistics is limited, however detailed study of systematic uncertainties from trigger biases on lifetime was performed and resulted in $c\tau(B_s) = (479 \pm 29(stat) \pm 5(syst))\mu m$ again in agreement with w.a. Cross checks of the whole method was further performed on similar hadronic decays $B_d^0 \to D^+\pi^-$ and $B^+ \to D^0\pi^+$ with higher statistics around 8000 and 7000 events respectively. Results $c\tau(B_d) = (453 \pm 7(stat) \pm 4(syst))\mu m$ and $c\tau(B_u) = (498 \pm 8(stat) \pm 4(syst))\mu m$ are in perfect agreement with w.a. lifetimes.

We can conclude that proper time biases coming from displaced track trigger SVT are well under control and that this samples are suitable for mixing analysis.

Flavour Tagging

Several methods can be used to determine the b quark flavour at production time. These can be divided in two categories one of which studies the opposite b in the event. In fact $b\bar{b}$ pairs are produced at collider and when one of those quarks (b) hadronize in a B_s the opposite ($\bar{b}$) will hadronize in a B meson or b-barion with opposite b flavour. The second category relies on the hadronization processes of the b quark: $b + (q\bar{q})_{vacuum} \to B_q + hadron(s)$, where the charge and the species of the hadronization particle close to the B meson (Same Side) will be a tag for the b flavour at production. For this analysis only opposite side flavour tagging methods are used, perspectives on how to include Same Side tagging for Δm_s analysis are discussed in the last section.

Three opposite side tagging where exploited: soft muon, soft electron and jet charge tagging. The first two soft lepton tags relies on the presence in the opposite b jet of a lepton coming from prompt $b \to l^-$ decay. the charge of the lepton reflects the b flavour. However other processes can give a lepton in the final state such as cascade decays $b \to c \to l^+$ resulting in wrong sign tags. For soft muon tag the candidate lepton is matched with muon detector identification while for soft electron both calorimeter information and dE/dx in the drift chamber are used for lepton identification. The jet charge tag exploits the fact that the momentum weighted sum of the particle charges in the opposite b jet is the same of the charge of the b quark producing the jet. All tagging algorithms provide a predicted Dilution D on an event by event basis, which has initially been studied on a large inclusive control sample based on lepton + SVT trigger data. Soft lepton tags predicted Dilution are parametrized as a function of p_T^{rel} w.r.t. jet direction while jet charge tag as a function of total charge. Taggers information has been used exclusively and ordered in decreasing Dilution. A total tagging power εD^2 (see eq.4) of around 1.1% and 1.4% is predicted for hadronic and semileptonic samples respectively. Each tagger Dilution parametrization has been tested on B_d^0 mixing analysis both on semileptonic and hadronic samples and consequently further adjusted by scale factors. The predicted Dilution are then used as a event by event input in Δm_s analysis.

SYSTEMATIC UNCERTAINTIES

Before performing the final amplitude scan for Δm_s, systematic uncertainty $\sigma_A(syst)$ on the amplitude has been evaluated for both samples using a realistic toy montecarlo reproducing data features. Varying the main systematic effects, amplitude variation was estimated using the formula $\sigma_A(syst) = \Delta A + (1 - A) \cdot \Delta\sigma_A/\sigma_A$ [6].

Main systematic error in the hadronic analysis comes from the tagger Dilution parameterizations and their Scale Factors because the relatively small statistics calibration sample of hadronic B^0 and B^+ decays. This error will scale with luminosity increase. In the semileptonic analysis main systematic contribution are from the prompt and double charm backgrounds fraction and Dilution knowledge. For the present analysis for both samples systematic uncertainties are actually negligible with respect to statistical. We expect that their importance will rise with increasing collected statistics and consequent higher Δm_s range reach.

AMPLITUDE SCAN RESULTS

Finally the results of the Δm_s amplitude scans on semileptonic and hadronic samples in terms of sensitivity and actual measured 95% C.L. limit are:

- $\Delta m_s > 7.7 ps^{-1}$ and sens. of $7.4 ps^{-1}$ (semileptonic)
- $\Delta m_s > 0.0 ps^{-1}$ and sens. of $0.4 ps^{-1}$ (hadronic)

A straightforward way to combine the two results considering them two independent measurements of the same physics quantity is possible in the amplitude scan method and hence combined CDF winter 2005 result for Δm_s is:

- $\Delta m_s > 7.9 ps^{-1}$ at 95% C.L., Sensitivity of $8.4 ps^{-1}$

Combined Amplitude scan is showed in Fig. 3 where statistical only Amplitude error band (yellow), systematic included band (green) are used to set the limit and mea-

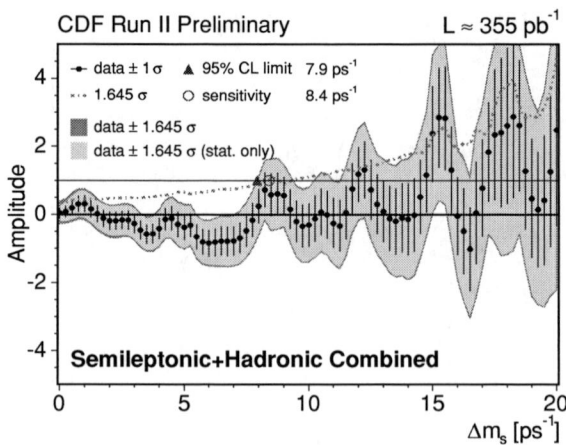

FIGURE 3. CDF combined Δm_s Amplitude scan result.

sured sensitivity curve is superimposed. The actual CDF result is dominated by the semileptonic sample, however the hadronic analysis will lead our sensitivity with increasing statistics.

PERSPECTIVES

With the result presented here CDF marginally improves the world average Δm_s sensitivity and do not increase the average limit. However we expect in the short time range to perform world best analysis on B_s oscillation frequency. In fact major improvements to the Δm_s search are expected from several sides. CDF has at this time collected more than a factor of two new data and much more is coming before the LHC starts. We are working in the inclusion of Same Side falvour tag algorithm for the B_s. This tag has already been used for the B_d and we foresee a Same Side tagging power for B_s at least comparable to all the opposite tags together. Moreover additional B_s signals from different decay modes and from different triggers are going to be included in the Δm_s search. CDF should start to probe for Δm_s in the SM preferred range for the next analysis round next year.

REFERENCES

1. M. Bona et al., *http://utfit.roma1.infn.it/* (2005).
2. H.F.A.G., *http://www.slac.stanford.edu/xorg/hfag/* (2005).
3. The CDF II detector: Technical Design Report, *FERMILAB-TM-2198* (2003).
4. A. Bardi et al., *NIM A*, **485** (2002).
5. D. Acosta et al., *NIM A*, **518** (2004).
6. H.G. Moser, A. Roussarie, *NIM A*, **384** (1997).

Branching Fractions and CP Asymmetries in Hadronic Charmless Decays of B Mesons at CDF

Massimo Casarsa

INFN Trieste, Italy
e-mail: Massimo.Casarsa@ts.infn.it

Abstract. We present an overview of branching ratio and direct CP asymmetry measurements for charmless b-hadron decays into two-body hadronic final states using 180 pb^{-1} of data collected by the CDF II experiment at the Fermilab Tevatron collider. The following modes are treated: $B^0_{d,s} \to h^+h'^-$, $\Lambda^0_b \to ph^-$, where h and h' stand for K or π, $B^+_u \to \phi K^+$, and $B^0_s \to \phi\phi$.

Keywords: CDF, B Physics, charmless B decays, hadronic B decays.
PACS: 13.25.Hw, 13.30.Eg, 14.40.Nd, 14.20.Mr

INTRODUCTION

Charmless hadronic decays of B mesons are rare decays, characterized by branching ratios (BR) of the order of or smaller than 10^{-5}, which occur through the $b \to u, d, s$ quark-level transitions and are mediated by both tree and penguin diagrams. From measurements of branching ratios and CP asymmetries of such decays the Cabibbo-Kobayashi-Maskawa matrix parameters can be determined. Furthermore, charmless B decays dominated by one-loop penguin transitions are potentially sensitive to new CP-violating phases from Physics beyond the Standard Model.

Hadronic machines like the Tevatron are characterized by a large $b\bar{b}$ production cross-section and provide large yields of B mesons. In addition, they give access to B^0_s and b-baryon states. On the other hand, the total inelastic cross-section is one thousand times larger than $b\bar{b}$ production cross-section, resulting in a messy environment with large combinatorics. The CDF experiment exploits a dedicated trigger on displaced secondary vertices to select enriched samples of b hadrons.

In this paper we present an overview of measurements of two-body hadronic charmless B decays, which use 180 pb^{-1} of $p\bar{p}$ collision data at $\sqrt{s} = 1.96$ TeV collected by the upgraded Collider Detector (CDF II) at the Fermilab Tevatron between February 2002 and September 2003.

THE EXPERIMENTAL APPARATUS

The components of the CDF II detector pertinent to the measurements reported herein are briefly outlined. A more complete description can be found elsewhere [1]. We use tracks in the pseudorapidity range $|\eta| \lesssim 1$ reconstructed by a silicon microstrip vertex detector (SVX II) and the Central Outer Tracker (COT), which are immersed in a 1.4 T solenoidal magnetic field. The SVX II detector consists of double-sided sensors arranged in five cylindrical layers. Surrounding the SVX II is the COT, an open cell drift chamber with 96 sense wires. The integrated charge collected by each wire provides a measurement of the specific ionization (dE/dx) for charged particles, allowing a separation equivalent to 1.4 Gaussian σ between π and K for $p_T > 2$ GeV/c. A set of planar drift chambers, located outside the calorimeters and additional steel absorbers, is used to detect muons within $|\eta| \leq 1$ with high purity.

A sample enriched with heavy flavor particles is collected by a three-level trigger which selects events with track pairs originating from a common displaced vertex. At Level 1, charged tracks are reconstructed in the COT by the eXtremely Fast Tracker (XFT). The trigger requires two oppositely charged tracks with transverse momenta $p_T \geq 2$ GeV/c and the scalar sum $p_{T_1} + p_{T_2} \geq 5.5$ GeV/c. At Level 2, the Silicon Vertex Tracker (SVT) associates SVX II r-φ position measurements with XFT tracks, providing a precise measurement of the track impact parameter (d_0), the distance of closest approach of the track trajectory to the beam axis in the transverse plane. Decays of heavy flavor particles are identified by requiring two tracks with $120\,(100)\,\mu$m $\leq |d_0| \leq 1.0$ mm and an opening angle $2°\,(20°) \leq |\Delta\phi| \leq 90°\,(135°)$ in the case of multi-track (two-track) final states. A requirement $L_{xy} > 200\,\mu$m is also applied for both topologies, where the two-dimensional decay length, L_{xy}, is calculated as the transverse distance from the beam axis to the two track intersection projected onto the total transverse momentum of the track pair.

A complete event reconstruction is performed at Level 3, where the Level 1 and Level 2 trigger requirements are confirmed.

HADRONIC B CHARMLESS DECAYS

First we treat issues which are common to all the analyses, then we present the highlights of each measurement and quote the results.

In all BR measurements, in order to cancel the uncertainty in the B hadron production cross section and to reduce the systematic uncertainties on detector efficiencies, the branching fractions are extracted from ratios of the decay rates of interest normalized to well established decay modes, which are characterized by the same number of decay vertices and charged tracks in the final state as in the signal modes.

Candidate reconstruction and selection are based on a common procedure and a similar set of requirements is used to reject physical and combinatorial backgrounds. Requiring a good vertex fit χ^2 reduces background from mismeasured tracks. Combinatoric background is reduced by exploiting several variables sensitive to the long lifetime and relatively hard p_T spectrum of B mesons and the isolation of B hadrons inside b-quark jets. For this purpose we define the quantity I_R as the ratio of the B candidate p_T over the total transverse momenta of all tracks within a cone of radius $R = \sqrt{\Delta\eta^2 + \Delta\varphi^2} = 1$ around the B flight direction. Requiring the B flight direction to extrapolate back to the beam axis decreases background from partially reconstructed decays. The cut values on the discriminating variables are optimized by maximizing $\mathcal{S}/\sqrt{\mathcal{S}+\mathcal{B}}$ for the already observed $B^0_{d,s} \to h^+h'^-$ and $B^+_u \to \phi K^+$ signals and $\mathcal{S}/(1.5+\sqrt{\mathcal{B}})$ for $B^0_s \to \phi\phi$ and $\Lambda^0_b \to ph^-$ whose branching ratios are unknown. The latter choice is equivalent to maximizing the potential to reach a 3σ observation of a new signal [2]. The signal ($\mathcal{S}$) is derived from a Monte Carlo (MC) simulation of the CDF II detector and trigger, whereas the background ($\mathcal{B}$) is represented by appropriately normalized sideband data.

We now describe the $B^0_{d,s} \to h^+h'^-$ analysis, where $h, h' = \pi, K$. All events with two-prong displaced secondary vertices are reconstructed with a pion mass assignment to the final state tracks. From the MC simulation four major modes are expected to show up: $B^0_d \to \pi^+\pi^-$, $B^0_d \to K^+\pi^-$, $B^0_s \to K^-K^+$, and $B^0_s \to K^-\pi^+$ [1]. The selection requirements are: impact parameters of the two tracks $|d_0^{(1)}|, |d_0^{(2)}| > 150$ μm and $d_0^{(1)} \times d_0^{(2)} < 0$, B^0-meson decay length $L_{xy} > 300$ μm, B^0 impact parameter $|d_0^B| < 80$ μm, and B^0 isolation $I_B > 0.5$. In the two-track invariant mass distribution the four signals overlap to form a single unresolved bump, which

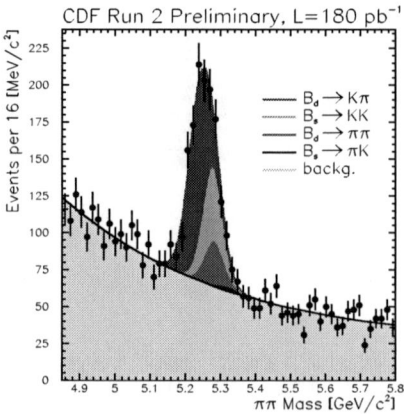

FIGURE 1. $M_{\pi\pi}$ distribution of the $B^0_{d,s} \to h^+h'^-$ candidates after all selection cuts. The fit result is overlaid.

counts 893 ± 47 events. Analysis strategy consists in disentangling the single modes by means of an unbinned maximum-likelihood fit, which combines information on the kinematics of each decay and on the particle identification (PID) of the final state products. It can be shown that all the relevant kinematic properties of the decay are summarized in the two-track invariant mass ($M_{\pi\pi}$) and the signed momentum imbalance $q_{\min}(1 - p_{\min}/p_{\max})$, where the subscripts min (max) refers to the smaller (larger) momentum track. PID is based on the dE/dx, measured in the drift chamber. The fit returns the relative contributions of each component to the signal and the direct CP asymmetry of the self-tagging $B^0_d \to K^+\pi^-$ mode. Figure 1 shows the invariant mass distribution with the fit result overlaid. In order to extract relative branching ratios from the raw fit results, it is necessary to correct the fractions by 5-10% to accont for different selection efficiencies for each channel. A $\sim 1\%$ correction is also applied to A_{CP} to compensate for the effect of the CDF tracking charge asymmetry. Corrections are estimated both from the complete CDF detector simulation and from data control samples. Main contributions to the total systematics are 16% from dE/dx calibration and 10% from isolation cut efficiency. Both are due to statistical uncertainties in the calibration samples and are expected to decrease with sample size.

Results [2] for B^0_d modes are:

$$\frac{\text{BR}(B^0_d \to \pi^+\pi^-)}{\text{BR}(B^0_d \to K^+\pi^-)} = 0.24 \pm 0.06 \pm 0.04,$$

$$A_{CP}(B^0_d \to K^+\pi^-) = -0.04 \pm 0.08 \pm 0.01,$$

---

[1] Charge conjugate modes are implied throughout this paper unless otherwise stated.

[2] Throughout this paper, the first quoted error is statistical, the second systematic.

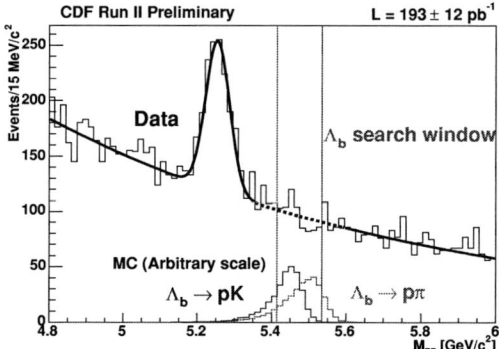

FIGURE 2. $M_{\pi\pi}$ invariant mass distribution after all selection criteria are applied. The peak is given by the $B \to hh'$ events. The dotted line indicates the region that was blinded during the cut optimization. The solid line indicated the region used to determine the background level.

where the time-integrated direct CP asymmetry is defined as $(N_{\bar{B}_d^0} - N_{B_d^0})/(N_{\bar{B}_d^0} + N_{B_d^0})$ with $N_{\bar{B}_d^0}$ ($N_{B_d^0}$) being the number of $\bar{B}_d^0$ (B_d^0). Both results are in agreement with the world average [3].

For which concerns the B_s^0 sector, we measure:

$$\frac{f_s}{f_d} \frac{\mathrm{BR}(B_s^0 \to K^-K^+)}{\mathrm{BR}(B_d^0 \to K^+\pi^-)} = 0.50 \pm 0.08 \pm 0.07,$$

$$\frac{f_d}{f_s} \frac{\mathrm{BR}(B_d^0 \to \pi^+\pi^-)}{\mathrm{BR}(B_s^0 \to K^-K^+)} = 0.48 \pm 0.12 \pm 0.07.$$

The above results are obtained under the assumption that $\Gamma_s = \Gamma_d$, the $B_s^0 \to K^-K^+$ mode is dominated by the short-lived component, and that $\Delta\Gamma_s/\Gamma_s = 0.12 \pm 0.06$. Using the values for f_s/f_d and $\mathrm{BR}(B_d^0 \to K^+\pi^-)$ quoted in [3], we calculate the absolute branching ratio $\mathrm{BR}(B_s^0 \to K^-K^+) = (34.3 \pm 5.5 \pm 5.2) \times 10^{-6}$ finding it in good agreement with the theoretical prediction [4]. No evidence is found for the $B_s^0 \to K^-\pi^+$ decay:

$$\frac{f_s}{f_d} \frac{\mathrm{BR}(B_s^0 \to K^-\pi^+)}{\mathrm{BR}(B_d^0 \to K^+\pi^-)} < 0.11 \text{ @ } 90\% \text{ C.L.},$$

where it is assumed that both modes have the same average lifetime.

Furthermore, a search was performed for two $B_{d,s}^0$ rarer decay modes, whose BR are expected to lie in the range 10^{-8} to 10^{-7}. The following limits are set:

$$\frac{\mathrm{BR}(B_d^0 \to K^-K^+)}{\mathrm{BR}(B_d^0 \to K^+\pi^-)} < 0.17 \text{ @ } 90\% \text{ C.L.},$$

$$\frac{\mathrm{BR}(B_s^0 \to \pi^-\pi^+)}{\mathrm{BR}(B_s^0 \to K^-K^+)} < 0.10 \text{ @ } 90\% \text{ C.L.}.$$

The latter is derived under the assumption that both modes have the same average lifetime.

The same data as for the $B_{d,s}^0 \to h^+h'^-$ analysis have been used in a separate search for the charmless $\Lambda_b^0 \to p\pi^-/pK^-$. In that case a blind analysis was performed: the data in the signal mass window were hidden and the background was calculated by fitting the invariant mass spectrum and interpolating in the blinded signal region. After all selection criteria were fixed and the systematic uncertainty estimated, the signal region was unblinded and the number of events counted and compared with the expected background. The optimized selection requirements are $|d_0^\Lambda| < 50$ μm, $L_{xy} > 400$ μm, and $\min(d_0^{(1)}, d_0^{(2)}) > 180$ μm. The search window which maximizes the sensitivity is 5.415-5.535 GeV/c^2. Figure 2 shows the invariant mass distribution after all selection cuts are applied. No evidence for signal is found. Using $B_d^0 \to K^+\pi^-$ as a normalization channel, a Bayesian upper limit, based on uniform prior, is set for the sum of the two branching fractions:

$$\mathrm{BR}(\Lambda_b^0 \to ph^-) < 2.3 \times 10^{-5} \text{ @ } 90\% \text{ C.L.},$$

where $h = \pi, K$. This result improves significantly the previously published limits of 5×10^{-5} at 90% C.L. for both decay modes [3]. Theoretical predictions lie in the range $(0.9 - 1.2) \times 10^{-6}$ for $p\pi$ and $(1.4 - 1.9) \times 10^{-6}$ for pK [5].

As for the $B_u^+ \to \phi K^+$ analysis, ϕ is reconstructed in the K^+K^- final state. The optimization results in the following requirements: vertex $\chi^2 < 8$, B^+ decay length $L_{xy} > 350$ μm, B^+ reconstructed impact parameter $|d_0^B| < 100$ μm, isolation $I_R > 0.5$, non-trigger track transverse momentum $p_T^{soft} > 1.3$ GeV/c and impact parameter $|d_0^{soft}| > 120$ μm. The yield and the CP asymmetry are extracted

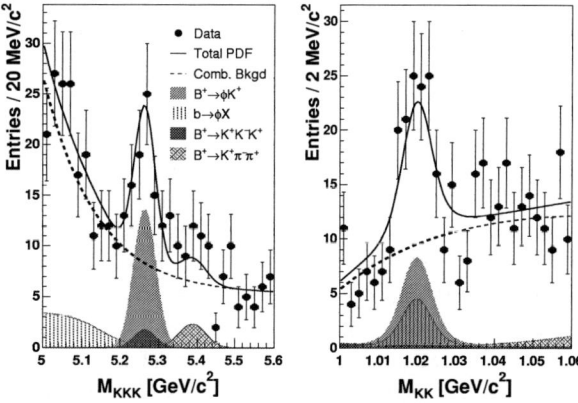

FIGURE 3. M_{KKK} (left) and M_{KK} (right) distributions for $B_u^+ \to \phi K^+$ with projections of the total likelihood function and its components. $B^+ \to K^+K^-K^+$ is the sum of non-resonant and $B^+ \to f_0(980)K^+$ contributions, while $B^+ \to K^+\pi^-\pi^+$ includes non-resonant and $B^+ \to K^{*0}(892)\pi^+$.

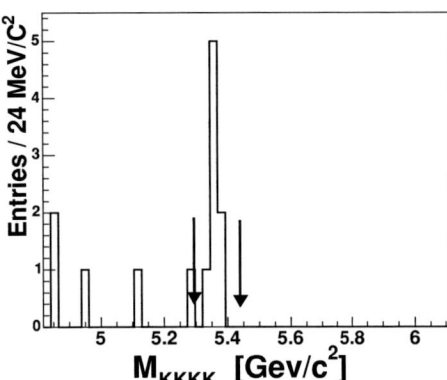

FIGURE 4. Invariant mass distribution for $B_s^0 \to \phi\phi$ candidates. The arrows indicate the signal region for the $B_s^0 \to \phi\phi$ search.

simultaneously from an extended unbinned maximum-likelihood fit on the three-track invariant mass (M_{KKK}), the ϕ invariant mass (M_{KK}), the ϕ helicity (defined as the angle between the K^+ momentum in the parent ϕ rest frame and the momentum of the ϕ in the B rest frame), and the measured dE/dx deviation from the expected value for pions for the lowest momentum trigger track. The fit discriminating variables are assumed to be uncorrelated. The fit yields $47.0 \pm 8.4 \pm 1.4$ events. Figure 3 shows the M_{KKK} and M_{KK} distributions with projections of the total likelihood function and its components. The BR is measured relative to $B_u^+ \to J/\psi K^+$ with $J/\psi \to \mu^+\mu^-$. A similar fit on $M_{\mu\mu K}$ and $M_{\mu\mu}$ yields 439 ± 22 events and $A_{CP}(B_u^+ \to J/\psi K^+) = 0.046 \pm 0.050$ (statistical error only). Absolute BR is calculated by estimating the total efficiencies from MC and using PDG [3] BR's for $\phi \to K^+K^-$ and $J/\psi \to \mu^+\mu^-$:

$$BR(B_u^+ \to \phi K^+) = (7.6 \pm 1.3 \pm 0.6) \times 10^{-6},$$
$$A_{CP}(B_u^+ \to \phi K^+) = -0.07 \pm 0.17 {}^{+0.03}_{-0.02},$$

where A_{CP} is defined as $(N_{B^-} - N_{B^+})/(N_{B^-} + N_{B^+})$. The main sources of systematic uncertainty come from the fit modeling (3%) and, for the BR only, the MC efficiencies (5.6%). Results are in agreement with the previously published measurements [3].

A "blind" search for $B_s^0 \to \phi\phi$ decays was performed fixing the selection requirements and evaluating the combinatoric background from independent samples before examining the signal region in the data. ϕ's are reconstructed in the K^+K^- final state. The signal is selected requiring two pairs of kaons having an invariant mass within 15 MeV/c^2 of the world average ϕ mass [3]. The optimized selection criteria are the following: vertex $\chi^2 < 10$, B_s^0 decay length $L_{xy} > 350 \mu$m, B_s^0 reconstructed impact parameter $|d_0^B| < 80 \mu$m, minimum ϕ meson transverse momentum $p_T^\phi > 2.5$ GeV/c and minimum of the two kaons' impact parameters from each ϕ meson candidate $|d_{0\,min}^{\phi_1}| > 40 \mu$m, $|d_{0\,min}^{\phi_2}| > 110 \mu$m, where ϕ_1 is the lower momentum ϕ candidate. The $B_s^0 \to \phi\phi$ candidate mass distribution is shown in Fig. 4. In a region of ± 72 MeV/c^2 around the world average [3] B_s^0 mass, corresponding to a window three times the expected mass resolution, we observe 8 events.

Two sources of background are expected in the B_s^0 signal region: combinatoric background and $B^0 \to \phi K^{*0}$ decays with the pion from the K^{*0} decay treated as a kaon. The contribution of the combinatoric background in the signal region is estimated using a background enriched sample where both ϕ meson candidates have invariant mass lying in the ϕ mass sideband region: 0.35 ± 0.37 events (combining statistical and systematic uncertainties). The $B^0 \to \phi K^{*0}$ background contribution, derived from simulation, is 0.37 ± 0.18 events where the error includes both statistical and systematic uncertainties. The resulting signal yield is $7.3 {}^{+3.2}_{-2.5} \pm 0.4$ events.

For the determination of BR($B_s^0 \to \phi\phi$), a normalization sample of $B_s^0 \to J/\psi\phi$ decays is selected: $N_{J/\psi\phi} = 69 \pm 10 \pm 5$ events. We subtract 3.7 ± 1.7 background events from $B^0 \to J/\psi K^{*0}$ decays, with the pion treated as a kaon, as estimated from simulation. The result is:

$$BR(B_s^0 \to \phi\phi) = (14 {}^{+6}_{-5} \pm 6) \times 10^{-6},$$

where the systematic uncertainty includes a 36% contribution due to the uncertainty on BR($B_s^0 \to J/\psi\phi$), 8% from $B_s^0 \to J/\psi\phi$ yield and background, and 4% from polarization of the decay vector particle and the value assumed for $\Delta\Gamma_s$. The measurement is in agreement with the estimate in [6].

CONCLUSION

An overview of first generation measurements of BR's and CP asymmetries for hadronic charmless decays of b-hadrons on 180 pb^{-1} has been presented. At the time of this writing ~ 1 fb^{-1} of integrated luminosity is available on disk. Therefore, a series of more accurate and precise results is expected in the next future.

REFERENCES

1. D. Acosta et al., Phys. Rev. D **71**, 032001 (2005).
2. G. Punzi, eConf C030908, MODT002 (2003).
3. S. Eidelman et al. [Particle Data Group], Phys. Lett. B **592**, 1 (2004).
4. A.J. Buras et al., Nucl. Phys. B **697**, 133 (2004).
5. R. Mohanta et al., Phys. Rev. D **63**, 074001 (2001).
6. X.Q. Li et al., Phys. Rev. D **68**, 114015 (2003); **71**, 019902(E) (2005).

Parallel Session IV:

Neutrinos and Cosmic Rays

Conveners:

D. Montanino
University of Lecce and INFN, Italy

G. Riccobene
INFN LNS, Italy

Neutrino Mass Phenomenology

A. Marrone

Dipartimento di Fisica and Sezione INFN di Bari, Via Amendola 173, 70126 Bari, Italy

Abstract. The present knowledge about neutrino mass and mixing, in the framework of three active neutrino mixing, is reviewed. All available data on solar, atmospheric, reactor, and accelerator oscillation neutrino experiments are analysed and combined with data on absolute neutrino masses, coming from beta decay and neutrinoless double beta decay experiments and astrophysical and cosmological researches.

Keywords: neutrino mass and mixings
PACS: 14.60Pq

INTRODUCTION

In the last ten years a number of experiments on neutrinos did provide a robust evidence for neutrino oscillations. In this work we analyse and combine the results of the solar neutrino experiments Homestake [1], SAGE [2], GALLEX-GNO [3, 4], Super-Kamiokande (SK) [5], and Sudbury Neutrino Observatory (SNO) [6, 7], the long-baseline reactor neutrino experiment KamLAND [8], the atmospheric neutrino experiment Super-Kamiokande [9], the long-baseline accelerator neutrino experiment K2K [10] and the short-baseline reactor experiment CHOOZ [11]. All current available data, with the possible exception of the LSND experiment [12] can be explained by neutrino oscillation in a three-generation framework [13], where the flavor neutrino eigenstate ν_α ($\alpha = e, \mu, \tau$) are a linear superposition of three mass eigenstates ν_i ($i = 1, 2, 3$) through a mixing matrix U, depending on three mixing angle ($\theta_{12}, \theta_{13}, \theta_{23}$) and a CP-violating phase δ.

Neutrino oscillations depend on neutrino masses m_i through two squared mass differences, δm^2 and Δm^2, defined as follows:

$$\delta m^2 = m_2^2 - m_1^2, \qquad (1)$$
$$\Delta m^2 = \left| m_3^2 - \frac{m_1^2 + m_2^2}{2} \right|. \qquad (2)$$

We use the convention that δm^2 is always positive while Δm^2 can be either positive (normal hierachy, NH) or negative (inverted hierarchy, IH). Moreover, only the two CP-conserving cases $\cos \delta = \pm 1$ will be analysed, the two being formally related by the replacement $\sin \theta_{13} \to -\sin \theta_{13}$. While oscillation experiments are only sensitive to neutrino mass differences, beta decay [14] and neutrinoless double beta decay ($0\nu 2\beta$) experiments [15] are sensitive to absolute neutrino masses through the two effective parameters m_β and $m_{\beta\beta}$, respectively:

$$m_\beta = \left[c_{13}^2 c_{12}^2 m_1^2 + c_{13}^2 s_{12}^2 m_2^2 + s_{13}^2 m_3^2 \right]^{\frac{1}{2}}, \qquad (3)$$
$$m_{\beta\beta} = \left| c_{13}^2 c_{12}^2 m_1 + c_{13}^2 s_{12}^2 m_2 e^{i\phi_2} + s_{13}^2 m_3 e^{i\phi_3} \right|. \qquad (4)$$

The effective Majorana mass $m_{\beta\beta}$ probed in $0\nu 2\beta$ experiments depends on two Majorana phases, ϕ_2 and ϕ_3, that are currently unobservable and so marginalised away in the global analysis. We also include in our global analysis the upper bounds on the sum of neutrino masses ($\Sigma = m_1 + m_2 + m_3$), coming from astrophysical and cosmological searches [16, 17, 18, 19, 20, 21], namely from data on large scale structure, Cosmic Microwave Background and Lyman α forest data.

SOLAR NEUTRINOS AND KAMLAND

The combination of solar neutrino and KamLAND data can be explained by a well-defined and unique solution in the parameter space, the so called large mixing angle solution (LMA). In the case of two-neutrino mixing, solar neutrino oscillations, as well as long-baseline reactor neutrino oscillations, depend only on the parameters $(\delta m^2, \theta_{12})$, while in the three-generation mixing scheme the mixing angle θ_{13} must be also considered. Figure 1 shows the bounds obtained by our analysis in the parameter space $(\delta m^2, \theta_{12}, \theta_{13})^1$. The bounds on δm^2 are dominated by the KamLAND results and will be improved with higher statistics, while the uncertainty on θ_{12} is dominated by the solar neutrino data, in particular by the SNO measure of the charged-to-neutral current ratio (CC/NC). The final SNO-II data set shifts the previous LMA solution to slightly higher values of $\sin^2 \theta_{12}$,

---

[1] We use the convention to define allowed regions at $n\sigma$ the part of the parameter space where $\Delta\chi^2 \leq n^2$, so that the projection onto each parameter gives the $n\sigma$ range on that parameter.

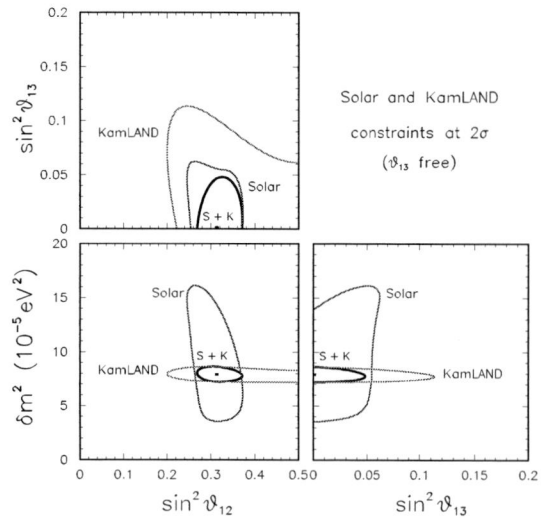
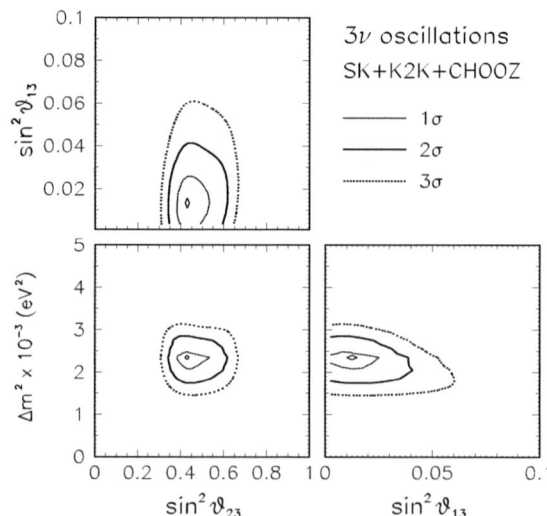

FIGURE 1. Three falvor analysis of solar and KamLAND data.

FIGURE 2. Three flavor analysis of SK+K2K+CHOOZ data, including subleading LMA effects. The contours represent allowed regions after marginalization with respect to the four cases $[\cos\delta = \pm 1]\otimes[\pm\Delta m^2]$.

since the measured CC/NC ratio (0.34) is larger than the previous central value (0.31). Moreover the SNO-II CC spectral data prefer slightly higher values of δm^2, as compared to older results. KamLAND data show a weak anticorrelation between $\sin^2\theta_{13}$ and $\sin^2\theta_{12}$, due to the fact that a total rate reduction can be achieved both with a higher θ_{13} or a lower θ_{12}. Both solar and KamLAND data, as showed in right bottom panel, are consistent with a small θ_{13} value, the combined upper limit being $\sin^2\theta_{13} \sim 0.05$ at two sigma level. This upper limit is not much weaker than the one placed by the combination of CHOOZ+SK+K2K data.

SK, K2K AND CHOOZ

The dominant parameters in the atmospheric neutrinos oscillations are Δm^2, θ_{23}, and θ_{13}, for three-generation mixing. The two remaining parameter $(\delta m^2, \theta_{12})$ induce subleading effects on the distribution of the SK theoretical expectations, and in the analysis are fixed at the solar best fit ($\delta m^2 = 8 \times 10^{-5}$ eV^2, $\sin^2\theta_{12} = 0.314$). In the three-generation framework, with $\theta_{13} > 0$, SK+K2K data are sensitive to the sign of Δm^2 and to the value of δ. However whitin the limits on θ_{13} set by the negative results of CHOOZ, the allowed regions depend very weekly on the hierarchy and δ, for the two cases $\cos\delta = \pm 1$ studied (not shown). The K2K data are octant-symmetric and relatively weak in $\sin^2\theta_{23}$, while they significantly contribute in constraining the allowed Δm^2 range, especially in the upper part. SK data, instead, are effective in constraining both Δm^2 and $\sin^2\theta_{23}$. When the effect of the subleading parameters δm^2 and θ_{12} is included, the allowed region begins asymmetric in the mixing angle θ_{23} and move to the left, toward smaller values of the angle. That happens because, when θ_{23} is not maximal ($\theta_{23} \neq \pi/4$) the LMA subleading effect induces an excess of sub-GeV electron events in SK that helps the global fit. However, given the present statistical and systematic errors data cannot distinguish between the first and the second octant of θ_{23}. In Figure 2 the global χ^2 is marginalised with respect to the two hierarchies and the two values of δ considered in this work ($\cos\delta = \pm 1$). The combination of SK, K2K and CHOOZ data prefer a value of θ_{13} slightly different from zero (even if with small statistic significance, since $\theta_{13} = 0$ is allowed at less than one sigma), mostly due to a small excess in the low-energy ν_e sample.

GLOBAL CONSTRAINTS ON OSCILLATION PARAMETERS

As mentioned before, current analyses of oscillation experiments cannot distinguish between normal and inverted neutrino mass hierarchy, nor can give substantial information on the CP phase δ, so that in the results presented in Figures 3 and 4, the four cases ($\pm\Delta m^2$, $\cos\delta = \pm 1$) are marginalized away. The two figures show the $(\chi^2)^{1/2}$ as a function of the two mass squared differences and of the three mixing angles. Figure 3 shows that the accuracy in the determination of δm^2 and θ_{12} is already enough to provide almost linear errors (the curves are approximately straight lines), while it is better to quote two or three sigma errors on Δm^2 and θ_{23} since the re-

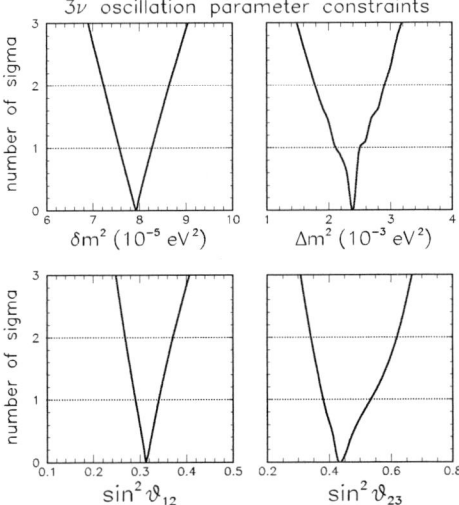

FIGURE 3. Global three-neutrino analysis of oscillation data. The bounds on δm^2, Δm^2, $\sin^2\theta_{12}$, and $\sin^2\theta_{23}$ are show in terms of $n\sigma = \sqrt{\Delta\chi^2}$.

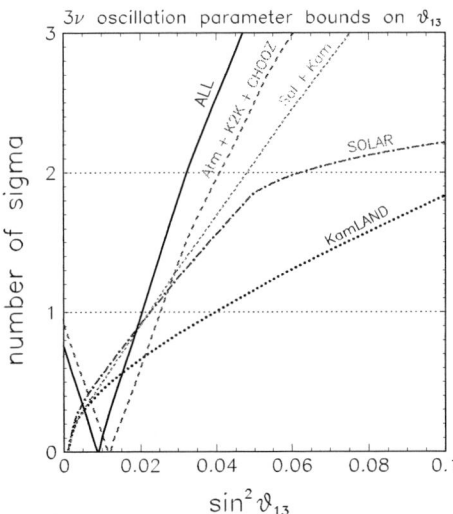

FIGURE 4. Global three-neutrino analysis of oscillation data. Bounds on $\sin^2\theta_{13}$ are shown in terms of $n\sigma = \sqrt{\Delta\chi^2}$ for KamLAND, solar, solar+KamLAND, SK+K2K+CHOOZ and all data combined. The parameters $(\Delta m^2, \delta m^2, \sin^2\theta_{12}, \sin^2\theta_{23})$ are marginalized away.

sults of the analysis is less "clear" near the best-fit point. Figure 4 shows the constraints on the less known of the oscillation parameters, θ_{13}. It is interesting to notice that solar+KamLAND data give a limit on θ_{13} comparable to the dominant one, provided by the combination of SK, K2K and CHOOZ data, partly due to the preference of $\theta_{13} \neq 0$ of the latter. The two sigma bounds on the parameter can be summarized as follows:

$$\sin^2\theta_{13} = 0.9^{+2.3}_{-0.9} \times 10^{-2}, \quad (5)$$
$$\delta m^2 = 7.92(1 \pm 0.09) \times 10^{-5}\,\text{eV}^2, \quad (6)$$
$$\sin^2\theta_{12} = 0.314(1^{+0.18}_{-0.15}), \quad (7)$$
$$\Delta m^2 = 2.4(1^{+0.21}_{-0.26}) \times 10^{-3}\,\text{eV}^2, \quad (8)$$
$$\sin^2\theta_{23} = 0.44(1^{+0.41}_{-0.22}). \quad (9)$$

The lower uncertainty on $\sin^2\theta_{13}$ is purely formal corresponding only to the condition $\sin^2\theta_{13} \geq 0$. The correlation among the parameters is not reported, being small as can been graphically seen from Figures 1 and 2.

ABSOLUTE MASSES

The parameters $(m_\beta, m_{\beta\beta}, \Sigma)$ that depend on absolute neutrino masses, are strongly correlated, once the bounds on the oscillation parameters given in the previous section are taken into account. In general, both in the case of normal and inverted hierarchy they are positively correlated so that an upper bound on one of them translate into an upper bound on the other two. However current limits on m_β from beta decay experiments are not as compelling as bounds on $m_{\beta\beta}$ and Σ from $0\nu 2\beta$ experiments and cosmology. In our analysis we use the combined upper limit of the Mainz and Troitsk experiments:

$$m_\beta < 1.8\,\text{eV at } 2\sigma. \quad (10)$$

Bounds at the level of few eV on m_β are still too weak to contribute significantly to the global fit in the $(m_\beta, m_{\beta\beta}, \Sigma)$ parameter space. Concerning the $0\nu 2\beta$ analysis, we use the two following inputs on $m_{\beta\beta}$:

$$\log_{10}(m_{\beta\beta}/\text{eV}) = -0.23 \pm 0.14\ (\text{signal}), \quad (11)$$
$$\log_{10}(m_{\beta\beta}/\text{eV}) = -0.23^{+0.14}_{-\infty}\ (\text{no signal}). \quad (12)$$

We assume no prior knowledge on the majorana phases and let them vary into the range $[0, \pi]$. The inputs for astrophysical and cosmological data are described in [?]. Figure 5 shows the two sigma bounds in the plane $(\Sigma, m_{\beta\beta})$ from all the available oscillation and non-oscillation data. The horizontal dashed band is the region allowed by the positive claim of a part of the Heidelberg-Moscow $0\nu 2\beta$ collaboration [22]. The two continuos curves represent the allowed region for normal and inverted hierarchy, when the positive claim is rejected. The upper limit on Σ from astrophysical and cosmological data implies that there is a non negligible tension between all the available oscillation and non-osscilation data and the positive $0\nu 2\beta$ claim, even though, given the current uncertainties on the sistematics of the Lyman α forest data and on the nuclear decay rates for neutrinoless double beta decay, more experimental and theoretical investigation is needed to definitively asses the data interpretation.

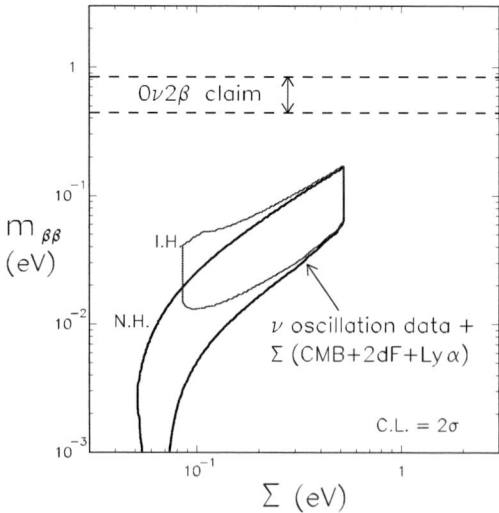

FIGURE 5. Analysis of oscillation and non-oscillation data in the plane $(m_{\beta\beta}, \Sigma)$. The two sigma horizontal dashed band is allowed by the positive $0\nu2\beta$, while the two regions below are preferred by all the other data.

CONCLUSIONS

Neutrino oscillations driven by non zero mass and mixing are an experimental established fact. All oscillation data (excepted the LSND result) can be explained in three-neutrino mixing framework. It is not currently possible to determine the sign of Δm^2 and the value of the phase δ. The less known of the oscillation parameters is the mixing angle θ_{13}, which nonetheless is constrained to be small by the combination of all the available data. The bounds on the absolute masses are dominated by the limits on the sum of neutrino masses Σ from cosmology at the eV/sub-eV level. If the claim of a positive signal in the neutrinoless double beta decay experiment is accepted, then a strong tension emerges between this results and all other data.

ACKNOWLEDGMENTS

This work is supported by the Italian Ministero dell'Istruzione, Università e Ricerca (MIUR) and Istituto Nazionale di Fisica Nucleare (INFN) through the "Astroparticle Physics" project.

REFERENCES

1. Homestake Collaboration, B.T. Cleveland *et al.*, Astrophys. J. **496**, 505 (1998).
2. SAGE Collaboration, J.N. Abdurashitov *et al.*, J. Exp. Theor. Phys. **95**, 181 (2002).
3. GALLEX Collaboration, W. Hampel *et al.*, Phys. Lett. B **447**, 127 (1999).
4. GNO Collaboration, M. Altmann *et al.*, Phys. Lett. B **616**, 174 (2005).
5. SK Collaboration, M.B. Smy *et al.*, Phys. Rev. D **69**, 011104 (2004).
6. SNO Collaboration, S.N. Ahmed *et al.*, Phys. Rev. Lett. **92**, 181301 (2004).
7. SNO Collaboration, B. Aharmim *et al.*, nucl-ex/0502021.
8. KamLAND Collaboration, K. Eguchi *et al.*, Phys. Rev. Lett. **94**, 081801 (2005).
9. Super-Kamiokande Collaboration, Y. Ashie *et al.*, hep-ex/0501064.
10. K2K Collaboration, E. Aliu *et al.*, Phys. Rev. Lett. **94**, 081802 (2005).
11. CHOOZ Collaboration, M. Apollonio *et al.*, Eur. Phys. J. C **27**, 331 (2003).
12. LSND Collaboration, A. Aguilar *et al.* Phys. Rev. D **64**, 112007 (2001).
13. G.L. Fogli, E. Lisi, A. Marrone, and A. Palazzo, hep-ph/0506083, invited review submitted to Progress in Particle and Nuclear Physics (2005).
14. K. Eitel, in *Neutrino 2004*, Nucl. Phys. B (Proc. Suppl.) **143** (2005), 197.
15. S.R. Elliott and J. Engel, J. Phys. G **30**, R183 (2004).
16. WMAP Collaboration, C.L. Bennett *et al.*, Astrophys. J. Suppl. **148**, 1 (2003).
17. 2dF GRS Collaboration, W.J. Percival *et al.*, MNRAS **327**, 1297 (2001).
18. SDSS Collaboration, P. McDonald *et al.*, astro-ph/0405013.
19. W. Hu, D.J. Eisenstein, and M. Tegmark, Phys. Rev. Lett. **80**, 5255 (1998).
20. SDSS Collaboration, M. Tegmark *et al.*, Phys. Rev. D **69**, 103501 (2004).
21. U. Seljak *et al.*, astro-ph/0407372.
22. H.V. Klapdor-Kleingrothaus *et al.*, Phys. Lett. B **586**, 198 (2004).

Long-Baseline Neutrino Oscillation Experiments in Europe

F. Terranova

Laboratori Nazionali di Frascati dell' INFN, Via E.Fermi 40, I-00044 Frascati (Rome), Italy.

Abstract. Long-baseline experiments will play a crucial role in the precision era of neutrino oscillation physics. In this paper we review the ongoing European programme, which is focused on CNGS, and we discuss the opportunities for new facilities based on the CERN acceleration complex.

Keywords: Neutrino oscillations, long baseline experiments
PACS: 14.60.Pq, 14.60.Lm

INTRODUCTION

Over the past seven years, an impressive sequence of experimental results have put on a solid ground the 40-year old hypothesis that neutrinos change flavor [1]. In particular, a clear evidence of disappearance of atmospheric neutrinos has been reported by the Super-Kamiokande experiment and is corroborated by Soudan2 and Macro data. Furthermore, Super-Kamiokande provided evidence in favor of a sinusoidal pattern of disappearance as a function of L/E. The most likely explanation for these data is an almost pure $\nu_\mu \to \nu_\tau$ transition, connected with the m_2 and m_3 mass eigenstates. This result is supported by K2K, a long-baseline experiment reporting a deficit of artificial ν_μ produced at KEK and sent to Super-Kamiokande through a path-length of ~ 250 km. Atmospheric neutrinos offer the unique possibility of measuring in a direct manner the appearance of new flavors, while this measurement is impossible with solar neutrinos. The Δm^2 driving solar oscillations is of the order of $\sim 8 \times 10^{-5}$ eV^2. Therefore, for any realistic baseline the energy of the neutrino is too small to overcome the kinematic threshold for μ or τ on-shell production. The CNGS [2] beam has been tuned to provide the first direct evidence for appearance of new flavors (ν_τ) from a pure ν_μ beam. The CNGS and the corresponding far detectors (ICARUS [3] and OPERA [4]) are described in the next section. Moreover, if the next round of accelerator experiments (MINOS, MiniBOONE, OPERA and ICARUS) confirms the present scenario, long-baseline experiments will offer the opportunity to access the presently unknown 1-3 sector of the leptonic mixing matrix. This fascinating issue and the role that future European facilities can play is also discussed. Emphasis is on the special role played by the mixing angle between the first and third ν generation (θ_{13})[1] and on the opportunities offered by novel neutrino sources for the study of $\nu_e \to \nu_\mu$.

THE CNGS PROJECT

Given the distance of 732 Km between the CERN accelerator complex and the existing underground laboratories at Gran Sasso, the CNGS ("CERN neutrinos to Gran Sasso") beam has been designed maximizing the number of ν_τ charged current interactions occurring at the detector location. Therefore, the beam energy is *not* at the peak of the oscillation probability ($E_\nu \simeq 1.5$ GeV, i.e. below the τ kinematic threshold) but at the optimal value of 17 GeV. Two experiments will exploit this beam. The first one (OPERA) is able to identify the τ decay topology on event-by-event basis, thus profiting of an outstanding background discrimination.

The OPERA experiment will search for ν_τ appearance by detecting the final state τ and identifying its decay kink topology. This method requires detectors with μm-scale granularity. The apparent conflict of the high resolution requirements with the large target mass needed to collect enough neutrino interactions, is solved in a modular way thanks to the so-called "brick". The brick is an implementation of the ECC (Emulsion Cloud Chamber), validated for τ search by the DONUT experiment. It represents the basic unit of OPERA and consists of a sandwich of 56 lead plates (1 mm thick) alternated with 56 nuclear emulsion sheets.

The lead constitutes the target mass for neutrino interactions while nuclear emulsion sheets, with angular resolution of about 2 mrad and a space resolution of 0.5 μm, allow for the reconstruction of the vertex and decay kink. Moreover, this basic ECC unit provides the momenta of

---

[1] There is currently no evidence that $\theta_{13} \neq 0$. Best limits come from reactor disappearance experiments [5] and they constrain $\theta_{13} \lesssim 11°$.

FIGURE 1. Layout of the OPERA detector.

FIGURE 2. Installation of the OPERA magnetic spectrometers (March 2005).

charged particles produced in the interactions from their multiple scattering in the lead plates; it allows the identification of electrons and photons and the measurement of their energy by sampling electromagnetic showers with emulsions sheets; moreover, it provides pion to muon separation from dE/dx in the proximity of the track end of range.

Being a completely passive device, the brick has to be integrated with electronic detectors. In order to provide a trigger and localize in which brick the neutrino interacted, scintillator plane hodoscopes have to be inserted in the target section in between the bricks. The latter are arranged in walls and 31 walls + scintillator planes constitute the actual ν target. A muon spectrometer installed at the end of the target is needed for identification and measurement of the charge and momentum of penetrating tracks. OPERA is made of two such targets, each followed by a spectrometer, as shown in Fig.1. The brick selected as the one in which the neutrino interacted is hence extracted from the target and its emulsion sheets are scanned with automatic microscopes. The installation of the two magnetic spectrometers has been completed in spring 2005 and the installation of the hodoscopes and the support structure for the wall is in progress (see Fig.2). About 50% of the emulsion are already stored underground at LNGS and brick production will start in 2006.

The number of expected signal and background events is presented in Tab.1 for full mixing and various values of Δm^2. 5 years of data-taking with the nominal beam (4.5×10^{19} protons on target per year) or with the beam upgrade ($\times 1.5$) are assumed.

Another detector, developed by the ICARUS Collaboration, will exploit the CNGS beam. It consists of a large vessel of liquid Argon filled with three planes of wires strung on different orientations. The device allows tracking, dE/dx measurements and a full-sampling

TABLE 1. A summary of expected events and background in five years of data taking for nominal and upgraded (in parenthesis) performance of the CNGS and various values of Δm^2 (units are eV^2).

1.9×10^{-3}	2.4×10^3	3.0×10^3	Bkgd.
6.6 (10)	10.5 (15.8)	16.4 (24.6)	0.7 (1.1)

electromagnetic and hadronic calorimetry. Furthermore, the imaging provides excellent electron and photon identification and electron/hadron separation. The energy resolution for electromagnetic showers is about $(3/\sqrt{E(\text{GeV})} \oplus 1)\%$ and for contained hadronic showers $\sim 30/\sqrt{E(GeV)}\%$. ICARUS can gain evidence for ν_τ appearance through a kinematic analysis of neutrino interactions and, for a detector of ~ 3 kton mass, it has a sensitivity comparable with OPERA. At present, a 600 ton module has been successfully operated at surface [6] and it has been installed underground in winter 2004.

The θ_{13} dilemma

Current experimental results indicate that the ratio $\Delta m_{21}^2/|\Delta m_{32}^2|$ of the neutrino squared mass differences driving the solar and atmospheric oscillations is of the order of $\mathcal{O}(10^{-2})$. This measurement has a enormous impact in the design of future experiments posed to thoroughly determine the leptonic mixing matrix (PMNS [7]). In particular, given a relatively large $\Delta m_{21}^2/|\Delta m_{32}^2|$ ratio, the determination of the currently unknown 1-3 sector of the PMNS, i.e. the mixing between the first and third generation and the CP violating Dirac phase, can be accomplished by long baseline experiments. It is done by measuring the contamination of $\nu_\mu \to \nu_e$ transitions in the bulk of $\nu_\mu \to \nu_\tau$ oscilla-

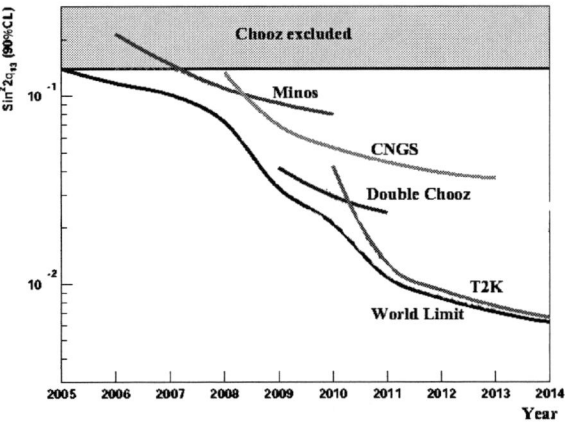

FIGURE 3. $\sin^2 2\theta_{13}$ sensitivity (90% CL) versus time after the startup of the next generation experiments. The darkest line represents the world overall sensitivity (courtesy of M.Mezzetto).

TABLE 2. Facilities needed to access CP violation in the leptonic sector for different values of θ_{13}. Abbreviations are SB=SuperBeams [11], BB=Beta Beams [12], NF=Neutrino Factories [13].

θ_{13}	Detector mass (kton)	ν source	Cost (M$)
$\sim 10°$	50	SB	<500
$\sim 3°$	1000	SB/BB	500-1000
$\sim 1°$	50	NF	>1000

tions at the atmospheric scale. Therefore, long baseline experiments will play a leading role in neutrino physics if $\nu_\mu \to \nu_e$ is experimentally accessible. The size of this sub-dominant contribution depends on the mixing angle between the first and third neutrino generation (θ_{13}) and an experimental determination of this angle is mandatory to establish to what extent future facilities are able to address CP violation in the leptonic sector or fix the neutrino mass hierarchy (sign of Δm_{32}^2) exploiting matter effects. The discovery of $\theta_{13} \neq 0$ has not only a scientific relevance but also a high practical value. The commissioning and running of an apparatus to observe CP violation in the leptonic sector at the atmospheric scale is a major technical and economical challenge; since most of its physics reach - in particular the measurement of leptonic CP violation and the determination of U_{e3} in the PMNS matrix - depends crucially on the size of θ_{13}, the latter should be determined by "Phase I" experiments tuned to maximize their θ_{13} sensitivity. Otherwise, the physics case of the new detectors should be drawn independently of their neutrino physics reach. In Tab.2 an estimate is done of the type and cost of the new facilities for different value of θ_{13}.

The CNGS [8] and other experiments running in the near future have some sensitivity to the θ_{13} angle but significant improvements will be done by dedicated long baselines experiments. Proposals have been put forward both in Japan [9] and in US [10]. In Fig.3 the evolution of sensitivities on $\sin^2 2\theta_{13}$ as a function of time is depicted. The world overall sensitivity along the time is also displayed (darkest line).

Should the size of θ_{13} be large enough to allow the observation of $\nu_\mu \to \nu_e$ oscillations at the atmospheric scale in the forecoming experiments ($\theta_{13} \gtrsim 3°$), new facilities would be needed to close up the PMNS. They should perform precision measurements of the 1-3 sector and particularly of the CP violating phase.

Long-term prospects in Europe

Europe has currently neither a powerful proton driver nor a massive detector for ν_e appearance. Accessing the T2K sensitivity in a timescale comparable with this japanese facility would be unlikely even with a very aggressive funding profile. In a longer timescale, an interesting way out is offered by the beam technologies that allow the detectors to operate in ν_μ appearance mode (i.e. search for $\nu_e \to \nu_\mu$ transitions and their CP conjugate) as the Beta Beams[12] or the Neutrino Factories[13]. In particular, the Beta Beam concept offers a strong synergy with nuclear physics (high intensity radioactive beams) and low background contamination in the $\nu_e \to \nu_\mu$ channel. The small energy of the parent ions requires, anyway, the construction of a 1M ton water Cherenkov to compensate for the smallness of the cross section at $E_\nu < 1$ GeV. The latter has been proposed as the main detector for a new underground laboratory at Frejus, which could start data taking ~ 2020 [14]. Physics performances, however, would be very similar to a SuperBeam since the improved background rejection is compensated by the lower neutrino energy (~ 300 MeV). An increase of the beta beam energy[15] could be envisaged by a fast cycling superconducting SPS ("Super-SPS") at CERN. This machine is currently considered as an option for the luminosity upgrade and, possibly, the energy upgrade of the LHC [16]. In this case, the detector mass could be significantly reduced and the baseline would match the CERN-LNGS distance. In this configuration, denser detectors operating in ν_μ appearance mode can be exploited [17] and therefore smaller experimental halls than Frejus (e.g. LNGS) are needed. It can be shown [18] that these setups have the sensitivity to address CP violation in the leptonic sector for any value of θ_{13} that allows evidence for $\nu_\mu \to \nu_e$ oscillations in the forecoming experiments (up to T2K). Finally, the Super-SPS would simplify ion injection into the LHC and provide, in a longer timescale, very high gamma Beta

Beams[19]. Hence, a possible synergy with the machine upgrade of the LHC is extremely appealing, particularly due to the central role of the Large Hadron Collider and its upgrades in European high energy physics.

ACKNOWLEDGMENTS

I wish to express my gratitude to A. Guglielmi, M. Mezzetto and P. Migliozzi for many useful insights on the future of long baseline experiments in Europe. A special thank to F. Ferroni and W. Scandale for several interesting discussions during and after this Workshop.

REFERENCES

1. For a review and reference to original literature see G. L. Fogli, E. Lisi, A. Marrone and A. Palazzo, arXiv:hep-ph/0506083, to appear in *Progr. Part. Nucl. Phys.*
2. G. Acquistapace et al., CERN 98-02, INFN/AE-98/05; CERN-SL/99-034(DI), INFN/AE-99/05 Addendum.
3. F. Arneodo et al. [ICARUS Coll.], ICARUS-TM/2001-08 LNGS-EXP 13/89 add.2/01.
4. M. Guler et al. [OPERA Coll.], CERN-SPSC-2000-028.
5. M. Apollonio et al. [CHOOZ Coll.], *Eur. Phys. J. C* **27**, 331–374 (2003).
6. S. Amerio et al., *Nucl. Instr. and Meth. A* **527**, 329–410 (2004).
7. B. Pontecorvo, *Sov. Phys. JETP* **6**, 429 (1957); Z. Maki, M. Nakagawa, S. Sakata, *Prog. Theor. Phys.* **28**, 870 (1962); B. Pontecorvo, *Sov. Phys. JETP* **26**, 984 (1968).
8. M. Komatsu, P. Migliozzi and F. Terranova, *J. Phys. G* **29**, 443–452 (2003).
9. Y. Itow et al., [T2K Collaboration], arXiv:hep-ex/0106019. Y. Hayato [T2K Collaboration], *Nucl. Phys. Proc. Suppl.* **143**, 269–276 (2005).
10. D. S. Ayres et al. [NOvA Collaboration], arXiv:hep-ex/0503053.
11. T. Kobayashi, *Nucl. Phys. Proc. Suppl.* **143**, 303–308 (2005).
12. P. Zucchelli, *Phys. Lett. B* **532**, 166–172 (2002).
13. S. Geer, *Phys. Rev. D* **57**, 6989–6997 (1998); [Erratum-ibid. D **59**, 039903 (1999)].
14. J. Bouchet, "Project of a Megaton-scale experiment at Frejus", talk at the *Workshop on the Next Generation of Nucleon Decay and Neutrino Detectors (NNN05)* 7-9 April 2005, Aussois, France.
15. J. Burguet-Castell, D. Casper, J. J. Gomez-Cadenas, P. Hernandez and F. Sanchez, *Nucl. Phys. B* **695**, 217 (2004).
16. R. Garoby and W. Scandale, *Nucl. Phys. Proc. Suppl.* **147**, 16–19 (2005); O. Bruning et al., CERN-LHC-PROJECT-REPORT-626, 2002; W. Scandale, "High Intensity Injection Chain for LHC", talk at the *High Intensity Frontier Workshop (HIF05)*, La Biodola, Italy, 2005.
17. F. Terranova, *Nucl. Phys. Proc. Suppl.* **149**, 185–187 (2005).
18. A. Donini et al., in preparation.
19. F. Terranova, A. Marotta, P. Migliozzi and M. Spinetti, *Eur. Phys. J. C* **38**, 69–77 (2004).

Solar Neutrinos: Present and Future

A. Ianni

I.N.F.N. Gran Sasso Laboratory, S.S. 17 bis km 18+910, 67010
Assergi (AQ), Italy

Abstract. In this paper, present and future experimental techniques to search for solar neutrinos are discussed. A number of reasons to measure sub-MeV solar neutrinos in the future are presented. This opportunity offers a unique possibility to search for a new physics case, besides neutrino oscillations and mixing, which may play a sub-dominant role in the solar neutrino phenomenology, in particular below 1 MeV. On one side non-standard neutrino interactions can be probed through sub-MeV solar neutrinos; on the other, the astrophysics of the sun can be studied at the level of a few %'s.

Keywords: neutrino interactions, solar neutrinos, neutrino detectors.
PACS: 13.15.+g, 14.60.Pq, 26.65.+t, 26.40.Ka, 26.40.Mc

INTRODUCTION

The Sun is a huge source of electron neutrinos, v_e. The flux of these neutrinos on earth is of the order of 60×10^9 cm^{-2} s^{-1}. The energy distribution of the solar neutrinos flux is shown in Fig. 1. The first attempt of measuring solar neutrinos started more than 35 years ago with the Chlorine experiment [2]. This early experiment (radiochemical) performed an integrated neutrino flux measurement with a threshold energy of 814 keV. The experimental method is based on a charge current interaction of neutrinos on ^{37}Cl target atoms. The ^{37}Ar atoms produced are extracted by He flushing. The radioactivity of the few Ar atoms collected in about a month of exposure gives a measure of the solar neutrino flux above the energy threshold. The first live counting for solar neutrinos was developed by the Kamiokande [3] detector, a water Cherenkov based on neutrino-electron elastic scattering. The radioactivity in U and Th of the water did not allow a measurement below about 7 MeV. Above this energy only ^{8}B can be measured as the hep neutrinos are much smaller in number (see Fig. 1). However, the directionality of the elastic scattering gave for the first time a clear indication of ^{8}B solar neutrinos. After this two pioneering experiments the radiochemical technique was used by GALLEX and SAGE [4] with a threshold of 223 keV (using ^{71}Ga as target atoms), and a new much larger water Cherenkov was built, the Super-Kamiokande (SK) [5]. In SK the radioactivity of U and Th in the water (~10^{-14} g/g) was about 10 times better than in Kamiokande. This allowed to lower the detection threshold at 5 MeV. These five experiments sampled the solar neutrino flux over a wide range in energy and showed an important deficit with respect to the predictions of the Solar Standard Model (SSM) [1]. This deficit is known as the Solar Neutrino Problem (SNP). A fundamental boost to the interpretation of the SNP came after the first results of the SNO experiment [6]. SNO is a water Cherenkov based on "heavy water". The deuterium allows to detect solar neutrinos via a charged current (CC) reaction

$$v_e + d \to e^- + 2p \qquad (1)$$

or a neutral current (NC) reaction

$$v_x + d \to v_x + p + n. \qquad (2)$$

The NC measurement reported by SNO shows that the total solar neutrino flux is in agreement with the SSM prediction. On the contrary the CC measurement shows a clear deficit. In particular, it turns out that $R_{CC/NC} = 0.305 \pm 0.033$. In a model independent way the SNO results show that the v_e suppression is due to a flavor-changing process. At present the solar neutrino measurements are explained by matter-enhanced neutrino oscillations, the so called MSW-LMA solution. In Fig. 2 a summary of solar neutrino measurements together with theoretical expectations is reported. An important ingredient to the interpretation of the solar neutrino phenomenology is the measurement performed by the KamLAND experiment [7]. KamLAND is a liquid scintillator detector which aims to measure electron anti-neutrinos from reactors with an effective baseline of about 180 km. Data collected by KamLAND are explained by using the hypothesis of neutrino oscillations in a vacuum. The δm^2 parameter which fits the experimental KamLAND data is the MSW-LMA predicted one ($\delta m_{12}^2 \sim 8 \times 10^{-5}$ eV^2, $\sin^2\theta_{12}=0.31$). Moreover, matter effects inside the sun ensure that $\delta m_{12}^2 > 0$ and that maximal mixing is ruled out at 6σ.

At present the SNO experiment is taking data and is planned to be stopped in 2006. SK is taking data with only about 46% of the PMTs used before 2001 incident (11200) and a coverage of 19% (40% in 2001). The SK collaboration has scheduled a new re-commissioning of the detector by the end of 2005. So, in 2006 SK will resume operations with the full size as in 2001.

In the following of the paper the main goals of solar neutrino research in the near and far future will be discussed together with the experimental methods planned to be used.

THE FUTURE OF SOLAR NEUTRINOS: WHAT WE NEED TO DO

As stated in the previous Section the favorite interpretation of the solar neutrino phenomenology is the MSW-LMA solution. As a matter of fact, the null hypothesis of no MSW transitions can be rejected at more than 5σ [9]. The oscillation parameters which drive the solar neutrino phenomenology (δm_{12}^2 and $\sin^2\theta_{12}$) are, at present, known with an accuracy of 12% and 22%, respectively. In Figure 3 it is shown the predicted electron neutrino survival probability on earth as obtained using the MSW-LMA solution. This probability shows an important feature: a transition from high energy (>1 MeV) to low energy. At high energy the value of the survival probability has been measured by SNO. In SNO the detection threshold is 5.5 MeV. Taking into account that the reaction in (1) has a Q value of 1.44 MeV, one can conclude that SNO measures solar neutrinos above about 7 MeV. So, in the energy region probed by SNO neutrino oscillations are dominated by matter effects. On the contrary, below 1 MeV neutrinos do not experience the MSW effect and the survival probability is given by averaged vacuum oscillations (for $E_\nu < 1$ MeV)

$$P_{ee}(E_\nu) = 1 - \frac{1}{2}\sin^2\theta_{12}. \quad (3)$$

The value of this probability below 1 MeV can be extracted by the GALLEX/GNO and SAGE measurements (0.58±0.04) [10]. A second important prediction of the MSW-LMA is the so-called day-night asymmetry: a lower suppression of ν_e's should be seen during the night. The day-night effect is expected at the level of 1-2% in SK and 2-3% in SNO.

So, a direct evidence of the MSW-LMA mechanism should be searched by a spectral distortion (up-turn of the survival probability) and a positive day-night asymmetry. At present, both effects have not seen yet. Therefore, a main goal of future experiments will be a direct test of the MSW-LMA solution by increasing the statistics, lowering the systematics and, in particular, going to sub-MeV neutrinos.

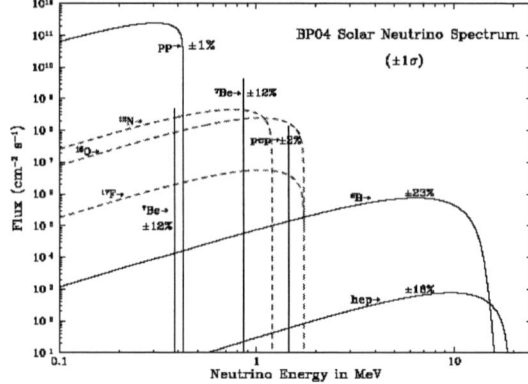

FIGURE 1. Energy distribution of the flux of solar neutrinos as predicted by the Solar Standard Model [1].

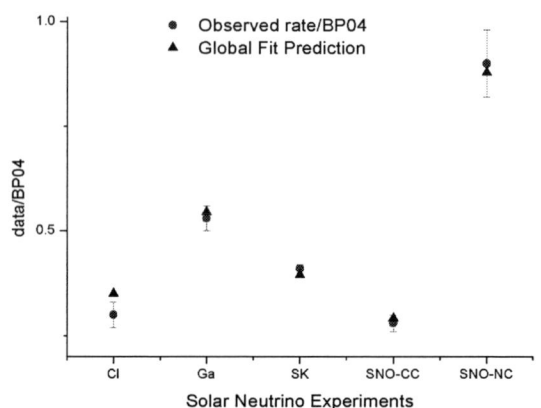

FIGURE 2. Experimental data (filled circles) on solar neutrinos together with theoretical expetations (filled triangles) according to the MSW-LMA [8].

The first attempt will be performed by SK starting next year. However, in order to measure sub-MeV neutrinos one has to wait for a new generation of experiments based on new experimental techniques. This matter will be discussed in the next Section.

Besides searching for MSW-LMA effects, in the future, solar neutrino observations aim to check a possible sub-leading mechanism, such as non-standard neutrino interactions driven by small flavor universality violations of the order of 0.1-0.2 G_F [11,12]. A non-zero neutrino magnetic moment of the order of 10^{-12} μ_B [13]. An extra light sterile neutrino [14]. These hypothesis can be probed with sub-MeV solar neutrinos. The reason being the fact that high energy solar neutrinos are almost pure ν_2, the neutrino mass eigenstate with mass m_2. Moreover, the possibility to search for differences between the photon luminosity and the neutrino one is also amplified once sub-MeV solar neutrinos have been measured with a precision at the level of 5% [15].

As far as neutrino mixing parameters, solar neutrinos can improve significantly the present results only with a pp neutrinos measurement is performed with a precision in the rate better than 4% [16].

So, in the near and far future a wide range of interesting physics cases can be probed by the next generation of solar neutrino experiments. In particular, sub-MeV neutrinos are of great interest. How this measurement can be performed is the matter of next Section.

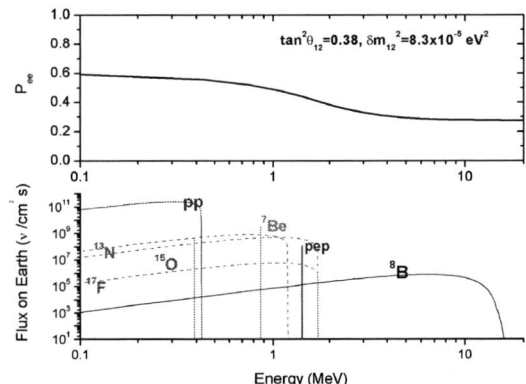

FIGURE 3. The predicted MSW-LMA electron-neutrino survival probability (upper plot) against the solar neutrino spectrum (lower plot).

TABLE 1. Future solar neutrino experiments. LS = liquid scintillator.

Experiment	Detection reaction	Target	Expected signal
Borexino	ES	300 tons LS	30 cpd/100tons only Be
KamLAND	ES	1 kton LS	180cpd/600tons only Be
CLEAN	ES	10 ton liquid Ne	2900 pp (1500 Be) 50 keV threshold
XMASS	ES	10 ton liquid Xe	2370 pp (1240 Be) 50 keV threshold
LENS	CC	20 tons LS In-loaded	2190 pp (511 Be)
MOON	CC	3.3tons Mo foils + plastic scintillators	240 pp (77 Be)

THE FUTURE OF SOLAR NEUTRINOS: THE EXPERIMENTAL METHODS

In this Section the current ideas on how to perform a measurement of sub-MeV solar neutrinos are discussed. In the sub-MeV region the natural radioactivity of U, Th and K becomes a dominant source of background. The level of U and Th in water it is too high to allow such a measurement. In SNO (water Cherenkov), as an example, the radiopurity of the water was pushed at 10^{-15} g/g in order to compensate the poor directionality of the CC interactions with respect to the elastic scattering. This, however, is not enough in spite of the much larger flux of ^{7}Be neutrinos with respect of the ^{8}B ones. For the first time the Borexino collaboration addressed the possibility to use an organic liquid scintillator, where the U and Th contamination was shown to be at the level of 10^{-16} g/g (~ 0.1 counts/day/ton) [17]. Moreover, Borexino has shown how a self-shielding design using a liquid scintillator with $\rho \sim 1$ g/cm^3 can reduce the external background at a negligible level with respect to the internal one, dominated by intrinsic radioactivity.

A detector based on an organic scintillator can measure ^{7}Be with a threshold at the level of 0.2 MeV. This because of the intrinsic ^{14}C activity. The ratio of ^{14}C/^{12}C ~ 10^{-18} as determined by the CTF [18] allows to measure ^{7}Be above 0.25 MeV but not pp neutrinos. In a water Cherenkov such as SK the number of photoelectrons/MeV is of the order of 6 and this gives an energy resolution around 13% at 10 MeV. In a liquid scintillator the number of photoelectrons/MeV can be as large as 300-400, and the energy resolution at the level of 5% at 1MeV. So, besides the radiopurity issue a liquid scintillator is more suited for sub-MeV energy than a water Cherenkov.

As far as pp neutrinos are concerned, a promising technique to overcome the ^{14}C problem is the possibility to use a liquefied noble gas as scintillator. At present, R&D projects based on liquid Xe [19] and liquid Ne [20] are underway. These experiments can perform a measurement of sub-MeV solar neutrinos via the elastic scattering interaction. A different approach is taken by LENS [21] and MOON [22]. In this case electron neutrinos are detected via a CC interaction

$$\nu_e + (A,Z) \rightarrow e^- + (A,Z+1)^*. \quad (4)$$

The CC transition is thought to have either an electromagnetic de-excitation with $\tau \sim 10^{-8} - 10^{-6}$ s or to be unstable and produce a fast delayed coincidence. In both cases the CC capture provides a delayed signal which is used to reduce the background. The energy threshold for the capture reaction must be such to allow a pp (<0.4 MeV) measurement. Unfortunately, only a few isotopes meet the requirements and some have intrinsic radioactivity problems: ^{115}In, ^{100}Mo, ^{82}Se, ^{160}Gd, ^{176}Yb. As an example, in LENS ^{115}In is loaded in the liquid scintillator. ^{115}In decays beta with an end-point energy of 0.496 MeV. For solar neutrino detection one uses the fact that with a threshold of 0.114 MeV neutrinos captured on ^{115}In produce a prompt signal (the electron) and a delayed one (de-excitation gamma's). However, the ^{115}In beta decay can mimic the neutrino signal. In fact, the emitted electron can be thought of as to come from a capture interaction (prompt) and Bremsstrahlung photons can be thought of as the delayed signal.

A second important source of background for these experiments (besides U, Th and K) comes from radioactive noble gases, such as ^{85}Kr and ^{39}Ar. In the liquid neon scintillator this will be the most important background.

In the near future (2-3 years) Borexino, under commissioning at the Gran Sasso laboratory, and KamLAND, after the upgrade to reduce the intrinsic radioactivity, seem the only possibility to measure in real time low energy solar neutrinos (^{7}Be). In the long term (5-10 years) there could be the possibility to measure pp neutrinos by one of the mentioned methods under study. In Tab. 1 a summary of the main characteristics of the future projects is reported. It is important to notice that a CC measurement together with an ES one is necessary to quantify the total flux of all active flavors. An NC measurement such as the one performed by SNO is not yet proposed for low energy solar neutrinos.

As far as high energy solar neutrinos are concerned, in the future there are a few possibilities to improve the present measurements. One is, as mentioned, SK after 2006. A second one is SK loaded with Gd proposed in particular to measure relic neutrinos from supernovae [23]. A third one is a megaton scale water Cherenkov to be built mainly to study a possible proton decay, neutrino mixing parameters from neutrino beams, supernovae. However, as pointed out in the previous Section, only low energy solar neutrinos offer a great and unique opportunity to search for new physics cases besides testing deeply the MSW-LMA scenario.

ACKNOWLEDGMENTS

I would like to thank the organizers of the neutrinos and cosmic rays parallel session, D. Montanino and G. Riccobene, for the invitation to the conference.

REFERENCES

1. J.N. Bahcall, et al., *Astrophys. J.*, 555, (2001), 990. J.N. Bahcall, hep-ph/0412068.
2. B.T. Cleveland, et al., *Astrophys. J.*, 496, (2005), 505.
3. Y. Fukuda, et al., *Phys. Rev. Lett.* 77, (1996), 1683.
4. GALLEX collaboration, W. Hampel, et al., *Phys. Lett.* **B447**, (1999), 127. SAGE collaboration, J.N. Abdurashitov, et al., Zh. Eksp. Teor. Fiz. **122**, (2002), 211 [*J. Exp. Theor. Phys.* **95**, (2002), 181].
5. Y. Fukuda, et al., *Phys. Rev. Lett.*, **81**, (1998), 1158.
6. Q.R. Ahmad, et al., *Phys. Rev. Lett.* **87**, (2001) 071301. Q.R. Ahmad, et al., *Phys. Rev. Lett.* **89**, (2002) 011301.
7. T. Araki, et al., hep-ex/0406035.
8. A. Bandyopadhyay, et al., *Phys. Lett.*, **B608**, (2005), 115-129.
9. G.L. Fogli, et al., *Phys.Lett.* **B583** (2004) 149-156.
10. A. Strumia and F. Vissani, hep-ph/0503246.
11. M.M. Guzzo, et al., hep-ph/0403134.
12. O.G. Miranda, et al., hep-ph/0406280.
13. O.G. Miranda, et al., *Phys.Rev.Lett.* 93 (2004) 051304.
14. A. Strumia and F. Vissani, *Nucl. Phys. Proc. Suppl.* **143** (2005) 144.
15. J.N. Bahcall and C. Pena-Garay, JHEP 0311 (2003) 004.
16. S. Goswami, Lepton-Photon 2005, June 30-July 5, Uppsala, Sweden.
17. Borexino collaboration, G. Alimonti, et al., *Astrop. Phys.*, **16** (2002) 205-234.
18. Borexino collaboration, G. Alimonti, et al., *Phys. Lett.* **B422** (1998) 349.
19. Y. Suzuki, LowNu workshop, Tokyo, 2000.
20. D.N. McKinsey and J.M. Doyle, *J. of Low Temp. Phys.* **118** (2000) 153.
21. R.S. Raghavan, *Phys. Rev. Lett.*, 78, 3618–3621 (1997).
22. H. Ejiri, et al., *Phys. Rev. Lett.* **85** (2000) 2917.
23. J.F. Beacom and M.R. Vagins, *Phys.Rev.Lett.* 93 (2004) 171101

Neutrinos from Supernovas and Supernova Remnants

M.L. Costantini* and F. Vissani[†]

*Università dell'Aquila and INFN, L'Aquila, Italia
[†]Laboratori Nazionali del Gran Sasso, INFN, Assergi (AQ) Italia

Abstract. Supernovae (SN) and supernova remnants (SNR) have key roles in galaxies, but their physical descriptions is still incomplete. Thus, it is of interest to study neutrino radiation to understand SN and SNR better. We will discuss: (1) The ~10 MeV thermal neutrinos that arise from core collapse SN, that were observed for SN1987A, and can be seen with several existing or planned experiments. (2) The 10-100 TeV neutrinos expected from galactic SNRs (in particular from RX J1713.7-3946) targets of future underwater neutrino telescopes.

Keywords: supernovas, supernova-remnants, neutrino and γ radiation, neutrino oscillations
PACS: 97.60.Bw; 98.58.Mj; 98.70.Sa; 13.15.+g; 14.60.Pq

1. INTRODUCTION

In the economy of a galaxy, the stars with mass $\geq 6-10\,M_\odot$ are known to play a prominent role. The last instants of their brief life (on ten million year scale) is a crucial moment. In fact, the gravitational collapse of the 'iron' core leads to supernovae of type II, IB, IC and eventually to pulsars, neutron stars and stellar black holes. The gas subsequently expelled by all supernovas including IA (that lead to the observed supernova remnants) is suspected to be the place where cosmic rays are accelerated. A complete theoretical understanding of these systems (and in particular of the supernova) is not yet available. However, it is known that neutrinos of ~10 MeV are emitted during and just after the gravitational collapse, and that neutrinos above 1 TeV can be produced in the supernova remnant at least in special circumstances. Therefore, neutrinos can be used as diagnostic tools of what happens in these interesting and complex systems. With these considerations in mind, we will discuss here briefly neutrinos from supernova and from supernova remnants, aiming to outline the main 'what', 'when', 'where', and 'how' questions.

Outline and references. In the rest of this section, we discuss a number of general facts pertaining to supernovas and supernova remnants; the references that we use are [1, 2]. In the second section we concentrate on supernova neutrinos; we use [1, 2], and also [3]. In the third section we describe the calculations of [4]. In these 4 papers, one can to find more details and a complete list of references. In the appendix, we show in details a simple way to calculate neutrinos from cosmic rays, adopted also in [4].

Generalities. Before passing to neutrinos, we would like to discuss a few general questions.

Guessing 'where' Let us recall that the only SN within the capability of existing neutrino detectors are those from our Galaxy (or some irregular galaxy around us). The best guess we can propose for next galactic core collapse SN is

$$\langle L \rangle = 10 \pm 4.5 \text{ kpc} \quad (1)$$

and it is motivated as follows:
1) We are $R = 8.5$ kpc from the galactic center.
2) The distribution of the matter that can go supernova is: $\rho \sim r e^{-r/r_0}$ with r=distance from the center and $r_0 = 3$ kpc, possibly summing a $\delta(r)$ to describe the 'bar'
3) We calculate the distribution in function of L=distance from us, integrate over the galactic azimuth θ and get the result above.

Guessing 'when' The rate of core collapse SN in the Milky Way is expected to be

$$R_{SN} = 1/(30\text{-}70 \text{ years}) \quad (2)$$

The most reliable method is: count SN in other galaxies, and correlate with galactic type.[1] A similar rate expected for SN Ia. Possibly, we missed several galactic SN due to dust.[2] For the future, we will have better coverage of galactic SNs with ν's, IR and perhaps with gravitational waves. From the absence of neutrino bursts, one derives

$$R_{SN} > 1/(21 \text{ years}) \text{ at } 1\sigma \quad (3)$$

We assume Poisson statistics (namely, $\exp(-TR_{SN}) = 1 - C.L.$) with $T = 24$ years. To obtain the last number,

---

[1] The Padova-Asiago database includes several thousand SN. The Milky Way could be Sb or Sb/c, which means a factor 2 uncertainty.
[2] Indeed, with about 10 supernovae seen in last 2000 years, in order to have a SN each 25 years we need to admit we see just one SN each 8, that is about the right figure on accounting the presence of the dust and the possibility that a supernova explodes on the day sky.

we note that till 1986 only Baksan worked with 90 % DAQ livetime, then we assume 100 % coverage.

We conclude by listing the number of SN precursors and descendants. We estimated these populations by assuming $R_{SN}^{tot} = 1/(25 \text{ years})$.

Object	Lifetime	Number
Pre-SN with ν	20 million y	400.000
Pulsars	2 million y	40.000
SNR	100.000 y	2×2000
young SNR	2000 y	2×40

By 'Pre-SN with ν' we mean core collapse SN. Recall that core collapse SN produce neutron stars (NS) or stellar black holes, and that pulsars are 'active' NS. Type Ia SN make white dwarfs, instead, but are also supposed to produce SNR (which explains the factor 2 above).

2. νS FROM CORE COLLAPSE SNS

We begin recalling the astrophysics of core collapse. The giant stars–**pre-SN**–burn in sequence H, He, C and Si, Ne, Mg, Na *etc*, and form an "onion structure", with a inert 'iron' core. Violent stellar winds occasionally modify external **mantle** in latest stages; apparently, this happened for SN1987A that was a $\sim 20\,M_\odot$ *blue* giant. What happens in the **core**? The gravitational pressure is balanced by e^- degeneracy pressure, and the core grows. As demonstrated by Chandrasekhar, the pressure of free electrons is unable to supply an equilibrium configuration when e^- become relativistic. Now, the iron core mass is $\sim 1.4\,M_\odot$ and the **collapse** begins. The sequence of the events becomes uncertain. More on the reference picture, the so-called "delayed scenario". What about the energetic of the collapse? The total energy of the collapse is very large, and 99 % of this energy is carried away by neutrinos. With $M_{ns}/M_\odot = 1-2$, $R_{ns} = 15\,\text{km}(M_\odot/M_{ns})^{1/3}$, we estimate

$$\mathcal{E} \simeq \frac{3G_N M_{ns}^2}{7R_{ns}} = (1-5) \times 10^{53}\,\text{erg}$$

The delayed scenario. This is a pictorial summary of the 'delayed scenario' by Wilson & Bethe:

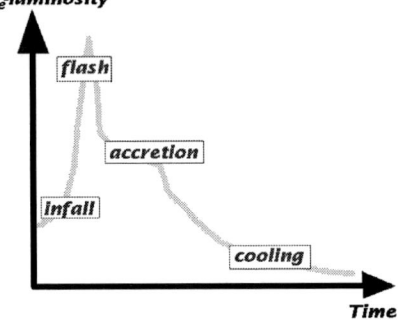

The energy radiated in any neutrino species $e, \bar{e}, \mu, \bar{\mu}, \tau, \bar{\tau}$ is expected to be the same within a factor of two (Janka et al.):

$$\mathcal{E}_e \sim \mathcal{E}_{\bar{e}} \sim \mathcal{E}_x$$

where x denotes any among $\mu, \bar{\mu}, \tau, \bar{\tau}$. Indeed, in this picture non-electronic neutrinos and antineutrinos are produced in a similar amount

Note that neutrinos are mostly emitted in cooling (80-90 %) and accretion (10-20 %) phases. The latter one is the phase crucial to understand the nature of the explosion, whereas the first phase is almost thermal. In this sense, the ignorance of the detailed mechanism of the explosion do not preclude a description the bulk of neutrino radiation (to put it with a slogan, ignorance is self-consistent).

Now we come to a prescription for the fluence:

$$F_i(E) = \frac{\mathcal{E}_i}{4\pi L^2}\frac{N}{\langle E_i \rangle^2} z^\alpha e^{-(\alpha+1)z}, \quad z = \frac{E}{\langle E_i \rangle}$$

where $\langle E_i \rangle$ is the average energy of the neutrino species $i = e, \bar{e}, x$; N ensures that the total energy carried is $\mathcal{E}_i$. (If one needs to describe time dependent situations, $\mathcal{E}_i \to L_i(t) \equiv d\mathcal{E}_i/dt$, $\langle E_i \rangle \to \langle E_i(t) \rangle$, $\alpha \to \alpha(t)$.) The expectations for time integrated quantities are:

$$\langle E_{\bar{e}} \rangle = 12 - 18\,\text{MeV}, \quad \langle E_x \rangle / \langle E_{\bar{e}} \rangle = 1 - 1.2$$
$$\mathcal{E}_{\bar{e}} = (2 - 10) \times 10^{52}\,\text{erg} \quad \mathcal{E}_x / \mathcal{E}_{\bar{e}} = 1/2 - 2$$

One guesses $\mathcal{E}_e = \mathcal{E}_{\bar{e}}$ (not so important); the ν_e average energy can be estimated from the emitted lepton number.

Oscillations of supernova neutrinos. Oscillations of SN neutrinos are pretty simple to describe. To account for oscillations we need to assign just 2 functions, P_{ee} and $P_{\bar{e}\bar{e}}$:

- $F_e = F_e^0 P_{ee} + F_\mu^0 P_{\mu e} + F_\tau^0 P_{\tau e}$
  $= F_e^0 P_{ee} + F_x^0 (1 - P_{ee})$
- $F_e + F_\mu + F_\tau = F_e^0 + F_\mu^0 + F_\tau^0$

and similarly for antineutrinos (we will consider only oscillations of massive νs). The relevant densities to calculate P_{ee} and $P_{\bar{e}\bar{e}}$ are $\rho_{sol} \sim 10$ gr/cc (He) and $\rho_{atm} \sim 10^3$ gr/cc (C+O).

There is a great interest on the unknown parameter U_{e3}, and supernovas offer a chance to learn more on that. However, we should keep in mind the uncertainties mentioned above. Let us try to discuss this point further. We begin with the formula $F_e = F_x^0 - P_{ee}(F_x^0 - F_e^0)$. If we have normal mass hierarchy:

$$P_{ee} = \begin{cases} \sin^2\theta_{13} \sim 0 & \theta_{13}\ \text{'large'}, > 1° \\ \sin^2\theta_{12} \sim 0.3 & \theta_{13}\ \text{'small'}, < 0.1° \end{cases}$$

Now we can ask a precise question on U_{e3}:

Can we distinguish the two cases?

TABLE 1. Depending on the phase of neutrino emission, it is possible to learn more on U_{e3} studying of SN neutrinos. In the table, goods and bads of each specific phase are mentioned. (LSD refers to the observations of the Mont Blanc detector, that according to Imshennik and Ryazhskaya could be interpreted assuming an intense neutronization phase before the main ν burst.)

Emission	Good	Bad	Remarks
cooling	strong ν radiation	uncertainties, small effect	$F_x^0 \sim F_e^0$
accretion	strong ν radiation	uncertainties!	$F_x^0 \sim F_e^0/2$
neutronization	clean signal	weak ν radiation	$F_x^0 \sim 0$
neutroniz.++ with rotation	clean and strong sign.	uncertain!!	$F_x^0 \sim 0$ (LSD?)

Table 1 provides a check-list. Note that if the mantle is stripped off till densities $\rho > 10$ gr/cc (e.g., with SN Ic) we have vacuum oscillations, $0.3 \to 0.6$ (Selvi). This is rare, but not impossible.

SN1987A. The first detection of SN neutrino is of epochal importance. These observations fit into the 'standard' picture for neutrino emission (see next figure), but there are some puzzling aspects:
1. IMB and Kamiokande-II find forward peaked distributions; e.g., $\langle \cos\theta \rangle$ are $\sim 2\,\sigma$ above expectations.
2. $\langle E_{vis}^{KII} \rangle \sim 15$ MeV and $\langle E_{vis}^{IMB} \rangle \sim 30$ MeV (± 2.5 MeV) are quite different even correcting for efficiencies.
3. The time sequence of events looks different; when combined not so bad (but abs. time is unknown). 3. The 5 LSD events, occurred 4.5 hours before the main signal, cannot be accounted for.
In figure 1, will stress the consistency of the 'standard' interpretation, but the space for non-standard ones is not small (not only due to limited statistics). A reasonable agreement with expectations is obtained if:

$$\langle E_{\bar{e}} \rangle \equiv E_0 = 12 - 16 \text{ MeV}$$
$$\mathscr{E} = (2-3) \times 10^{53} \text{ erg}$$
zero or a few $\nu_e\, e \to \nu_e\, e$ events in KII

3. νS FROM SNR

The leftover gas is the SNR. The kinetic energy is a few times 10^{51} erg, which means that the gas is in free expansion with velocities of about ~ 4000 km/sec. There are various phases of the SNR, according to their age, and various shapes (shell, plerionic, or mixed).

An argument by Ginzburg and Syrovatskii suggests SNR as main source of galactic cosmic rays (CR):
1) The Milky Way irradiates CR. Take $V_{CR} = \pi R^2 H$ with $R \sim 15$ kpc, $H \sim 5$ kpc as the volume of confinement. Take $\tau_{CR} = 5 \times 10^7$ years as CR lifetime in the Galaxy. We get the CR luminosity:

$$\mathscr{L}_{CR} = \frac{V_{CR} \cdot \rho_{CR}}{\tau_{CR}} = 0.9 \times 10^{41} \text{erg/s}$$

2) We have a new SN each $\tau_{SN} \sim 25$ year, with about $\mathscr{E} \sim 10^{51}$ erg in kinetic energy, that is

$$\mathscr{L}_{SN} = \frac{\mathscr{E}}{\tau_{SN}} = 1.2 \times 10^{42} \text{erg/s}$$

Comparing the two formulae, we see that if a SNR is able to convert a fraction $f_{CR} \sim 5 - 10$ % of the injected energy into CR, we are home (the numbers quoted above shouldn't be taken too seriously, but this 40-years-old arguments maintains its appeal).

We proceed describing the system. In 2000 years, the SNR proceeds by ~ 10 pc. The density is ~ 0.2 protons/cm^3, too faint to permit a significant production of secondaries. In the galaxy, the most dense non-collapsed objects are the molecular clouds, whose density can reach 10^4 protons/cm^3. The 2 objects form:

A COSMIC BEAM DUMP

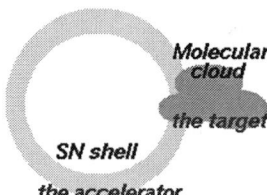

That is an ideal configuration, since

$$\text{CR collisions} \to \begin{cases} \pi^0 \to & \text{high energy } \gamma \\ \pi^\pm \to & \text{high energy } \nu_\mu, \nu_e \end{cases}$$

Thus, we expect to have γ and neutrinos.

There is some evidence that the system called RX J1713.7-3946 is a 'cosmic beam dump'. Indeed,
1) It is seen in X-rays, with many details
2) It is in Chinese Annales, 393 A.D.
3) A molecular cloud seen in CO and 21 cm H line
4) And most interestingly, CANGAROO (since 2000, see [5]) and H.E.S.S. (since 2004, see [6]) do see TeV γ rays. The distance is $L = 1$ kpc, the angular size about $1°$, the density of the cloud ~ 100 part/cm^3. The source is transparent to gamma rays, so that the neutrino flux can be calculated easily and reliably as shown in the appendix.

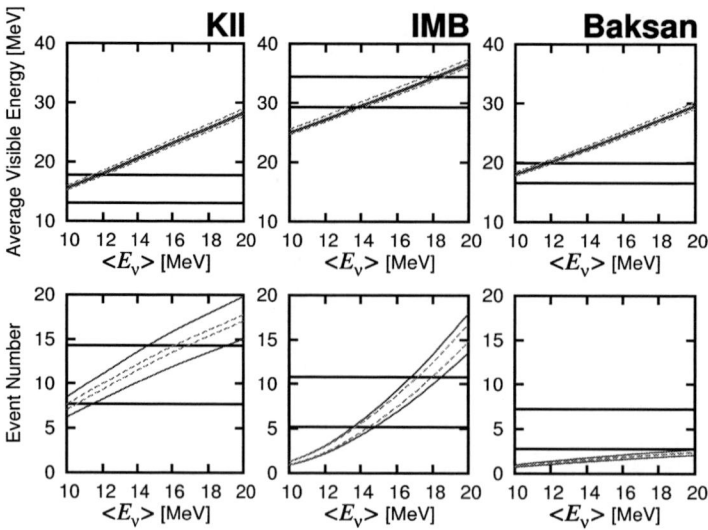

FIGURE 1. Average energies and number of events for three detectors. The horizontal lines describe the measured values, the oblique lines describe the theoretical expectations.

(However, no item above seems to be waterproof, what we discuss is just an appealing interpretation.)

We assume that the cosmic ray spectrum:

$$F_p = KE^{-\Gamma}, \quad \Gamma = 2 - 2.4$$

interacts with the molecular cloud and is the main source of visible gamma rays, through the reaction $p \to \pi^0 \to \gamma$. Neutrinos originate from similar processes, e.g., the reaction $p \to \pi^+ \to \mu^+ \to \nu_e$ will yield ν_e. Using the flux $F_\Gamma = 1.7 \cdot 10^{-11} (E/\text{TeV})^{-2.2}$ TeV^{-1}cm^{-2}s^{-1} measured by H.E.S.S. in the range between 1-40 TeV we get:

$$F^0_{\nu_\mu} = 7.3 \times 10^{-12} \left(\frac{E}{\text{TeV}}\right)^{-2.2} \frac{1}{\text{TeVcm}^2\text{s}}$$
$$F^0_{\bar\nu_\mu} = 7.4 \times 10^{-12} \left(\frac{E}{\text{TeV}}\right)^{-2.2} \frac{1}{\text{TeVcm}^2\text{s}}$$
$$F^0_{\nu_e} = 4.7 \times 10^{-12} \left(\frac{E}{\text{TeV}}\right)^{-2.2} \frac{1}{\text{TeVcm}^2\text{s}}$$
$$F^0_{\bar\nu_e} = 3.0 \times 10^{-12} \left(\frac{E}{\text{TeV}}\right)^{-2.2} \frac{1}{\text{TeVcm}^2\text{s}}$$

3 flavor oscillations take the simplest form (vacuum averaged or Gribov-Pontecorvo) and can be included easily. This leads to an important modification of the fluxes.

Finally we pass to the signals of neutrinos. ν interactions are due to deep elastic scattering. The simplest and most traditional observable is induced muons, that can be correlated to the source by mean of an angular cut. It is important to recall that high energy ν_μ are to some extent absorbed from the Earth, and that when the source is above the horizon, it is impossible to see anything due to the background from atmospheric μ. Along with oscillations, these effect decrease the observable neutrino signal. For an ideal detector, located in the Mediterranean, with Area=1 km^2, Data taking=1 year, $E_{thr.}$ = 50 GeV we find that the number of expected events is about 10 (this was 30 if oscillations, absorption and μ-background were ignored, or even 40 if the slope was $\Gamma = 2.0$).

4. DISCUSSION AND PERSPECTIVES

We do not have a clear understanding of how SN explode. Perhaps, because it is a very difficult problem, or there is some missing ingredient, perhaps the answer is not unique (a combination of various mechanisms?), ... the confusion could persist even after next galactic SN.

νs from next galactic SN have an impressive potential to orient our understanding. The hypothesis of an "accretion phase", implying 10-20 % of the emitted energy, can certainly be tested. SN1987A does not contradict the 'delayed scenario' seriously (but does not help much either). There are chances to learn on oscillations and in general on particle physics. The possibility to use ν_e and neutral current events deserves consideration.

νs from SNR are an uncharted territory. Recent results from H.E.S.S. motivated us to consider one specific SNR (but we expect new results and surprises with γ rays). Sure enough, CR acceleration in SNR is not fully understood, and there are other possible sources for TeV ν astronomy, however the number of events we found (about 10/km^2 y) suggests the need of large exposures. It is important to improve our theoretical tools to describe SNR γ and ν.

APPENDIX: Cosmic rays and neutrinos: We report here on the details of the calculation of (anti) neutrino spectra from the decays of secondary particles produced in pp interaction in astrophysical beam-dumps (table 1 of [4]). As a first approximation, we add the contributions to v-emission from two chains: (1) $p \to m^{\pm} \to (anti)v$ and (2) $p \to m^{\pm} \to \mu^{\pm} \to (anti)v$, where $m^{\pm} = \{\pi^{\pm}, K^{\pm}\}$, $v = \{v_e, \bar{v}_e, v_{\mu}, \bar{v}_{\mu}\}$.

If we assume scaling, and take a power-law proton spectrum $F_p(E_p) = cost. E^{-\Gamma}$, we can express the (anti) neutrino spectra from (1) and (2) as:

$$F_v^{(1),(2)}(E_v) = \frac{\Delta X}{\lambda_p} \cdot \psi_v^{(1),(2)} \cdot F_p(E_v) \quad (4)$$

where ψ_v indicate the neutrino emissivity coefficient, defined in [7], ΔX is the column density traversed by the protons and λ_p is the interaction length of CR. The total spectra are:

$$F_v = F_v^{(1)} + F_v^{(2)} = \frac{\Delta X}{\lambda_p} \cdot \psi_v \cdot F_p(E_v) \quad (5)$$

where $\psi_v = \psi_v^{(1)} + \psi_v^{(2)}$. Similarly, the γ-spectrum from $p \to \pi^0 \to \gamma$ is:

$$F_{\gamma}(E_{\gamma}) = \frac{\Delta X}{\lambda_p} \cdot \psi_{\gamma} \cdot F_p(E_{\gamma}), \text{ where } \psi_{\gamma} = \frac{2Z_{p\pi_0}}{\Gamma} \quad (6)$$

In equations (4) and (6) we assume a thin target. We note that $F_v/F_{\gamma} = (\psi_v/\psi_{\gamma})F_p$, and so defining $k_v = \psi_v/\psi_{\gamma}$:

$$F_v = k_v F_{\gamma}(E_v) \quad (7)$$

When we assume scaling, following Gaisser (see [8], formulas 4.1 and 4.2 in approximation 3.31) we have:

$$F_{v_{\mu}}^{(1)} = \Delta X \sum_{m^+ = \pi^+, K^+} \int_{\frac{E_{v_{\mu}}}{1-r_m}}^{\infty} \frac{dn_{v_{\mu}m^+}}{dE_{v_{\mu}}} D_{m^+}(E_{m^+}) dE_{m^+} \quad (8)$$

with:

$$\frac{dn_{v_{\mu}m^+}}{dE_{v_{\mu}}} = \frac{B_{m^+}}{1-r_m} \frac{1}{E_{m^+}}, \quad D_{m^+} = \frac{Z_{pm^+}}{\lambda_p} F_p(E_{m^+}) \quad (9)$$

where B_{m^+} is the branching ratio for $m^+ \to \mu^+ v_{\mu}$, $r_m = m_{\mu}^2/m_m^2$ (subscript $m = \{\pi, K\}$), and Z_{pm^+} are the spectrum-weighted momenta, defined in [8]. We recall that in the scaling hypothesis the spectrum-weighted momenta depend only from the spectral index Γ. We can get the $\bar{v}_{\mu}$ spectrum from the same formula, replacing m^+ with m^-. Evaluating the integrals, we get:

$$F_{v_{\mu}}^{(1)} = \frac{\Delta X}{\lambda_p} F_p(E_{v_{\mu}}) \sum_{m^+=\pi^+,K^+} B_{m^+} \frac{Z_{pm^+}}{\Gamma} (1-r_m)^{\Gamma-1} \quad (10)$$

and so:

$$\psi_{v_{\mu}}^{(1)} = \left[B_{\pi^+} \frac{Z_{p\pi^+}}{\Gamma} (1-r_{\pi^+})^{\Gamma-1} + B_{K^+} \frac{Z_{pK^+}}{\Gamma} \cdot (1-r_{K^+})^{\Gamma-1} \right] \quad (11)$$

Similarly, for $\psi_{v_{\mu}}^{(2)}$ we have (see [8], formula (7.14), in which we replace ϕ_{π} with the factor $[\Delta X Z_{pm^-} B_{m^-} F_p(E_{v_{\mu}})]/\lambda_p$, and then divide by $\Delta X F_p/\lambda_p$):

$$\psi_{v_{\mu}}^{(2)} = \sum_{m^-=\pi^-,K^-} Z_{pm^-} B_{m^-} \frac{1-r_m^{\Gamma}}{\Gamma(1-r_m)} \left\{ \langle y_{0v_{\mu}}^{\Gamma-1} \rangle + \right.$$
$$\left. + \left[1 + r_m - \frac{2\Gamma r_m}{1-r_m^{\Gamma}} \frac{1-r_m^{\Gamma-1}}{\Gamma-1} \right] \frac{\langle y_{1v_{\mu}}^{\Gamma-1} \rangle}{1-r_m} \right\} \quad (12)$$

where the moments:

$$\langle y_{iv}^{\Gamma-1} \rangle = \int_0^1 y^{\Gamma-1} g_{iv}(y) dy \quad (13)$$

for (anti)v_e and (anti)v_{μ} are given in table 7.3 from [8] (the $g_{iv}(y)$'s are given in table 7.2 from [8]). We can obtain $\psi_{\bar{v}_{\mu}}^{(2)}$ from eq. 12, replacing Z_{pm^-} with Z_{pm^+}. In our approximation, the (anti)v_e emissivity coefficients are given by eq. 12, inserting respectively Z_{pm^+} for ψ_{v_e}, and Z_{pm^-} for $\psi_{\bar{v}_e}$ and the appropriated moments. The numerical values of k_v for several values of Γ, obtained using the spectrum-weighted momenta from figure 5.5 by [8], are reported in table 1 by [4].[3]

REFERENCES

1. F. Cavanna, M. L. Costantini, O. Palamara and F. Vissani, Surveys High Energ. Phys. 19 (2004) 35
2. M. L. Costantini, A. Ianni and F. Vissani, Nucl. Phys. Proc. Suppl. 139 (2005) 27.
3. M. L. Costantini, A. Ianni and F. Vissani, Phys. Rev. D 70 (2004) 043006
4. M.L. Costantini and F. Vissani, Astrop. Phys. 23 (2005) 5
5. CANGAROO Collaboration, H. Muraishi et al., Astron. Astroph. 354 (2000) L57
6. H.E.S.S. Collaboration, F.A. Aharonian et al., Nature 432 (2004) 75
7. V.S. Berezinsky and V.V. Volinsky, in Proc. 16th Int. Cosmic Ray Conf. (Kyoto, 1979), vol. 10, page 326
8. T.K. Gaisser, *Cosmic rays and particle physics*, (1990) Cambridge Univ. Pr.
9. V.S. Berezinsky and V.V. Volinsky, in Proc. 16th Int. Cosmic Ray Conf. (Kyoto, 1979), vol. 10, page 332

---

[3] In our calculation, we omitted the contribution of v's produced in other decay modes of π and K, as well as in K^0's decay chains, and so we introduce and error on ψ_v at several % level. In the calculation made in [7], those contributions are taken into account. Moreover, the values of Z_{pm} used in [7] (reported in appendix of [9]) differ from our values at several % level. We note that in the calculation made in [7], only the first addend in the parenthesis, $\langle y_0^{\Gamma-1} \rangle$, is taken into account, that introduces an error at the few % level.

Seesaw and Leptogenesis

Pasquale Di Bari

Max-Planck-Institut für Physik (Werner-Heisenberg-Institut), Föhringer Ring 6, 80805 München

Abstract. Leptogenesis is a cosmological consequence of the seesaw mechanism and provides a link between the observed baryon asymmetry and neutrino masses. We show which information can be inferred from leptogenesis on neutrino masses and on those high energy seesaw parameters that represent a sort of 'dark side' for conventional experiments. We also report on a new scenario of leptogenesis that opens new opportunities for leptogenesis to be tested.

Keywords: Neutrino physics, early universe, baryogenesis, seesaw mechanism
PACS: 14.60.Pq,14.60.St,98.80.Cq,13.35.Hb

INTRODUCTION

Current cosmological observations support the idea that the observed matter-anti matter asymmetry of the Universe is the relic trace of a dynamical generation process [1], so called baryogenesis, occurred during the very early Universe history. The asymmetry is measured in the form of a *baryon to photon number ratio* at the time of recombination. A combination of WMAP and SLOAN data gives [2]

$$\eta_B^{CMB} = (6.3 \pm 0.3) \times 10^{-10}. \quad (1)$$

Even though all three Sakharov's conditions for successful baryogenesis are satisfied in the Standard Model, the final predicted asymmetry would fall by far below the observed value. Therefore, an explanation of the observed asymmetry requires 'new physics'. The discovery of neutrino masses is a strong indication of physics beyond the Standard Model. It is thus remarkable that the most elegant way to understand neutrino masses, the seesaw mechanism, can also provide a simple explanation of the matter-anti matter asymmetry of the Universe known as Leptogenesis [3]. Conversely Leptogenesis can be also regarded as a powerful cosmological tool to test the seesaw mechanism, together with the other more conventional phenomenologies, such as neutrino mixing and, hopefully to be observed in the next years, $\beta\beta 0\nu$ decay and CP violation in neutrino mixing. Here we will concentrate our attention on leptogenesis, showing the information that can be derived on the seesaw parameters.

FROM SEESAW TO LEPTOGENESIS

Adding to the Standard Model Lagrangian three right handed (RH) neutrinos with Yukawa couplings h and a Majorana mass term M, after spontaneous symmetry breaking, a Dirac neutrino mass term, $m_D = h\mathsf{v}$, is generated by the vacuum expectation value of the Higgs boson and the whole neutrino mass term can be written [1] as

$$\mathscr{L}^\nu_{\text{mass}} = -\frac{1}{2}\left[(\bar{\nu}_L^c, \bar{\nu}_R)\begin{pmatrix} 0 & m_D^T \\ m_D & M_R \end{pmatrix}\begin{pmatrix} \nu_L \\ \nu_R^c \end{pmatrix}\right] + h.c. \quad (2)$$

Assuming that the eigenvalues of M are much higher than those of m_D, then one has a splitting between 3 heavy eigenstates with masses $M_1 \leq M_2 \leq M_3$, approximately coinciding with the eigenvalues of M, and 3 light eigenstates with masses $m_1 \leq m_2 \leq m_3$ given by the seesaw formula

$$m_\nu = -m_D \frac{1}{M} m_D^T. \quad (3)$$

All matrices are in general complex and this provides a natural source of CP violation, one of the three Sakharov necessary conditions for successful baryogenesis. Indeed, considering the decays of the RH neutrinos N_i, one has that these can proceed with a different rate into leptons ($N_i \to l\phi^\dagger$) and anti-leptons ($N_i \to \bar{l}\phi$). The difference can be expressed in terms of the CP asymmetry parameter defined as

$$\varepsilon_i \equiv -(\Gamma_i - \bar{\Gamma}_i)/(\Gamma_i + \bar{\Gamma}_i). \quad (4)$$

If the mass differences $|M_i - M_j|$ are not too small compared to the rate differences, then ε_i can be calculated from the interference between tree level and one loop graphs, self energy plus vertex correction [4].

It has been pointed out [5] that, working in a basis where the Majorana mass term is diagonal, such that $M \to D_M \equiv \text{diag}(M_1, M_2, M_3)$, the seesaw formula is equivalent to an orthogonality condition for a matrix Ω through which the Dirac mass matrix can be parameterized as

$$m_D = U D_m^{1/2} \Omega D_M^{1/2}. \quad (5)$$

---

[1] We consider the case where a triplet Higgs term is absent.

Here U is the matrix that diagonalizes m_ν, such that $D_m \equiv \text{diag}(m_1, m_2, m_3) = -U^\star m_\nu U^\dagger$. It can be identified with the MNS neutrino mixing matrix in a basis where charged leptons are diagonal. This parametrization is particularly useful for leptogenesis. First of all it shows that there are 18 parameters: 9 'low energy parameters' (the 3 light neutrino masses m_i and the 6 parameters in the MNS mixing matrix U) and 9 'high energy parameters' (the 3 RH neutrino masses M_i and the 6 parameters in the orthogonal matrix Ω). From neutrino mixing experiments we have s information on some parameters in U, in particular we measure two mass squared differences: $(m_3^2 - m_1^2)^{1/2} = m_{\text{atm}} \simeq 0.05\,\text{eV}$ and $(m_{2(3)}^2 - m_{1(2)}^2)^{1/2} = m_{\text{sol}} \simeq 0.009\,\text{eV}$ for a normal (inverted) scheme. We still miss a determination of the *absolute neutrino mass scale*, that we will indicate in terms of the lightest neutrino mass m_1. The CP asymmetries ε_i depend only on the $m_D^\dagger m_D$ entries, and from the (5) one can see that the mixing matrix U cancels out, implying that the CP responsible for leptogenesis stems uniquely from Ω and is *in general* not dependent on U. Vice versa the high energy parameters do not enter the neutrino mass matrix m_ν and thus CP in neutrino mixing can only arise from U. Thus, the orthogonal parametrization shows a full disentanglement of CP in neutrino mixing and in leptogenesis. An important consequence is that it is not possible to prove or disprove leptogenesis by measuring or constraining CP in neutrino mixing.

On the other hand many possible connections between the leptogenesis CP asymmetry and CP in neutrino mixing have been worked out within specific frameworks: this is not in contradiction with the previous conclusion. Indeed suppose one adds, to the general framework, some extra theoretical input or phenomenological information that specifies or measures a subset **X** of the seesaw parameters in terms of which one can replace part or all the Ω parameters through a non trivial transformation $\Omega = \Omega(m_i, M_i, U; \mathbf{X})$. This brings to a new parametrization $(M_i, m_i, U, \mathbf{X})$, where now the **X**'s have to be regarded as known parameters. In this way one has a reduction of the number of parameters that allows to express Ω as a (non trivial) function $\Omega(m_i, M_i, U)$. When this is plugged into the ε_i's, the CP asymmetry responsible for leptogenesis becomes a function of U. Thus a measurement of CP in neutrino mixing, more than a test of leptogenesis, will provide a test of the specific 'X-framework'. However, *common sense* suggests that 'if there are phases in U, then why not also in Ω' and so the existence of such a framework is quite reasonable, Therefore, a detection of CP in neutrino mixing, if not a smoking gun, will certainly provide an additional piece in support of leptogenesis.

The orthogonal parametrization (cf. (5)) is also a useful technical tool to solve different problems in leptogenesis. The final asymmetry can be in general written as the sum of the contributions from the decays of all three N_i,

$$\eta_B = d \sum_i \varepsilon_i \kappa_i, \quad (6)$$

where $d = a_{\text{sph}}/f \simeq 10^{-2}$ takes into account both that only a fraction $a_{\text{sph}} \simeq 1/3$ of the $B - L$ asymmetry ends up into a baryon asymmetry through sphaleron conversion and the dilution of the asymmetry $f \simeq 35$ due to the photon production from the time of leptogenesis till recombination. Each κ_i is the efficiency factor associated to the asymmetry production from the decays of N_i. In general the baryon asymmetry will thus depend on 10 unknown seesaw parameters: the absolute neutrino mass scale m_1, the three M_i and the 6 parameters in Ω. However, assuming a mild heavy neutrino mass hierarchy, $M_2 \gtrsim 5 M_1$, and assuming that the inverse decays of the lightest RH neutrino N_1 strongly wash-out the asymmetry generated from the two heavier ones, then one has a simplified picture where the final asymmetry is produced only from N_1 decays and $\eta_B \simeq 10^{-2} \varepsilon_1 \kappa_1$. For values of $M_1 \ll 10^{14}\,\text{GeV}\, m_{\text{atm}}^2 / \sum_i m_i^2$, the main contribution to the wash-out comes from inverse decays [6, 7] and κ_1 is a function just of the *effective neutrino mass* $\widetilde{m}_1$ defined as $\widetilde{m}_1 \equiv (m_D^\dagger m_D)_{11}/M_1$. The assumption of strong wash-out holds for $\widetilde{m}_1 \gtrsim m_\star \simeq 10^{-3}\,\text{eV}$, where $m_\star$ is the *equilibrium neutrino mass*. The effective neutrino mass can be conveniently expressed in terms of the Ω matrix as $\widetilde{m}_1 = \sum_i m_i |\Omega_{j1}^2|$ and from the Ω orthogonality it easily follows that $\widetilde{m}_1 \geq m_1$ [8]. This is the only model independent information on $\widetilde{m}_1$, quite relevant if $m_1 \gtrsim m_\star$, since in this case one can conclude that the strong wash-out condition is always realized. If $m_1 \lesssim m_\star$, then $\widetilde{m}_1$ can also lie in the weak-wash out regime. However now, for fully normal (inverted) hierarchical neutrinos, one has $\widetilde{m}_1 \simeq m_1 |\Omega_{11}^2| + m_{\text{sol}} (m_{\text{atm}}) |\Omega_{21}^2| + m_{\text{atm}} |\Omega_{31}^2|$, and one can easily understand that the condition $\widetilde{m}_1 \ll m_\star \ll m_{\text{sol}}$ holds only when [9]

$$\Omega \simeq \begin{pmatrix} 1 & 0 & 0 \\ 0 & \Omega_{22} & \sqrt{1 - \Omega_{22}^2} \\ 0 & -\sqrt{1 - \Omega_{22}^2} & \Omega_{22} \end{pmatrix}, \quad (7)$$

such that $|\Omega_{21}^2| \ll m_1/m_2$ and $|\Omega_{31}^2| \ll m_1/m_{\text{atm}}$. If $M_1 \gtrsim 10^{14}\,\text{GeV}\, m_{\text{atm}}^2 / \sum_i m_i^2$, then off-shell $\Delta L = 2$ processes gives a non negligible contribution to the wash-out such that

$$\kappa_1 \simeq \kappa_1(\widetilde{m}_1) e^{-\frac{M_1}{10^{14}\,\text{GeV}} \frac{\sum_i m_i^2}{m_{\text{atm}}^2}}. \quad (8)$$

Another nice consequence of assuming a hierarchical heavy neutrino spectrum is that it allows to write down an approximate expression for ε_1,

$$\varepsilon_1 \simeq \varepsilon_1(M_1, m_1, \widetilde{m}_1, \Omega_{j1}^2) \equiv \overline{\varepsilon}(M_1)\, \beta(m_1, \widetilde{m}_1, \Omega_{j1}^2), \quad (9)$$

where $\overline{\varepsilon}(M_1) \equiv 3 M_1 m_{\text{atm}}/(16\pi v^2)$. Therefore, we can see that within the simplified picture where the dominant contribution to the final asymmetry arises from N_1 decays, one has that $\eta_B = \eta_B(M_1, m_1, \widetilde{m}_1, \Omega_{j1}^2)$ depends just on 6 parameters. One can find interesting constraints imposing that the predicted asymmetry explains the observed value. Interestingly, the function $\varepsilon_1(M_1, m_1, \widetilde{m}_1, \Omega_{j1}^2)$ has an upper bound that can be found maximizing β over the Ω_{j1}^2's [10, 11]. This is saturated for fully hierarchical neutrinos ($m_1 = 0$) and when $\Omega_{21}^2 = \text{Re}(\Omega_{31}^2) = 0$, such that $\beta = 1$ and $\varepsilon_1 = \overline{\varepsilon}(M_1)$. The upper bound on ε_1 implies a lower bound on M_1 [11, 12]

$$M_1 \gtrsim 4.2 \times 10^8 \,\text{GeV} \times [k_1(\widetilde{m}_1)]^{-1} \quad (M_1 \ll 10^{14}\,\text{GeV}), \tag{10}$$

and consequently a lower bound on the reheating temperature [7] $T_{\text{reh}} \gtrsim M_1/5$. For non-zero values of m_1, the upper bound on ε_1 becomes more restrictive [11, 13],

$$\varepsilon_1 \leq \overline{\varepsilon}(M_1, m_1, \widetilde{m}_1) \equiv \overline{\varepsilon}(M_1) \frac{m_{\text{atm}}}{m_1 + m_3} f(m_1, \widetilde{m}_1). \tag{11}$$

The bound is still saturated for $\Omega_{21}^2 = 0$ but this time $X \equiv \text{Re}(\Omega_{31}^2) \neq 0$ [14]. The function $f(m_1, \widetilde{m}_1) \in [0, 1]$ vanishes for $\widetilde{m}_1 = m_1$ and is equal to 1 in the limit $m_1/\widetilde{m}_1 \to 0$. For generic values of m_1 it can be calculated simply finding the maximum of $Y \equiv \text{Im}(\Omega_{31}^2)$ for fixed $\widetilde{m}_1$ and with the orthogonality implying $\Omega_{11}^2 + \Omega_{31}^2 = 1$ and then $f = (m_1 + m_3) Y/\widetilde{m}_1$ [9]. For $m_1 \ll m_{\text{atm}}$ the function f is well approximated by [13, 9]

$$f(m_1, \widetilde{m}_1) = \left(m_3 - m_1 \sqrt{1 + \frac{m_3^2 - m_1^2}{\widetilde{m}_1^2}} \right) / (m_3 - m_1), \tag{12}$$

while in the limit of quasi-degenerate neutrinos one has [14, 9] $f = \sqrt{1 - (m_1/\widetilde{m}_1)^2}$. It is quite remarkable that, accounting also for the suppression of the efficiency factor (cf.(8)), the final baryon asymmetry is strongly suppressed for $m_1 \gg m_{\text{atm}}$. This results into a stringent upper bound on the neutrino masses $m_i \lesssim 0.1\,\text{eV}$ [13]. For a generic choice of the seesaw parameters, the CP asymmetry can be written as

$$\varepsilon_1 = \overline{\varepsilon}(M_1, m_1, \widetilde{m}_1) \sin \delta_L(m_1, \widetilde{m}_1, \Omega_{j1}^2), \tag{13}$$

where the *effective leptogenesis phase* $\sin \delta_L$ takes into account the suppression of the asymmetry compared to the case when the upper bound is saturated and $\sin \delta_L = 1$. The exponential suppression of κ_1 for large M_1 (cf. (8)) yields a lower bound on $\sin \delta_L$ given by [9]

$$\sin \delta_L \gtrsim 4 \times 10^{-2} (\widetilde{m}_1/\text{eV}) \quad (\widetilde{m}_1 \gtrsim 10^{-3}\,\text{eV}). \tag{14}$$

All the derived bounds are valid under a set of minimal assumptions and it is actually non trivial that the final asymmetry can be explained within such a minimal picture. In particular it is interesting that this is possible only because the atmospheric neutrino mass scale lies in the range $10^{-3}\,\text{eV} \lesssim m_{\text{atm}} \lesssim 1\,\text{eV}$ [15], a result to be regarded as a successful test for the minimal leptogenesis scenario.

On the other hand the existence of the constraints on neutrino masses, the lower bound on the reheating temperature of the Universe and the upper bound on the neutrino masses, can potentially disprove the minimal scenario. In particular the first is a problem when leptogenesis is embedded within the minimal supersymmetric standard model scenario where an upper bound on T_{reh} is necessary to avoid a gravitino over-production. The second is of course a problem if neutrino masses larger than $0.1\,\text{eV}$ will be found. At the moment it is intriguing that the most restrictive upper bound from a combined analysis of the CMB acoustic peaks and of large scale structure gives $m_1 < 0.14\,\text{eV}$ and thus it is in agreement with the bound from leptogenesis. Therefore, at the moment, the minimal picture cannot be ruled out.

A NEW SCENARIO OF LEPTOGENESIS

In the weak wash-out regime, for $\widetilde{m}_1 \lesssim m_\star$, the minimal picture where the asymmetry is produced by the N_1's, encounters two serious obstacles. The first is that the CP asymmetry ε_1 vanishes in the limit $\widetilde{m}_1 \to m_1$, implying that one has to tune $\widetilde{m}_1$ such to be lower than $m_\star \simeq 10^{-3}\,\text{eV}$ but much larger than m_1. The second problem is that, in the weak wash-out regime, the final asymmetry depends on the initial asymmetry and, more importantly, on the initial N_1 abundance and a calculation of the final asymmetry becomes very model dependent [7].

These two problems are both solved considering another possibility: that the asymmetry is produced from the decays of N_2, the next-to-lightest RH neutrino. Indeed if $\widetilde{m}_1 \lesssim m_\star$, as we are assuming, then necessarily $\widetilde{m}_2 \gtrsim m_{\text{sol}} \gg m_\star$ and thus the generation occurs in the strong wash-out regime, solving the problem of the initial conditions. Moreover now if $\widetilde{m}_1 = m_1$, when Ω is given by the eq. (7), the CP asymmetry ε_2 does not vanish. Therefore one does not need to fine tune Ω, it is enough that this is close to the form given by the eq. (7) and of course that the phase of Ω_{22} is large enough.

Another attractive feature is that N_1 does not play any role and thus the lower bound on M_1 disappears and is replaced by an analogous lower bound on M_2, still implying a lower bound on $T_{\text{reh}} \gtrsim M_2/5$. It is intriguing to speculate about *possible phenomenological implications of a light N_1*. The first is to think whether this can be light enough to be produced in the Large Hadron Collider. However, the N_1 Yukawa coupling upper bound, $U^\dagger h_{11} \lesssim 10^{-7} \sqrt{M_1/1\,\text{TeV}}$, gives no hopes, unless new extra-gauge interactions are assumed. Another intriguing possibility is that N_1 can play the role of cold Dark Matter. Indeed preliminary calculations [16] seem to suggest that for $M_1 \sim 1\,\text{GeV}$ it is possible to have both long lived N_1 and the right abundance $\Omega_{N_1} h^2 = \Omega_m h^2 \sim 0.1$.

BEYOND THE MINIMAL PICTURE

The new scenario described in the previous section is an interesting logical completion of the minimal picture. However, the possibility to evade the lower bound on the reheating temperature and the upper bound on the neutrino masses requires more drastic departures from the minimal scenario. A large variety of models have been proposed. They can be classified in three categories.

Leptogenesis from type II seesaw formula [17]. A first class of models is based on modifications of the minimal seesaw formula. The most popular is the account of the triplet Higgs term in the neutrino mass term (cf. (2)). In this was the simple seesaw formula (cf. (3) gets generalized into

$$m_\nu = -m_D \frac{1}{M} m_D^T - h^T \frac{v^2}{M_T}. \qquad (15)$$

The dominant wash-out processes constraints mainly the first term but not the second and in this way the upper bound on the neutrino masses can be easily evaded. Moreover the CP asymmetry receives additional contributions making possible also to evade the lower bounds on M_1 and on T_{reh}.

Degenerate heavy neutrino spectrum [13, 14, 9, 18]. If $M_2 \stackrel{<}{\sim} 5 M_1$ one expects deviations from the constraints obtained in the minimal picture, even in the strong wash-out regime for $\widetilde{m}_1 \gg m_\star$. There are two effects to consider: one is that the asymmetry and the wash-out from the $N_{2,3}$ decays and inverse decays can give a non negligible contribution and so one has to consider the general expression (6) for the final asymmetry. The second effect is that there can be significant deviations from the approximate expression for ε_1 (cf. (9)). Defining $\xi_\varepsilon \equiv \varepsilon_1/\overline{\varepsilon}(M_1)$, one has that for $M_2 \stackrel{<}{\sim} 5 M_1$ values $\xi_\varepsilon \stackrel{>}{\sim} 1.1$ are possible, with a consequent relaxation of the neutrino mass constraints. However, the possibility of a significant evasion of the bounds, especially the upper bound on the neutrino masses, is possible if the CP asymmetry undergoes a resonant enhancement, such that $\varepsilon_1 \sim 0.1$ independently on the values of M_1 and m_1 [18, 14].

Non thermal leptogenesis [19]. If the temperature $T_{\text{reh}} \stackrel{<}{\sim} M_1/5$, a thermal production of the N_1 is inefficient. Different mechanisms of non thermal production have been proposed, all associated to the occurrence of an inflationary stage. The bounds of the minimal picture can be easily evaded.

Within all these three categories of non minimal models, neutrino mass bounds can be easily evaded. However, at the same time the correlation between the neutrino mixing mass scales and the baryon asymmetry, the *leptogenesis conspiracy*, gets lost or in other words is like 'to throw away the baby with the bath water'. Therefore, until data will not disprove the minimal picture, this appears the most appealing way to realize leptogenesis. In a near future the most important experimental tests for the minimal picture are represented by the measurement of the absolute neutrino mass scale to be compared with the upper bound $m_i < 0.1\,\text{eV}$ and a test of the supersymmetric models in LHC, that could confirm the gravitino upper bound on T_{reh} at variance with the minimal leptogenesis lower bound.

REFERENCES

1. A. D. Sakharov, Pisma Zh. Eksp. Teor. Fiz. **5** (1967) 32.
2. WMAP Collaboration, D. N. Spergel *et al.*, Astrophys. J. Suppl. **148** (2003) 175; M. Tegmark *et al.* [SDSS Collaboration], Phys. Rev. D **69** (2004) 103501.
3. M. Fukugita and T. Yanagida, Phys. Lett. B **174** (1986) 45.
4. M. Flanz, E. A. Paschos, U. Sarkar, Phys. Lett. B **345** (1995) 248; Phys. Lett. B **384** (1996) 487 (E); L. Covi, E. Roulet, F. Vissani, Phys. Lett. B **384** (1996) 169; W. Buchmüller, M. Plümacher, Phys. Lett. B **431** (1998) 354.
5. J. A. Casas and A. Ibarra, Nucl. Phys. B **618** (2001) 171.
6. G. F. Giudice, A. Notari, M. Raidal, A. Riotto and A. Strumia, Nucl. Phys. B **685** (2004) 89.
7. W. Buchmuller, P. Di Bari and M. Plumacher, Annals Phys. **315** (2005) 305.
8. M. Fujii, K. Hamaguchi, T. Yanagida, Phys. Rev. D **65** (2002) 115012.
9. P. Di Bari, arXiv:hep-ph/0502082.
10. T. Asaka, K. Hamaguchi, M. Kawasaki and T. Yanagida, Phys. Lett. B **464** (1999) 12; K. Hamaguchi, H. Murayama and T. Yanagida, Phys. Rev. D **65** (2002) 043512.
11. S. Davidson, A. Ibarra, Phys. Lett. B **535** (2002) 25.
12. W. Buchmüller, P. Di Bari, M. Plümacher, Nucl. Phys. B **643** (2002) 367.
13. W. Buchmüller, P. Di Bari, M. Plümacher, Nucl. Phys. B **665** (2003) 445.
14. T. Hambye, Y. Lin, A. Notari, M. Papucci and A. Strumia, Nucl. Phys. B **695** (2004) 169.
15. W. Buchmuller, P. Di Bari and M. Plumacher, New J. Phys. **6** (2004) 105.
16. P. Di Bari, in preparation.
17. S. Antusch and S. F. King, Phys. Lett. B **597** (2004) 199.
18. A. Pilaftsis, Phys. Rev. D **56** (1997) 5431.
19. G. Lazarides, Q. Shafi, Phys. Lett. B **258** (1991) 305; H. Murayama, T. Yanagida, Phys. Lett. B **322** (1994) 349; G. F. Giudice et al. JHEP **9908** (1999) 014.

Neutrinoless Double Beta Decay and ν–Mass Determination

M. Pedretti

Università dell'Insubria, Via Vallegio 11, 22100 Como, Italy

Abstract. The search for Neutrinoless Double Beta Decay could improve our knowledge on neutrino properties. After a brief discussion on the implications of the observation of this rare process, I will introduce the experimental approaches and review the prospects of the search for this nuclear transition.

Keywords: neutrino, rare decay, double beta decay, particle proprieties
PACS: 14.60.Pq, 14.60.St

INTRODUCTION: OPEN QUESTIONS

The results on neutrino oscillations have stimulated the scientific community to study neutrino properties. In fact, neutrino physics is a sensitive test for theoretical models beyond Standard Model. The observation of neutrino oscillations demonstrated that neutrino mass eigenstates don't coincide with the neutrino flavor eigenstates and that these particles have non-zero masses. However there are still many open questions in the physics of massive and mixed neutrinos. For example the neutrino oscillations indicate the differences among the squared neutrino mass eigenvalues but not their absolute value. The masses M_i of the neutrino mass eigenstates ν_i remain therefore unknown. Up to now all the three mass patterns are possible: the normal hierarchy, where $m_1 \ll m_2 \ll m_3$, the inverted hierarchy with $m_3 \ll m_1 < m_2$ and the quasi–degenerate one with $m_1 \sim m_2 \sim m_3$.

The second open question regards the neutrino nature: are the neutrino and the antineutrino the same particle? And why do neutrinos and antineutrinos produce respectively negative charged leptons and positive charged leptons via charged current weak interaction with matter?

There are two possible explanations of this phenomenology:

- in the Standard Model leptons differ from their antiparticles for the sign of the lepton number, as

$$\mathbf{L}(\nu_e, e^-) = -1 \quad \mathbf{L}(\bar{\nu}_e, e^+) = +1 \quad (1)$$

It is thus possible explain the matter behavior stating that, like electric charge, lepton number is conserved in weak interactions. In this case, the neutrino is classified as a **Dirac particle** and

$$\nu_e \neq \bar{\nu}_e \quad (2)$$

- the second possibility is to look at neutrino and antineutrino as the same particle, but characterized by different helicities:

$$\mathbf{H}(\nu_e) = -1 \, (L) \quad \mathbf{H}(\bar{\nu}_e) = +1 \, (R) \quad (3)$$

When the different neutrino and antineutrino phenomenology is accounted for by their different helicity, neutrino is said to be a **Majorana particle**, and then

$$\nu_e = \bar{\nu}_e. \quad (4)$$

As neutrinos have mass, then helicity depends on the reference frame and the neutrino nature question is not merely academic. In this scenario the search for the Double Beta Decay (DBD) plays a the crucial role as it can give information on the neutrino nature and the neutrino mass scale.

NEUTRINOLESS DOUBLE BETA DECAY

The Double Beta Decay (DBD) is a rare spontaneous nuclear transition [1, 2] where a nucleus (A,Z) changes the nuclear charge of two units maintaining the same mass number, so it becomes $(A,Z\pm2)$ nucleus. Let's focus the attention on the case in which a $(A,Z+2)$ nucleus appears in the final state. Normally DBD is not favored with respect to the single beta decay, and it is possible to observe it only for those nuclei for which the single β decay is either energetically forbidden or suppressed by a large change of the nuclear spin–parity state.

Two different DBD modes are usually considered: first, the decay with two neutrinos, where lepton number is conserved and so it is allowed by the SM, that is described by the reaction

$$2\nu\text{DBD}: \quad (A,Z) \to (A,Z+2) + 2e^- + 2\bar{\nu}_e \quad (5)$$

and, second, the neutrinoless decay given by

$$0\nu\text{DBD}: \quad (A,Z) \to (A,Z+2) + 2e^- \quad (6)$$

where the emission and the re–absorption of a virtual neutrino mediate the decay. An experimental confirmation of this decay mode will thus constitute an important

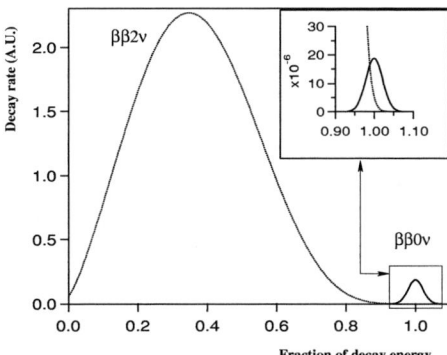

FIGURE 1. Expected spectrum of the sum of the energy of the electrons for DDB2ν (dashed line) and DDB0ν (solid line). In the inset the contribution of the 2ν decay to the 0νDDB background is underlined.

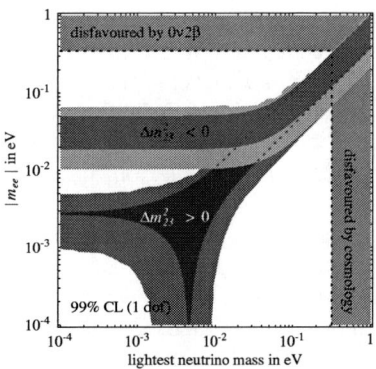

FIGURE 2. Plot of the effective mass $\langle m_\nu \rangle$ as a function of the lightest neutrino mass (on a log-log scale). The darkness region are the possible neutrino mass range in case that neutrino oscillation experiment results have negligible errors. This plot is taken from [4].

step in the study of elementary particle physics beyond the SM as the decay requires Majorana neutrinos with a non–zero mass.

As shown in fig. 1, the 2νDBD is a four body decay and so the spectrum is a continuum with a maximum value around one third of the Q value. On the contrary, in 0νDBD, the two electrons retain all the available kinetic energy (neglecting the nuclear recoil). For this reason, the expected spectrum is just a spike at the transition energy.

In both cases, the DBD is a semileptonic second–order weak interaction and thus is characterized by a very long lifetime: $\tau_{\beta\beta} \sim 10^{18} - 10^{22}$ years for the 2 neutrino channel. Thus, the experimental observation of this decay turns out to be a great challenge, because very rare events have to be detected in the presence of unavoidable traces from other radioisotopes with similar transition energies but with decay times that are even 10 orders of magnitude shorter. Presently, 2νDBD has been observed for ~ 10 nuclei [3].

0νDBD probability is usually expressed using the general relation derived from Fermi's golden rule:

$$\left[T_{1/2}^{0\nu}\right]^{-1} = G^{0\nu} \left|M^{0\nu}\right|^2 \langle m_\nu \rangle^2 \qquad (7)$$

where $G^{0\nu}$ is the phase space integral and can be estimated exactly, $\left|M^{0\nu}\right|^2$ is the decay matrix element and $\langle m_\nu \rangle$ the effective neutrino mass:

$$\langle m_\nu \rangle = \sum_i \phi_i M_i |U_{ei}|^2 \qquad (i = 1,2,3) \qquad (8)$$

with $|U_{ei}|$ the neutrino mass matrix elements. Let's observe that the ϕ_k phases that appear in the last equation are the intrinsic neutrino CP parities. Their presence implies that cancellations are possible.

From eq (8) it is clear that 0νDBD could help to solve the questions on the absolute value of neutrino masses and to disentangle the hierarchy scheme as shown in fig. 2 where the neutrino oscillation results have been considered.

In particular a sensitivity of $\sim 10 - 50$ meV on $\langle m_\nu \rangle$ could definitely exclude the inverted hierarchy and the and quasi–degenerate pattern [4].

From eq. (7), it is also clear that the evaluation of $\langle m_\nu \rangle$ from an experimental measure of $T_{1/2}^{0\nu}$ will require the exact knowledge of the nuclear matrix elements. From a theoretical point of view, this is the main limit when describing 0νDBD. Several models have been proposed and is not simple to evaluate their correctness and accuracy, especially because they are not connected with other nuclear processes that allow a simple verification. Even the comparison with the two neutrino decay presents unclear points, mostly because of the different neutrino role in the two processes. In this situation, one possibility is to consider the spread of the theoretical values of the nuclear matrix elements as a measure of their uncertainty.

EXPERIMENTAL APPROACHES

The experimental signature in the case of neutrinoless double beta decay is in principle very clear: one should expect a peak (at the $Q_{\beta\beta}$ value) in the summed two–electron energy spectrum, whereas a continuous spectra will feature the two–neutrino decay mode. In spite of such characteristics, their identification is very difficult. The search for the 0νDBD decay is a search for a peak superimposed on a continuum.

To measure double beta decays, three general approaches have been followed: geochemical, radiochemical, and direct counting measurements.

As both geochemical and radiochemical approaches are not able to distinguish the 2ν contribute from the 0ν one, we focus our attention on the direct counting measurements.

This approach allows measurement of the $\beta\beta$–electron energy, their spectral shape and, sometimes also permit event reconstruction. Direct experiments are based on two different methods. In the source$\neq$detector approach a double beta active material is inserted, normally in form of thin sheets, in a suitable detector. In the source=detector or "calorimetric experiments" the detector itself is made of a material containing the double beta active nuclei. Well–known examples of $\beta\beta$ emitters measured in direct counting experiments are ^{48}Ca, ^{76}Ge, ^{96}Zr, ^{82}Se, ^{100}Mo, ^{116}Cd, ^{130}Te, ^{136}Xe, ^{150}Nd.

Calorimeters can achieve good energy resolution and almost 100% efficiency. They normally lack the tracking capabilities to identify the background on an event–by–event basis but they have, in their favor, that their sharp energy resolution does not allow the leakage of too many counts from ordinary double beta decay into the neutrinoless region, and so the sensitivity intrinsic limitation is not too severe. In the other case, the identification capabilities make them very well suited for 2νDBD experiment, but, having low energy resolution their 0νDBD regions can be dominated by the 2νDBD background events. For this reason, at least in principle, calorimetric experiments are preferred to search for 0νDBD.

In order to understand the general strategy to follow to perform a neutrinoless double beta decay experiment, I report the sensitivity to the effective neutrino mass (1σ)

$$m_{ee} \propto \frac{1}{\sqrt{G^{0\nu}}|M^{0\nu}|} \times \left(\frac{B\cdot\Gamma}{M\cdot t}\right)^{1/4} \qquad (9)$$

where B is the background rate (in c/(keV kg y)), M the mass of the $\beta\beta$ emitter (in kg), Γ energy resolution at the energy transition of the searched decay, and t the running time measurement in years.

Thus in order to obtain powerful 0νDBD experiments it is necessary

- to work with very large source masses, of order of one ton or larger in the next generation experiments;
- to use good energy resolution and high efficiency detectors; good energy resolution is a must, not only to improve the signal to background in the peak search, but also because poor resolution would result in the 2νDBD tail extending up into the peak region to become a background itself.

- to perform the experiment in "clean" conditions from the point of view of radioactivity background; this implies that detectors must be shielded from the environment and its associated radioactivity. Usually cosmic rays contribution are suppressed by performing experiments in underground laboratories. To perform an effective background suppression strategy, it is useful that the detectors provide much information. For example if detector permits an event topology reconstruction it is possible to perform an event–by–event background rejection.
- to have long data taking periods.

PRESENT EXPERIMENTS

Complete reviews on 0νDBD experiments could be found in Refs. [5, 1] Here more attention is devoted to the Heidelberg–Moscow (HM) experiment, that presently gives the best limits on the effective neutrino mass, and on the two running experiment which have the potential to improve the HM results: Neutrino Ettore Majorana Observatory (NEMO) and Cuoricino experiments.

The Heidelberg–Moscow experiment

The best limit on 0νDBD comes from the Heidelberg–Moscow (HM) experiment [6, 7] using ^{76}Ge source. This experiment operates five enriched (to 86%) high–purity Ge detectors, with an active mass of 10.96 kg of ^{76}Ge. The experiment is installed in the Gran Sasso Underground Laboratory (LNGS) under heavy shields for gamma and neutron environmental radiation. Extremely low background levels are achieved thanks to a careful selection of the setup materials and further improved by the use of Pulse Shape Discrimination (PSD) techniques. In the 0νDBD region the background level is $\sim$ 0.06±0.02 c/keV/kg/year. The limit obtained is $T^{0\nu}_{1/2} = 1.9 \times 10^{25}$ years at 90% of C.L., corresponding to $< m_\nu > < 0.35$ eV.

In January 2002, a few members of the HM collaboration claimed evidence for 0νDBD peak [8]. The authors quoted in a recent paper [9] with $T^{0\nu}_{1/2} = 1.19\ 10^{25}$ years as a best value corresponding to a $< m_\nu >$=440 eV.

This claim raised skepticism in most of the scientific community and a controversy followed which is still open [10, 11, 12]. In particular, a strong criticism is due mainly to the fact that the line identification is still weak and to the width of the energetic window chosen to calculate the background. It is probable therefore that a definite answer to the correctness of the claim will be given only by the very sensitive next generation 0νDBD projects.

The NEMO experiment

The NEMO collaboration is working with NEMO–3 detector at Frejus underground labs (4800 m.w.e.) with the primary goal of studying different isotopes DDB0ν (decay) with large $Q_{\beta\beta}$ values. The NEMO–3 detector can contain $\sim$ 10 kg of different decaying $\beta\beta$ isotopes. The biggest effort has been dedicated to ^{100}Mo ($Q_{\beta\beta}$ = 3034 keV) and ^{82}Se ($Q_{\beta\beta}$ = 2995 keV).

This experiment is based on the direct detection, by a tracking set–up, of the two electrons resulting from the decay and on the direct measurement of their energy with a calorimeter. The tracking capability of NEMO–3 detector permits a good background rejection, but, on the contrary, the modest energy resolution makes 2νDBD an important source of background.

The present lower limits for the half-lives are 3.5 $\times 10^{23}$ years at 90% C.L for ^{100}Mo and 1.9 $\times 10^{23}$ years for ^{82}Se. The sensitivity that the NEMO–3 experiment will reach after five years of data taking has been estimated [13] corresponding to $\langle m_\nu \rangle$ = 0.2 eV. A next generation experiment, named Super NEMO, has been proposed by NEMO collaboration as an expansion of the present set-up, in order to reach limits on neutrino mass of the order of few tens of meV.

Cuoricino Experiment

With its 40 kg of TeO$_2$ in form of bolometric calorimeters, Cuoricino is the biggest running 0νDBD experiment. It uses low temperature true calorimeters proposed [14] and developed by Milano group. This technique measures the energy deposited by particles as an increase of temperature of a crystalline absorber (in the case of Cuoricino experiment, TeO$_2$ crystals). The source of 0νDBD is ^{130}Te that, thanks to its high natural abundance, permits to work with large active masses also without isotopic enrichments.

Cuoricino detector has reached good energy resolution in the DBD energy region (3.9 keV at 2615 keV at the best) and is charcterized by good efficiency (86%). The current background in 0νDBD region (0.19$\pm$0.02 c/keV/kg/y) is dominated by degraded alphas coming from surfaces of the TeO$_2$ crystals and of the copper structures holding the detectors. Cuoricino is presently giving a limit of 1.8 10^{24} y at the 90 % of C.L. on $T_{1/2}^{0\nu}$, corresponding to a bound on $\langle m_\nu \rangle$ between 0.2 and 1.1 eV. Cuoricino can reach a sensitivity on neutrino mass between 0.13 and 0.31 eV in a few years [15]. Unfortunately, due to the uncertainties on nuclear matrix elements, it will not be able to exclude totally the range indicated by part of the H-M collaboration for effective neutrino mass. Cuoricino is also a test for a next generation experiment, CUORE (Cryogenic Underground Observatory for Rare Events), that has been proposed and partially funded [16]. A dedicated R&D is going on in order to obtain a relevant background reduction by using passive and active techniques [17]. This experiment will be able to explore the neutrino mass pattern and, thanks to its sensitivity of a few tens of meV on the Majorana neutrino mass, has a large discovery potential in case of inverted mass hierarchy.

CONCLUSIONS

In the next years the running experiments will test part of the region indicated by the authors of [9]. This controversy remarks the necessity to have reliable calculations on nuclear matrix elements. It is also clear that experiments that study different 0νDBD sources, with different techniques and so different systematics, are needed. Anyway, to have a definitive answer regarding the H–M claim, we have to wait for the results coming from the next generation experiments. that will be able also to test the inverted hierarchy pattern (until few tens of meV of majorana neutrino mass).

REFERENCES

1. S. R. Elliott, and P. Vogel, *Ann. Rev. Nucl. Part. Sci.*, **52**, 115–151 (2002), `hep-ph/0202264`.
2. A. Faessler, and F. Simkovic, *J. Phys.*, **G24**, 2139–2178 (1998), `hep-ph/9901215`.
3. M. K. Moe, *Nucl. Phys. Proc. Suppl.*, **38**, 36–44 (1995).
4. A. Strumia, and F. Vissani (2005), `hep-ph/0503246`.
5. O. Cremonesi, *Nucl. Phys. Proc. Suppl.*, **118**, 287–296 (2003), `hep-ex/0210007`.
6. H. V. Klapdor-Kleingrothaus, et al., *Eur. Phys. J.*, **A12**, 147–154 (2001), `hep-ph/0103062`.
7. A. M. Bakalyarov, A. Y. Balysh, S. T. Belyaev, V. I. Lebedev, and S. V. Zhukov (2003), `hep-ex/0309016`.
8. H. V. Klapdor-Kleingrothaus, A. Dietz, H. L. Harney, and I. V. Krivosheina, *Mod. Phys. Lett.*, **A16**, 2409–2420 (2001), `hep-ph/0201231`.
9. H. V. Klapdor-Kleingrothaus, A. Dietz, I. V. Krivosheina, and O. Chkvorets, *Nucl. Instrum. Meth.*, **A522**, 371–406 (2004), `hep-ph/0403018`.
10. C. E. Aalseth, et al., *Mod. Phys. Lett.*, **A17**, 1475–1478 (2002), `hep-ex/0202018`.
11. H. V. Klapdor-Kleingrothaus (2002), `hep-ph/0205228`.
12. H. L. Harney (2001), `hep-ph/0205293`.
13. R. Arnold, et al., *JETP Lett.*, **80**, 377–381 (2004), `hep-ex/0410021`.
14. E. Fiorini, and T. O. Niinikoski, *Nucl. Instr. Meth.*, **A224**, 83 (1984).
15. C. Arnaboldi, et al. (2005), `hep-ex/0501034`.
16. R. Ardito, et al. (2005), `hep-ex/0501010`.
17. L. Foggetta, et al., *Appl. Phys. Lett*, **86**, 134106 (2005).

Neutrinos and Cosmology: An Update

Ofelia Pisanti* and Pasquale D. Serpico[†]

*Dipartimento di Scienze Fisiche, Università di Napoli Federico II, and INFN, Sezione di Napoli,
Complesso Universitario di Monte Sant'Angelo, Via Cintia, I-80126 Napoli, Italy
[†]Max Planck Institut für Physik, Werner-Heisenberg-Institut,
Föhringer Ring 6, 80805, München, Germany

Abstract. We review the current cosmological status of neutrinos, with particular emphasis on their effects on Big Bang Nucleosynthesis, Large Scale Structure of the universe and Cosmic Microwave Background Radiation measurements.

Keywords: neutrinos, big bang nucleosynthesis, large scale structure of the universe, cosmology
PACS: 14.60.Lm, 14.60.St, 26.35.+c, 98.65.Dx, 98.80.-k

INTRODUCTION

The Standard Cosmological Model predicts the existence of a neutrino background (CνB) filling the universe with densities of the order $n_\nu \approx 110\,\text{cm}^{-3}$ per flavor. Neutrino properties are rather difficult to be probed experimentally, due to the weakness of neutrino interactions which, especially at low energies, makes hopeless at present any perspective of direct detection of the CνB. Nevertheless, neutrinos are one of the most abundant relics of the primordial universe and played a key role in different stages of its evolution. Several cosmological observables are then sensitive to neutrinos, and can be used to put bounds on their properties.

Given their extremely low interaction rate, the natural out-of-equilibrium driving force of the expansion of the Universe pushed neutrinos to decouple from the thermal bath very early, when the temperature was $\mathcal{O}(1\,\text{MeV})$. This temperature is close to the electron mass m_e, setting the scale of the electron/positron annihilation, and both are close to the $\mathcal{O}(0.1\,\text{MeV})$ scale of the synthesis of the light nuclei via thermonuclear fusion. So, Big Bang Nucleosynthesis (BBN) is a privileged laboratory for the CνB studies. In particular, it is sensitive to the ν (weak) interactions as well as to the shape of the $\nu_e - \bar{\nu}_e$ phase space distributions entering the $n \leftrightarrow p$ inter-conversion rates. Apart from the energy density due to the extra (*i.e.* non electromagnetic) relativistic degrees of freedom, the BBN tests the *dynamical* properties of the neutrinos in a thermalized (almost) CP-symmetric medium.

Other cosmological probes are Cosmic Microwave Background (CMB) anisotropies or the Large Scale Structure (LSS) of the universe, which are, however, sensitive only to the CνB *gravitational* interaction. The role of neutrinos as the dark matter (DM) particles has been widely discussed since the early 1970s. For values of neutrino masses much larger than the present cosmic temperature one finds a contribution in terms of the critical density $\Omega_\nu \simeq 0.0108\,h^{-2}\sum m_\nu/\text{eV}$, h being the Hubble parameter in units of 100 Km s^{-1} Mpc^{-1}. Nowadays, we know that neutrinos cannot constitute all the DM ($\Omega_{\text{DM}} \approx \Omega_m \sim 0.3$ [1]), and the main question is how large the contribution of neutrinos can be, deducing Ω_ν from their contribution to cosmological perturbations. In fact, neutrino background erases the density contrasts on wavelengths smaller than a mass-dependent free-streaming scale. Neutrinos of sub-eV mass behave almost like a relativistic species for CMB considerations and therefore the power spectrum suppression can be seen only in LSS data. Even if neutrino mass influences only slightly the spectrum of CMB anisotropies, it is crucial to combine CMB and LSS observations, because the former give independent constraints on the cosmological parameters, and partially remove the parameter degeneracy that would arise in an analysis of the LSS only.

NEUTRINOS AND COSMOLOGY

At temperatures above $\mathcal{O}(1\,\text{MeV})$, neutrinos are in thermal equilibrium with the thermal bath and their distribution is a perfect Fermi-Dirac one,

$$f_{\nu_\alpha}(y) = \frac{1}{e^{y-\xi_\alpha}+1}, \quad (1)$$

where $y \equiv p/T_\nu$ and $\xi_\alpha \equiv \mu_\alpha/T_\nu$ (here μ_α is the chemical potential of the flavor α, which is neglected in the standard scenario). As the temperature goes down, the universe expansion prevents weak interactions from maintaining neutrinos in equilibrium and they decouple. As a first approximation, the neutrino decoupling can be described as an instantaneous process taking place around 2-4 MeV, without any overlap in time with $e^+ - e^-$ annihilation. All flavors would then keep Fermi-

Dirac distributions (both neutrino momenta and temperature redshift identically with the universe expansion), but the neutrino temperature T_ν will not benefit of the entropy release from $e^+ - e^-$ annihilations. The asymptotic ratio T/T_ν for $T \ll m_e$ can be evaluated in an analytic way, and turns out to be $(11/4)^{1/3} \approx 1.401$.

More accurate calculations by solving the kinetic equations have been performed, and they show a partial entropy transfer to the neutrino plasma. As a consequence, the neutrino distributions get distorted, since this transfer is more efficient for larger neutrino momenta. In [2, 3] it was shown that with a very good approximation the distortion in the α-th flavor can be described as

$$f_{\nu_\alpha}(x,y) \simeq \frac{1}{e^y + 1}\left(1 + \sum_{i=0}^{3} c_i^\alpha(x) y^i\right), \quad (2)$$

with $x \equiv m_e/T_\nu$. The electron neutrinos get a larger entropy transfer than the μ and τ, since they also interact via charged currents with the $e^\pm$ plasma. The effective ratio $T/T_\nu \approx 1.3984$ is slightly lower than the instantaneous decoupling estimate.

The incomplete decoupling of neutrinos also induces a modification in the contribution of neutrinos to the energy density. By fully consistently including order α QED corrections to the photon and $e^\pm$ equation of state, in [2] the energy density in the neutrino fluid is found to be enhanced by 0.935% (for ν_e) and 0.390% (for ν_μ and ν_τ). A refined treatment, also including the effects of three-flavor neutrino oscillations, has been recently provided in [4].

The contribution of neutrinos to the total relativistic energy density of the universe is usually parameterized via the "effective number" of neutrinos, N_{eff},

$$\rho_\nu = N_{eff} \frac{7}{8}\left(\frac{T_\nu}{T}\right)^4 \rho_\gamma. \quad (3)$$

N_{eff} measures neutrino energy density in "units" of the energy density of massless neutrinos with zero chemical potential, but it can in principle receive a contribution from other (relativistic) relics. For three massless neutrinos with zero chemical potential and in the limit of instantaneous decoupling, $N_{eff} = 3$. The inclusion of entropy transfer between neutrinos and the thermal bath modifies this number to about 3.04 at the CMB epoch.

Shortly after neutrino decoupling the temperature reaches the value of the neutron-proton mass difference, and weak interactions are no longer fast enough to maintain equilibrium among nucleons: a substantial final neutron fraction survives, however, down to the phase of nucleosynthesis where all neutrons become practically bound in ^{4}He nuclei. The predicted value of the ^{4}He mass abundance, Y_p, is poorly sensitive to the nuclear network details and has only a weak, logarithmic dependence on

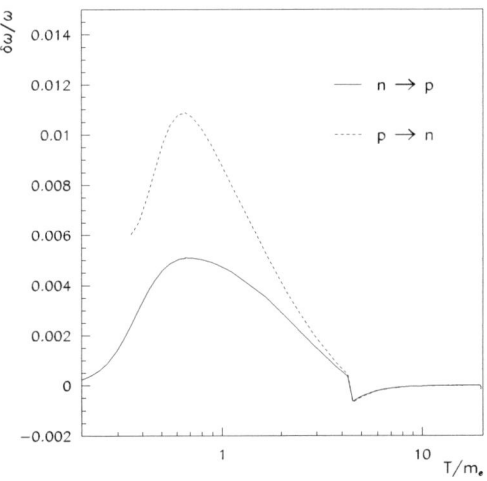

FIGURE 1. The relative correction to the $n \to p$ (solid line) and $p \to n$ (dashed line) total rates, due to the neutrino distortion (see Ref. [7] for details).

the baryon fraction of the universe, $\omega_b = \Omega_b h^2$, being fixed essentially by the ratio of neutron to proton number density at the onset of nucleosynthesis. This in turn crucially depends on the weak rates and on the (standard or exotic) neutrino properties.

In several papers (see [5, 6] and references therein) the value of Y_p has been computed by improving the evaluation of the weak rates including electromagnetic radiative corrections, finite mass corrections and weak magnetism effects, as well as the plasma and thermal radiative effects. In particular in [7] it has been also considered the effect of the neutrino spectra distortions and of the process $\gamma + p \leftrightarrow \nu_e + e^+ + n$, which is kinematically forbidden in vacuum, but allowed in the thermal bath. The latter is shown to give a negligible contribution, while neutrino distortions have a significant influence on the rates for different reasons: a) the larger mean energy of ν_e induces a δT_{ν_e} and (indirectly, through a decrease in $\rho_{e.m.}$) a δT; b) the ratio T/T_ν, which enters the weak rates, is changed. Moreover, the time-temperature relationship is changed and so the time at which BBN starts.

The total effect on the rates are shown in Figure 1. Even though one would expect effects up to $\mathscr{O}(1\%)$, the spectral distortion and the changes in the energy density and $T_\nu(T)$ conspire to almost cancel each other, so that Y_p is changed by a sub-leading $\mathscr{O}(0.1\%)$. This effect is of the same order of the predicted uncertainty coming from the error on the measured neutron lifetime, $\tau_n = 885.7 \pm 0.8$ s [8], and has to be included in quoting the theoretical prediction, $Y_p = 0.2481 \pm 0.0004$ (1σ, for $\omega_b = 0.023 \pm 0.001$).

Apart from the uncertainty on ω_b, the ^{2}H, ^{3}He and ^{7}Li abundance predictions are mainly affected by the nuclear reaction uncertainties. An updated and critical review of the nuclear network and a new protocol to perform the nuclear data regression has been widely discussed in [7], to which we address for details.

A different scenario, the Degenerate BBN (DBNN) one, has received a new attention in the last few years, especially when the first data on CMB seemed to indicate a tension between the determination of ω_b from CMB and standard BBN [9]. Such a tension could, in fact, be relaxed assuming that the number of ν_α and $\bar{\nu}_\alpha$ be different, i.e. $\mu_\alpha \neq 0$ in Eq.(1). In this degenerate case, changes are expected: a) in the weak rates (a reduction in Y_p), since a positive ξ_e enhances $n \to p$ processes with respect to the inverse ones, and modifies the initial condition on the n/p ratio at $T \gg 1$ MeV; b) in the expansion rate (an increase in Y_p), since non zero ξ_α's contribute to total N_{eff} as

$$N_{\text{eff}}^{\text{DBBN}} \simeq N_{\text{eff}} + \sum_\alpha \left[\frac{30}{7} \left(\frac{\xi_\alpha}{\pi} \right)^2 + \frac{15}{7} \left(\frac{\xi_\alpha}{\pi} \right)^4 \right]. \quad (4)$$

When several neutrino species are degenerate, both the effects might combine and particular values of ξ_α's exist for which the predictions of DBBN are still in good agreement with the observational data on the abundances of primordial elements. Notice that BBN is more sensitive to neutrino degeneracy than CMB or LSS, due to the further effect played by the $\nu_e - \bar{\nu}_e$ distributions in the weak rates, while the latter are only sensitive to the extra energy density present in the $\xi \neq 0$ case.

Earlier claims of discrepancies between BBN and CMB have been largely overcome by new data, and it was recently realized [10] that the flavor oscillations in the primordial plasma induced by (presently determined) mass differences and mixing angles from atmospheric and solar neutrinos almost equalize the three asymmetry parameters ξ_α. Still, exotic models, where both a common relatively large ξ and a $N_{\text{eff}} \neq 3.04$ exist, have been considered ("hidden relativistic degrees of freedom" [11, 12]), and were shown to be compatible with the data. However, if one sticks to the scenario with N_{eff} fixed by standard Physics, introduces no sterile species, and assumes the CMB prior on ω_b, BBN turns into a powerful "leptometer", constraining the common ξ to an unprecedent accuracy even under conservative assumptions for Y_p [13]. This provides an indirect consistency check for the sphaleron mechanism at electroweak phase transition, predicting baryon and lepton asymmetries of the same order.

Let us come to neutrino role in structure formation. In general, neutrinos tend to stream freely across gravitational potential wells, and to erase density perturbations. Free-streaming is efficient on a characteristic scale λ_J called the Jeans length, corresponding roughly to the distance on which neutrinos can travel in a Hubble time.

For ultra-relativistic neutrinos, λ_J is by definition equal to the Hubble radius c/H, but for non-relativistic ones it is lower than c/H. Neutrinos with masses smaller than approximately 0.3 eV are still relativistic at the time of last scattering, and their direct effect on the CMB perturbations is identical to that of massless neutrinos. In the intermediate mass range from 10^{-3} eV to 0.3 eV, the transition to the non-relativistic regime takes place during structure formation, and the matter power spectrum will be directly affected in a mass-dependent way. Wavelengths λ smaller than the current value of the neutrino λ_J are suppressed by free-streaming. The largest wavelengths, which remain always larger than the neutrino λ_J, are not affected. Finally, there is a range of intermediate λ which become smaller than the neutrino λ_J for some time, and then encompass it again: these scales smoothly interpolate between the two regimes. The net signature in the matter power spectrum $P(\lambda, \tau)$ is a damping of all wavelengths smaller than the Hubble scale at the time τ_0 of the transition of neutrinos to a non-relativistic regime.

Then, for $\lambda \ll \lambda_J$, if $\Omega_\nu \ll \Omega_m$ the suppression is given roughly by the factor

$$\frac{\Delta P}{P} \simeq -8 \frac{\Omega_\nu}{\Omega_m}, \quad (5)$$

that is by the ratio between neutrino and matter energy densities.

Notice that one can somehow play with both N_{eff} and $\sum m_\nu$ and find models which give excellent fits of the data. In fact, models with massive neutrinos have suppressed power at small scale. Adding relativistic energy further suppresses power at scales smaller than the horizon at matter-radiation equality. For the same matter density such a model would therefore be even more incompatible with data. However, if the matter density is increased together with m_ν and N_{eff}, data can be described very nicely (see, for example Figure 3 in [14]).

COMPARISON WITH DATA

Still few years ago, the BBN theory together with the observations of the abundances of primordial nuclides were used to determine the baryon fraction of the universe, ω_b. Nowadays $\omega_b \approx 0.023$ is fixed to better than 5% accuracy by detailed CMB anisotropies analysis [1], thus leaving the BBN as an over-constrained and very predictive theory. Once $\omega_b = 0.023 \pm 0.001$ is plugged into the BBN theory, the prediction for the deuterium, which is the nuclide most sensitive to ω_b, nicely fits the range of the observed values in high redshift, damped Ly-α QSO systems [15], thus offering a remarkable example of internal consistency of the current cosmological scenario. Moreover, the predictions of other light nuclei

TABLE 1. Bounds on N_{eff} (2σ) from different analyses

Ref.	Bound on N_{eff}	Data used
[7]	$1.8 \leq N_{eff} \leq 3.7$	CMB, BBN
[12]	$1.3 \leq N_{eff} \leq 6.1$	CMB, BBN(D)
[12]	$1.6 \leq N_{eff} \leq 3.6$	BBN(D+Y_p)
[17]	$1.4 \leq N_{eff} \leq 6.8$	CMB, LSS, HST
[18]	$1.9(2.3) \leq N_{eff} \leq 7.0(3.0)$	CMB, LSS, (+BBN)
[19]	$1.7 \leq N_{eff} \leq 3.0$	CMB, BBN
[20]	$N_{eff} \leq 4.6$	CMB, BBN
[21]	$1.90 \leq N_{eff} \leq 6.62$	CMB, LSS, HST

TABLE 3. Bounds on $\sum m_\nu$ (2σ) from different analyses

Ref.	Bound on $\sum m_\nu$	Data used
[1]	≤ 0.69	CMB, LSS, Lyα
[18]	≤ 1.01	CMB, LSS, HST, SNIa
[22]	≤ 0.65	CMB, LSS, HST, Lyα
[23]	$= 0.56^{+0.30}_{-0.26}$	CMB, LSS, f_{gas}, XLF
[24]	≤ 1.7	CMB, LSS
[25]	≤ 0.75	CMB, LSS, HST
[26]	$\leq 1.0(0.6)$	CMB, LSS (+HST, SNIa)
[27]	≤ 0.47	CMB, LSS, HST, SNIa, Lyα

which at least qualitatively agree with the observed values are likely to put constraints on the Galactic chemical evolution (^{3}He) or on the temperature scale calibration or depletion mechanisms in PopII halo stars (^{7}Li).

The determination of Y_p is usually performed by extrapolating to zero metallicity the measurements done in dwarf irregular and blue compact galaxies. The typical statistical errors are of the order of 0.002 (*i.e.*, at the 1% level), but the systematics are such that in the recent reanalysis [16] the authors argue for the conservative range $0.232 \leq Y_p \leq 0.258$, *i.e.* a 1σ error of $\mathcal{O}(5\%)$.

In Table 1 the present bounds on the effective number of neutrinos from various analyses are presented, together with the type of data used. The most stringent bounds come from BBN alone (Deuterium+Helium), while CMB and BBN-Deuterium are less effective in constraining N_{eff}. Some differences are due to (slightly) different databases or assumptions.

Table 2 shows the bounds on ξ_e. The interval from Ref. [12] is broader than what in Ref. [11], since in the first case only a prior from CMB instead of all data is used in the analysis. The third line shows the bound obtained in [13] assuming only standard physics, while the previous two bounds assume no prior on N_{eff}.

Finally, Table 3 shows, with the same notations, the upper bound on the neutrino mass. As can be gauged from this table, a fairly robust bound on the sum of neutrino masses is at present somewhere around 1 eV, depending on the specific priors and data sets used.

In conclusion, the complementarity among different fields of cosmology (BBN, CMB, LSS) can be used to test the role of neutrinos from very early epochs (redshift $z \simeq 10^{10}$) down to relatively recent history ($z \simeq$ a few).

TABLE 2. Bounds on ξ_e (2σ) from different analyses

Ref.	Bound on ξ_e	Data used
[11]	$-0.10 \leq \xi_e \leq 0.25$	CMB, BBN
[12]	$-0.13 \leq \xi_e \leq 0.31$	BBN(D+Y_p)+prior on ω_b
[13]	$-0.05 \leq \xi_e \leq 0.07$	BBN(Y_p)+priors on ω_b, N_{eff}

New nuclear rate measurements and a better understanding of possible systematics affecting primordial abundance determination (BBN) and more data together with an increased precision (CMB and LSS) will give us the opportunity to constrain standard neutrino properties as well as to test new physics in the neutrino sector, gaining at the same time a deeper insight on the physics of the early universe.

REFERENCES

1. D. N. Spergel *et al.*[WMAP Collaboration], *Astrophys. J. Suppl.*, **148**, 175 (2003).
2. G. Mangano *et al.*, *Phys. Lett.*, **B534**, 8 (2002).
3. S. Esposito *et al.*, *Nucl. Phys.*, **B590**, 539 (2000).
4. G. Mangano *et al.*, hep-ph/0506164.
5. R. E. Lopez and M. S. Turner, *Phys. Rev.*, **D59**, 103502 (1999).
6. S. Esposito *et al.*, *Nucl. Phys.*, **B568**, 421 (2000).
7. P. D. Serpico *et al.*, *JCAP*, **0412**, 010 (2004).
8. S. Eidelman *et al.*[Particle Data Group], *Phys. Lett.*, **B592** (2004) 1.
9. S. Esposito *et al.*, *JHEP*, **0009**, 038 (2000).
10. A. D. Dolgov *et al.*, *Nucl. Phys.*, **B632**, 363 (2002).
11. V. Barger *et al.*, *Phys. Lett.*, **B569**, 123 (2003).
12. A. Cuoco *et al.*, *Int. J. Mod. Phys.*, **A19**, 4431 (2004).
13. P. D. Serpico and G. G. Raffelt, *Phys. Rev.*, **D71**, 127301 (2005).
14. S. Hannestad, *New J. Phys.*, **6**, 108 (2004).
15. D. Kirkman *et al.*, *Astrophys. J. Suppl.*, **149** (2003) 1.
16. K. A. Olive and E. D. Skillman, *Astrophys. J.*, **617**, 29 (2004).
17. P. Crotty *et al.*, *Phys. Rev.*, **D67**, 123005 (2003).
18. S. Hannestad, *JCAP*, **0305**, 004 (2003).
19. V. Barger *et al.*, *Phys. Lett.*, **B566**, 8 (2003).
20. R. H. Cyburt *et al.*, *Astropart. Phys.*, **23**, 313 (2005).
21. E. Pierpaoli, *Mon. Not. Roy. Astron. Soc.*, **342**, L63 (2003).
22. S. Hannestad, hep-ph/0409108.
23. S. W. Allen *et al.*, *Mon. Not. Roy. Astron. Soc.*, **346**, 593 (2003).
24. M. Tegmark *et al.*[SDSS Collaboration], *Phys. Rev.*, **D69**, 103501 (2004).
25. V. Barger *et al.*, *Phys. Lett.*, **B595**, 55 (2004).
26. P. Crotty *et al.*, *Phys. Rev.*, **D69**, 123007 (2004).
27. G. L. Fogli *et al.*, *Phys. Rev.*, **D70**, 113003 (2004).

Anti-GZK Effect in UHECR Spectrum

R. Aloisio[1] and V.S. Berezinsky

INFN - Laboratori Nazionali del Gran Sasso, Assergi (AQ), Italy

Abstract. In this paper we discuss the anti-GZK effect that arises in the framework of the diffusive propagation of Ultra High Energy (UHE) protons. This effect consists in a jump-like increase of the maximum distance from which UHE protons can reach the observer. The position of the jump is independent of the Intergalactic Magnetic Field (IMF) strength and depends only on the energy losses of protons, namely on the transition energy from adiabatic and pair-production energy losses. The Ultra High Energy Cosmic Rays (UHECR) spectrum presents a low-energy steepening approximately at this energy, which is very close to the position of the observed second knee. The dip, seen in the universal spectrum as a signature of the proton interaction with the Cosmic Microwave Background (CMB) radiation, is also present in the case of diffusive propagation in magnetic fields.

Keywords: cosmic rays, diffusion, intergalactic magnetic field
PACS: PACS numbers: 01.30.Cc, 13.85.Tp, 98.62.Ra, 95.85.Sz

INTRODUCTION

Recently a very interesting phenomenon, determined by the UHE proton propagation in IMF, has been found [1, 2]. It consists in a low energy steepening of the proton spectrum that occurs at energy below 10^{18} eV. This steepening is caused by an increase of the diffusive propagation time, that rapidly exceeds the age of the universe, and can be explained trough the diffusive propagation of UHECR in IMF. The position of the steepening energy $E_s = 1 \times 10^{18}$ eV, is determined only by the proton energy losses on CMB and coincides with a good accuracy to the position of the 2nd knee observed in the CR spectrum [3]. In this paper we will discuss the main features of the diffusive propagation of UHECR in IMF, focusing our attention on the steepening energy scale E_s. Before entering the details of the diffusive propagation of UHECR protons let us review the main experimental evidences related to IMF.

The presence of an IMF is still an open question, the most reliable observations of this field are based on the measurement of the Faraday rotation (RM) of polarized radio emission [4]. The upper limit obtained with these measurements is $RM < 5$ rad/m^2, it implies an upper limit on the IMF that depends on the assumed scale of coherence length. For instance, according to [5], in the case of an inhomogeneous universe $B_{l_c} < 4$ nG with a scale of coherence of about $l_c = 50$ Mpc. In general, as follows from the observations of Faraday rotation, the magnetic field is high, of the order of 1 μG with a coherence length $l_c = 1$ Mpc, in clusters of galaxies and radio lobes of radio galaxies [4]. Apart from observations,

the IMF can be predicted, in principle, trough Magneto-hydrodynamics (MHD) simulations. The main ambiguities in these simulations are related to the assumed seed magnetic field and to the capability of simulations to reconstruct the local Universe as we observe it (i.e. constrained [6] and unconstrained simulations [7]). Unfortunately, because of these uncertainties, MHD simulations are not completely conclusive, there are at least two opposite results in literature with predicted magnetic field in voids (filaments) that vary from 10^{-3} nG (10^{-1} nG) [6] up to 10^{-1} nG (10 nG) [7].

While a direct evaluation of the IMF strength is still challenging, indirect informations about the UHECR propagation mode can be inferred from UHECR data. The analysis of the arrival directions of UHECR at energies $E > 10^{19}$ eV shows a small angle clustering within the angular resolution of the detectors. The AGASA detector has found 3 doublets and 1 triplet among 47 detected events [8]. This analysis is confirmed also by the combined data of different detectors [8] in which 8 doublets and 2 triplets are found in 92 collected events. This evidence can be well understood in terms of a rectilinear propagation of protons at the highest energies ($E > 10^{19}$ eV) with a random arrival of two (three) particles from the same source and a source number density of about $n_s \simeq 10^{-5}$ Mpc^{-3} [9]. However, the small angle clustering may survive in the case of UHE protons propagation in IMF [7, 10].

Another remarkable evidence of an almost rectilinear propagation of UHECR, in the energy range $2 - 8 \times 10^{19}$ eV, has been found by Tinyakov and Tkachev [11]. These authors have found a correlation between arrival directions of UHE particles in the AGASA and Yakutsk detectors and the directions of several BL-Lac objects, (i.e. AGNs with jet directed toward us). The combined evi-

---
[1] Talk presented by R. Aloisio

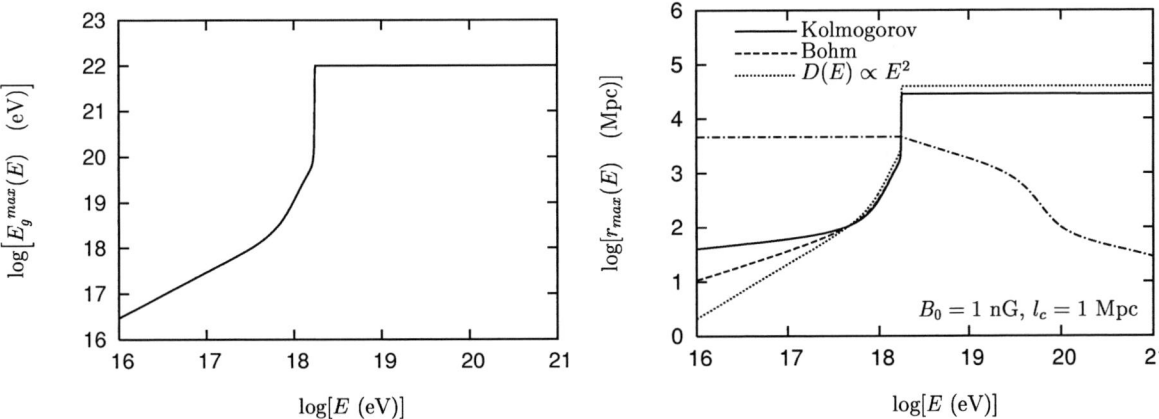

FIGURE 1. [**Left Panel**] Maximum generation energy E_g^{max} defined as $\min[E_g(E,t_0), E_{max}^{acc}]$, where E_{max}^{acc} is the maximal acceleration energy and t_0 is the age of the universe (see text). [**Right Panel**] Maximal distance $r_{max}(E)$ to the contributing sources as function of the observed energy E. Three merging curves in the left-low corner give $r_{max}^{diff}(E)$ and the dash-dotted curve gives $r_{max}^{rect}(E)$, which numerically is very close to the energy-attenuation length $l_{att}(E) = [(1/cE)dE/dt]^{-1}$.

dences of small angle clustering and correlation with BL-Lacs favor a scenario with a quasi-rectilinear propagation of protons at energies larger than 10^{19} eV.

UHECR DIFFUSIVE PROPAGATION

In order to describe the diffusive propagation in IMF of UHECR protons we will use the Syrovatsky [12] solution to the diffusive equation. Following [2], we will also assume a distribution of sources on a lattice; under this hypothesis the diffuse flux can be calculated as the sum over the fluxes from the discrete sources at distances r_i:

$$J_p^{diff}(E) = \frac{c}{4\pi} \frac{L_p K(\gamma_g)}{b(E)E_{min}^2} \sum_i \int_E^{E_g^{max}} dE_g \left(\frac{E_g}{E_{min}}\right)^{-\gamma_g} \times \\ \times \frac{exp\left[-\frac{r_i^2}{4\lambda(E,E_g)}\right]}{(4\pi\lambda(E,E_g))^{3/2}}, \quad (1)$$

where $b(E) = dE/dt$ is the proton energy losses (we have used $b(E)$ as computed in [13]) and $Q_{inj} = (L_p K(\gamma_g)/E_{min}^2)(E_g/E_{min})^{-\gamma_g}$ is the particle generation rate per unit energy with L_p the source luminosity (here we assume identical sources with the same luminosity) and $K(\gamma_g) = \gamma_g - 2$ (if $\gamma_g > 2$) a normalization constant. The function $\lambda(E,E_g)$ is

$$\lambda(E,E_g) = \int_E^{E_g} d\varepsilon \frac{D(\varepsilon)}{b(\varepsilon)} \quad (2)$$

with $D(E)$ the diffusion coefficient. The quantity in (2) describes the squared distance traversed by a proton in the direction of the observer, while its energy decreases from E_g to E. The Syrovatsky parameter $\lambda(E,E_g)$ poses a natural cut to the flux contributing sources, it is clear from equation (1) that a source at distance $r > 2\sqrt{\lambda(E,E_g)}$ gives a negligible (exponentially suppressed) contribution to the flux.

The Syrovatsky solution (1) formally includes all propagation times up to $t \to \infty$ and the generation energy of particle is limited only by the maximum energy that the source can provide E_{max}. Nevertheless, the propagation time should be smaller than the age of the universe t_0, this poses an additional limit to the generation energy represented by $E_g(E,t_0)$, which can be computed evolving backward in time the proton energy from E at $t=0$ up to E_g at $t=t_0$. Therefore, to limit the propagation time, the upper limit of integration $E_g^{max}(E)$ in equation (1) is fixed at the minimum between E_{max} and $E_g(E,t_0)$. It is interesting to note that at energies $E < E_s = 1 \times 10^{18}$, where only adiabatic energy losses are relevant, the upper limit of integration in (1) is fixed by $E_g^{max}(E) = E_g(E,t_0)$ while at higher energies, where pair-production and photo-pion-production energy losses become relevant, the upper limit of integration is $E_g^{max} = E_{max}$. This behavior of $E_g^{max}(E)$ is responsible for the low energy steepening of the UHECR diffusive spectrum. In Figure 1 (right panel) we have reported the E_g^{max} as function of the observed energy E, with $E_{max} = 1 \times 10^{22}$ eV.

Let us now concentrate on the diffusion coefficient that enters in $\lambda(E,E_g)$ function, see equation (2). Following [2], we assume diffusion in a random magnetic field with a strength B_0 on the coherence length l_c. This assumption determines the diffusion coefficient $D(E)$ at the highest energies when the Larmor radius of protons $r_L(E) = 1(B_0/nG)^{-1}(E/10^{18}eV)$ Mpc becomes larger than l_c, namely $D(E) = D_0(r_L(E)/l_c)^2$ with $D_0 = cl_c/3$.

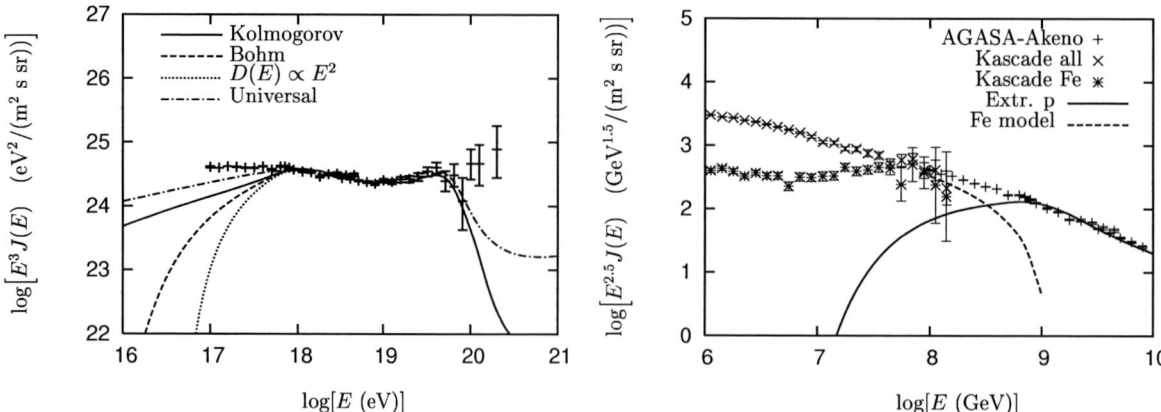

FIGURE 2. [Left Panel] Energy spectrum in the case of $B_0 = 1$ nG, $l_c = 1$ Mpc and for the diffusion regimes: Kolmogorov (continuous line), Bohm (dashed line) and $D(E) \propto E^2$ (dotted line). The separation between sources is $d = 30$ Mpc and the injection spectrum index is $\gamma_g = 2.7$ (see text). The AGASA-Akeno data with the universal spectrum (dash-dotted line) are also reported. [Right Panel] The galactic iron-nuclei spectrum computed by subtracting the extragalactic proton spectrum from the Akeno-AGASA data. The extragalactic proton spectrum is taken for the case $B_0 = 1$ nG, $l_c = 1$ Mpc, $d = 30$ Mpc, $\gamma_g = 2.7$ with the Bohm diffusion at $E < E_c$.

At low energy (i.e. when $r_L(E) \lesssim l_c$) we have considered three different cases: (i) the Kolmogorov diffusion $D_K(E) = D_0(r_L(E)/l_c)^{1/3}$; (ii) the Bohm diffusion $D_B(E) = D_0(r_L(E)/l_c)$; (iii) an arbitrary case $D(E) \propto E^\alpha$ with $\alpha = 2$ as the extreme possibility, also in this case $D(E)$ is normalized by D_0 at $r_L = l_c$, that corresponds to an energy $E_c \simeq 10^{18}(B_0/nG)(l_c/Mpc)$ eV. The diffusion length can be evaluated through the interpolation formula $l_{diff} = \Lambda_d + r_L^2/l_c$, with: $\Lambda_d = r_L$ and $\Lambda_d = l_c(r_L/l_c)^{1/3}$ in the Bohm and Kolmogorov cases respectively. At distances $r < l_{diff}(E)$ the proton propagation becomes rectilinear and the fluxes from individual sources on the lattice are computed in the rectilinear propagation limit. In this limit, taking into account the cosmological evolution of the Universe, the diffuse flux in our lattice model will be:

$$J_p^{\text{rect}}(E) = \frac{L_p K(\gamma_g)}{(4\pi E_{\min})^2} \sum_i \left(\frac{E_g(E,z_i)}{E_{\min}}\right)^{-\gamma_g} \times \quad (3)$$

$$\times \frac{1}{r_i^2(1+z_i)} \frac{dE_g(E,z_i)}{dE}$$

where z_i is the red shift that, according to the standard cosmology, is associated to the source at distance r_i, the two quantities $E_g(E,z_i)$ and $dE_g(E,z_i)/dE$ can be computed according to [13].

Following [2] we fix a reasonable strength of the IMF: namely $B_0 = 1$ nG on the coherence scale $l_c = 1$ Mpc. This choice of (B_0, l_c) is compatible with the limits that follows from the Faraday rotation measures and with a rectilinear propagation regime at the highest energies $E > 10^{19}$ eV. In order to reproduce, in our lattice model, the density of sources as follows from small angle clustering, we have chosen a separation between sources, i.e. a lattice spacing, of $d = 30$ Mpc that corresponds to a source space density of $n_s = 1/d^3 = 3.7 \times 10^{-5}$ Mpc^{-3}. In the computation of the fluxes we have assumed an injection spectrum with a single power law $\gamma_g = 2.7$, starting from the minimum energy $E_{\min} = 1$ GeV.

As follows from equation (1) the particles that, as a whole, contribute to the diffusive flux are those particles produced inside a sphere of radius $r_{\max}(E) = 2\sqrt{\lambda(E, E_g^{\max})}$, the contribution to the flux is exponentially suppressed for particles produced at higher distances. In Figure 1 (left panel) we have reported, fixing $l_c = 1$ Mpc and $B_0 = 1$ nG the behavior of the maximum contributing distance $r_{\max}(E)$ in the three cases Kolmogorov, Bohm, and $D(E) \propto E^2$. The dot-dashed line represents the proton attenuation length $l_{\text{att}} = E(dE/dl)^{-1}$, that can be interpreted as the maximum contributing distance in the rectilinear propagation regime. Figure 1 (right panel) clearly illustrates the steepening of the diffusive spectrum at low energies. While the energy-attenuation length l_{att} diminishes with energy and has the GZK steepening at energy $E \simeq 5 \times 10^{19}$ eV, the diffusive maximum distance $r_{\max}(E)$ increases with energy and has a sharp jump at energy $E_{\text{eq}} \simeq 2 \times 10^{18}$ eV. The position of this jump is related only to the proton energy losses, it does not depend on the magnetic field configuration (i.e. on the diffusion parameters). The energy E_{eq} corresponds to the proton energy at which pair-production energy losses become equal to adiabatic energy losses. At this energy the upper limit of integration in the Syrovatsky solution changes abruptly from $E_g(E,t_0)$ to the maximum energy $E_{\max}$ that the source can provide.

Let us consider now the UHECR spectra shown in Figure 2 (left panel). In this figure we report the diffusive and universal spectra, the latter corresponds to the rectilinear propagation (see equation (3)) in the case of a homogeneous distribution of sources (see [14] for a detailed discussion). The effect of the low energy steepening is clearly seen in the spectra, with the steepening energy E_s being independent of the diffusive regime chosen. According to the results presented in Figure 1 (right panel) for $r_{max}(E)$ the flux below the steepening energy E_s is largest for the Kolmogorov diffusion and lowest in the case $D(E) \propto E^2$, with the Bohm diffusion between them. The source luminosity L_p needed to provide the observed spectrum is very high, for a distance between sources of $d = 30$ Mpc the required luminosity is $L_p = 3.0 \times 10^{48}$ erg/s. To reduce the required emissivity one can assume that the acceleration mechanism works only starting from a somewhat higher minimum energy E_{min} (see [2, 3] for a detailed discussion).

It is also worthwhile to stress that the feature of the dip, that signals in the experimental data for a proton dominated spectrum, is not washed out by the IMF in the diffusive approximation. The validity of this approximation is related to the validity of the Syrovatsky solution at energies $E < 10^{19}$ eV. The diffusive equation, and hence its solution, are valid only in the case in which the energy losses and diffusion coefficient are time independent. At energies lower than 10^{19} eV, this is not the case, because the propagation time of protons approaches the age of the universe and the effect of the CMB temperature variation with time (red-shift) becomes important. However, our computations show a good agreement between the quasi-rectilinear regime of propagation and the exact rectilinear propagation, this agreement shows the approximate validity of the Syrovatsky solution at the discussed energies.

Following [1, 2, 15] we shall conclude by discussing shortly the transition from galactic to extragalactic cosmic rays. The remarkable feature of the diffusive spectra is the low-energy steepening at the fixed energy $E_s \sim 1 \times 10^{18}$ eV, which provide the transition from extragalactic to galactic CR. This energy coincides approximately with the position of the 2nd knee E_{2K} and gives a non-trivial explanation of its value as $E_{2K} \sim E_s$. Like in the above-mentioned works we shall assume that at $E \gtrsim 1 \times 10^{17}$ eV the galactic spectrum is dominated by iron nuclei and calculate their flux by subtracting the calculated flux of extragalactic protons from all-particle Akeno spectrum. For these calculations we shall fix the spectrum with magnetic configuration (1 nG, 1 Mpc), the Bohm diffusion and a separation between sources on the lattice $d = 30$ Mpc The calculated spectrum of galactic iron is shown in Figure 2 (right panel) by the dashed curve. This prediction should be taken with caution because obtained with a model-dependent calculations (assumption of the Bohm diffusion) and uncertainties involved in the Syrovatsky solution. However, it is interesting to note that the iron-nuclei spectrum in Figure 2 (right panel) is well described by the Hall diffusion [16] in the galactic magnetic field at energies above the knee.

CONCLUSIONS

We have analyzed the anti-GZK effect in the diffusive propagation of UHE protons. This effect consists in a sharp increase of the maximum distance $r_{max}(E)$ from which UHE protons can arrive (Figure 1 (left panel)). The observational consequences of the anti-GZK effect is a low-energy steepening of the diffuse spectrum. While the shape of the steepening depends on the magnetic field configuration, the steepening energy E_s is practically independent of it. The steepening of the spectrum at $E_s \sim 1 \times 10^{18}$ eV coincides with the 2nd knee observed in the CR spectra by most of the detectors and provides a natural transition from galactic iron nuclei to extragalactic protons.

REFERENCES

1. M. Lemoine, Phys. Rev. **D71** 083007 (2005).
2. R. Aloisio and V.S. Berezinsky, Astrop. J. **625** (2005) 249.
3. R. Aloisio, V.S. Berezinsky, P. Blasi, A.Z. Gazizov and S.I. Grigorieva, *in preparation*.
4. P.P. Kronberg, Rep. Progr. Phys. **57**, 325 (1994); J.P. Vallee, Fund. Cosm. Phys. **19**, 1 (1997); C.L. Carilli and G.B. Taylor, Annual Rev. Astr. Astroph., **40**, 319 (2002).
5. P. Blasi, S. Burles, A.V. Olinto, Ap.J. **514**, 79 (1999).
6. K. Dolag, D. Grasso, V. Springel, I. Tkachev, JTEP Lett. **79** 583 (2004); Pisma Zh. Eksp. Teor. Fiz. **79** 719 (2004).
7. G. Sigl, F. Miniati and T.A. Enßlin, Phys. Rev. **D68** 043002 (2003).
8. M. Takeda [AGASA collaboration], Astrop. J. **522**, 225 (1999); C. B. Finley and S. Westerhoff, Astrop. Phys. **21**, 359 (2004); Y. Uchihori et al., Astrop. Phys. **13**, 151 (2000).
9. S. L. Dubovsky P. G. Tinyakov and I. I. Tkachev, Phys. Rev. Lett. **85** 1154 (2000); Z. Fodor and S. Katz, Phys. Rev. **D63**, 23002 (2001); P. Blasi and D. De Marco, Astrop. Phys. **20**, 559 (2004); M. Kachelriess and D. Semikoz, *preprint* astro-ph/0405258.
10. H. Yoshiguchi, S. Nagataki, S. Tsubaki and K. Sato, Astrophys.J. **586** (2003) 1211-1231.
11. P.G. Tinyakov and I. Tkachev, JETP Lett. **74**, 445 (2001).
12. S.I. Syrovatskii, Sov. Astron. **3** 22 (1959).
13. V.S. Berezinsky and S.I. Grigorieva, Astron. Astrophys. **199** 1 (1988); V.S. Berezinsky, A.Z. Gazizov and S.I Grigorieva, *preprint* hep-ph/0204357.
14. R. Aloisio and V.S. Berezinsky, Astrop. J. **612** 900 (2004).
15. V.S. Berezinsky, S.I. Grigorieva and B.I. Hnatyk, Astrop. Phys. **21**, 617 (2004).
16. V.S. Ptuskin et al., Astron. Astroph. **268**, 726 (1993).

Ultrahigh Energy Cosmic Rays Detection

Carla Aramo

Istituto Nazionale di Fisica Nucleare - Sezione di Napoli
Complesso Universitario di Monte Sant'Angelo, Via Cintia, 80126 - Napoli

Abstract. The paper describes methods used for the detection of cosmic rays with energies above 10^{18} eV (UHECR, UltraHigh Energy Cosmic Rays). It had been anticipated there would be a cutoff in the energy spectrum of primary cosmic rays around $3 \cdot 10^{19}$ eV induced by their interaction with the 2.7 °K primordial photons. This has become known as the GZK cutoff. However, several showers have been detected with estimated primary energy exceeding this limit.

Keywords: Cosmic Ray, Ground Array, Fluorescence Detectors, P. Auger Observatory
PACS: 96.40.Pq, 96.40.De, 95.55.Vj

INTRODUCTION

UHECRs form the tail of the cosmic-ray spectrum, which extends from 1 GeV to beyond 10^{20} eV [1]. Their energy is equivalent to that of a tennis ball moving at 100 km/h and their flux is about once a year every 100 km^2 of the earth's surafce. Because of their rarity we know relatively little about them; in particular, we do not understand how or where these particles gain their remarkable energies. The prediction of the existence of the Greisen-Zatsepin-Kuzmin (GZK) cut-off [2] for particles with energy above $3 \cdot 10^{20}$ eV has been faulted with the detection of several showers generated by a primary particles with energy well beyond 10^{20} eV [3].

In this paper methods used for UHECR detection will be described. Moreover first results of P. Auger Observatory (PAO) will be shown. This experiment [4] has been conceived to measure the properties of the highest-energy cosmic rays with unprecedented statistical precision. The PAO detects ultra-high energy cosmic rays by implementing two complementary airshower techniques: the combination of a large ground array and fluorescence detectors. This hybrid observation of events allows a rich variety of measurements on a individual shower, providing much more information than with either detector alone.

The complete observatory will consist of two instruments, constructed in the northern and southern hemispheres, each covering an area of 3000 km^2.

DETECTION TECHNIQUES

Currently UHECRs are investigated using two different detection methods. The first method consists on the distribution of a number of particle counters across a large area allowing detection of particles which survive to the detection level. The other method exploits the excitation of nitrogen molecules by the particles in the shower and the associated fluorescence emission of light in the 300-400 nm band. The light is detected by photomultipliers and the profile of the shower in the atmosphere can be inferred rather directly.

UHECR detection with ground arrays

An air shower produces a large number of particles spread out over a large area at the observation level. Particles are detected with an array of detectors deployed over an appropriate area of many square kilometers. The separation of detectors is typically many hundreds of meters. The density of charged particle and their arrival time are measured at each detector location. These informations allow the reconstruction of shower axis and shower core (the impact point of the axis on the ground) by fitting the station signal size to expected lateral distribution function (LDF). The primary energy is estimated by a local charged particle density at fixed distance from the core in meters S(core distance) which depend by the array size. For example in the AGASA array was used $S(600)$ [5, 6] and in the PAO is used $S(1000)$ [7]. All of the arrays built to detect cosmic rays above 10^{19} eV have been located between 800 g cm^{-2} and sea level. This is appropriate, as the average maximum depth of showers of these primaries is about 750 g cm^{-2} and it is effective to study showers close to or beyond shower maximum. The shower disc has a thickness that increases from a few nanoseconds close to the shower core up to several microseconds at distances beyond 1 km. The accuracy of the timing measurement is only one of the factors that limit the directional precision: a second is the area of the detector. With giant arrays the arrival direction has been measured to an accuracy of between 0.5° and 5°.

Detection with fluorescence detectors

The first successful implementation of the fluorescence technique was obtained by the Fly's Eye group [8]. The fluorescence detector follows the trajectory of an extensive air shower and measures the energy dissipated by shower particles in the atmosphere that acts as an air calorimeter of more than 10^{10} tons. For this purpose, the whole sky is viewed by many segmented mirrors focusing the collected fluorescence light emitted isotropically along the trajectory of the shower on a photomultiplier matrix. Correlation between the light intensity and light arrival time detected in each PMT provides unambiguous information on energy released and shower path in the atmosphere. The shower detector plane (SDP), defined in Figure 1, is constructed from sequence of hit photomultipliers. Then the distance to the shower axis (impact parameter) Rp and incident angle ψ in the plane are determined by fitting the time sequence of several photomultiplier signals. Once the track geometry is determined, the number of photons N_γ received by a photomultiplier is calculated. Since there is also contamination from direct and scattered Cerenkov light, the longitudinal size $N_e(x)$ of the extensive air shower for each angular bin is calculated via an iterative process to remove those contributions which depend upon the viewing angle between the pointing direction of the photomultiplier and the shower axis. The resultant photoelectrons are directly proportional to the number of charged particles in the angular bin. From the integration, $\int N_e(x)dx$, the total track length is estimated. If a shower is seen simultaneously by two fluorescence detectors (stereo event), a shower detector plane for each one can be determined and the intersection of these planes defines the shower trajectory without timing information. The total track length can also be determined independently by each detector.

AGASA and HiRes Experiment

As an example of the two techniques above described AGASA [9] and HiReS [10] Experiment are discussed. AGASA was operated at the Akeno Observatory (Tokyo), the operation had been started in February 1990 and was closed in January 2004 with an $\sim$ 95% overall live ratio. In an $\sim$ 100 km^2 area 111 detector stations were deployed. Each station was equipped with 2.2 m^2 surface detector with a 5 cm-thick scintillator viewed by a PMT. At 27 southern stations muon detectors were built. They consisted of 14–20 proportional counters aligned below an absorber (30 cm-thick iron or 1 m-thick concrete: 0.5 GeV threshold energy for vertical incidence). The primary energy was estimated by a local charged particle density at 600 m from the core,

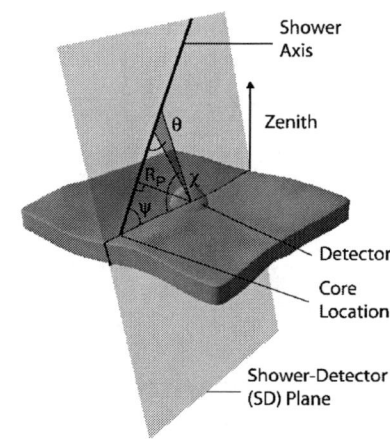

FIGURE 1. The geometry of the reconstruction for a monocular air fluorescence detector

known as $S(600)$: $E[\text{eV}] = 2.03 \times 10^{17} \cdot S(600)$. The error of primary energy determination is ± 30% at $10^{19.5}$eV and ± 25% at 10^{20}eV. The exposure of AGASA is almost constant above $\sim 7 \times 10^{18}$ eV and is 5.8×10^{16} m^2 sr s. Figure 2 shows the energy spectrum of UHECRs above $10^{18.5}$ eV [11]. The vertical axis denotes the differential flux multiplied by E^3. Poisson bounds are given at a confidence level (CL) of 68%. Upper limits are given at a 90% CL. The dashed curve represents the expected flux by the GZK hypothesis for the uniform source distribution [12]. The most noticeable feature is the observation of cosmic rays beyond the GZK cutoff energy. Eleven events were detected above 10^{20} eV against the expected ~ 1.9 events [9].

HiRes Experiment have two detectors located atop desert mountains in west central Utah. The detectors consist of mirrors that collect fluorescence light and focus it on arrays of 256 hexagonal photomultiplier tubes (PMT's). Each PMT subtends about one degree of sky. The HiRes-I detector consists of 21 mirrors arranged to look from 3 to 17 degrees in elevation and almost 360 degrees in azimuth. The HiRes-II detector, located 12.6 km SW of HiRes-I, consists of 42 mirrors, which cover 3 to 31 degrees in elevation and almost 360 degrees in azimuth. The spectrum of the HiRes-I and HiRes-II detectors, observing in monocular mode are shown in Figure 2 where a marked deficit of events above $10^{19.8}$ eV can be noted. This energy is the threshold for pion production in interactions between cosmic ray protons and the average photon of the CMBR; i.e., the deficit occurs at the energy of the GZK cutoff. On the contrary the results of the AGASA experiment seem to indicate that the spectrum continues above the ankle at a constant power law. To test whether HiRes data are consistent with this interpretation of the Agasa result, the HiRes data were fit, from the ankle to the pion production threshold, to a power law, then

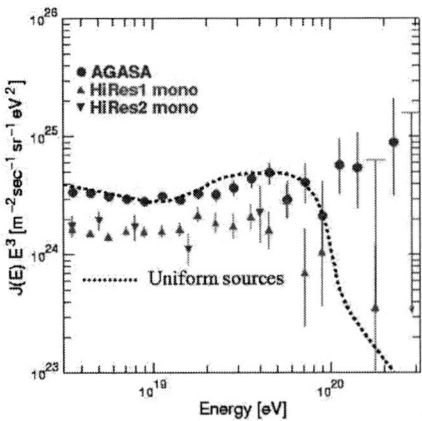

FIGURE 2. UHECR energy spectrum

continue the power law to higher energies. This tests the hypothesis that the GZK cutoff is absent, as the AGASA data seem to show. The power law index is 2.8 ± 0.1. If the cutoff were absent HiRes expect to see 29.0 events above $10^{19.8}$ eV, while only 11 event have been seen. The Poisson probability of seeing 11 or fewer events, with a mean of 29.0, is 1×10^{-4} [10].

PIERRE AUGER OBSERVATORY

The PAO was designed to observe, in coincidence, the shower particles at ground and the associated fluorescence light generated in the atmosphere. This is achieved with a large array of water Cherenkov detectors coupled with air-fluorescence detectors (eyes) that overlook the surface array. It is not simply a dual experiment. Apart from important cross-checks and measurement redundancy, the two techniques see air showers in complementary ways. The surface detector (SD) measures the lateral structure of the shower at ground level, with some ability to separate the electromagnetic and muon components. On the other hand, the fluorescence detector (FD) records the longitudinal profile of the shower during its development through the atmosphere. A hybrid event is an air shower that is simultaneously detected by the fluorescence detector and the ground array. Data are recovered from both detectors whenever either system is triggered. The Observatory started operation in hybrid production mode in January, 2004. Surface stations have a 100% duty cycle while fluorescence eyes can only operate on clear moonless nights. Both surface and fluorescence detectors have been running simultaneously 14% of the time. The SD accumulated aceptance is larger than 1600 km^2 sr yr while the FD aceptance is $\sim 14\%$ of the SD due to the limited duty cycle [13]. The number of hybrid events represents 10% the statistics of the surface array data. The southern site of the Pierre Auger Cosmic Ray Observatory in Argentina now covers an area of approximately 1500 km^2 with an explosure of 1750 km^2 sr yr and a full efficiency above 3 EeV for zenith angles less than 60° [14]. Two of the Auger fluorescence detector sites (Los Leones and Coihueco) have been operating in a stable manner since January 2004 and a third site (Los Morados) began operation in March 2005.

The Hybrid Performance and Measurements

A hybrid detector has excellent capability for studying the highest energy cosmic ray air showers. Much of its capability stems from the accurate geometrical reconstructions it achieves. Timing information from even one surface station can much improve the geometrical reconstruction of a shower over that achieved using only eye pixel information. The reconstruction accuracy is better than the ground array counters or the single eye could achieve independently. A core location resolution of 50 m is achieved. The resolution for the arrival direction of cosmic rays is 0.6° [15]. Due to the much improved angular accuracy, the hybrid data sample is ideal for anisotropy studies and, in particular, for point source searches. The combination of the air fluorescence measurements and particle detections on the ground provides an energy measurement almost independent of air shower simulations. The fluorescence measurements determine the longitudinal development of the shower, whose integral is proportional to the total energy of the electromagnetic particle cascade. At the same time, the particle density at any given distance from the core can be evaluated with the ground array. It is important to note that both techniques have different systematics, and results are preliminary at this stage while the Observatory is under construction. The hybrid analysis benefits from the calorimetry of the fluorescence technique and the uniformity of the surface detector aperture. Operation started in January, 2004 and over 16000 hybrid events have been successfully reconstructed up to now [16]. An example of hybrid and stereo (two FD eyes) event with an energy of ~ 21 Eev is reported in Figure 3.

The first AUGER energy spectrum and Anisotropy Studies

The methods to calculate the cosmic ray energy spectrum are simple and robust, exploiting the combination of FD and SD. The methods do not rely on detailed numerical simulation or any assumption about the chemical composition. The spectrum in Figure 4 is only a first es-

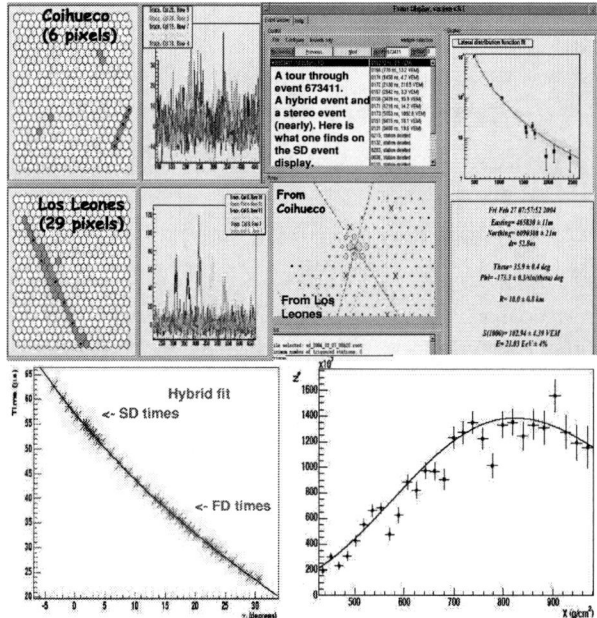

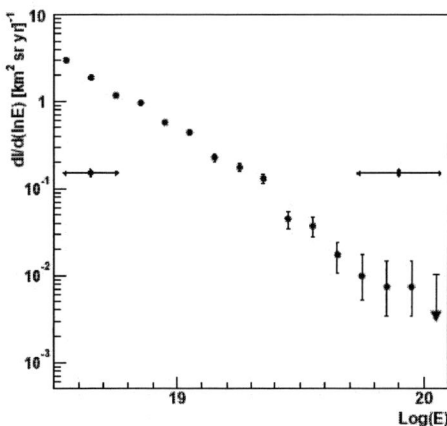

FIGURE 4. Estimated Auger energy spectrum [17]. Error bars on points indicate statistical uncertainty. Systematic uncertainty is indicated by double arrows at two different energies.

FIGURE 3. Example of hybrid stereo event (21 EeV). **Top**: at left, green line is the hit PMTs in FD camera of Coihueco eye (up) and Los Leone eye (down) with on the right the PMT signals of marked pixels. At right the SD informations: on the left the core position (10 km) as cross point of two lines that are the SDP from Coihueco and Los Leones and on the right the LDF function. **Down**: at left the hybrid time fit is shown with blue (red) points related to tanks (pixels in the FD camera). On the right the number of electrons as function of slant depth in the atmosphere is shown. The solid line is the Gaisser-Hillas function that best fit the data.

timate and is related with data from 1 Jan 2004 through 5 Jun 2005 [17]. Events are included for zenith angle 0-60° and for energies above 3 EeV, in total 3525 events. It has significant systematic and statistical uncertainties. The indicated statistical error for each point comes directly from the Poisson uncertainty in the number of measured showers in that logarithmic energy bin. There is larger systematic uncertainty in the conversion of S(1000) to energy. Part of that comes from the FD energies themselves. The accuracy is limited by the available statistics, and the total systematic energy uncertainty grows with energy: from 30% at 3 EeV to 50% at 100 EeV. This uncertainty is indicated by horizontal double arrows in Figure 4, and a 10% systematic uncertainty in the exposure is indicated by vertical arrows.

The Auger data have been analyzed to search for excesses of events near the direction of the galactic center in several energy ranges around EeV energies [18]. In this region the statistics accumulated by the Observatory are already larger than that of any previous experiment. Using both the data sets from the surface detector and hybrid data sets any significant excess is find. These results do not support the excesses reported by AGASA and SUGAR experiments. An upper bound on the flux of cosmic rays arriving within a few degrees from the galactic center in the energy range from 0.8-3.2 EeV is set: $\Phi_s < 2.8\,\xi\,10^{-15}\,\text{m}^{-2}\,\text{s}^{-1}$ @ 95% CL with ξ in [1.0-4.0] [18]. Also the search for correlations of cosmic ray arrival directions with the galactic plane and with the super-galactic plane at energies in the range 1-5 EeV and 12 above 5 EeV found no significant excess [18].

ACKNOWLEDGMENTS

I wish to thank all the P. Auger Collaboration, in particular the Napoli group. I thank the organizers of IFAE 2005 for an exciting workshop and for their hospitality.

REFERENCES

1. M. Nagano and A.A. Watson, Rev. Mod. Phy., 72 (2000)
2. Greisen, K., Phys. Rev. Lett., 16 (1966) 748; Zatsepin, Z.T. and Kuz'min, V.A., Pisma Zh. Eksp. Teor. Fiz. 4 (1966)
3. C. Aramo(ed.) et al.,"GZK and Surroundings" CRIS 2004 Nucl. Phys. B (Proc. Suppl.) 136 (2004) 1-453
4. The AUGER Coll., NIM A523 (2004) 50-95
5. Hillas, A.M. et al., 12th ICRC (Australia) 3 (1971) 1001
6. H.Y. Dai et al., J. Phys. G 14 (1988) 793
7. P. Ghia for AUGER Coll., 29th ICRC (India) (2005)
8. D. Bird et al., NIM A349 (1994) 592
9. Agasa Coll., Nucl. Phys. B (Proc. Suppl.) 136 (2004) 18
10. G. Thomson, Nucl. Phys. B (Proc. Suppl.) 136 (2004) 28
11. Takeda, M. et al., Astropart. Phys. 19 (2003) 447
12. Glushkov, A.V. et al., 28th ICRC (Japan) (2003) 389
13. J. Matthews for AUGER Coll., 29th ICRC (India) (2005)
14. E. Parizot for AUGER Coll., 29th ICRC (India) (2005)
15. C. Bonifazi for AUGER Coll., 29th ICRC (India) (2005)
16. M. Mostafa for AUGER Coll., 29th ICRC (India) (2005)
17. P. Sommers for AUGER Coll., 29th ICRC (India) (2005)
18. A. Selvon for AUGER Coll., 29th ICRC (India) (2005)

Gamma-Ray Astronomy with Cherenkov Telescopes

A. Stamerra

Dipartimento di Fisica, Università di Siena, via Roma 1, Siena, Italy and INFN Sez. Pisa, Italy

Abstract.
The γ-astronomy with Cherenkov telescope has recently made great progress coming to the 2nd generation of instruments. The higher sensitivity and lower energy threshold have brought to discover new sources. The determination of their spectra, necessary for developing models for acceleration, emission and absorption of the γ-rays, will shed new light on the century-old problem of the origin of galactic cosmic rays.

A short summary of the current physics objectives in the γ-astronomy and a description of the Cherenkov telescopes technique is given. Some representative observations from the two operative telescopes, HESS and MAGIC are described.

Keywords: Cherenkov Telescopes; gamma-rays observations;
PACS: 95.55.Ka 95.85.Pw

INTRODUCTION

The importance of the γ-ray astronomy at very high energies (VHE) is steadily grown in the last decade. It is believed it should play a crucial role in solving the problem of origin of galactic cosmic rays (CR)[1]. Due to the deviation of the charged particles in the interstellar magnetic fields, localized accelerators cannot be identified by observations of such particles near the Earth. Individual CR sources may be found amongst the regions of localized γ-emission. In fact most energetic objects like pulsars, supernovae remnants, microquasars, active galaxies, emit radio and X-ray synchrotron radiation, indicating that electrons of tens of TeV energy range are present in these objects. It is a common belief that also CR nucleons should have been accelerated. The inverse Compton effect (IC) of electrons on the photon field and the γ-ray emission from nucleon-induced π^0-decay produce the observed VHE emission.

The increasing knowledge of high energy emission processes in astrophysical sources, has also shown the importance of this emission in the global energetic budget of these sources.

CR ORIGIN AND THE HADRONIC-LEPTONIC DILEMMA

Since a full discussion of the physics objectives of ground-based γ-astronomy goes beyond the limit of this contribution and can be found elsewhere[2], only a simple description of the problem of the origin of CR is given.

Although the supernovae remnants (SNRs) are the favourites candidates as CR sources up to 10^{15} eV, no clear hadronic acceleration site has been found among them. The diffusive shock acceleration (DSA) mechanism provides the theoretical framework and allows observation predictions to probe the content of CRs in the SNRs[7]. The alternate scenario to the emission related to the shock-accelerated protons and nuclei, through π^0-decay γ-rays, is the inverse Compton (IC) emission of the synchrotron electrons.

In figure 1 the broad-band spectral energy distribution of the TeV J2032+41 source is shown, together with the theoretical prediction for IC and hadronic emission, superimposed on the HEGRA measured spectrum. It is clear that a discrimination between the two models will come from high precision spectrum measurements and from observations carried on at lower energies.

The actual generation of Cherenkov telescopes could be able to probe the SNRs emission of γs originated from π^0-decay and thus to provide a decisive test for the SNRs origin of galactic CRs.

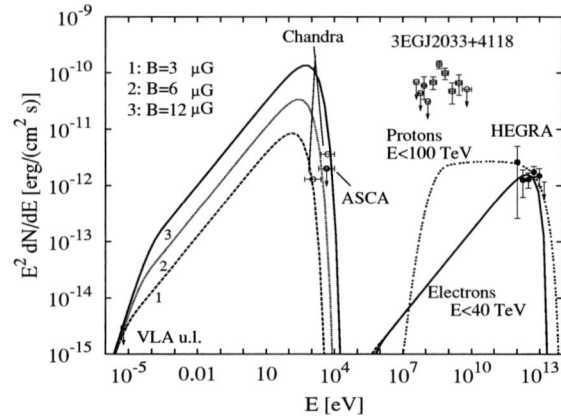

FIGURE 1. The integrated broad band SED of TeV J2032+41 source[8].

THE IACT TECHNIQUE

The Earth's atmosphere (28 radiation length thickness) is opaque to high-energy photons. Nevertheless, one can observe from ground secondary effects when the VHE γs hit the atmosphere, producing the so-called extended air showers. The Cherenkov radiation generated by the energetic charged particles in the shower are collected by the optical reflector of the Cherenkov telescope on a pixelated camera that records the images of the shower. Cherenkov photons from a single shower arrive on the ground as a short pulse of few nanoseconds duration and spread over a large area, roughly a circle with several hundred meter diameter. Thus the effective area of a Cherenkov telescope is given by the size of the Cherenkov light pool -roughly $10^4 m^2$ - and is much higher than the geometric area of the instrument.

The other critical point of gamma-ray telescopes is their efficiency to reject the cosmic-ray background, which is orders of magnitude larger than the observed gamma-ray signal. The successful detection of gamma-ray signals by Cherenkov telescopes, is based on the imaging technique[6], that can remove over $\sim$ 99.7% of the cosmic-ray background exploiting the characteristic differences of the hadronic and purely electromagnetic showers, as shown in fig. 2. These two points, the large effective area and the background rejection, constitute the basis of the most successful technique of ground-based gamma-ray astronomy, the Imaging Atmospheric Cherenkov Telescope (IACT) technique.

A BIT OF HISTORY

The observation time necessary to detect the Crab Nebula can be used to measure the performance improvement of the Cherenkov technique in the last decades.

The first camera mounted on the Whipple telescope belongs to the the so-called 0th generation of Cherenkov telescopes (1968-1976). It consisted of a single 12.5 cm phototube and had no imaging properties. The background was estimated pointing to a sky position where no source was expected. After three years of observations a γ-ray detection of the Crab Nebula above 250 GeV at 3 standard deviations above background was reported [3].

In 1982 the first imaging camera was built and implemented on the 10m Whipple reflector. The ability to distinguish between background hadronic showers and VHE γ-ray showers, on the basis of the shape of their image, allowed the Crab detection at 9 standard deviations above background after 90 hours of data taking at an energy threshold of 700 GeV [4]. This was one of the most important milestones in the development of γ-ray astronomy.

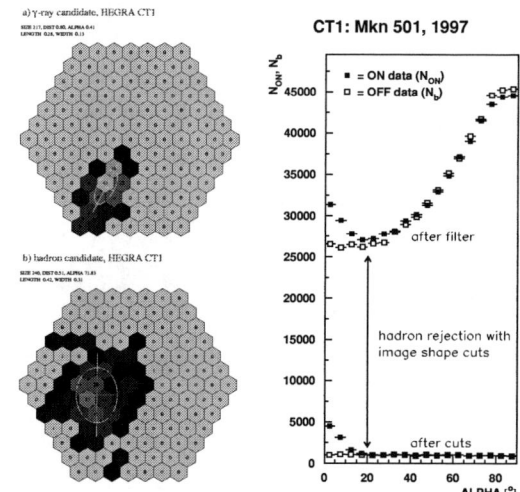

FIGURE 2. Typical images of a gamma ray (top-left) and a hadron-initiated shower (bottom-left). The major axis of the gamma image is oriented toward the camera center, where the source is located. The angle between the major axis of the image and the line connecting the source location with the center of gravity of the image is the ALPHA parameter. (right) ALPHA distribution of the data from the 1997 Markarian 501 flare observed by HEGRA CT1. ALPHA is small for showers coming from the direction of the source. Cuts in ALPHA and the shape parameters can achieve a background suppression factor of about 200.

In the following decade the IACT technique flourished in many projects employing high resolution cameras and the new stereoscopic observation mode. Fifteen galactic and extragalactic sources have been established as VHE emitters by Whipple, HEGRA, CAT, CANGAROO and other groups. The Crab could be detected in 15 minutes at 5 standard deviations. The 1st generation of γ-ray telescopes has firmly established ground based γ-ray astronomy as a highly regarded discipline.

SENSITIVITY AND ENERGY THRESHOLD

The 1st generation of IACTs could not fill the gap between the VHE sky seen at $E > 300$ GeV and the HE region explored by EGRET at $E < 100$ MeV, where 271 point sources were discovered.

The actual generation of Cherenkov telescope has improved the IACT technique following two paths. HESS, VERITAS and CANGAROO has raised the sensitivity employing the stereoscopic observation mode and a high resolution camera. MAGIC has reduced the energy threshold to few tenth of GeV using a very large reflecting surface (225 m^2) and introducing new challenging techniques[5].

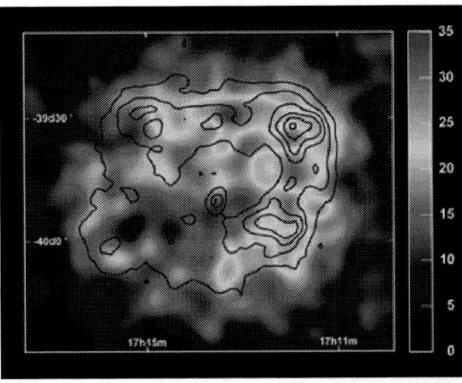

FIGURE 3. TeV gamma-ray image of .RX J1713.7-3946 obtained with the H.E.S.S. telescopes[9]. No background subtraction has been applied. The superimposed contours show the X-ray surface brightness as seen by ASCA in the 1-3 keV range. The energy threshold is 800 GeV.

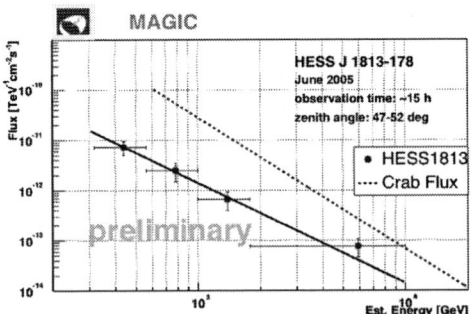

FIGURE 4. Spectrum of HESS1813-178 measured by MAGIC in the Northern Emisphere. Despite the large zenith angle observation the energy threshold is 300 GeV.

Some up-to-date results of the two operative telescopes, HESS and MAGIC, are here reviewed.

HESS

HESS (High Energy Stereoscopic System) consists of four 13 m diameter optical dishes, each equipped with a 960-pixel PMT camera with a field of view of 5°. The array was completed in December 2003 and is located in the Southern Emisphere (Namibia). It operates at an energy threshold of 100 GeV, with an angular resolution of few arcminutes and an energy resolution ≈15%.

HESS has a high resolution and sensitivity to conduct morphological and spectrometric studies of extended galactic sources. The observations of the SNR RX J1713.7-3946 have first provided a γ-ray image of a TeV source on arcminute scales (fig.3). The new HESS data obtained in 2004 with the full four-telescope system, will allow significant extension of the energy spectrum which could provide conclusions concerning the emission processes of this source[10].

The wide field of view allowed HESS to scan the inner part of the Galactic plane[11]. Eight unknown VHE sources were detected at the level of 6 standard deviations. For two of these sources, HESS J1813-178 and HESS J1614-518, no counterparts in other wavelength regimes were found in the literature. Shortly after the discovery of HESS J1813-178 was published, counterparts were located in existing but unpublished multi-wavelength data. In radio data a shell-type supernova remnant was located[12], and an X-ray emission is seen by ASCA[14] and INTEGRAL[13]. MAGIC has confirmed the VHE emission from HESS J1813-178 at $E >$ 300 GeV, providing its first spectrum[15].

The recent HESS discoveries are contributing to populate the γ-sky and to reveal new galactic source populations emitting in the VHE region.

MAGIC

The Major Atmospheric Imaging Cherenkov telescope (MAGIC]) is currently the largest single dish Imaging Air Cherenkov Telescope (IACT). Located on the Canary Island La Palma (N, W, 2200m a.s.l), the telescope has a 17m diameter tessellated parabolic mirror, supported by a light weight carbon fiber frame. It is equipped with a high efficiency 576-pixel FOV photomultiplier camera. It operates at an energy threshold of 30 Gev with an angular resolution of $\approx 0.1°$.

MAGIC aims for the lowest energy threshold possible for a Cherenkov telescope. A low energy threshold allows the detection of sources with a soft spectrum and sources where a cut-off is expected.

Figure 5 shows the Crab spectrum measured down to energies of 100 GeV[16].

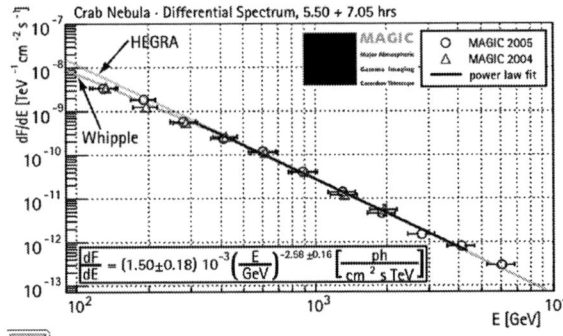

FIGURE 5. Observed differential crab spectra for the 2004 and 2005 datasets. A power law fit to the combined data between 300GeV and 2TeV as well as the fit to HEGRA data and the parametrization of Whipple data are shown.

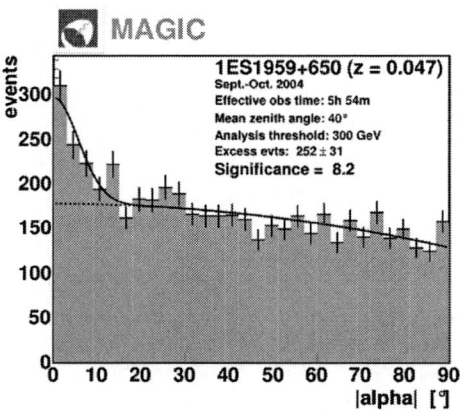

FIGURE 6. ALPHA plots after γ-hadron separation for 1ES1959. The background has been determined with a fit in the region of ALPHA > 20, with a polynomial function of second order.

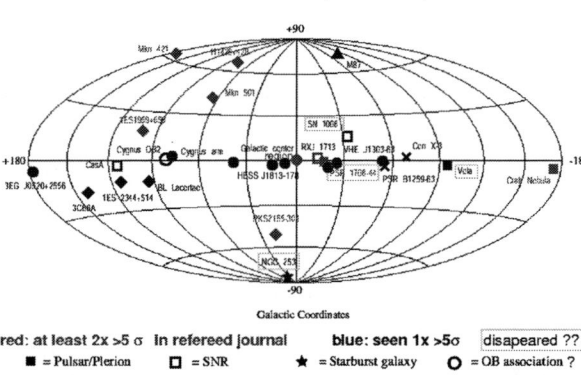

FIGURE 7. Sources discovered in the VHE region. Status on March 2005.

The advantages of a low energy threshold are demostrated by the observation of the AGN 1ES1959+650 in quiet state [17]. The evidence of a steep spectrum (differential index $\alpha \approx -3$) favors the detection at low energies. 1ES1959 is at $\sim 20\%$ of the Crab level at about 200 GeV, while the flux drops at $\sim 6\%$ Crab at ~ 2 TeV. As a consequence this AGN could be detected by MAGIC in 6 hrs, whereas HEGRA needed 94 hrs!

The effect of absorption of γ-rays on the infrared background for pair production suppresses the detection of distant sources at high energies. The so-called γ-horizon is a function of the energy[18] and at redshift $z > 1$ 100 GeV γs are completely absorbed. With an energy threshold of 50 Gev MAGIC can observe sources at $z \sim 2$.

MAGIC has been also designed to catch the GRB bursts. On July 13th 2005 the GRB050713a has been promptly observed by MAGIC 40 seconds after the burst, overlapping for 70 seconds with the SWIFT observation. The presence or absence of a signal will provide strong constraints on the actual GRB models on the prompt emission.

CONCLUSIONS

The γ-ray astronomy with Cherenkov telescope has been observationally motivated in the last decade, and the recent exciting discoveries of new γ-emitters are providing lots of results to be understood, and new sources to be investigated. The new generation of IACTs has fulfilled the expected explosion in the number of VHE γ-ray sources (see fig. 7), making this field an exciting area to work in.

ACKNOWLEDGMENTS

The support of the italian INFN is gratefully aknowledged. I am grateful to the organizers of the XVII IFAE meeting for inviting me to give this talk.

REFERENCES

1. F.A. Aharonian, *Proc. 22nd Texas Symposium, Stanford*, p.43 (2004)
2. R. Bock et al., *Astrop. Phys.*, **23** 493 (2005)
3. G.C. Fazio et al, *Ap. J.*, **175** L117 (1972)
4. T.C. Weekes et al, *Ap. J.*, **342** 379 (1989)
5. A. Stamerra [MAGIC Collaboration], *Proc. Workshop on Science with the New Generation of High-Energy Gamma-ray Experiments, Perugia* (2003)
6. A.M. Hillas, *Proc. 19th ICRC, La Jolla*, **3** 445 (1985)
7. M.A. Malkov, L.O'C. Drury, *Rep. Prog. Phys.*, **64** 429 (2001)
8. F.A. Aharonian et al. (HEGRA coll), *A&A*, **431** 197A (2005)
9. F.A. Aharonian et al. (HESS collaboration), *Nature*, **416** 823 (2004)
10. F.A. Aharonian et al. (HESS collaboration), in preparation (2005)
11. F.A. Aharonian et al. (HESS collaboration), *Science*, **307** 1938 (2005)
12. Helphand et al., *astroph/0505392*, (2005)
13. Ubertini et al., *astroph/0505191*, (2005)
14. Brogan et al., *astroph/0505145*, (2005)
15. H. Bartko et al. (MAGIC collaboration), *Proc. 29. ICRC, Pune*, to be published (2005)
16. R. Wagner et al. (MAGIC collaboration), *Proc. 29. ICRC, Pune*, to be published (2005)
17. N. Tonello et al. (MAGIC collaboration), *Proc. 29. ICRC, Pune*, to be published (2005)
18. T.M. Kneiske, K. Mannheim, D.H. Hartmann, *A&A*, **386** (2002)

Experimental High Energy Neutrino Astrophysics

Carla Distefano

Laboratori Nazionali del Sud, Catania, Italy

Abstract.
Neutrinos are considered promising probes for high energy astrophysics. More than four decades after deep water Čerenkov technique was proposed to detect high energy neutrinos. Two detectors of this type are successfully taking data: BAIKAL and AMANDA. They have demonstrated the feasibility of the high energy neutrino detection and have set first constraints on TeV neutrino production astrophysical models. The quest for the construction of km^3 size detectors have already started: in the South Pole, the IceCube neutrino telescope is under construction; the ANTARES, NEMO and NESTOR Collaborations are working towards the installation of a neutrino telescope in the Mediterranean Sea.

Keywords: high energy neutrino astronomy, underwater/ice telescopes
PACS: 95.55.Vj, 95.85.Ry, 96.40.Tv

INTRODUCTION

The detection of high energy neutrinos may solve several problems connected to the physics of cosmic rays. Many indications suggest, indeed, that cosmic objects, where acceleration of charged particles takes place, e.g. GRBs and AGNs, are the sources of the detected UHE-CRs. Accelerated hadrons, interacting with ambient gas or radiation, can produce HE neutrinos [1, 2]. Unlike charged particles and gamma rays having $E_\gamma >$ TeV, neutrinos can reach the Earth from far cosmic accelerators, travelling in straight line and therefore carrying direct information on the source. Theoretical models indicate that a detection area of $\simeq 1$ km^2 is required for the measurement of HE cosmic neutrino fluxes. The underwater/ice Čerenkov technique is widely considered the most promising experimental approach to build high energy neutrino detectors.

High energy neutrinos are detected by observing the Čerenkov radiation from secondary particles produced by neutrinos interacting inside large volumes of water/ice instrumented with a lattice of photomultiplier tubes (PMTs). In the case of muon neutrino interactions, muons range out over kilometres at TeV energy, to ten of kilometres at EeV energy, generating showers along their track by bremsstrahlung, pair production and photo-nuclear interactions. These are the sources of the Čerenkov light, together with the direct Čerenkov radiation emitted by the muons, detected by the lattice of PMTs. The Čerenkov wave-front is then reconstructed using the information of photon hits and PMT positions. Apart from muon tracks, cascades can also be detected. With a typical size of tens of meters in ice or water, cascades can be considered as quasi point-like compared to the spacing of OMs. All three neutrino flavors contribute to this signature and the possibility to detect all neutrino flavors allows to study neutrino oscillations. A cosmic accelerator is, in fact, believed to produce neutrinos from the decay of pions in the flux ratio $\nu_e : \nu_\mu : \nu_\tau = 1 : 2 : 0$. Neutrino oscillations change the ratio at the detection point into $1 : 1 : 1$, because approximately one half of the muon neutrinos convert to tau neutrinos over cosmic baselines.

The first generation of underwater/ice neutrino telescopes, BAIKAL and AMANDA, despite their limited size, have already set first constraints on TeV neutrino production astrophysical models. The quest for the construction of km^2 size detectors have already started: in the South Pole the IceCube neutrino telescope is under construction; the ANTARES, NEMO and NESTOR Collaborations are working towards the installation of a neutrino km^3 telescope in the Mediterranean Sea.

THE OPERATING DETECTORS: BAIKAL AND AMANDA DETECTORS

After the pioneering work carried out by DUMAND offshore Hawaii Island [3], Baikal was the first collaboration which installed an underwater neutrino telescope and, after more than ten years of operation, it is still the only neutrino telescope located in the Northern Hemisphere. The BAIKAL NT-200 is an array of 200 PMTs, moored between 1000 and 1100 m depth in lake Baikal (Russia) [4]. BAIKAL is a high granularity detector with a threshold $E_\mu \simeq 10$ GeV and an estimated effective detection area $\leq 10^5$ m^2 for TeV muons. The limited depth and the qualities of lake water (light transmission length of $15 \div 20$ m, high sedimentation and bio-fouling rate, optical background due to bioluminescence) limit the detector performances as a neutrino telescope. The Collaboration has successfully measured the atmospheric neutrino flux and set a 90% c.l. upper limit for diffuse astrophysical neutrino fluxes of $E_\nu^2 \Phi_\nu < 4 \times 10^{-7}$ cm^{-2} s^{-1} sr^{-1} GeV [5].

AMANDA is currently the largest neutrino telescope installed [6]. In the present stage, named AMANDA II, the detector consists of 677 optical modules (OM) pressure resistant glass vessel hosting downward oriented PMTs and readout electronics. OMs are arranged in 19 vertical strings, deployed in holes drilled in ice between 1.3 and 2.4 km depth. Vertical spacing between OMs is 10÷20 m, horizontal spacing between strings is 30÷50 m. The ice optical properties have been mapped as a function of depth: at detector installation depth the average light ($\lambda = 400$ nm) absorption length is $L_a \simeq 100$ m, the effective light scattering length is $L_b \simeq 20$ m. This makes AMANDA a good calorimeter for astrophysical events, with a resolution 0.4 in log(E) for muons and 0.15 in log(E) for electron cascades. The detector angular resolution is between 1.5° and 3.5° for muons and $\simeq 30°$ for cascades. AMANDA data have permitted to measure for the first time the upgoing atmospheric neutrino spectrum in the energy range from few TeV to 300 TeV (see figure 1). The atmospheric neutrino spectrum recorded by

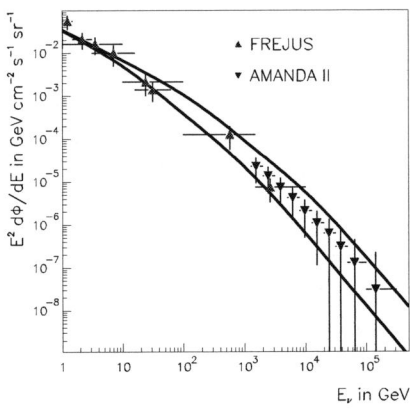

FIGURE 1. Atmospheric neutrino energy spectrum (preliminary) from AMANDA data, compared to the Frejus spectrum [7] at lower energies. The two solid curves indicate model predictions [8] for the horizontal (upper) and vertical (lower) flux.

AMANDA-II was then used to set a 90% c.l. upper limit on a diffuse muon neutrino flux of $E_{\nu_\mu}^2 \Phi_{\nu_\mu} < 2.6 \times 10^{-7}$ GeV cm^{-2} s^{-1} sr^{-1} ($100 < E_\nu < 300$ TeV). AMANDA can also detect cascades originated by the leptonic vertex of electron and tau neutrino charged current interactions, and hadronic vertex cascades from all-flavor neutral current interactions. The measured 90% c.l. upper limit with respect to the flux of all three flavors is $\Phi_{\nu_e+\nu_\mu+\nu_\tau} E_\nu^2 = 8.6 \cdot 10^{-7}$ cm^{-2}s^{-1}sr^{-1}GeV (50 TeV $< E_\nu <$ 5 PeV) [9]. At energies above 1 PeV, the Earth is opaque to neutrinos except for tau neutrinos which undergo regeneration. The search for astrophysical neutrinos is therefore concentrated on events close to the horizon and even from above, since the atmospheric muon background is low at these high energies. No excess above background is observed and the measured 90% c.l. upper limit on all flavors neutrino flux is $\Phi_{\nu_e+\nu_\mu+\nu_\tau} E_\nu^2 < 9.9 \times 10^{-7}$cm^{-2}s^{-1}sr^{-1}GeV (1 PeV $< E_\nu <$ 3 EeV) [10].

Search for neutrino point sources requires good pointing resolution and are thus restricted to the ν_μ channel. For point sources the detector reached a sensitivity $E_{\nu_\mu}^2 \Phi_{\nu_\mu} \simeq 7 \times 10^{-8}$ GeV cm^{-2} s^{-1} calculated over 807 days live time (years 2000-2003) [11]. The search for 33 pre-selected sources yielded no evidence for extraterrestrial point sources. The strongest excess was observed from the direction of the Crab nebula, with 10 events where 5 were expected [12]. A much larger array like a km^3 scale detector is therefore necessary to detect astrophysical sources.

THE FUTURE KM3 NEUTRINO TELESCOPES

AMANDA and BAIKAL have demonstrated the possibility to use the Čerenkov technique to track muons induced by $E_\nu >$ 100 GeV neutrinos. This success opened the way to the construction of km^3 neutrino telescopes such as IceCube [13], that will extend the AMANDA detector at South Pole.

The contemporary observation of the full sky with at least two neutrino telescopes in opposite Earth Hemispheres is an important issue for the study of transient phenomena. Moreover neutrino events detection from the Northern Hemisphere is required to observe the Galactic Centre region (not seen by IceCube), already observed by HESS as an intense TeV gamma source [14]. In the Northern Hemisphere a favourable region is offered by the Mediterranean Sea, where several abyssal sites (> 3000 m) close to the coast are present and where it is possible to install the detector near scientific and industrial infrastructures. Three Collaborations, ANTARES, NEMO and NESTOR, which are building demonstrator detectors and prototypes, aim at the construction of the km^3 detector in the Mediterranean Sea.

IceCube

The IceCube telescope will be the natural extension of AMANDA to the km^3 size. When completed (expected in 2010), it will consist of 4800 downward looking PMTs arranged in 80 strings. The distance between PMTs will be 16 m with a spacing of 125 m between the strings. IceCube will be able to identify muon tracks from muon neutrinos with $E_{\nu_\mu} > 100$ GeV. Simulations run by the IceCube Collaboration show that in three years of live time, the detector will reach a sensitivity $E^2_{\nu_\mu}\Phi_{\nu_\mu} = 4.2 \times 10^{-9}$ GeV cm^{-2} s^{-1} sr^{-1} for a diffuse E_ν^{-2} neutrino spectrum and a sensitivity $E^2_{\nu_\mu}\Phi_{\nu_\mu} = 2.4 \times 10^{-9}$ GeV cm^{-2} s^{-1} for a point-like source [13]. Besides, IceCube will be able to detect cascades from electron neutrinos with $E_{\nu_e} > 10$ TeV and from tau neutrinos with $E_{\nu_\tau} > 1$ PeV.

It is also worthwhile to mention that the under-ice detectors are not affected by radioactive and biological optical noise which is present in natural sea or lake waters. This makes them suitable for the search of low energy neutrino fluxes from Galactic SuperNova explosions.

The Mediterranean km^3

An underwater detector offers, compared to IceCube, the possibility to be recovered, maintained and/or reconfigured. Moreover detector installation at depth ≥ 3000 m will reduce the atmospheric muon background by a factor ≥ 5 with respect to 2000 m depth, where IceCube is located. As discussed above the Mediterranean sea offers several abyssal sites where a km^3 telescope could be installed at these depths. Moreover the long light scattering length of the Mediterranean abyssal seawater preserve the Čerenkov photons directionality and will permit excellent pointing accuracy (order of $0.1°$ for 10 TeV muons). On the other hand the light absorption length in water is shorter than in ice, then it reduces the photon collection efficiency for a single PMT. The selection of a marine site with optimal oceanographic and optical parameters is, therefore, a major task for the Mediterranean Collaborations involved in the km^3 project. In the following we discuss the three collaborations operating in the Mediterranean sea.

NESTOR. NESTOR [15], the first Collaboration that operated in the Mediterranean Sea, proposes to deploy a modular detector at 3800 m depth in the Ionian Sea, near the Peloponnese coast (Greece). Each module is a semi-rigid structure (*the NESTOR tower*), 360 m high and 32 m in diameter, equipped with $\simeq 170$ PMTs looking both in upward and downward directions. After a long R&D period, in March 2003 NESTOR has successfully deployed 1 tower floor, with a diameter of 12 m and equipped with 12 PMTs. This module was deployed at a depth of 3800 m in order to test the overall detector performance and particularly that of the data acquisition systems. It acquired, on-shore, underwater optical noise and cosmic muon signals for about 1 month. From data acquired during this time, 745 atmospheric muon events have been reconstructed, making it possible to measure the cosmic ray muon flux as a function of the zenith angle [16]. In the next future the Collaboration aims at the deployment of the first tower with 2×10^4 m^2 effective area for $E_\mu > 10$ TeV muons.

ANTARES. ANTARES will be a *demonstrator* neutrino telescope with an effective area of 0.1 km^2 for astrophysical neutrinos [17]. It will be located in a marine site near Toulon (France), at 2400 m depth. ANTARES will be a high granularity detector consisting of 12 strings, each equipped with 25 equidistant stories made of 3 PMTs (total 75), placed at an average distance of 60 m. The PMT are $45°$ downward oriented, in order to avoid their obscuration by sediments and bio-fouling. The whole detector installation is scheduled to be completed in 2007 [18]. After a first operation of two prototypal lines in spring 2003, a new improved version of a short line instrumented with oceanographic sensors and optical modules has been put in operation in April 2005 and it is taking data since then. Data recorded by optical modules show an unexpectedly high optical background ranging from 60 to several hundreds kHz, well above the one produced by ^{40}K decay, then probably due to bioluminescence.

NEMO. The NEMO Collaboration was formed in 1998 with the aim to carry out the necessary R&D towards the km^3 neutrino telescope. The activity has been mainly focused on the search and characterization of an optimal site for the installation and on the development of a feasibility study of the detector [19]. NEMO has intensively studied the oceanographic and optical properties in several deep sea (depth ≥ 3000 m) sites close the Italian coast. Results indicate that a large region located 80 km SE of Capo Passero (Sicily) is excellent for the installation of the km^3 detector. The bathymetric profile of the region is extremely flat over hundreds km^2, with an average depth of $\simeq 3500$ m. Deep sea currents are, in the average, as low as 3 cm s^{-1}, and never stronger than 12 cm s^{-1}. The average value of blue ($\lambda = 440$ nm) light absorption length is $L_a = 66 \pm 5$ m. The same device measured $L_a(440\text{nm}) \simeq 48$ m in Toulon site [21] and $L_a(488\text{nm}) = 27.9 \pm 0.9$ m in Baikal lake [22]. The optical background noise was also measured at 3000 m depth in Capo Passero. Data show that optical background induces on 10" PMTs (0.5 s.p.e.) a constant rate of $20 \div 30$ kHz (compatible with the one expected from ^{40}K decay), with negligible contribution of bioluminescence bursts. These results were confirmed by biological analysis that show, at depth> 2500, extremely small concentration of dissolved bioluminescent organisms [21].

Concerning the detector design, NEMO proposes an innovative structure to host OMs: the NEMO *tower*. It is designed to deploy, during a single operation, a large number of PMTs (≥ 60) arranged in a 3-dimensional shape in order to locally permit event trigger and track reconstruction [23]. The structure is mechanically flexible, being composed by a sequence of 16 ÷ 20 stories hosting OMs, interlinked by a net of syntectic fiber ropes. Each storey is 15 ÷ 20 m long and hosts two optical modules (one downward looking and one looking horizontally) at each storey end. The vertical inter-spacing between stories is ≃ 40 m. One of the detector geometries proposed by the Collaboration is the NEMO140dh. It consists of a squared of 9 × 9 NEMO towers equipped with 5832 optical modules (10" diameter PMTs). The distance between the towers is 140 m. This configuration reaches an effective area of 1 km^2 at muon energy of about 10 TeV, with an angular resolution lower than 0.1 degrees [24]. Figure 2 shows a preliminary result of the expected sensitivity to a E_v^{-2} muon neutrino spectrum coming from a point-like source, obtained for a search bin of 0.3 degree [25]. The same figure reports, as a comparison, the sensitivity of the IceCube detector, obtained for a search bin of 1 degree [13]. The NEMO140dh detector has, therefore, a better sensitivity and smaller search bin, allowing a better identification of point-like sources.

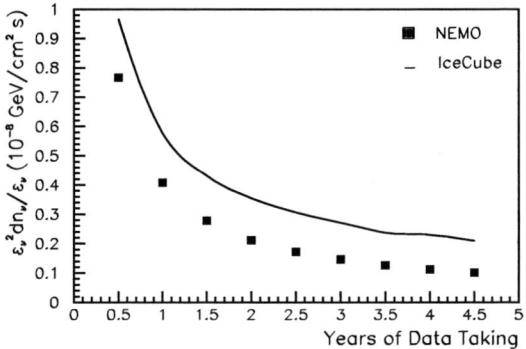

FIGURE 2. Preliminary result of the NEMO140dh sensitivity to a E_v^{-2} muon neutrino spectrum coming from a point-like source [25], as a function of years of data taking, and comparison with the IceCube detector [13].

As an intermediate step towards the km^3, the Collaboration has decided to realize a technological demonstrator including most of the critical elements of the proposed detector. This project is called NEMO Phase 1 [19] and it is under realization at the Underwater Test Site of the Laboratori Nazionali del Sud in Catania [20]. The completion of this project is foreseen by the end of 2006.

CONCLUSIONS

The first generation of underwater/ice Čerenkov neutrino telescopes, BAIKAL and AMANDA, have successfully measured the atmospheric neutrino spectrum, demonstrating the feasibility of the high energy neutrino detection. Besides, despite their limited size, they have already set first constraints on TeV neutrino production astrophysical models. The forthcoming km^3 neutrino telescopes will be *discovery* detectors that could widen the knowledge of the Universe. In the South Pole the IceCube detector is extending AMANDA to the km^3 size and it is planned to be completed within year 2010, being probably the first km^3 telescope running. In the Mediterranean Sea, ANTARES is going to install a 0.1 km^2 neutrino telescope demonstrator, NESTOR has deployed 12 PMTs in deep water, aiming at a 2×10^4 m^2 tower and NEMO is installing a technological demonstrator for km^3 detector and proposed Capo Passero as an optimal installation site. The research and development activities conducted by the three Mediterranean collaborations can represent a valuable experience in the construction of the underwater km^3 detector, which will be the result of their efforts.

REFERENCES

1. T.K. Gaisser, et al., *Phys. Rep.*, **D258**, 173, (1995).
2. E. Waxman & J. Bahcall, *Phys. Rev.*, **D64**, 023001, (1999).
3. A. Roberts et al., *Rev. of Mod. Phys.*, **64**, 259 (1992).
4. I. Belolaptikov et al., *Astrop. Phys.*, **7**, 263, (1997).
5. R. Wischnewski for the BAIKAL Collaboration, Proc. of *28th ICRC*, Tsukuba, Japan, 2003.
6. E. Andres et al., *Nature*, **40**, 441, (2001). AMANDA web page http://amanda.uci.edu.
7. K. Daum et al., *Z. Phys.*, **C66**, 417, (1995).
8. L.V. Volkova, *Sov. J.Nucl.Phys.*, **31**, 784, (1980).
9. M. Ackermann et al., *Astrop. Phys.*, **22**, 127, (2004).
10. M. Ackermann et al., *Astrop. Phys.*, **22**, 339, (2005).
11. M. Ackermann et al., *Phys. ReV.*, **D077102**, (2005).
12. C. Spiering, Proc. of the Nobel Symposium on Neutrino Physics, Haga Slott, Sweden, (2004).
13. J. Ahrens et al., *Astrop. Phys.*, **20**, 507, (2004).
14. F. Aharonian et al., *A&A*, **425**, L13, (2004).
15. NESTOR web page http://icecube.wisc.edu.
16. G. Aggouras et al., *Astrop. Phys.*, **23**, 377, (2005).
17. *ANTARES: A Deep Sea Telescoper for High Energy Neutrinos*, astro-ph/9907432.
18. E.V. Korolkova for the ANTARES Collaboration. Proc. of the *CRIS 2004*, Catania, Italy, 2004.
19. E. Migneco et al. Proc. of the CRIS 2004, Catania, Italy, 2004.
20. G. Riccobene for the NEMO Collaboration. Proc. of *29th ICRC*, Pune, India, 2005.
21. G. Riccobene for the NEMO Collaboration. Proc. of *VLVvT Workshop*, Amsterdam, 2003.
22. V. Balkanov et al. *NIM*, **A498**, 37231, (2003).
23. M. Musumeci for the NEMO Collaboration. Proc. of *VLVvT Workshop*, Amsterdam, 2003.
24. P. Sapienza et al., *LNS Activity Report*, (2005).
25. C. Distefano et al., *LNS Activity Report*, (2005).

Cosmogenic Neutrinos and New Physics Signal at Neutrino Telescopes

J. I. Illana*, M. Masip* and D. Meloni[†]

CAFPE and Depto. de Física Teórica y del Cosmos, Universidad de Granada, 18071 Granada, Spain
[†]*INFN and Dipto. di Fisica, Università degli Studi di Roma "La Sapienza", 00185 Rome, Italy*

Abstract. Cosmogenic neutrinos reach the Earth with energies around 10^9 GeV, and their interactions with matter will be measured in upcoming experiments (Auger, IceCube). Models with extra dimensions and the fundamental scale at the TeV could imply signals in these experiments. In particular, besides the production of microscopic black holes by cosmogenic neutrinos, gravity-mediated interactions at larger distances (that can be calculated in the eikonal approximation) can take place. In these processes a neutrino of energy E_ν interacts elastically with a parton inside a nucleon, loses a small fraction y of its energy, and starts a hadronic shower of energy $yE_\nu \ll E_\nu$. We show that for the expected fluxes of cosmogenic neutrinos these elastic processes give a stronger signal than black hole production in neutrino telescopes and that the energy distribution of contained hadronic showers can help to distinguish between eikonal and black hole or Standard Model events. On the other hand, the absence of any signal at IceCube would imply a bound of $M_D \gtrsim 5$ TeV.

Keywords: Extra-dimensions, Neutrinos
PACS: 04.50.+h, 13.15.+g, 96.40.Tv

INTRODUCTION

Cosmogenic neutrinos appear in any scenario proposed to explain the most energetic cosmic rays. In particular, if the observed air showers of up to 10^{11} GeV [1] are produced by primary protons, in their way to the Earth these protons will interact with the cosmic microwave background (CMB) photons and produce pions:

$$p + \gamma_{2.7K} \rightarrow \Delta^+ \rightarrow n + \pi^+ \, (p + \pi^0) \, . \quad (1)$$

The flux of cosmogenic neutrinos is created in the decay of the charged pions, and it will appear correlated with observable fluxes of nucleons and photons.

Cosmogenic neutrinos are of great interest as probes of new TeV physics because of two generic reasons. First, they provide large center of mass energies. Second, the relative effect of new physics on the weakly interacting neutrinos is larger than on quarks or charged leptons, making it easier to see deviations. The signals of new physics could be detected in deeply penetrating air showers and neutrino telescopes.

In particular, in models with extra dimensions and the fundamental Planck scale (M_D) at the TeV [2] the gravitational interactions are unsuppressed in the transplanckian regime ($s \gg M_D^2$). The possibility of black hole (BH) formation [3] by cosmogenic neutrinos has been discussed in several papers [4]-[8]. Here we will consider the gravitational interaction at larger distances, where it can be calculated using the eikonal approximation [5, 9, 10]. We will show that these elastic processes are more frequent than BH formation and Standard Model events and that neutrino telescopes are the ideal place to observe them.

TEV GRAVITY

The simplest picture of TeV gravity includes only two free parameters: the value of the higher-dimensional Planck scale M_D, and the number n of compact dimensions where gravity propagates. A third parameter, the (common) length $2\pi R$ of the n dimensions, could be deduced from the 4-dimensional Newton constant $G_N \equiv M_P^{-2}$:

$$G_D = (2\pi R)^n G_N = \frac{(2\pi)^{n-1}}{4 M_D^{n+2}} \, . \quad (2)$$

At processes below M_D the model-independent signature of extra dimensions is graviton emission. The amount of energy radiated would be proportional to the accessible phase space or, in the Kaluza-Klein (KK) picture, to the number of KK modes of mass below the center of mass energy. In this type of experiments for a given n one sets bounds on R and then deduces the limits on M_D. From collider experiments one obtains $M_D \geq 1.4 \, (1.0)$ TeV for $n = 2 \, (\geq 3)$ [11].

The bounds obtained from transplanckian collisions are complementary in the sense that given n they are a direct probe of M_D, and R is then adjusted in order to reproduce G_N. At energies above M_D and impact

---

[1] Talk given by D. Meloni at IFAE, Incontri di Fisica delle Alte Energie, Catania (Italy), March 30th-April 2nd

parameters smaller than R the collision is a pure higher-dimensional process independent of the compactification details that fix the value of the effective Newton constant. The transplanckian collision does not *see* that the extra dimensions are compact, they could be taken infinite with no effect on the cross section.

Neutrino-parton amplitude

The TeV gravity model should be embedded in a string theory, which would relate M_D with the string scale M_S. In the simplest set-up [12] the standard model (SM) fields (open strings) would be attached to a 4d brane, whereas gravity (closed strings) would propagate in the whole Dd space. In this case

$$M_D^{n+2} = \frac{8\pi}{g^4} M_S^{n+2}, \quad (3)$$

with g the string coupling. The transplanckian regime corresponds then to energies above the string scale, where any tree-level amplitude becomes very *soft*. In the ultraviolet string amplitudes go to zero exponentially at fixed angle and, basically, only the forward (long distance) contribution of the graviton survives. This is precisely the regime where the eikonal approximation is valid.

Let us consider the elastic collision of a neutrino and a parton that exchange D-dimensional gravitons (see [5, 10] for details). The eikonal amplitude $\mathscr{A}_{eik}(s,t)$ resums the infinite set of ladder and cross-ladder diagrams. Essentially, $\mathscr{A}_{eik}$ (reliable as far as the momentum carried by the gravitons is smaller than the center of mass energy) is the exponentiation of the Born amplitude in impact parameter space:

$$\mathscr{A}_{eik}(s,t) = \frac{2s}{i} \int d^2b\, e^{i\mathbf{q}\cdot\mathbf{b}} \left(e^{i\chi(s,b)} - 1 \right), \quad (4)$$

where $\chi(s,b)$ is the eikonal phase and **b** spans the (bidimensional) impact parameter space. $\chi(s,b)$ can be deduced from the Fourier transform to impact parameter space of $\mathscr{A}_{Born}(s,t)$. Our Born amplitude comes from the t-channel exchange of a higher dimensional graviton:

$$\mathscr{A}_{Born} = -\frac{s^2}{M_D^{n+2}} \int \frac{d^n q_T}{t - q_T^2}, \quad (5)$$

where the integral over momentum q_T along the extra dimensions (equivalent to the sum over KK modes) gives an UV divergence. However the contributions from large q_T introduce corrections to the phase $\chi(s,b)$ only at small b ($\approx 1/q_T$), but this small b region gives a negligible contribution to $\mathscr{A}_{eik}$ in the transplanckian regime.

From $\mathscr{A}_{eik}$ one obtains $\chi(s,b) = (b_c/b)^n$, with

$$b_c^n = \frac{(4\pi)^{\frac{n}{2}-1}}{2} \Gamma\left(\frac{n}{2}\right) \frac{s}{M_D^{2+n}}. \quad (6)$$

The amplitude in Eq. (4) can [5, 10] then be written as $\mathscr{A}_{eik}(s,q) = 4\pi s b_c^2 F_n(b_c q)$, where

$$F_n(y) = -i \int_0^\infty dx\, x J_0(xy) \left(e^{ix^{-n}} - 1 \right), \quad (7)$$

$q = \sqrt{-t}$, and the integration variable is $x = b/b_c$. For $q < b_c^{-1}$ this integral is dominated by impact parameters around b_c, and for $q > b_c^{-1}$ by a saddle point at b_s. As q (or $y = q^2/s$) grows nonlinear corrections (H diagrams) become important [10]. The differential cross section $d\sigma_{eik}/dy$ grows as y decreases [5]. For example, taking $M_D = 1$ TeV and $E_\nu = 10^{10}$ GeV, for $n = 2$ (6) it is a factor of 265 (62) larger at $y = 10^{-3}$ than at $y = 0.1$. The small y region corresponds to long distance processes, with typical impact parameter distances larger than R_S, where the neutrino interacts with a parton and transfers only a small fraction of its energy (we take $y_{max}=0.2$). This region is less important for a larger number of extra dimensions, since then gravity *dilutes* faster and becomes weaker at long distances.

For $-t/s \approx 1$, b_s approaches [5] the Schwarzschild radius R_S of the system:

$$R_S = \left(\frac{2^n \pi^{\frac{n-3}{2}} \Gamma\left(\frac{n+3}{2}\right)}{n+2} \right)^{\frac{1}{n+1}} \left(\frac{s}{M_D^{2n+4}} \right)^{\frac{1}{2(n+1)}}. \quad (8)$$

At $b \leq R_S$ one expects an inelastic collision, with a significant emission of gravitons, and black hole (BH) formation. The latter possibility has been considered in several analyses [3]-[8], where is it also shown that a number of factors (angular momentum, charge, geometry of the trapped surface, radiation before the collapse) make a quantitative estimate difficult. For this kind of *inelastic* processes, we adopt a geometrical cross section, $\sigma_{BH} = \pi R_S^2$, including BH formation and *hard* scatterings with important graviton emission where the neutrino loses most of its initial energy.

SIGNALS AT NEUTRINO TELESCOPES

The flux of cosmogenic neutrinos depends on the production rate of primary nucleons of energy around and above the GZK cutoff E_{GZK}. It will always appear correlated with proton and photon fluxes that should be consistent, respectively, with the number of ultrahigh energy events at AGASA and HiRes [1] and with the diffuse γ-ray background measured by EGRET [13].

We will base our analysis on the two neutrino fluxes described in [14] (solid and dashed lines in Fig. 1). The

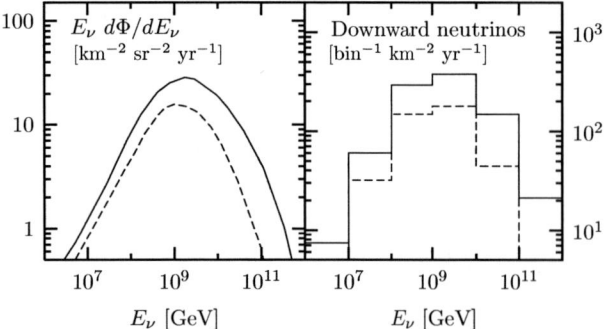

FIGURE 1. Left: cosmogenic neutrino fluxes referred in the text as *higher* (solid) and *lower* (dashed). We plot the fluxes for one flavor $\Phi = \phi_{\nu_\ell} + \phi_{\bar\nu_\ell}$ and assume that all flavors have the same frequency. Right: corresponding number of downward ($0 \leq \theta_z < \pi/2$) cosmogenic neutrinos of each flavor.

first one saturates the observations by EGRET, whereas for the second one the correlated flux of γ-rays contribute only a 20% to the data, with the nucleon flux normalized in both cases to AGASA/HiRes.

To observe a hadronic event inside the telescope, a cosmogenic neutrino must first *survive* as it crosses the atmosphere and the ice (or water) above the detector. Its typical interaction length in a medium of density ρ is $L_0 = [\rho N_A \sigma^{\nu N}]^{-1}$ where $\sigma^{\nu N} = \sigma^{\nu N}_{SM} + \sigma^{\nu N}_{BH}$ is the total cross section to have an interaction that *destroys* the neutrino. Notice that we include in $\sigma^{\nu N}$ both the SM and the short distance gravitational interactions, but we ignore the soft (eikonalized) gravitational interactions because they take from the neutrino just a small fraction of its energy.

A neutrino from a zenith angle θ_z must cross a column density of material

$$x(\theta_z) = \int_{\theta_z} dl \, \rho(l, \theta_z) \,. \quad (9)$$

The probability that it does not interact before reaching the detector is then

$$P_{\text{surv}}(E_\nu, \theta_z) = e^{-x/x_0} \,. \quad (10)$$

Once in the detector, the probability of an event is

$$P_{\text{int}}(E_\nu) \approx 1 - e^{-L \rho N_A \sigma^{\nu N}_{\text{int}}} \,, \quad (11)$$

where L is the linear dimension of the detector and $\sigma^{\nu N}_{\text{int}}$ the total cross section. Therefore, the total number of events in the telescope in an observation time T is

$$N = 2\pi A T \int dE_\nu \sum_{\nu_i, \bar\nu_i} \frac{d\phi_{\nu_i}}{dE_\nu} \int d\cos\theta_z P_{\text{surv}} P_{\text{int}} \,, \quad (12)$$

where A is the detector's cross sectional area and ϕ_{ν_i} the neutrino flux.

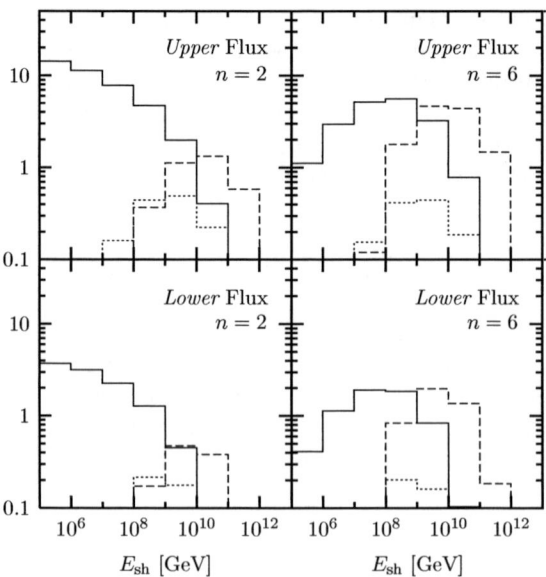

FIGURE 2. Energy distribution (events per bin) of the eikonal (solid), BH (dashed) and SM (dotted) events in IceCube per year for the *higher* and the *lower* cosmogenic fluxes, $M_D = 2$ TeV and $n = 2, 6$.

Cosmogenic neutrinos at AMANDA and IceCube

Let us now study the total number of events [15] at AMANDA (0.03 km^2 and a length of 700 m) and IceCube (1 km^3) [16] for the neutrino fluxes in Fig. 1.

In the SM, for the *higher* flux (910 downward cosmogenic neutrinos of each flavor per year and km^2), we would expect 1.32 contained events per year in IceCube. Of those, 0.38 would come from a neutral current and 0.94 from a charged current (one third of the events of each lepton flavor). The distribution of energy of these events is given in Fig. 2 (we show it despite the low statistics just for a comparison purpose). For the *lower* flux (around 410 cosmogenic neutrinos of each flavor per year and km^2), we would expect just 0.50 SM events per year inside the detector. The numbers for AMANDA can be easily obtained just multiplying by a volume factor $V_{AM}/V_{IC} \approx 0.02$, namely 0.03 SM events per year for the *higher* flux and 0.01 for the *lower* flux.

In a scenario with $n = 2$ (6) extra dimensions, for the *higher* flux we obtain a signal above the SM background (1.32 contained events per year in IceCube) if $M_D \leq 5.6$ (4.9) TeV, whereas for the *lower* flux we have a signal above the 0.50 SM events if $M_D \leq 4.8$ (4.5) TeV.

The event rate at IceCube and AMANDA for different values of M_D and a minimum energy of the shower of 100 TeV is plotted in Fig. 3. We give the number of short distance (BH) and of soft (eikonal) events; the largest total rate due to elastic interactions is clearly visible.

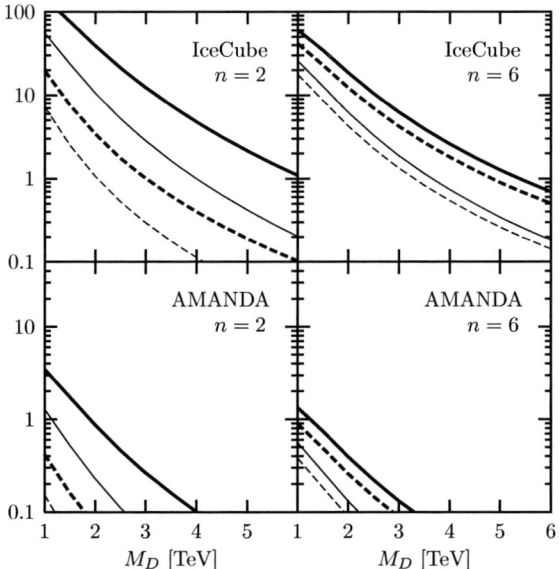

FIGURE 3. Contained events per year in IceCube and AMANDA for the *higher* (thick) and the *lower* (thin) cosmogenic fluxes and $n = 2, 6$. We show eikonal (solid) and BH (dashed) events.

The energy distributions of contained hadronic showers in IceCube for both fluxes, $M_D = 2$ TeV and $n = 2$ (6) are also shown in Fig. 2. There is a clear difference between the energy distribution of eikonal and BH or SM events: while these have a shape similar to the cosmogenic flux, eikonal events are typically of much lower energies, specially for $n = 2$, since $E_{sh} = y E_\nu \ll E_\nu$ and the eikonal cross section is larger for smaller values of y.

In summary, we think that elastic eikonalized interactions provide a clear (distinguishable from possible SM events) and model-independent signal of TeV gravity. Being at impact parameter distances larger than R_S, these interactions have a cross section that is larger than the geometric cross section to produce a BH. The eikonal event would be much less energetic than a SM or a BH event, but neutrino telescopes are sensitive to showers of energy up to four orders of magnitude below the average energy of cosmogenic neutrinos. The values of the fundamental scale of gravity that IceCube could reach, around 5 TeV, are comparable to those to be explored at the LHC or the ILC [10].

ACKNOWLEDGMENTS

D.M. is indebted with the Organizer Commitee for the stimulating atmosphere of the IFAE Conference.

REFERENCES

1. L. Anchordoqui, T. Paul, S. Reucroft and J. Swain, Int. J. Mod. Phys. A **18** (2003) 2229.
2. N. Arkani-Hamed, S. Dimopoulos and G. Dvali, Phys. Lett. B **429** (1998) 263; I. Antoniadis, N. Arkani-Hamed, S. Dimopoulos and G. Dvali, Phys. Lett. B **436** (1998) 257.
3. P. C. Argyres, S. Dimopoulos and J. March-Russell, Phys. Lett. B **441** (1998) 96; R. Emparan, G. T. Horowitz and R. C. Myers, Phys. Rev. Lett. **85** (2000) 499; D. M. Eardley and S. B. Giddings, Phys. Rev. D **66** (2002) 044011. S. Dimopoulos and G. Landsberg, Phys. Rev. Lett. **87** (2001) 161602; S. B. Giddings and S. Thomas, Phys. Rev. D **65** (2002) 056010.
4. J. L. Feng and A. D. Shapere, Phys. Rev. Lett. **88** (2002) 021303; L. A. Anchordoqui, J. L. Feng, H. Goldberg and A. D. Shapere, Phys. Rev. D **65** (2002) 124027; Phys. Rev. D **66** (2002) 103002; Phys. Rev. D **68** (2003) 104025.
5. R. Emparan, M. Masip and R. Rattazzi, Phys. Rev. D **65** (2002) 064023; M. Masip, arXiv:hep-ph/0210143.
6. A. Ringwald and H. Tu, Phys. Lett. B **525** (2002) 135; M. Kowalski, A. Ringwald and H. Tu, Phys. Lett. B **529** (2002) 1; S. I. Dutta, M. H. Reno and I. Sarcevic, Phys. Rev. D **66** (2002) 033002.
7. J. Álvarez-Muñiz, J. L. Feng, F. Halzen, T. Han and D. Hooper, Phys. Rev. D **65** (2002) 124015.
8. E. J. Ahn, M. Cavaglia and A. V. Olinto, arXiv:hep-ph/0312249.
9. G. 't Hooft, Phys. Lett. B **198** (1987) 61; I. J. Muzinich and M. Soldate, Phys. Rev. D **37** (1988) 359; D. Amati, M. Ciafaloni and G. Veneziano, Phys. Lett. B **197** (1987) 81; D. Kabat and M. Ortiz, Nucl. Phys. B **388**(1992)570.
10. G. F. Giudice, R. Rattazzi and J. D. Wells, Nucl. Phys. B **630** (2002) 293.
11. E. A. Mirabelli, M. Perelstein and M. E. Peskin, Phys. Rev. Lett. **82** (1999) 2236; G. F. Giudice, R. Rattazzi and J. D. Wells, Nucl. Phys. B **544** (1999) 3; for a review, see F. Feruglio, arXiv:hep-ph/0401033.
12. S. Cullen, M. Perelstein and M. E. Peskin, Phys. Rev. D **62** (2000) 055012; F. Cornet, J. I. Illana and M. Masip, Phys. Rev. Lett. **86** (2001) 4235.
13. P. Sreekumar *et al.*, Astrophys. J. **494** (1998) 523.
14. D. V. Semikoz and G. Sigl, JCAP **0404** (2004) 003.
15. J. I. Illana, M. Masip and D. Meloni, Phys. Rev. Lett. **93**, 151102 (2004) [arXiv:hep-ph/0402279]; D. Meloni, Acta Phys. Polon. B **35**, 2781 (2004); J. I. Illana, M. Masip and D. Meloni, Phys. Rev. D **72** (2005) 24003 [arXiv:hep-ph/0504234].
16. J. Ahrens [IceCube Collaboration], arXiv:astro-ph/0305196; see also http://icecube.wis.edu/.

Parallel Session V:

Detectors and New Technologies

Conveners:

C. Civinini
INFN Firenze, Italy

S. Miscetti
INFN LNF, Italy

Detector R&D for the International Linear Collider

Erika Garutti

DESY, Notkestr. 85, 22607 Hamburg, Germany
tel: 0049-40-8998-3779
e-mail: erika.garutti@desy.de

Abstract. To probe physics of interest a decade after the LHC begins to take data, the Linear Collider will need to have the capability of performing precision measurements of masses and couplings. The demands that this places in its associated detectors will be shortly sketched in the following. The status and plans for detector R&D necessary to meet these demands will be discussed.

Keywords: international Linear Collider, detector R&D
PACS: 29.17.+w

A DETECTOR FOR ILC

There is the worldwide consensus that the next particle physics project after the proton-proton collider LHC, is to be the International Linear Collider (ILC) for electron and positron able to deliver energies at the interaction point ranging from 91 to about 1000 GeV, and capable of providing high luminosity (the measure of the collision rate) above 10^{34} cm^{-2}s^{-1}. The ILC will extend the discoveries made at the LHC and provide a wealth of measurements that are essential for giving deeper understanding of their meaning, and pointing the way to further evolution of particle physics in the future.

Hardware wise moving from a LHC to a ILC detector means to shift the emphasis of R&D studies from radiation hardness and high rate capabilities, to precision measurements with high resolution.

At the ILC, given the precise knowledge of the initial state, the Standard model (SM) Higgs can be measured as the recoil mass of processes like $e^+e^- \to ZH$ where the Z bosons decays leptonically into l^+l^-. The resolution of this measurement is limited by the intrinsic width of the Z, if the resolution of the transverse momentum of charged particles in the trackers is $(\delta p_t/p_t)^2 \leq 5 \times 10^{-5}$. If the Higgs is found, the determination of its branching rations and couplings will allow to distinguish between different underlying physics models, i.e. SM or MSSM. For this an excellent capability of flavor tag is required, which imposes limits for the charged-particle impact parameter resolution to be better than 3 μm at high momentum and better than 10 μm at p=1 GeV/c.

If the Higgs mechanism is not the method chosen by Nature to obtain electroweak symmetry breaking, a precise measurement of the WW scattering will become essential for the understanding of new physics. In order to disentangle the two concurring processes of $WWvv$ and $ZZvv$ from their dijet final states, a jet energy resolution of $30\%/\sqrt{E(GeV)}$ is necessary. This can be achieved by combining the potentials of the particle flow approach with high granularity in transverse and longitudinal directions for both electromagnetic (ECAL) and hadronic (HCAL) calorimeters.

With these very strong requirements in mind, three concepts for an ILC detector are emerging from the already advanced R&D studies carried out over the past years. The three detector designs, sketched in Fig. 1, differ mainly in the choice of the technology for the tracking chamber, a five-layers solid-state tracker and a large-volume Time Projection Chamber (TPC) are proposed. The other main difference is the detector size, which has a direct influence on the required magnetic field of the solenoid (between 3 to 5 T from the most compact to the largest design).

The main news on R&D activities on various detector components are reviewed, a complete overview of ILC detector R&D status is found in [1].

THE VERTEX DETECTOR

The baseline vertex detector concept (sketched in Fig. 2) consists of 5 layers of 22x22 μm^2 CCD pixels, with a space point resolution of 5 μm, and with an overall material budget of 0.12% X_0 per layer, the inner one being at 1.2 cm from the beam trajectory.

The main issue for vertex detectors at the ILC is to maintain the occupancy below 0.1% during the foreseen $\sim$3000 bunch crossings (of 300 ns duration) per train. Given the proposed pixel size and the bunch structure fixed by the chosen cold RF-cavity technology in the ILC machine, the occupancy during a bunch train would be approximately 20 times the required one. Solutions have being investigated and prototypes have already shown very encouraging results, please refer to [2] for more de-

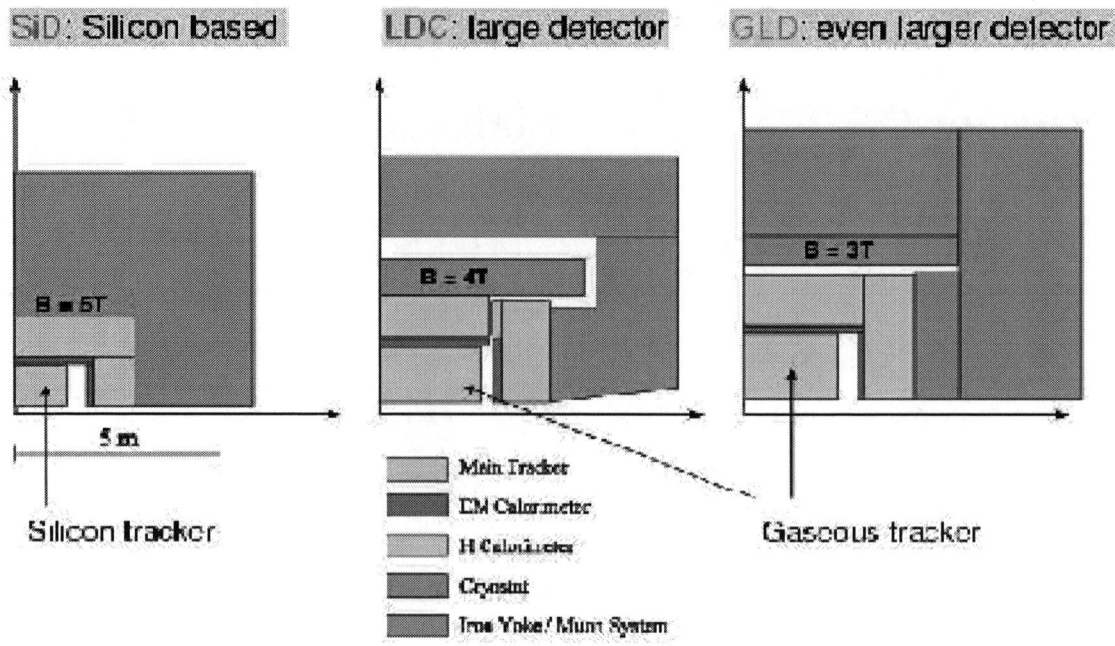

FIGURE 1. The three proposed design concepts for an ILC detector.

tails. One approach to keep the occupancy below 1% is to readout the pixels 20 times during a bunch train. Examples of this technology are the Charge-Coupled Devices (CCD) with column parallel readout[1], the Monolithic Active Pixel Sensors[2] (MAPS), the DEpleted P-channel Field Effect Transistor[3] (DEPFET) and the Silicon on Insulator [4] (SoI). MAPS and DEPFET are both examples of monolithic approach in which part of the readout circuitry is deposited directly onto the pixel sensor. In this case the charge released by ionizing radiation is collected on an n-well at the gate of the CMOS transistor situated between the epitaxial and the metal layer. The potential difference generated by the stored charge induces a current flow through the transistor, proportional to the charge itself.

A different approach is offered by the possibility to store charge in situ during the bunch train and subsequently read out the storage cells in between bunches. This technology is represented by the Imagine Sensor with In-situ Storage[5] (ISIS) and the Hybrid Pixel Sensors (HAPS). The charge collected during fixed time periods is shifted through subsequent n-channel storage wells on the pixel surface, where it remains stored till the next clock cycle. An elegant alternative to the complication of multiple readout during a bunch train is to reduce the pixel area by about a factor of 20. A matrix of Fine Pixel CCD (FPCCD) has been industrially produced with pixel size of 5x5 μm^2 and is under study.

It should be said that in spite the large R&D effort and the positive results obtained on various prototypes (some of which are also tested at test beam facilities) a substantial period of development is required before the possible application of the present technologies to the LC Detector.

THE TRACKING DETECTOR

The main difference in the various detector concepts proposed for the ILC is driven by the choice of the central tracking detector technology. The main advantages of a large-volume TPC over a Si-tracker are the fully three-dimensional track information and the large number of space points, which make the track reconstruction

---

[1] Pioneered by the Rutherford and the International Linear Collider Flavor ID (LCFI) groups.
[2] One example being the sequence of MIMOSA chips produced by LEPSI electronics and IRES (Strasburg).
[3] Project involving: MPI Munich, MPI Halle, U. Bonn, U. Mannheim
[4] Project involving: AGH-University of Science and Technology, Krakow, Poland; Universita' dell'Insubria, Como, Italy; Institute of Electron Technology, Warsaw, Poland

[5] Pioneered by W. Kosonocky et al., IEEE SCC 1996, Digest of Technical Papers, 182; and T Goji Etoh et al., IEEE ED 50 (2003) 144.

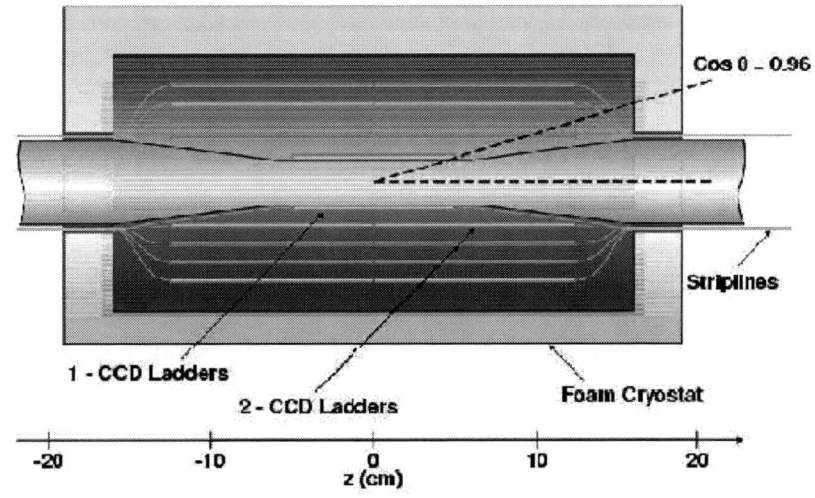

FIGURE 2. Baseline design of the 5-layers ILC vertex tracker.

straight-forward. Also, the minimum material budget in the barrel detector makes it possible to extend tracking to large radii.

The large magnetic field of ~4T, in which the ILC detector has to be operated to reduce e^+e^- background, will also reduce the transverse diffusion in the TPC improving the momentum resolution. Various TPC prototypes have been tested in magnetic fields up to 5.3T to verify the improvement in space point-resolution[6].

To achieve best point-resolution new gas amplification techniques as Gas Electron Multipliers [3] (GEM) and Micromegas [4] are being optimized. The very compact and localized electron avalanche produced in the micro holes of the gas amplifiers are sensed by conducting pads with a typical pitch of 1-3 mm [7]. The possible problem of overlap of events from several bunch crossings due to the slow drift time (~ 100 μsec) can be overcome by associating tracks in the TPC with time correlated track segments in the vertex detector, recovering in this way the single bunch-crossing information.

An alternative to gaseous detectors is a central tracker composed of five or more layers of silicon strips detectors. The key issue in this design is to minimize the material budget of sensors plus electronics, mechanical support, power and data traces, and cooling lines to below 1.5% of a radiation length per layer. Various proposals exist in this direction. Power cycling during the 200 msec gap between bunch trains reduces the power dissipation eliminating the need for active cooling, and avoiding the associated cooling system material. Studies are also undergoing[8] to produce single silicon ladder of half the total detector length, i.e. 167 cm for the outer most layer. In this way the readout electronics could be placed outside the barrel reducing the material in the tracking volume. Such long sensors with low noise are a serious challenge for the silicon technology and have not been produced before.

Fig. 3 shows the momentum resolution curve obtained with a conventional solid-state tracker (SD Thick, ~1.5% X_0/layer) compared to that of an ideal solid-state tracker (SD Thin) where only the sensors material is considered (from 100 to 300 μm from the inner- to the outer-most layer). While for high momentum particles the silicon tracker is always better than the gaseous detector (L), at low momenta only a thin tracker is competitive with the TPC. In addition for the comparison of the two technologies it should be added that pattern recognition for non-pointing tracks is easier in TPC (important for particle flow) but, on the other hand,

---

[6] Projects by Saclay, Orsay, U. Berkeley; U. Victoria, U. Aachen, DESY; and by Triumf.

[7] An independent approach to the electronic readout of micro-pattern gas detectors is a 55×55 μm^2 CMOS pixel sensor, Medipix2, proposed by NIKHEF and Saclay.

[8] U. of California (Santa Cruz) and LPNHE (Paris).

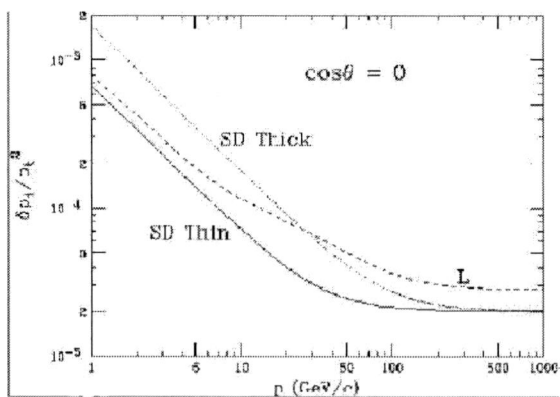

FIGURE 3. Momentum resolution curve for a conventional (SD Thick) and idealized low-mass (SD Thin) solid-state trackers and for a gaseous (L) tracker.

the large amount of material in front of ECAL endcap needed for the TPC endplate is a point in favour of Si-trackers.

THE CALORIMETER SYSTEM

In view of the demanding requirements of the particle-flow approach both ECAL and HCAL need to have unprecedented longitudinal segmentation and transverse granularity. A sandwich structure alternating high-Z absorber material to active layers is proposed for both calorimeters. Tungsten and steal are respectively the chosen absorbers for ECAL and HCAL.

To exploit the small Moliere radius of Tungsten, a very compact readout structure is required with a very high segmentation. Active layers with 1×1 cm^2 readout cells are proposed. This can be achieved with silicon pads according to the broadly accepted Si-W design[9], or eventually with small scintillating tiles[10], or with a hybrid structure alternating silicon and scintillator layers[11]. A Si-W prototype has been already operated in low-energy electromagnetic test beam at DESY

In the HCAL the minimum required segmentation is on the order of 3×3 cm^2 to match the Moliere radius of steel. Two possible designs are being pursued: a steel-scintillator structure with analogue readout of 3x3 cm^2 tiles; and a steel-radiator structure with gaseous signal amplification and digital readout of 1x1 cm^2 pixels.

The major R&D activities on calorimetry are pioneered by the CALICE collaboration [5], which has started the construction of ECAL and HCAL prototypes to be tested starting next year. Such a highly segmented calorimeter has never been built before and no precise knowledge of neither electromagnetic nor hadronic showers exist on such a fine scale. The data collected at test beam will serve as a benchmark to discriminate between existing shower models.

OUTLOOK

To exploit the full potential of the ILC, it is necessary to design and construct a detector with performance characteristics that exceed those of all existing and proposed cylindrical geometry detectors. Following the machine time schedule reasonable proposals of detector concept should appear by the end of 2007. To meet this goal a huge R&D effort has already started, but more studies on prototype detectors are crucial and should get a growing momentum. A challenging future for detector development is ahead of us in the next few years.

REFERENCES

1. J. Brau *et al.*, "Linear Collider Detector R&D" http://blueox.uoregon.edu/ lc/randd.html
2. "Vertex Detector R&D for the Future Linear Collider" http://hepwww.rl.ac.uk//damerell/International_VTX_R&D.htm
3. F. Sauli, *Nucl. Instrum. Methods* **A386** (1997) 531.
4. Y. Giomataris *et al.*, *Nucl. Instrum. Methods* **A412**(1998) 47.
5. J.-C. Brient *et al.*, "The CALICE project" http://polywww.in2p3.fr/flc/calice.html

---

[9] Pionered by: CALICE and BNL, SLAC, Oregon
[10] R&D studies in Japan by KEK, Kobe, Konan, Niigata, Shinshu, Tsukuba and JINR.
[11] A project involving: U. Como,ITE-Warsaw,LNF,U. Padova, U. Trieste

Modern Detectors for Astroparticle Physics

Oscar Adriani

*Università degli Studi di Firenze – INFN Sezione di Firenze
Via Sansone 1
I-50019 Sesto Fiorentino (Fi) – ITALY
adriani@fi.infn.it*

Abstract. This paper focus on the necessary requirements for a modern astroparticle physics detector based either on stratospheric balloons, either on satellite. The main technical solutions used to build a reliable detector are described. Finally, the most relevant experiments that have been developed with the INFN contribution and that will be ready in the near future (both for γ and charged cosmic rays detection) are described.

Keywords: cosmic rays, detector, astroparticle physics.
PACS: 95.55.Vj, 95.85.Ry, 13.85.Tp

INTRODUCTION

In the last ten years, a big effort has been made by INFN and other scientific institutions all around the world to promote the cosmic ray (CR) field, that is very interesting for many different aspects, related both to the elementary particle study and to the understanding of fundamental mechanisms in the universe.

In this paper I will mainly focus on the aspects related to the technologies that have been developed to allow the construction of reliable scientific experiments devoted to the direct cosmic rays detection, to be installed either on the higher part of the atmosphere (on a stratospheric balloon), either in space (on a dedicated satellite or on the International Space Station – ISS).

Finally, I will try to describe the most relevant and innovative aspects of the various experiments that have been developed in the last few years for the direct detection of charged cosmic rays and γ with the contribution of INFN, and that are expected to continuously take data in the next few years.

PROBLEMATICS OF DIRECT CR DETECTION

In order to have a better understanding of the CR phenomena, direct measurements are mandatory, to avoid the unwanted and 'catastrophic' screening effect of the atmosphere. However, since the charged CR flux has a very strong dependence on the energy ($\Phi \propto E^{-2.7}$), direct measurements are currently feasible only up to energies of the order of 10^{16} eV. Above this value, the only effective way to study CR is through the analysis of the secondary particles produced in the development of the atmospheric showers. This topic is not covered in my report.

Similar arguments are also valid to support the necessity of direct measurement in γ ray studies. In this field, moreover, the necessity of a very good pointing precision (to identify the sources) is another important point to support the requirement of direct detection.

The necessity of doing direct measurements implies that the CR detectors should be placed either in the higher parts of the atmosphere (by using stratospheric balloons) or in orbit around the earth (by using rockets and satellites/ISS).

Both these techniques have some serious drawbacks that I will try to explain in the next sections.

Stratospheric Balloons

Stratospheric balloons are a 'standard' technique that can be used to take CR data with a detector placed in the higher part of the atmosphere. The float height range between 30 and 40 km above the sea level; the residual density (≈ 5 g/cm^2) is small enough to allow direct measurements of CR, without big contamination from the secondary particles produced by the interaction of primary cosmic rays with the atmosphere.

One of the main advantage of this technique is that it is possible to lift quite heavy detectors (mass ≈ few tons) at a reasonable cost. However, the duration of the flight is usually limited to a few days by the day-night effect. To solve this problem, in the last few years almost all the agencies involved in this activity made a huge effort to flight from Antarctica, during the summer period, where, avoiding the day-night excursion, long duration flight can be accomplished. Just as a reference, the CREAM detector (described later in the paper) had a very successful flight in December 2004 that was lasting for 42 days.

Moreover, at the end of the flight period normally the payload can be recovered through the use of parachutes, allowing successive flights with the same detector.

Satellite/ISS

The other possibility to do direct measurements of CR is to install detectors completely outside the atmosphere, by putting them in orbit around the earth, on a dedicated satellite or on the ISS.

The use of this technique is very challenging and attractive; a space based detector can take data for many years (typically at least 3 years) and it can record data all around the earth (extremely useful for γ ray source searches).

However, there are many drawback related to the use of this technique. One of the biggest concerns is the cost of the mission, related both to the launch phase and to the construction of a space qualified reliable detector. In addition, the mass and power budgets are critical parameters. These strong limitations prevented in the past the use of complex and big CR detectors in space.

Anyway, things changed quite a lot in the last 10 years; an enormous R&D effort was done to build up reliable experiment to be operated in space, with a complexity typical of the high energy physics experiments.

The results of this effort are very encouraging; many complex detectors were developed in this way, and they are almost ready to be launched in the next few years.

In the next paragraph I will describe the main technological problems that came out during this development phase, and the way that were found to solve them. Most of theses aspects are also relevant for the balloon technique briefly described before.

THE SPACE ENVIRONMENT

Any detector to be put in orbit around the earth should operate under extreme conditions, related to the hostile space environment and to the vehicle used for the launch. I will focus on few of these problems, and I will try to present some of the solutions that were adopted by the CR physicists involved in the space projects.

1. Mechanical shocks and vibration during the launch phase of the rocket and during the fall phase of the balloon borne experiments: the detector and associated electronics should not be damaged by vibration with power spectral density up to 10 g_{rms}, and to repeated shocks up to 50 g for few msec. The solutions for this problem are mainly related to the mechanical projects and to the use of proper materials. The first resonance frequency of the structure is raised as much as possible (above 50 Hz) and many repeated tests under conditions similar to the real ones are conducted before the start of the mission by using dedicated facilities.

2. Thermal effects: the temperature control of the system is particularly complex, due to both the absence of convective heat exchange and to the extremely different temperature conditions in the absence of atmosphere. The solution for this is normally the use of an active cooling system, supported by a special care in the design and tests of the thermal heat exchangers. The problem is reduced for devices that operate in pressurized containers.

3. Vacuum effect: special care should be paid to avoid High Voltage discharge under low pressure conditions, by properly coating and potting the electronics circuits.

4. Effect of radiation: all the electronics should not be damaged by the radiation environment, especially during the crossing of the radiation belts and the South Atlantic Anomaly. These kinds of effects can be grouped into three different categories: Total Dose, Single Event Upset on the memory cells and Single Event Latch-up on the CMOS devices, that can be potentially destructive if the circuits are not properly protected. The traditional solution adopted in the past was the use of space qualified components, with the drawbacks of slowness, big power consumption and costs. However, in the last few years the research institutes decided to adopt a novel approach, based on custom made qualification of the proposed components, by using the available facilities all around the world. In particular, heavy ions beams were used to test the devices against

heavily ionizing particles. My personal opinion is that this is one of the most important and significant development connected to this research field.

5. Reliability and redundancies: the system should be developed trying to reduce the failures probability. To cope with this fact, normally the most critical parts of the systems are hot/cold redunded, and proper protection systems (based on the majority logic) are inserted in the critical logic circuits and memories.

6. Reduction of the power consumption. The use of custom qualified CMOS components, with low voltage power supply, allowed the handling of many thousands electronics channels, hence giving the possibility, for example, to operate very complex tracking devices with an acceptable power supply consumption.

Given these developments, in the last few years it became possible to design very complex CR detectors to be operated in space or to be accommodated on balloon, profiting of the technologies developed for the high energy particle physics experiment. The last generation CR experiments are very sophisticated devices, that make extensive use of silicon trackers and magnets to measure CR momentum, Cherenkov and Transition Radiation Detectors for particle identification, Time of Flight systems to trigger and measure CR velocity, calorimeters to measure energy and to separate hadrons from electromagnetic particles. This is really a breakthrough with respects to the previous CR experiments.

CHARGED CR AND γ EXPERIMENTS

In this paragraph I will focus the attention on five of the experiments that are approved by INFN to study charged CR (CREAM, AMS02 and PAMELA) and γ (AGILE and GLAST), trying to emphasize their most relevant aspects from the technological point of view.

CREAM

CREAM [1] is a balloon borne detector; its main scientific goals are the study of the charged CR in the region between 1 TeV and 1 PeV, by directly measuring the energy and the elemental composition (from H to Fe) of the cosmic rays. CREAM was launched for the first time from Antarctica on December, 16, 2004. The flight was very successful: it went over for 42 days, allowing collection of very high statistics, and the payload was recovered without damages. NASA already approved yearly flights from 2005 up to 2008.

CREAM is composed by a Timing Charge Detector to measure the charge of the incoming particle, by a TRD (complemented by a threshold Cherenkov detector) to measure the γ factor of particles with Z>3, by a Tungsten – Scintillating Fibers calorimeter for energy measurements, and by a pixel Silicon Charge Detector to identify the Z of the CR (up to Z<28).

Simultaneous measurement of the energy for Z>3 particles allows a cross calibration of the calorimeter and the TRD. The energy resolution for the TRD alone is $\approx$ 15% for C and 7% for Fe at $\gamma=3.10^3$.

AMS02

AMS02 [2] (Fig. 1) is a very complex detector that will be installed on the ISS in 2008 by using the Space Shuttle. The main scientific objectives of AMS02 are the search for Antinuclei and Dark Matter, and the precise measurements of the CR spectra in the GeV – TeV range ($e^\pm$, $p^\pm$, ^{3}He, ^{4}He, ^{9}Be, ^{10}Be etc.).

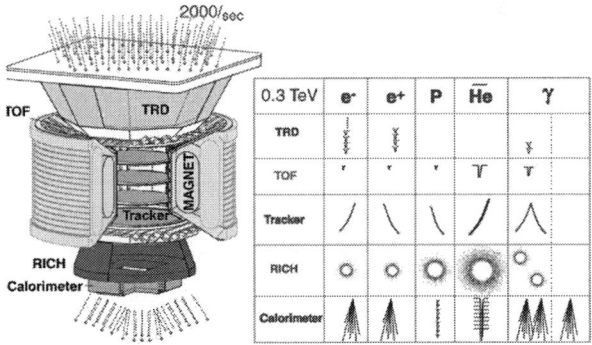

FIGURE 1. Exploded view of the AMS02 experiment. The table shows schematically how the CR are identified with the help of the various subdetectors.

To accomplish these tasks, the basic idea is to use many complementary subdetectors. The core of the experiment is a large superconducting magnet (with acceptance $\approx$ 0.5 m^{2}sr), that is a really a technological challenge. The maximum field (0.9 T) and the large lever arm determine a huge analyzing power (0.8 Tm^2), giving the possibility to do high statistics measurements up to very high energies. The total AMS02 weight is 7 ton, and the maximum allowed power consumption 2 kW. A wide acceptance silicon tracker (8 m^2) is inserted inside the magnetic field region to measure the momentum of the CR; the Maximum Detectable Rigidity (MDR) of the spectrometer is 2 TV/c, and the high dynamic range of the preamplifiers allows the measurement of the CR charge up Fe (Z=26).

The spectrometer is complemented by a series of particle identification detectors: a gas TRD, a Ring

Imaging Cherenkov detector, and a Tungsten / Scintillating fibers electromagnetic calorimeter that allows the precise identification and energy measurement of $e^{\pm}$ in a fraction of the total geometrical acceptance. A system for the Time of Flight measurement is important for low velocity particles, and is also used for the trigger.

PAMELA

PAMELA [3] (Fig. 2) is a satellite borne experiment, that will be launched by a Russian Soyuz launcher at the end of 2005, to precisely measure the CR spectra in the range between 100 MeV and few hundreds GeV. The instrument will be installed on the Resurs-DK1 satellite, on a high latitude orbit ranging between 350 and 600 km, allowing also the low energy CR measurements; the main purpose of the experiment is the identification of antiparticles in CR, mainly antiprotons and positrons, in an energy range never explored up to now.

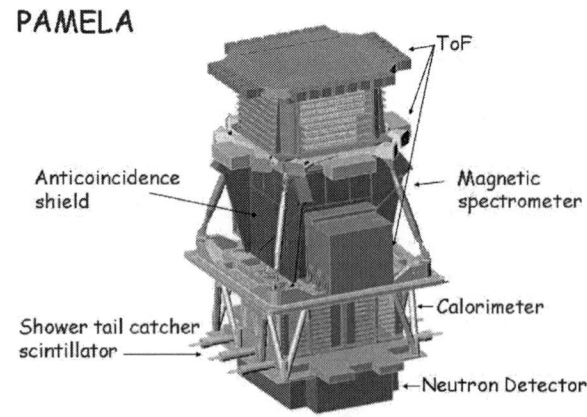

FIGURE 2. Schematic view of the PAMELA experiment. The total mass is 470 kg and the power consumption 345 W.

The instrument is based on a Nd-Fe-B permanent magnet, providing an uniform field B=0.45 T, with a geometrical acceptance of 20.5 cm^{2}sr. Inside the magnetic cavity 6 planes made by microstrip silicon sensors allow the precise reconstruction of the trajectory of the charged CR, hence measuring the momentum. The especially good spatial resolution of the silicon sensors (3 μm on the bending view) allows reaching an MDR greater than 1 TV/c.

The combined use of a fine segmented Silicon/Tungsten electromagnetic imaging calorimeter, giving a rejection factor greater than 3.10^4 between electromagnetic particles and hadrons, allows the clean identification of positrons and antiprotons up to ≈ 200/300 GeV, and of electrons up to 2 TeV.

AGILE

AGILE [4] is a small detector (120 kg total weight, 50 W total power) that will be placed in an equatorial orbit (550 km height) in 2006, to identify the γ rays in the CR.

It is composed by a silicon tracker system to measure the $e^{\pm}$ pairs created by the interaction of the γ with thin tungsten absorbers. The incident direction of the photons is hence precisely measured; the energy determination is realized by means of a 1.5 X_0 CsI homogeneous calorimeter. A peculiar feature of AGILE is the 'Super AGILE' Mask, placed above the detector, used to detect X rays correlated with γ rays.

GLAST

The GLAST experiment [5] is a sophisticated γ ray observatory to be installed on a 550 km height orbit in 2007. It is composed by a Large Area Telescope-LAT to identify γ with energy > 20 MeV, and a Gamma Ray Burst Monitor, optimized for lower energy photons (20 keV < E < 20 MeV). The basic concept of the LAT is the same as the AGILE one: the detection of the $e^{\pm}$ pair produced by the interaction of the γ with thin tungsten absorbers. The huge silicon tracker ($\approx 10^6$ channels, 160 W power consumption), coupled with a 8.4 X_0 CsI homogeneous calorimeter, allow a very precise survey of all the sky, looking for point γ sources, thanks to the very big effective area (1.8 x 1.8 m^2). One of the biggest challenges of GLAST is the reduction of the total power consumption to 650 W; the mass of the payload is ≈ 3 ton.

CONCLUSIONS

In this paper I have described the main technical solutions adopted to build reliable and complex detectors to be operated in space or in high atmosphere to measure CR, focusing on the last generation experiment developed with the INFN contribution and that will take data in the next few years.

REFERENCES

1. E.S. Seo et al., *Advances in Space Research* **30** (2002) 1263.
2. C. Lechanoine-Leluc, *Nuclear Instruments and Methods in Physics Research B* **214** (2004) 103.
3. M. Boezio et al., *Nuclear Physics B (Proc. Suppl.)* **134** (2004) 39.
4. C. Pittori et al., *Nuclear Physics B (Proc. Suppl.)* **134** (2004) 72.
5. L. Latronico, *Nuclear Instruments and Methods in Physics Research A* **511** (2003) 68.

Neutrino Detectors Review

Nicola D'Ambrosio

Istituto Nazionale di Fisica Nucleare, Sezione di Napoli, Via Cinthia, Napoli 80126 Italy

Abstract. The neutrino physics is one of the most important research field and there are several experiments made and under construction focused on it. This paper will present a review on some detectors used for Solar Neutrinos detection, Atmospheric Neutrinos detection and in Long Baseline Experiments.

Keywords: Neutrino physics, neutrino detector, neutrino oscillations.
PACS: 14.60.Lm, 14.60.Pq, 13.15.+g.

INTRODUCTION

The panorama of detectors used for neutrino physics is various and a review on all of them, even if short, would require a lot of pages and a big effort. In this paper will be presented a brief review with no intention to be exhaustive, some detectors will be described together, where possible, with their experimental results.

SOLAR AND ATMOSPHERIC NEUTRINO DETECTORS

Gallex/GNO and SAGE

The GALLium EXperiment (Gallex) and is successor Gallium Neutrino Observatory (GNO) [1] were located in Gran Sasso Underground Laboratory. The detector was made of 30.3 t of $GaCl_3$+HCl solution. They were set up to test the oscillation hypothesis at lower energy (threshold at 0.233 Mev). The gallium detectors could only detect electron type neutrino using the effect of the reaction:

$$v_e + {}^{71}Ga \rightarrow e^- + {}^{71}Ge \qquad (1)$$

A proportional counter has been used to count the germanium atoms looking for their radioactive decay.

The Soviet-American Gallium Experiment (SAGE) [2] was located in the Baksan Neutrino Observatory (northern Caucasus) 2,000 m deep. The detector was made of 50 t of metallic Ga.

Together with Chlorine experiment [3] they are still the only experiments with threshold energy below MeV. Both Gallex/GNO and SAGE found about half of the expected rate (see Table 1).

TABLE 1. Results of Gallex/GNO and SAGE compared with the expected rate in SSM.

	Gallex/GNO (SNU)	SAGE (SNU)
MEASURED	69.3 ± 5.5	$66.9^{+5.3}_{-5.0}$
EXPECTED	128 ± 9	128 ± 8

Sudbury Neutrino Observatory (SNO)

The Subdury Neutrino Observatory (SNO) experiment is located 2000 meters underground in Creighton mine in Canada [4]. The detector (Fig. 1) consists of a transparent acrylic vessel (12 m diameter) containing 1000 t of heavy water (D_2O). A geodesic structure surrounding of the acrylic vessel, holds about 9500 PMTs. To provide support and shielding the previous structure is contained inside a barrel-shaped cavity filled with purified water. The solar-neutrino detectors in operation prior to SNO were sensitive to the electron neutrino type. The SNO detector allows measuring the flux of all neutrino types. Observing the Cerenkov process, the neutrino from ^8B decay in the Sun are detected in SNO following these reactions:

(1) Charge Current (CC) reaction
$$d + v_e \rightarrow p + p + e^-$$
(2) Neutral current (NC) reaction
$$v_x + d \rightarrow n + p + v_x$$

(3) Elastic-scattering (ES) reaction

$$v_x + e^- \to v_x + e^-$$

The first reaction (CC) is sensitive only to electron-type neutrinos, the other two (NC and ES) are sensitive to all neutrino flavors.

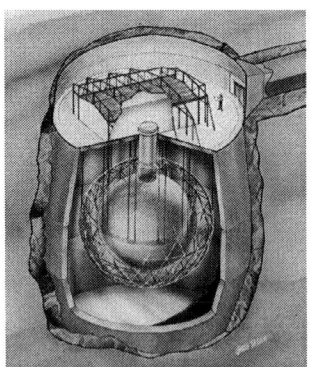

FIGURE 1. Schematic view of SNO detector. It is visible the acrylic vessel, the geodetic structure for PMTs and the barrel-shaped cavity.

The SNO experimental history is following three different phases. In the first phase (1999-2001) using pure heavy water, neutrons were observed through the Cerenkov light produced when they were captured on deuterium producing γ at 6.25 MeV. In the second phase (2001-2003) about 2 tons of NaCl were addend to D_2O to enhance neutron detection trough capture on Cl. During the third phase (2004-2006) the NaCl will be removed and will be installed an array of proportional counter filled with ^3He, this will provide direct detection of neutrons. The results of the second phase (*salt phase*) are summarized in Fig. 2, the ellipses represent the joint probability contours for ϕ_e and $\phi_{\mu\tau}$ in the case of 68%, 95% and 99%. There is evidence of solar neutrino flavor transformation at 5.3 σ level [5].

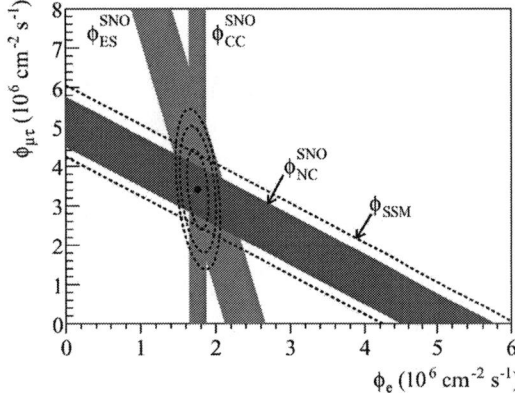

FIGURE 2. Flux of ^8B solar neutrinos which are of μ or τ flavor versus the flux of electron neutrinos deduced from the three neutrino reactions in SNO. The diagonal bands show the total ^8B flux as predicted by the SSM and that measured with NC reaction in SNO.

Super-Kamiokande

Super-Kamiokande is real time water Cerenkov detector [6]. It consists in a cylinder with a diameter of 39 m and height of 42 filled with about 50 kt of pure water (Fig. 3). The fiducial volume is about 22.5 kt, the water tank is divided into an inner part and an outer part. The read out of Cerenkov light is made using 13000 PMTs (11146 50-cm PMTs in the inner detector and 1885 20-cm PMTs in the outer detector. The detector is installed about 1000 m underground at the Kamioka zinc mine. The outer detector is used to identify muon events coming from outside and, at same time, it is a shield for the gamma rays and neutrons from the rock. Super-Kamiokande uses the elastic scattering ($v + e^- \to v + e^-$) of neutrinos coming from ^8B, this makes the experiment sensitive to all neutrino flavors. It is able to measure neutrino interactions with energy from 5 MeV to a few hundreds of TeV. The L/E analysis has confirmed the oscillatory patter, giving an experimental evidence for atmospheric neutrino oscillation. The SK data are consistent with $v_\mu \to v_\tau$ oscillation with $\Delta m^2 = 1.9 \div 3.0 \times 10^{-3} eV^2$ and $\sin^2 2\theta > 0.90$. K2K (KEK to Kamioka experiment, Long Baseline 250km) has confirmed the oscillation data using accelerator beam.

FIGURE 3. Super-Kamiokande detector

KamLAND

The KamLAND [7] detector consists in a 13 m diameter balloon filled with 1000 t of liquid scintillator (Fig. 4). It is surrounded by spherical tank (18 m) holding 1879 phototubes. Oil buffer fills the space between the tank and the balloon. The tank is immersed in a container filled with 3200 t of pure water instrumented with 225 PMTs (Cerenkov veto counter). It is located in the same place used for the original Kamiokande experiment. The aim of the experiment is to investigate the oscillation of anti-ν_e emitted from distant nuclear reactors through neutrons beta decay. It looks for a deficit of anti-ν_e and spectral distortion at distance L. The events are tagged by coincidence in time, space and energy of the neutron capture. The recent published results of KamLand data analysis [8] present an evidence of spectral distortion. In Fig. 5 the curves show the expectation for the best-fit oscillation, best-fit decay, and best-fit decoherence models taking into account the individual time-dependent flux variations of all reactors and detector effects. The data points and models are plotted with $L_0 = 180$ km, as if all antineutrinos detected in KamLAND were due to a single reactor at this distance [8]. The neutrino oscillation are much favoured with more than 99% CL, the decay and decoherence are excluded at the 95% and 94 % CL.

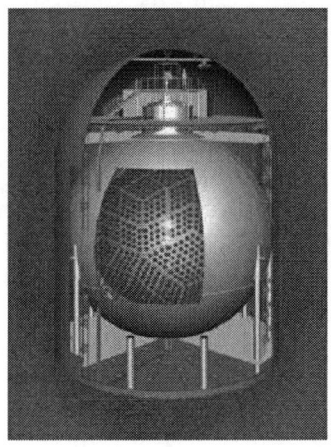

FIGURE 4. KamLAND detector

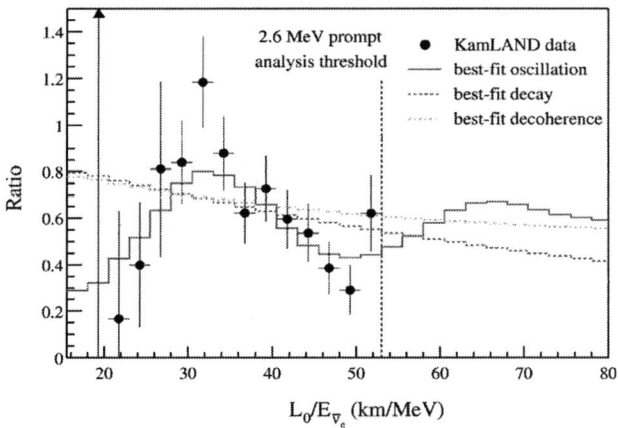

FIGURE 5. Ratio of the observed anti-ν_e spectrum to the expectation for no-oscillation versus L_0/E.

NEUTRINO OSCILLATIONS LONG BASELINE EXPERIMENTS

The discovery of neutrino oscillations by Super-Kamiokande in 1998 has pushed up the development of many projects using accelerator neutrinos and a *Long Baseline* to confirm that observation and measure the parameters of neutrino oscillations.

MINOS

The MINOS experiment will use a muon neutrinos beam produced at FNAL. The beam is tunable (1-20 GeV) and will be sampled by near detector at Fermilab and, after 735 km, by the far detector at Soudan mine. The MINOS experiment basic idea is to measure the ratio of neutrino energy spectrum in far detector (oscillated) to that in the near detector (unoscillated). It is a disappearance experiment. The near and far detector are similar [9], basically they are calorimeters made of planes of thick steel (2.54 cm) interleaved with sensitive planes of wide solid scintillator strips (1cm thick × 4 cm wide). The steel plates are made in octagonal shape and magnetised (about 1.5 T). In the far detector the octagons have a diameter of 8 m and forms two "supermodules" (Fig. 6). The total mass of far detector is 5.4 kt. The near detector has a mass of 0.98 kt. The sensitivity depends on number of protons on target. The physics goals of MINOS are: demonstrate oscillation behaviour, precise Measurement of Δm_{23}, search for sub-dominant $\nu_\mu \rightarrow \nu_e$ oscillations.

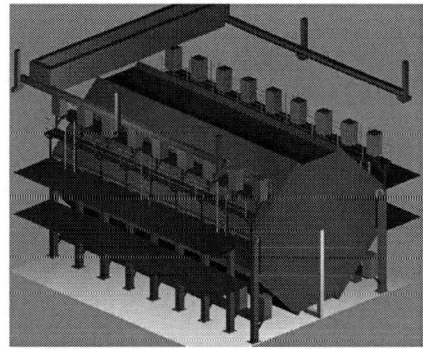

FIGURE 6. Schematic view of one "supermodule" of the MINOS far detector

OPERA

OPERA (Oscillation Project with Emulsion-tRacking Apparatus) is an appearance experiment. It will use the CNGS (CERN Neutrinos to Gran Sasso) beam (mean neutrino energy of 17 GeV).

The physics goal of OPERA [10] are: provide an unambiguous evidence for $\nu_\mu \to \nu_\tau$ oscillations in the region of atmospheric neutrinos by looking for ν_τ appearance in a pure ν_μ beam; search for sub leading $\nu_\mu \to \nu_e$ oscillations (measurement of Φ_1).

The detector is under construction in the Gran Sasso underground laboratory (730 km from CERN). The OPERA experiment is based on the direct observation of τ decay topology by means of nuclear photographic emulsions. It is a "hybrid" experiment based on an electronic detector and a massive target detector composed by a sandwich of lead and nuclear emulsion. The detector (Fig. 7) consist of two "supermodules", each of them is a sequence of 31 "modules". Each modules is composed by a "wall" of lead-emulsion "bricks" (about 200000 bricks in total) and two planes of orthogonal scintillator strips, a Spectrometer (magnetized iron dipoles, drift tubes and RPCs complete the electronic part of the detector. The total area of emulsion films is $\sim$ 170000 m². Each brick consist of 56 emulsion sheets (10 X 12.5 cm², two emulsion layers of 42 µm on a 200 µm of plastic support) for a total of $\sim$ 13 millions. The brick is based on the concept of Emulsion Cloud Chambers (ECC) used in the ν_τ discovery by DONUT experiment [11]

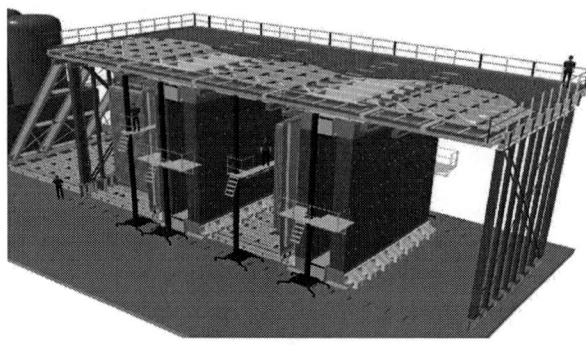

FIGURE 7. OPERA detector.

The data from electronic detectors will be used to select bricks where neutrino interactions occur, after the development the emulsions will be sent to a dedicated scanning machines [12] (based on automatized microscopes) and analyzed for vertex reconstruction. For τ candidate event a precision scanning measurement will be made to improve particle identification and momentum assignment. The expected brick extraction rate will be 30 bricks/day. The experiment will start with data taking during 2006 and will be operative for 5 years. The expected performance is shown in Table 2 for different values of Δm^2, in the all cases the background is 0.7(1.2).

TABLE 2. OPERA expected performances. Full mixing and 5 years of run. In parenthesis the expected number of events in case of CNGS beam upgrade (X 1.5)

$\Delta m^2 = 1.9 \times 10^{-3}$ eV$_2$	$\Delta m^2 = 2.4 \times 10^{-3}$ eV$_2$	$\Delta m^2 = 3.0 \times 10^{-3}$ eV$_2$
6.6 (10)	10.5 (15.8)	16.4 (24.6)

CONCLUSION

The combined results of the radiochemical experiments (Chlorine Gallex/GNO, SAGE) have established Solar Neutrino Oscillation. SNO provided a direct evidence of flavor conversion of solar electron neutrinos, an evidence of neutrino oscillations has been confirmed by KamLAND analysis using anti-ν_e emitted by nuclear reactor. Super-Kamiokande data analysis has provided an evidence for an oscillatory signature in atmospheric neutrino oscillation [13]. K2K has given an indication of neutrino oscillation in a long baseline experiment, but its sensitivity is limited, the new long baseline experiments MINOS and OPERA will give a significant proof that neutrino oscillations are responsible of atmospheric neutrino deficit. They will provide a precision measurements of neutrino oscillation parameters.

REFERENCES

1. W. Hampel *et al.*, *Phys. Lett. B* 447, 127 (1999) ; M. Altman *et al.*, *Phys. Lett.* 490, 16 (2000).
2. J. N. Abdurashitov *et al.*, *Astrophys. J.* B 496, 505 (1998)
3. R. Davis, Jr., *Phys. Rev. Lett.*, 302 (1964).
4. The SNO collaboration, *Nucl. Instr. and Meth.* A452, 115 (2000).
5. Q.R. Ahmad *et al.*, *Phys.Rev.Lett.*89:011301,2002.
6. M. Nakahata *et al.*, *Nucl. Inst. Meth.* A 421, (1999).
7. K. Eguchi *et al.*, *Phys. Rev. Lett.* 90, 021802 (2003).
8. T. Araki *et al.*, *Phys. Rev. Lett* 94, 081801 (2005)
9. MINOS Collaboration, Fermilab, NuMI-L-337
10. OPERA Collaboration, M. Guler *et al.*,*CERN-SPSC-2000-028*.
11. K. Kodama *et al.*, *Phys. Lett.* B504, 218 (2001).
12. N. D'Ambrosio, *Nucl. Inst. Meth.* A 525 (2004) 193-198; N. D'Ambrosio *et al.*, *Nucl. Phys. B Proc. Suppl.* 125 (2003) 22.
13. Super-Kamiokande Collaboration (Y. Ashie *et al.*), *Phys.Rev.Lett.* 93:101801, (2004)

Data Analysis Techniques at LHC

Tommaso Boccali

Scuola Normale Superiore and INFN Pisa
Piazza dei Cavalieri 7 - 56126 Pisa ITALY

Abstract. A review of the recent developments on data analysis techniques for the upcoming LHC experiments is presented, with the description of early tests ("Data Challenges"), which are being performed before the start-up, to validate the overall design.

Keywords: LHC Computing, LCG, GRID.
PACS: 29.85.+c

INTRODUCTION

The task of data analysis for the upcoming LHC Experiments ([1,2,3,4]) is unprecedented, due to both the computing power and the storage needed. In what follows I will try to sketch the problem, together with the proposed solutions.

THE PROBLEM

The size of the problem the LHC experiments must confront with can be understood from Table 1, where Data size and CPU power needed by current experiments and CMS are compared.

TABLE 1. Computing resources needed by current and future experiments

	ALEPH[5] (1995)	CDF[6] (2004)	CMS (2007)
Data Size	1 TB	1 PB	~10 PB
CPU power for analysis (SI2k)	< 100k	1.4 M	>25 M

The size of the data is exceeding at least by a factor 10 the needs by CDF, while the CPU estimates are even worse.

If the four LHC experiments are considered globally, the estimated needs are exceeding 100 MSI2k, with 40 PB a year on disk and at least the same quantity on tape. By current standards, these figures would translate in 25000 CPU (2007) and tens of thousand of hard drives, clearly not easy to handle. So, which are the possible proposed solutions for a problem of such scale? Let's try and figure out extreme approaches:

1. A supercomputer approach, in which (a number of) the most powerful available SC on the market are bought. As of today (top500.org[7]), the fastest machine is the BlueGene Computer, with an estimated cost of 250 M$ and with a power equivalent to ~10000 current CPUs. With these numbers, 10 such machines would be needed, with a cost comparable with LHC, without solving the data storage problem.
2. A mega farm approach, with a building stuffed with O(25000) PCs; only Google has proved up to now that such a technology is viable. It poses even more problems on the human side: all the people in LHC Collaborations working on computing should be full time at CERN. It is clear that in such a situation, national funding agencies are not willing to spend million of euros to finance an abroad laboratory, with scarce possibilities to use local staff.

The need for a different solution has been recognised long ago, and in late '90 the MONARC (MOdels of Networked Analysis at Regional Centres) Project[8] proposed a collaborative model based on the use of local resources, worldwide. In 2001, the Hoffman Review of Computing[9] defined MONARC as the baseline for LHC Computing.

THE MONARC MODEL

The MONARC Model divides computing resources in Tiers, where each Tier defines responsibilities and functions in LHC analyses. Figure 1 shows the proposed levels.

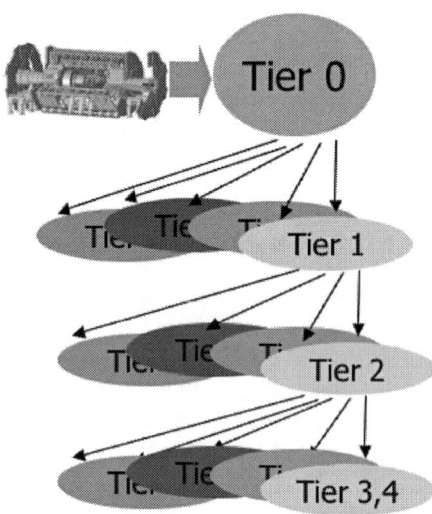

FIGURE 1. **The Tier structure as proposed by the MONARC architecture.**

The Tier0 centre is close to the apparatus, and is the place where raw data from the experiment is stored. A Tier0 should also be used for fast calibrations and reconstructions, mainly used to monitor the data taking.

Tier1 sites are big national or international centres, where a copy of the raw data is saved as backup, and where more accurate reconstruction is performed, producing reduced data sets ready for analyses.

Tier2 sites are hosted by big universities or institutions, and are the place where most of the simulation and analysis takes place. These should also host the activities of something like 50 users, allowing the production of physics results.

The role of smaller Tiers is somehow less defined, including small universities' clusters and even desktop machines.

The relative size of these centres is thought such that, for example, the sum of Tier2 resources, which refer to the same Tier1, should equal its size; moreover, the sum of Tier1 and Tier2 resources should be roughly equivalent.

The advantages of such an approach are clear: there is no centralized and unique resource (apart from the detector itself, obviously), while local expertises and staff are used at best.

Also some disadvantages show up, like the increased complexity of the system (think of replicating the data, handling of the load, synchronization of the code base), and an increased need of network bandwidth to move samples across the planet. While the first problem is easily sustainable using local staff at the various Tiers, the second is not too relevant, having in mind the current situation of the research network deployment. Figure 2 shows how Atlantic links exceeding 10 Gbit/s are already now a reality. With these figures, a month is sufficient to deliver more than 2.5 PB, so far more than the inclusive produced rate considering raw and reconstructed data.

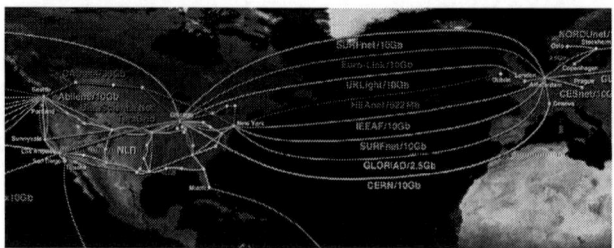

FIGURE 2. **Atlantic WAN connections.**

Since the Hoffman review big developments in GRID technology foresee a solution for the added complexity, allowing (yet to be proven on such a large scale) transparent data movement and resource allocation between large-scale facilities.

THE GRID

The GRID concept is defined in analogy with the power grid systems available in every house: a resource (CPU, disk) should be available regardless of its exact location and technology, provided we are authorized to use it.

This serves well as the infrastructure on which construct a MONARC-like architecture, since it allows the user to be less sensitive to the actual configuration of the computing resources. The user sees a site, whatever the internal configuration and the complexity of its architecture, simply as a service, providing computing and storage facilities with a coherent interface.

This is extremely useful when considering the standard use case when doing physics analysis: a user does not want to (and should not) know where data sits, and where he can find free CPUs for the analysis; his only interest is to have results in a controlled way, and in the least possible time. An automatic system for routing jobs and moving data as GRID promises would be ideal in this case.

The European Community recognized the usefulness of the GRID concept, far beyond the HEP needs. The EGEE[10] (Enabling Grids for E-sciencE) project has been funded to provide the software infrastructure (Middleware) upon which build a GRID system, useful for both research and industry.

On the other hand, CERN has launched the LCG[11] (LHC Computing Grid) to prepare the infrastructure and to be ready for the 2007 LHC start up.

Other communities have launched different and often incompatible GRID projects (there are tens of them); in what follows I will focus only on LCG, showing how the LHC experiments are preparing for data analysis.

DATA CHALLENGES

It is clear that it is not possible to wait 2007 to see whether the GRID concepts and infrastructure are up to the expectations.

For this reason, the LHC experiments are simulating in "Data Challenges" reduced scale day-to-day operations, testing the readiness of the LCG.

If a certain fraction of the load can be already sustained today, the natural improvements in CPU power and disk capacity in the next couple of years should guarantee the usability of the whole system.

CMS

In 2004, the CMS Collaboration tested Tier0 reconstruction, using samples of simulated data as if they were coming from the detector, processing the 25% of the expected rate in 2007 in real time.

In this way, 25 Hz were sustained for more than 24 consecutive hours, and 80 million events where processed (Figure 3). The second step attempted has been data transfer from Tier0 to various Tier1, using GRID infrastructures and some custom software.

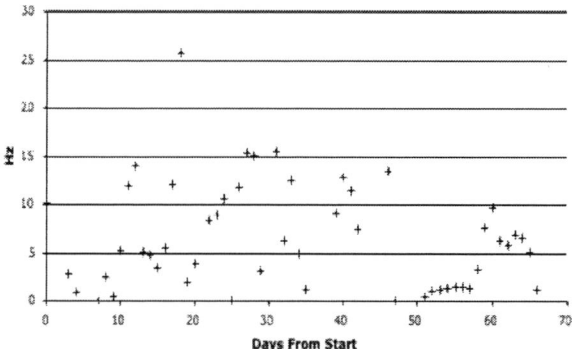

FIGURE 3. The event processing rate at Tier0 surpassed 25 Hz for 24 consecutive hours

Figure 4 shows how a sustained rate of more than 340 Mbit/s has been kept for over 5 hours; the numbers show that more than 4 TB/day where transferred without saturating the system.

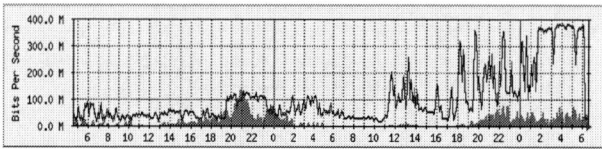

FIGURE 4. Sustained transfer rate from CERN to Italian T1, CNAF.

ATLAS

The ATLAS Collaboration has produced, with different GRID technologies operating in a sort of meta-GRID, more than 10 million events, and storing 35 TB of data. The GRIDs used are LCG, NorduGrid[12] (a collaboration of north European institutions) and the US GRID3[13], with a good share between the different technologies (see Figure 5).

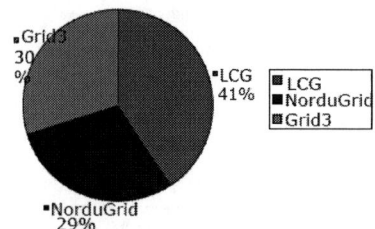

FIGURE 5. ATLAS production share between three different GRID technologies.

Tested operations are Tier0 reconstruction, at 10% of the nominal rate, using a total of more than 140.000 jobs submitted. A second phase, focused towards analysis, has to follow.

ALICE

The ALICE Collaboration had the very ambitious aim to produce and analyze realistically 10% of the data collected in the first year of data taking. This has been organized in three steps. The first step is the distributed production of Pb/Pb and proton/proton collision, for a total of 20 million events. The second step is the reconstruction of trigger responses, while the third will be distributed analysis. The ALICE Collaboration has implemented its own GRID system, called Alien, but that has integrated transparently also LCG resources, which appear under Alien as a single node (see Figure 6).

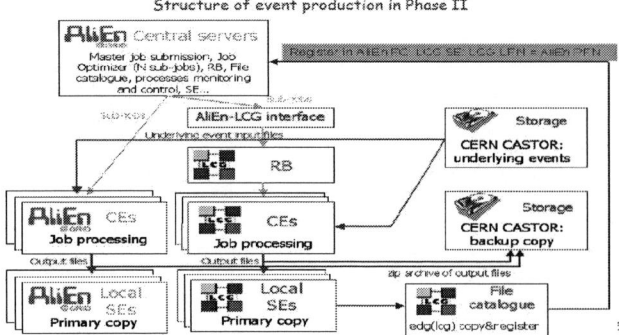

FIGURE 6. ALICE production and analysis system integrates native resources via Alien with LCG resources, seen as a single site.

LHCB

The LHCb Collaboration has shown the functionality of its software system with the GRID infrastructure, having has physics goals high-level trigger and background studies. In particular, the GRID has been used to simulate a big part of the 186 million events produced, while a smaller fraction has been produced with the proprietary system DIRAC. The use of LCG resources allowed a huge increase in

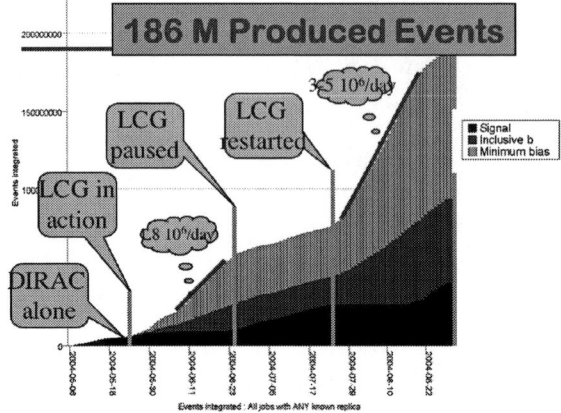

FIGURE 7. LHCb production system used both proprietary (DIRAC) and LCG resources; the second allowed a huge speed-up of the simulation.

the production capacity (Figure 7). A second step in the data challenge is the reconstruction of these samples, ongoing now, while a third analysis step has to follow.

CONCLUSIONS

The four LHC Collaborations completed or are completing in the last 12 months large-scale tests of production, reconstruction and analysis using GRID infrastructures. Even if various and important problems were encountered, the general feeling is that

the model can evolve to something really usable for the LHC start-up. Already with this series of data challenges, a sizeable fraction of the expected computing load has been reached. Overall, more than a billion events has been processed, on ~10000 CPUs and producing data for over half a Petabyte. Among the various problems encountered, storage has shown as the weak point, and already now the EGEE project is re-implementing part of it.

Since in a couple of years LHC should have its pilot run, it seems clear that all the experiments will be using, at least for the first part of data taking and analysis, systems based on GRID; the final tuning of such systems will require a new series of data challenges in the next years, already in preparation.

REFERENCES

1. The CMS Coll., http://cmsinfo.cern.ch
2. The ATLAS Coll., http://atlas.web.cern.ch/atlas
3. The ALICE Coll., http://aliceinfo.cern.ch
4. The LHCb Coll., http://lhcb.web.cern.ch
5. The ALEPH Coll., "Performance of the ALEPH Detector at LEP", Nucl. Instrum. Methods Phys. Res., A 360 (1995).
6. The CDF Coll., Phys. Rev. D71 (2005).
7. http://top500.org
8. The MONARC Project, http://monarc.web.cern.ch
9. The Hoffman Computing Review, CERN-LHCC-2001-04
10. EGEE, http://egee-intranet.web.cern.ch
11. LCG, http://lcg.web.cern.ch
12. NorduGrid, http://www.nordugrid.org
13. GRID3, http://www.ivdgl.org/grid3/

The CDF Analysis Farm

M.Casarsa[1]*, S.-C.Hsu[†], E.Lipeles[†], M.Neubauer[†], S.Sarkar**, I.Sfiligoi[‡] and F.Würthwein[†]

INFN, Trieste, Italy
†UCSD, La Jolla, CA, USA
**CNAF-INFN, Bologna, Italy
‡LNF-INFN, Frascati, Italy

Abstract. Since the beginning of Run II in 2001 the Fermilab Tevatron has been delivering proton-antiproton collisions to CDF with a steadily increasing performance. By the end of Run II up to 8 fb^{-1} of integrated luminosity are expected to be available on tape. In order to provide the CDF Collaboration members with adequate computing resources for analyzing those data, a globally distributed pool of dedicated PC farms has been commissioned and is currently being successfully deployed. This paper describes the main features of the CDF Analysis Farm with particular attention to the user perspective. The last part is devoted to a brief overview of the more relevant technical details on the farm implementation and, finally, the future developments toward its integration in the GRID framework are outlined.

Keywords: CDF, computing, data analysis, PC cluster.
PACS: 07.05.-t, 07.05.Kf

INTRODUCTION

The Collider Detector at Fermilab (CDF) [1] is an experiment at the Fermi National Accelerator Laboratory (FNAL or Fermilab), which investigates the ultimate structure of matter exploiting the proton-antiproton collisions provided by the Tevatron hadron collider. The CDF Collaboration consists of about eight hundred Physicists, affiliated to fifty-nine different Institutions from twelve Countries all over the World.

By the end of Run II up to 8 fb^{-1} of integrated luminosity are expected to be available for analyses. In order to fulfill the increasing Collaboration needs for computing resources, a dedicated PC farm was commissioned with the main goal of granting adequate CPU power and disk space to at least two-hundred users simultaneously every day, a rough estimate of the number of Physicists active in analyses. For their analyses CDF users are mainly concerned about accessing datasets of tens of TB's and hundred million events, whose processing usually takes several days at a rate of about 0.1-0.5 s/event, and producing large Monte Carlo (MC) samples, where the detector simulation represents a serious limiting factor in the execution time, ~10 s/event.

The CDF Analysis Farm (CAF) [2] was designed and assembled in 2002, originally as a PC cluster localized at FNAL. Successively, the CAF model was exported off-site and several decentralized CDF Analysis Farms arose in many sites worldwide. Currently, ~50% of CDF com-

[1] e-mail: Massimo.Casarsa@ts.infn.it

TABLE 1. Current CAF's resources around the World.

Cluster Name	Location	CPU [GHz]	Disk [TB]
FNAL CAF	FNAL	3200	370.0
TORCAF	Toronto (Canada)	576	10.0
CNAFCAF	CNAF (Italy)	480	32.0
MITCAF	MIT (USA)	322	3.2
SDSCCAF	San Diego (USA)	300	4.0
KORCAF	KNU (South Korea)	178	5.1
JPCAF	Tsukuba (Japan)	152	10.0
ASCAF	Academia Sinica (Tw)	134	3.0
HEXCAF	Rutgers (USA)	100	4.0
CANCAF	Cantabria (Spain)	50	1.5
Total		**5492**	**442.8**

puting power resides outside Fermilab: Tab. 1 lists the current CAF resources around the World. All CAF's are structurally identical, but, since data are stored at Fermilab, the FNAL CAF is mostly used for data analysis jobs, whereas the other CAF's are mainly devoted to MC production. Recently, dataset replicas allow users to run analysis jobs also at some remote sites.

In the following sections the main features of the CAF system will be outlined. In the first part, the CAF will be described from the outside, as it appears from the standpoint of a Physicist user. Then, the reader will be provided with a brief overview of the CAF implementation with the most relevant technical details. Finally, a first prototype of the CDF Analysis Farm integration within the GRID environment will be presented.

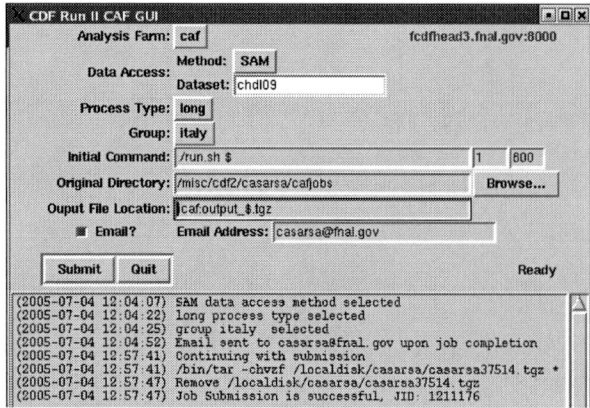

FIGURE 1. CAF graphical user interface.

THE CDF ANALYSIS FARM

The CDF Analysis Farm (CAF) is a portal which allows users to easily perform Physics analysis and MC production exploiting remote cluster resources: CPU power, scratch disk, massive data access. The computing model on which the CAF design is based is to delegate any interactive tasks, i.e. code developing and debugging, to the user desktop and to accept only batch jobs from remote. The result is a very versatile system, whose main features are ease of use and a large parallelization capability.

CAF Usage

CAF usage is very simple and straightforward. Users develop and debug their applications on their desktops or laptops anywhere in the World. Once the executable is produced, it is submitted to the CAF. Upon job completion user is notified by a comprehensive e-mail and can either retrieve the job output from a CAF scratch space or have it delivered wherever he/she decides. A series of command line tools allow users to interact with remote jobs as if they were running on a local machine and a web monitoring interface displays the job status and a lot of useful information on the system.

Job Submission

Jobs can be submitted to any CAF from any kerberized desktop or laptop running the CAF software. User has to provide an executable, all the necessary shared libraries, a startup script and any other configuration files needed by the job, which at submission are tarred up and sent to the CAF cluster.

Both a GUI interface and a command line application are available for job submission. An example of job submission via the CAF graphical user interface is shown in Fig. 1. In both cases user has to enter the following input arguments:

analysis farm: Using the same application, user can submit jobs to any CAF.

data access method and dataset name: For jobs running on data the dataset name has to be declared. Before submission the dataset status and availability are checked. The SAM data handling system [3] allows to run data analysis jobs also at those remote sites, which host dataset replicas.

process type: Depending on the time limit, different process types are available: short (6h), medium (12h), long (72h). Furthermore, a 2h-limited high-priority process type is available for quick tests.

group: Groups have been implemented in order to allow the members of Institutions, which contributed funds to the CAF, to have higher privileges on the fraction of the pool they own.

initial command: In principle any command can be executed on the remote host, the most flexible choice turns out to run a shell script which performs all the tasks users need, including running the data analysis executable.

section range: Many instances of the same job (sections) may be submitted at the same time. The differentiation of each parallel section is realized by passing an integer argument to the startup script from the range provided by the user at submission. The details of implementation are left to the user.

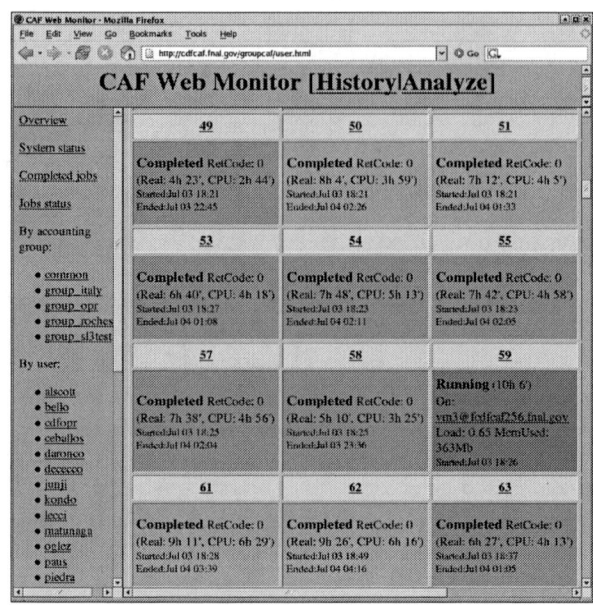

FIGURE 2. CAF Web Monitor.

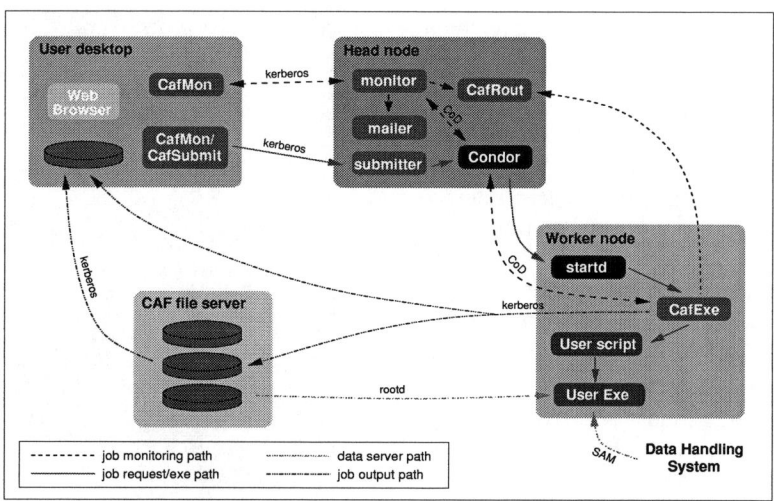

FIGURE 3. CAF structure overview.

original directory: The directory, which contains all files needed by the job, is archived at submission time and sent to the CAF.

output file location: At the end of the script execution, a tar file of the working directory on the remote node is created and copied to the location requested by the user, usually the CAF file server scratch disk. Files can be retrieved later from the CAF file server by means of kerberized `rcp` or `ftp` or accessed by other jobs via `rootd` [4].

e-mail address: Upon job completion a comprehensive report is e-mailed to the specified address, listing the exit codes and the waiting, CPU, and real times of all job sections, and an input/output data summary.

Job Monitoring and Job Management

Users have at their disposal different tools to monitor and manage their jobs running on the remote cluster. A snapshot of the system, taken every few minutes, is accessible via a web browser. Users can get information both on the status of their jobs and the status of the system as a whole: for example, the number of waiting, running, and completed sections, the waiting and execution times, the active processes, the CPU load and memory usage. Moreover, the information relative to all completed jobs is stored on the web server. Figure 2 shows a screen-shot of the web monitoring which displays the section status of a running job.

A command line application allows users to interact with remote jobs as if they were running locally to perform:

- job management tasks: killing jobs and sections, holding and releasing jobs, changing the priority and group of a running job;
- jobs and remote system interactive monitoring: listing the submitted jobs and sections, printing out the section log files, showing the status of the nodes running a specific section, showing the working directory of a specific section, and inspecting any text file in the section working directory;
- debugging: attaching a debugger session to any remote running process.

CAF Operation Overview

The CAF realization was driven by a balanced compromise between the Collaboration requirements on computing power and the effective constraints of a fixed budget. The final choice was a large batch-based cluster of commodity PC hardware: dual Intel/AMD processors and IDE disk, configured in RAID 50 arrays. CAF nodes are functionally divided in one head node, where the control and management daemons run, and many worker nodes, on which user jobs are executed. All nodes run Linux as an operating system and have access to the CDF software.

In order to ease the administration tasks, users have no direct access to CAF and user jobs run under generic accounts. The farm batch manager is Condor [5], configured with just a submission node, which plays also the role of manager node, and many worker nodes. Six virtual machines (VM) are allocated on every worker node, for each VM a generic user account is created. More details on Condor configuration can be found elsewhere [6].

On top of the farm batch manager lies the CAF software, which represents the interface between the user and the PC cluster. Figure 3 schematizes the elements which make up the CAF:

User desktop: On user desktop run the user front-end applications for jobs submission (`CafGui` and `CafSubmit`), which send the user submission parameters to the head node, and the user interactive tool to manage and monitor the running jobs (`CafMon`). Both communicate with the head node via a kerberos authenticated connection. A web browser is used to access the web monitoring pages.

Head node: The head node hosts all the CAF and Condor management processes. The submission request manager (`submitter`) grants kerberos authentication to users, transfers the CAF tar-ball from the user desktop, creates the Condor submission files and submits the CAF job wrapper (`CafExe`) to Condor. The monitoring daemon (`monitor`) provides the user with monitoring information and manages the user interactive requests. The CAF router (`CafRout`) allows communication between the `monitor` daemon and `CafExe`. The notification daemon (`mailer`) sends the summary mail upon job completion and cleans up the Condor files.

Worker node: Condor starts `CafExe`, which gets the user kerberos credentials, untars the user tar-ball, and runs the user initial command. At the end of the job execution it archives the working directory and transfers the tar-ball to the output location. Moreover, `CafExe` performs some monitoring tasks: creates job summary file and serves as interactive `CafMon` callback. The user executable can either get data from the data handling system or read data from the CAF file server via `rootd`. All outward connections are kerberized.

A FIRST STEP TOWARD THE GRID

Figure 4 gives an example of the CPU saturation of one of the CAF's during the last year: around 500 users were active with an average of 35000 completed sections per day, belonging to 50-100 different users. The CDF increasing volume of data pushes for more and more computing power in the next future. While it may be problematic to attain funds for CDF dedicated resources, GRID offers plenty of resources.

A first step toward the integration of the CDF computing in the GRID framework has been moved at CNAF in Italy, where a Tier1 is currently operative. A working prototype farm, based on `condor_glidein` [5], is exploiting Tier1 idle resources. `condor_glideins` provides a way to dynamically increase the Condor pool, by submitting Condor daemons to the GRID. Once the Con-

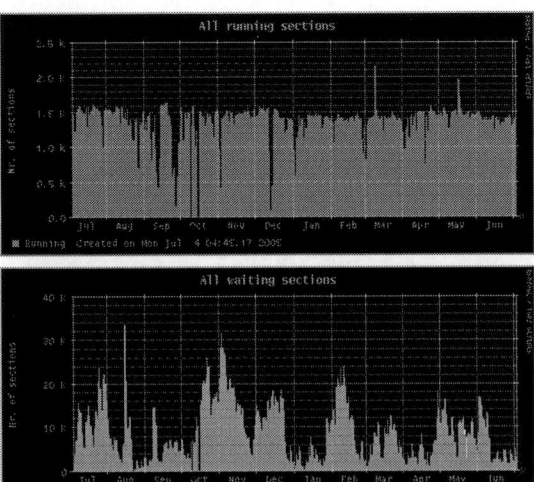

FIGURE 4. Number of running (upper) and waiting (lower) sections on FNAL CAF during the last year.

dor daemons start, the GRID worker nodes become a part of the Condor pool and are available for CDF users' jobs.

The final goal is to make the CAF software entirely compatible with the GRID framework, creating a CAF portal which can seamlessly exploit any GRID Computer Element in the world.

CONCLUSION

The need for computing resources has been driving the evolution of the CDF Analysis Farm from an old-fashioned farm, originally localized at FNAL, to a GRID oriented structure, distributed worldwide. Currently ∼50% of CDF computing power resides outside Fermilab. Future CAF developments aim to exploit the abundant reservoir of computing resources which GRID represents.

REFERENCES

1. CDF Collaboration, *"The CDF II Technical Design Report"*, FERMILAB-Pub-96/390-E (1996).
2. T.H.Kim *et al.*, *"The CDF Analysis Farm"*, IEEE Trans. Nucl. Sci. 51 (2004) 892.
3. The SAMGrid Project Home Page, http://projects.fnal.gov/samgrid/.
4. R.Brun and F.Rademakers, *"ROOT – An object Oriented Data Analysis Framework"*, Nucl. Inst. Meth. A 369 (1997) 81.
5. The Condor Project Home Page, http://www.cs.wisc.edu/condor/.
6. I.Sfiligoi *et al.*, *"The Condor based CDF CAF"*, in proceedings to CHEP04, Interlaken, Switzerland (2004).

The Silicon Drift Detector of the ALICE Experiment

G. Batigne for the ALICE collaboration

INFN, via Giuria, 1 10125 Torino Italy

Abstract. The ALICE experiment studies the properties of quark-gluon plasma and requires a good tracking system. This document presents the silicon drift detector which is part of the Inner Tracking System. Its principle and main features are given, especially its sensitivity to temperature variation and the effect of parasitic fields on measurement. Finally, the typical spatial resolution of this detector, which has been measured during beam tests, is shown.

Keywords: Silicon drift,ALICE, drift velocity, spatial resolution
PACS: 29.40.Wk, 29.40.Gx,61.72.Tt

INTRODUCTION

The purpose of the ALICE experiment[1, 2] is to study the quark gluon plasma (QGP) by using collisions of ultra-relativistic heavy ions. The direct observation of this state of matter is impossible. But the properties of the QGP can be extracted from signatures such as the modification in the production of strangeness, J/Ψ or bottomium. Moreover, the highest theoretical estimation of the number of particles created per unit of pseudo-rapidity (η) in Pb-Pb collision at 5.5 TeV is about 8000 at $\eta = 0$.

The two main detectors used for the particle tracking are the Time Projected Chamber (TPC) and the Inner Tracking System (ITS). They are embedded in a large solenoid magnet generating a field between 0.2 and 0.5 T. This tracking system covers, over the full azimuth, the polar angle range of $\pm 45°$ which corresponds to $|\eta| < 0.9$.

In the first section, a short presentation of the ITS is given. One will focus on the main functions and requirements of this system. The second section is dedicated to one of the three detectors of the ITS : the Silicon Drift Detector (SDD). The principle and the description of the SDD of the ALICE experiment will be presented, then the two main sensitive points of this kind of detector will be emphasized : the drift time calibration and the deviations in position. Finally, the spatial resolution measured during beam tests in 2000 will be shown.

INNER TRACKING SYSTEM

The main detector for particle tracking is the TPC. Its inner radius is fixed by the maximum acceptable hit density and its value is 90 cm[3]. The ITS is used to track particles between the beam pipe (radius of 3cm) and the TPC. Hence, the ITS allows to improve the efficiency of particle tracking. The specific functions of the ITS are two-fold[4] :

- determination of the primary vertex and secondary vertices necessary for the reconstruction of charm and hyperon decays.
- tracking and identification of low momentum particles ($p_T < 100$ MeV/c) which cannot reach TPC because of the magnetic field.

The ITS is an assembly of six concentric cylinders of silicon detectors whose the inner radius is 4 cm and the outer one is 43.6 cm with a total length of 97.6 cm. Three technologies are used. The two innermost cylinders are composed of Silicon Pixel Detectors (SPD) because, considering the high particle density close to the beam pipe, a very good resolution is necessary. The next two cylinders are made of Silicon Drift Detector. Finally, double sided Silicon Strip Detectors (SSD) with a small stereo angle (17.5 mrad) are used for the two outermost cylinders.

The ITS should be able to work in a stand alone mode for the measurement of low momentum particles. To apply the truncated-mean method for particle identification, a minimum of four measurement is mandatory. This explains the choice of the SDD for the two intermediate cylinders because it combines a good spatial resolution and an energy loss measurement. The SSD has a less good spatial resolution but has also an analog output to calculate dE/dx.

The table 1 gives some information about the different type of detectors among which the radii of the cylinders and the total number of detectors. The most important data are the spatial precision along the beam axis (z) and the transverse plan ($r\phi$). With such values, the precision on vertices position can reach 100μm.

TABLE 1. Some information about each kind of detector

Type	Mean radii of cylinders in cm	Total number of detectors	Spatial precision $r\phi$ in μm	Spatial precision z in μm
pixel	4 - 7	960	12	70
drift	14.9 - 23.8	260	38	28
strip	39.1 - 43.6	1770	20	830

SILICON DRIFT DETECTOR

Principle

When a charged particle passes through a silicon detector, some electrons go to the conduction band. In a drift detector, an electric field make these electrons migrate toward a line of anodes which are connected to the readout electronics. This kind of detector allows to reconstruct the position of the particle crossing on its surface. One coordinate is given by the centroid of the charges collected by anodes. The mean drift time of electrons from impact point to anodes is related to the other coordinate. The integral of charges is directly proportional to the energy loss, information required for particle identification.

This principal allows to reduce the number of readout channels but some systematic effects, such as electron mobility fluctuations or resistivity inhomogeneity have to be under control in order to achieve a good spatial precision. Moreover, the ratio signal over noise of the electronics should be as low as possible to make the energy loss resolution dominated by Landau fluctuation. After the description of the SDD, the drift time calibration and the systematic deviations in position reconstruction will be presented. Finally the spatial resolution of the SDD will be shown.

Description

Each SDD detector have a size of 7.25 cm by 8.76 cm with an active area of 7.02 cm by 7.53 cm. They were produced on 300 μm thick 5" Neutron Transmutation Doped (NTD) wafers with a resistivity of 3 kΩcm[5]. The figure 1 represents a sketch of a SDD detector structure. One can see that the active area is divided into two symmetrical part in order to decrease the maximum drift time and hence the data acquisition time. Each of these two parts is equipped with 256 anodes (pitch of 294 μm). The electric field of 667 V/cm is created by 291 cathodes (for each part) on both side of the wafer, parallel to the anode line and with a pitch of 120 μm. Under the anodes, there is an additional cathode used to "kick-up" the electrons to the anodes. In this way, the electric field acts like a toboggan. The electrons produced along the thickness of the wafer are focused in its center and drift toward the anodes.

The front-end electronic (FEE) boards are connected to anodes of one half SDD. They perform the pre-amplification, the analog to digital conversion, compression of data. FEE boards are based on two 64 channel ASICs, called PASCAL[7] and AMBRA[8]. The PASCAL chip provides the pre-amplification and digitalization and AMBRA chip the buffering of events and data compression from 10 bits to 8 bits. The clock frequency for these chips is 40 MHz. When a trigger is received, the data stored in 256 channel analog memory rings are frozen and digitalized. The sampling period of 25 ns is given by the clock. Hence each event represents an array of 256 anodes by 256 time bins. With a typical electron speed of 8 μm/ns, the maximum drift time (5 μs) fits in the time range of one event (6.4 μs). Then the data are transferred to AMBRA to be sent to the data acquisition system.

The SDDs and readout electronics are mounted on linear supports, called ladders, in carbon fiber (material chosen for its stiffness and low material budget). The inner cylinder is composed of 14 ladders of 6 SDDs each and the outer of 22 ladders of 8 SDDs. The detectors are mounted at different distances from the ladders in order to allow a small overlap of active areas. In this way, there is no dead zones in the acceptance of the SDD and all possible vertex positions can be measured.

Drift Time Calibration

The crucial parameter to control is the electron speed because it is the conversion factor between drift time and position. The electron speed is the product of the value of the electric field and the electron mobility within the crystal : $v = \mu_e E$. The variation in drift speed are mainly due to temperature fluctuations ($\mu_e \propto T^{-2.4}$). To achieve a spatial resolution of 30 μm, the temperature should not vary more than 0.1 °C. The SDDs and their FEE are cooled by water at ambient temperature[4]. This cooling system ensures to stabilize the temperature of the SDDs within the specifications. The first SDD cylinder is also thermally shielded from the heat produced by SPDs[4]. The drift speed can be monitored during the experiment by the mean of MOS charge-injectors. The

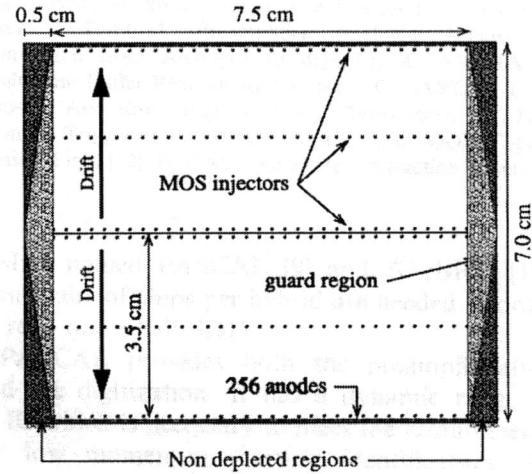

FIGURE 1. Sketch of the SDD detector (figure extracted from reference [6].

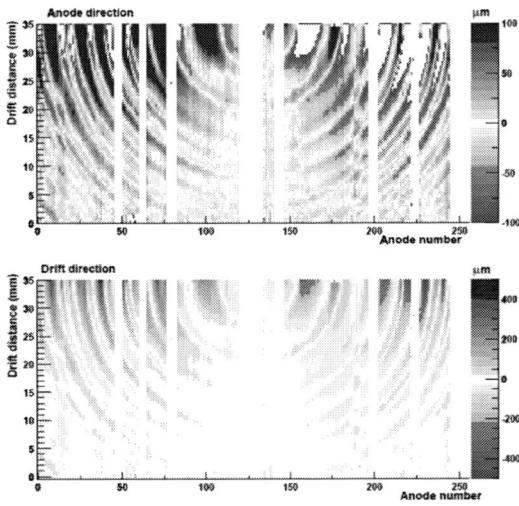

FIGURE 3. Differences between measured and real positions, in μm, along the anode direction (above) and the drift direction (below).

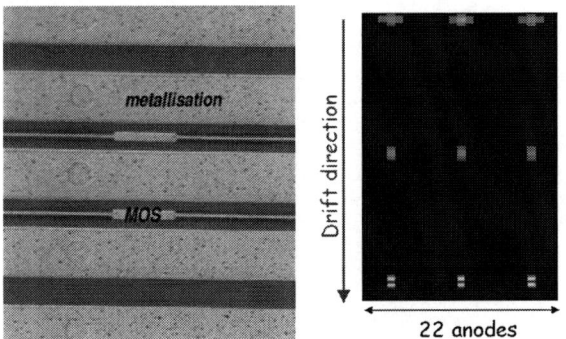

FIGURE 2. Left : picture of a MOS injector on the surface of the detector. Right : result of a measurement with injectors. This last picture is a zoom on an area of 22 anodes. One can also notice the effect of the diffusion on the width of the electron cloud which increases with drift time.

figure 1 shows the position of the three lines of 33 injector for each half SDD. These injectors released electrons in the silicon wafer when a specific trigger is given to the detector. The figure 2 represents on the left side one injector at the surface of a SDD and on the right side the data measured when an injector trigger has been given. The position of each injector on the surface is known with a very good precision (at the level of $1\mu m$). Any variation of temperature will be translated in a variation of the mean drift time of signals associated to injectors. Therefore this kind of measurement allows to correct the effect of temperature fluctuations.

Deviations in position

The main systematic effect linked to silicon drift detectors is due to impurities and dopant fluctuations in the silicon crystal. They introduce electrical parasitic fields which deviate the electrons from their straight trajectory toward the anodes. Depending of the amplitude of these parasitic fields, the electrons can be even collected by an other anode. Beam tests[9] performed at CERN have demonstrated that the error on the reconstructed position of the impact point can reach values as much as $700\mu m$. This number has to compared compared to the need resolution of $30\mu m$ on SDD to be able to reconstruct positions of vertices with a precision of $100\mu m$[4]. Therefore, the differences between reconstructed and real positions of impact points have to be measured all over the surface of each detector in order to correct them. An example of such deviation is given by the figure 3 ; the circular structure is explained by the process of the wafer production.

Spatial resolution

After the correction of data taken during beam tests, the spatial resolution can be extracted. The figure 4 shows one beam test result. One can see that the spatial resolution is mainly below 30 μm and its dependency with drift axis is the same along the drift and anode directions. In fact, the shape is the result of a mix of two effects : diffusion during the drift and anode bining. During the travel of the electron cloud toward the anodes,

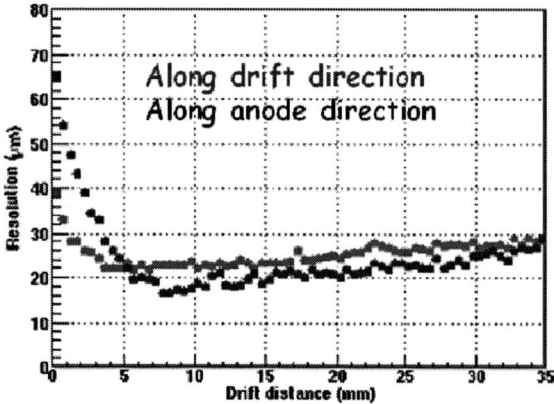

FIGURE 4. Spatial resolution along anode and drift axis in function of drift time.

its width increases with time because of the diffusion phenomenon. But meanwhile, the amplitude of the signal and therefore the ratio signal over noise decreases. This explains the slow increase in the value of the resolution with the drift distance. The quick decrease at low distance is due to the ratio between the width of the electron cloud and the anode pitch. When the cloud is created very close to one anode, all the electron are collected by this anode and therefore the error on the measurement is dominated by the anode pitch. When the drift distance increases, the diffusion makes the width of the electron cloud big enough to have signals on more than one anode. The behavior of the spatial resolution is also illustrated by the figure 2 where the signals of injectors at different drift distances can be seen.

CONCLUSION

The SDD is a detector which combines a good spatial resolution (at the order of 30μm) and the measurement of energy loss for particle identification. The number of readout channels is small but the price to pay is its slow response. The main systematic effects of this kind of detector are its sensitivity to temperature fluctuation and resistivity inhomogeneity. Nevertheless proper calibrations and monitoring allow to correct the consequences and therefore to achieve the required spatial resolution. Hence the SDD will fulfill its function in the ITS in order to determine vertices position and measure the low momentum particle during ALICE data taking.

REFERENCES

1. N. Antoniou, *et al.*, CERN/LHCC 93-16, 1993
2. ALICE web site : http://aliceinfo.cern.ch
3. TPC Technical Design Report, CERN/LHCC 00-001, 2000 or http://aliceinfo.cern.ch/Collaboration/Documents/TDR/index.html
4. ITS Technical Design Report, CERN/LHCC 99-12, 1999 or http://aliceinfo.cern.ch/Collaboration/Documents/TDR/index.html
5. A. Rashevsky, *et al.*, *Nucl. Instr. and Meth. A*, **461**, 133–138 (2001)
6. D. Nouais, *et al.*, *Nucl. Instr. and Meth. A*, **501**, 119–125 (2003)
7. A. Rivetti, *et al.*, CERN/LHCC 00-41
8. G. Mazza, *et al.*, CERN/LHCC 01-34
9. E. Crescio *et al.*, *Nucl. Instr. and Meth. A*, **539** 250–261 (2005)

The Tracker of the CMS Experiment

Ernesto Migliore[1]

Università di Torino and INFN, Via P. Giuria 1, 10125 Torino, Italy

Abstract. With more than 200 m^2 the Silicon Strip Tracker of the Compact Muon Solenoid (CMS) experiment will be the largest silicon detector ever built. In this contribution the main design considerations and the status of the construction, at about one and a half year after the begin of the production of the modules, are reviewed.

Keywords: CMS, tracking detectors, silicon strip detectors
PACS: 29.40.Gx,29.40.Wk

TRACKING AT LHC

The CMS experiment at the LHC collider at CERN will be equipped with an "all-silicon" detector for the tracking of the charged particles[1]. With the LHC running in pp mode, at the luminosity of 10^{34} cm^{-2}s^{-1} 20 minimum bias events will be produced on average at each bunch crossing. In the 4 T magnetic field of the CMS solenoid the expected average density of charged tracks is of about 1 track/cm^2 at 10 cm from the interaction point, decreasing to 0.1 track/cm^2 at 25 cm and to 0.01 track/cm^2 at 60 cm. Because of the high track density and of the short time interval between two bunch crossings, 25 ns, a fine granularity and fast response of the detector electronics is required. In these extreme conditions the tracker will have to measure precisely the points of all the tracks in a pseudorapidity region of $|\eta| \leq 2.5$ in order to determine with a good accuracy the particle transverse momentum, $\Delta p_T/p_T \sim 0.1 p_T$ (p_T in TeV) in the region of $|\eta| \leq 1.6$, and to provide an efficient tagging of long lifetime particles, b hadrons and τ's, $\varepsilon(b) \sim 60\%$ for a backgound contamination $\varepsilon(uds) \sim 1\%$. This has to be achieved with a limited number of points per track, about 10, and with a channel occupancy of 1-2%. As in the CMS magnetic field a 1 TeV p_T track has a sagitta of about 200 μm, the resolution required on a single measurement point is about 20-30 μm. This will be achieved by a combination of high resolution pixel detectors in the innermost region, $r < 15$ cm, and silicon strip detectors outside, $20 < r < 110$ cm. This paper describes the layout and the status of the construction of the latter, the Silicon Strip Tracker (SST).

---

[1] on behalf of the CMS Collaboration.

THE SILICON STRIP TRACKER

The SST is composed of 4 independent sub-systems. In the central region, $|z| < 120$ cm, there are an Inner Barrel (TIB) made of 4 cylindrical layers, complemented on each side by three Inner Disks (TID), and an Outer Barrel (TOB) made of 6 layers. In the forward region, $120 < |z| < 280$ cm, 9 Endcap Disks (TEC) are present on each side. A sketch of 1/4 of the CMS tracker in the rz projection is shown in Figure 1. The choice of an "all-silicon" layout has been made possible thanks to the achievements of the R&D activities on the read-out chip and on the silicon strip sensors.

The development of the APV25 read-out chip. The APV25[2] is the read-out chip used in the SST. The chip is built in 0.25 μm CMOS technology and it has been proved to be radiation tolerant up to a dose of 20 Mrad (the expected maximum dose in the CMS Tacker is about 10 Mrad). Each channel consists of a preamplifier coupled to a shaper which produces a 50 ns CR-RC pulse shape. The shaper output of each channel is then sampled at 40 MHz into a 192 cell deep pipeline which is in turn read out by an Analogue Pulse Shape Processor (APSP) circuit which can operate in two modes. In *peak* mode only one sample is read out while in *deconvolution* mode three consecutive samples are read and the output is a weighted sum of the three. The deconvolution operation results in a re-shaping of the analogue pulse shape to a narrowed one that peaks at 25 ns and returns rapidly to the baseline. The power required by one channel to operate in deconvolution mode is about 1.8 mW and the noise figure is ENC=396+59.4/pF both values smaller than those required by a simple CR-RC shaper with equivalent timing performances.

The R&D activity on the silicon strip sensors. The SST is equipped with p-on-n single sided AC coupled silicon strip sensors. Their design is the result of a R&D

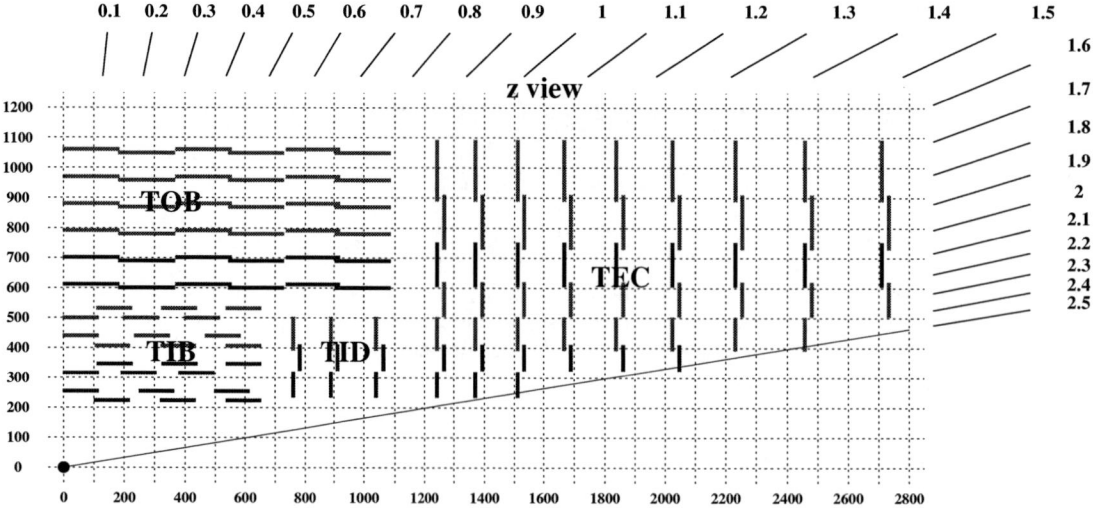

FIGURE 1. Sketch of one quarter of the SST (*rz* projection), with the different sub-detectors indicated.

study which brought to the following results[3]:

- the definition of a radiation[2] tolerant design of the sensor based on: the use of $\langle 100 \rangle$ silicon crystal, to guarantee stable interstrip capacitances after the exposure to irradiation; two different substrate resistivity values (higher for sensors to be used in regions where the expected fluence is lower and viceversa), and the adoption of a metal strip 4-8 μm wider than the p implant to improve the IV stability of the sensor when operating at the high bias voltage (400 V) required for an efficient charge collection after irradiation;
- a careful tuning of the layout of the strips to guarantee a capacity of about 1.25 pF/cm independently from the strip pitch and the sensor thickness. This allows to have similar granularity in both the inner and the outer layers with a similar signal/noise figure as the larger noise of longer strips, 19 cm in the TOB compared to 12 cm in the TIB, is compensated by the larger signal of thicker detectors, 500 μm in the TOB compared to 320 μm in the TIB.

---

[2] Fluences up to $1.5 \cdot 10^{14}$ 1 MeV n_{eq} cm^{-2} are expected in the SST volume.

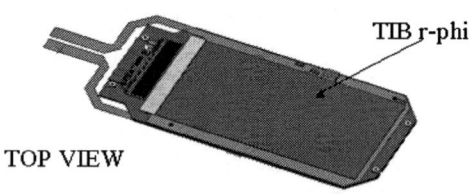

FIGURE 2. A TIB double sided module.

The availability of sensors obtained from 6" silicon wafers allowed a large scale production of silicon strip sensors and has been proved to have the same performance of the 4" sensors.

THE SST MODULES

The module is the basic mechanical and read-out unit of the SST. Its simplest layout consists of a single sided silicon strip sensor glued onto a carbon fiber frame together with a front-end hybrid (Figure 2). The hybrid is a 4 layers kapton circuit laminated onto a ceramic rigidifier and it hosts four or six APV25 chips, the additional ASICs and a glass pitch adapter to match the constant pitch of the APV's channels, 44 μm, with the pitch of the silicon

TABLE 1. Amount of the different components for the different SST sub-detectors: SS stands for single sided, DS for double sided modules.

	n. of modules	n. of sensors	n. of APVs
TIB	1188 SS	2724	14k
	768 DS		
TOB	3048 SS	10416	24k
	1080 DS		
TID	240 SS	816	4k
	288 DS		
TEC	4096 SS	10288	30k
	1152 DS		
Total	8572 SS	24244	73k
	3288 DS		

TABLE 2. Main acceptance criteria for the sensors.

I_{leak} at 450 V	<10 μA
I_{strip} at 400 V	<100 nA
I_{diel} with 10 V across	<1 nA
R_{poly} at 400 V	1.5$\pm$0.5 MΩ
Good strips	>99%

strips of the different geometries. Longer modules used in the TOB and TEC are made of two silicon sensors chained together while double sided modules are made of a sandwich of two single sided modules with the silicon sensors tilted by an angle of 100 mrad. Given the large amount of modules required to build the SST, see Table 1, the production of modules has been split in several steps and distributed in different centers.

Production and qualification of the silicon sensors. The sensors for the SST are produced by two companies: STM[3] and HPK[4]. A first set of tests, to guarantee that 99% of the delivered sensors will later pass the CMS acceptance criteria, are performed by the vendors. The sensors are then shipped, via CERN, to four Quality Test Centers (QTC) where after an optical inspection they undergo the following electrical tests[4]: an IV curve up to 550 V, a CV curve up 350 V at 1 kHz measurement frequency, a strip-by-strip test at 400 V bias voltage for measuring I_{strip}, I_{diel}, C_{AC}, R_{poly}. Tests in QTC were performed on all the sensors during the first 5% of the production while on a 5-10% sample basis for the remaining lots. The main acceptance criteria for a sensor are listed in Table 2.

Production of the front-end hybrids. The production of hybrids is made in two steps performed by two different companies: production of the substrate[5] and populating them with the ASICs and the passive components[6]. Hybrids are then shipped to CERN for gluing the pitch-adapter, bonding the APV channels and a fast qualification test.

---

[3] ST Microelectronics, Catania, Italy.
[4] Hamamatsu Photonics K. K., Hamamatsu-City, Japan.
[5] Cicorel, Boudry, Switzerland.
[6] Hybrid SA, Combamare, Switzerland.

Assembly of the modules. Qualified sensors and hybrids, together with a kapton cable for the connections, are glued onto the carbon fiber support by means of a robot ("gantry"). The robot guarantees micrometric movements and the achieved resolution in the positioning of the pieces is of 9 μm in the measurement coordinate, much better than the 39 μm required for the initial alignment of the detector. Typically 4 modules are produced on the same gantry plate and the full process lasts 1/2 day.

Bonding and qualification of the modules. Assembled modules are sent to production centers where the bias and read-out connections of the sensors are made by means of automated ultrasonic microwire bonding machines. The resistance of the bonds to a pull force is tested on a sample basis. The average resistance achieved in all bonding centers is well above the lower threshold of 5 g. The time required for the full bonding of a module on Delvotec 6400 machine is of about 30 minutes. Each module is then tested with a compact and standalone system called ARC[5, 6]. The ARC test consists of an IV curve up 450 V and a series of read-out tests (pedestal run, calibration run, runs with a pulsed and a continuous light source) to identify possible defects as noisy strips, shorts between neighbouring strips, open read-out channels, broken AC capacitances. The test requires about 30 minutes. A module is rejected if the leakage current at 450 V is more than 5 times larger that measured in the sensor test or if it has more than 2% of the channels identified as defective. To verify the resistance of the module to the temperature changes it will experience during its operation in the experiment and to spot possible weaknesses that may lead to an early mortality in the SST, TIB and TID modules are placed in a climatized box in which they are repeatedly cycled for three days between +20 °C and -20 °C while being continuously powered and read-out ("long term test"). Similar tests, but with the modules already mounted onto larger structures, are performed for the TEC and for the TOB modules.

At the time of this paper, April 2005, about 3000 modules have been produced corresponding to different fractions of the required amount for the different sub-detector systems (about 50% of the TIB and TID, 5% of the TOB and 15% of the TEC). The fraction of good modules is above 95%. In the case of the TIB modules, the largest sample produced so far, about 1700 units, the fraction of good modules is 97% with the main source of rejection being the bad IV behavior (2%) while the fraction of modules discarded for bad strips is about 0.5%.

Despite its excellent overall quality, the delivery of the modules suffers a delay because of two main problems encountered during the production. The first affected the front-end hybrids and consisted in fragile contacts in the vias, the metalized holes connecting different layers of the hybrid. The problem, affecting only some batches mainly of TEC and TOB hybrids, has been solved adopting a new hybrid design, with the diameter of the vias enlarged from 100 μm to 250 μm and with the rigidity of the support increased by the insertion of an additional kapton layer, and monitoring more closely all the production stages at the company. The second problem concerned the quality of the thick sensors provided by the STM. The main defects observed on these sensors were a large and unstable leakage current, an excessive charge concentration at the silicon-oxide interface and values of the interstrip resistance at 20 V bias voltage much lower than 1 GΩ. Because of the repeated problems the CMS management has decided in the Summer of 2004 to stop the contract with the STM and place an order for the remaining 85% of the required thick sensors to HPK.

THE SST INTEGRATION

At the time of this paper, the integration of the modules on the different sub-detectors is already started. Each sub-detector group has adopted its own procedure even if the major steps, which are related to how the SST is linked to the rest of the CMS experiment, are similar. The SST will be placed in CMS inside a cylindrical vessel of 5.8 m length and 2.5 m diameter where it will be kept at -10 °C to avoid the thermal runaway of the current of the sensors after irradiation. Digital controls and analogue outputs inside the SST will be in form of electrical signals while they will be transmitted to the CMS control room via optical fibers[7]. The digital control is based on the Communication and Control Unit (CCU)[8] which takes the control of a group of modules. Several CCUs are chained together to form a control ring. The CCU receives inputs from a Digital OptoHybrid (DOH). On the other side analogue outputs from the module are converted to optical signals by an Analogue OptoHybrid (AOH). Finally the SST should be supplied with the 15 kA current required by the front-end electronics and connected to the cooling system which should take away the 35 kW power dissipated[9]. As a consequence the integration consists of the following steps: the fixing of the cooling pipes on the mechanical structures and the test of the gas tightness, the mounting of the cards housing the DOH and the AOH, the mounting of the cables housing the CCUs which the modules will be connected to, and finally the mounting of the modules. The structural unit for the integration in the case of the TIB is a layer, corresponding to one half of a cylindric shell, for the TOB is a rod, a string of 6/12 modules, and for the TEC is a petal, a sector of 22.5°. The first integrated layer of the TIB is shown in Figure 3.

FIGURE 3. One half a the forward layer 3 of the TIB with mounted modules and services.

REFERENCES

1. CMS Collaboration, CERN-LHCC/98-6 (1998); CMS Collaboration, CERN-LHCC/2000-16 (2000).
2. M. J. French et al., *Nucl. Instr. and Meth.*, **A466**, 359–365 (2001).
3. S. Braibant et al., *Nucl. Instr. and Meth.*, **A485**, 343–361 (2002).
4. J.-L. Agram et al., *Nucl. Instr. and Meth.*, **A517**, 77–93 (2004).
5. M. Axer et al., CMS Note 2001/46 (2001).
6. M. Meschini et al., CMS IN Note 2004/18 (2004); C. Civinini et al., CMS IN Note 2004/50 (2004).
7. http://cms-tk-opto.web.cern.ch/cms-tk-opto/.
8. C. Ljuslin et al., in *Proc. of the 8th Workshop on Electronics for LHC Experiments*, CERN-LHCC/2002-034 (2002).
9. A. Bocci et al., in *Proc. of the 10th Workshop on Electronics for LHC Experiments and Future Experiments*, CERN-LHCC/2004-030 (2004).

The Monitored Drift Tube Chambers of Atlas

S. Ventura

INFN - Laboratori Nazionali di Frascati

Abstract. The Atlas experiment has been designed to explore the high energy physics frontier at the TeV energy scale and to investigate on the physics of the Standard Model and beyond at the Large Hadron Collider (LHC) at Cern. The Muon Spectrometer represents the most part of the Atlas detector. It has been designed to provide standalone measurement of the transverse muon momenta with a relative accuracy of 3% over a wide momentum range up to 10% for momenta of 1 TeV. This high accuracy is provided by the Monitored Drift-Tube chambers (MDT) which can determine the track trajectory with a precision of 40 μm.

Keywords: Drift chamber, gas detector, precision drift tubes
PACS: 29.40Gx,29.40Cs

Introduction

Most of the processes of great interest in the Standard Model physic and beyond produce one or more high p_t muons in the finale state. For this reason the Atlas collaboration has chosen to design a Muon Spectrometer [1] able to provide standalone measurement of the muon transverse momentum with a relative accuracy from 3% up to 10% in a momentum range from few GeV to 1 TeV. The Atlas Muon Spectrometer is made up of tracking detectors (Monitored Drift-Tube chambers) and trigger detectors (Resistive Plate Chambers in the barrel and Thin Gap Chambers in the end-caps). In this paper the main features of the tracking detectors are described.

The test system of the Muon Spectrometer at the beam line H8 at Cern is also described. In particular the method to measure the muon sagitta and momentum at H8 is reported in more detail.

The ATLAS Muon Spectrometer

The main requirements for the Atlas Muon Spectrometer are dictated by some benchmark reactions that produce one or more high momentum muons. The most important are:

- $H \rightarrow ZZ^* \rightarrow 4\mu$ [2]: discovery channel for the Standard Model Higgs boson in the mass range from 130 to 170 GeV (in this range a mass resolution at the level of 1% is needed);
- $H/A \rightarrow \mu\mu$: discovery channel for the Minimal SuperSymmetric Model Higgs;
- Z', W': new vector bosons production, in their leptonic decays muons with transverse momentum up to several TeV are produced.

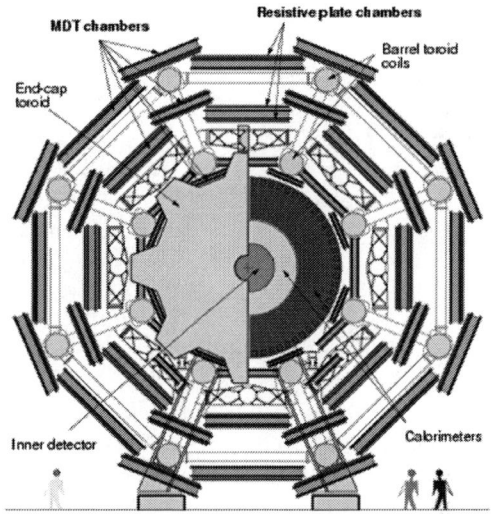

FIGURE 1. Section of the Muon Spectrometer in the plane transverse to the beam direction.

In order to reach a very high precision in the muon momentum determination the Atlas Muon Spectrometer has been designed with peculiar features.

In the barrel region the magnetic bending is provided by a large toroidal magnet in air composed by eight coils surrounding the hadron calorimeter, while in the end-caps the muons are bent by two smaller end-cap magnets. This configuration allows to minimize the multiple scattering.

The momentum measurement is provided by high precision tracking detectors, the Monitored Drift-Tube chambers (MDT). They are organized in three cilindrical layers for the central barrel region and in three wheel shaped planes in the end-caps. In the innermost ring of the end-cap inner station the Cathod Strip Chambers (CSC) are imployed instead of MDT, in order to sustain the very high background rate near the interac-

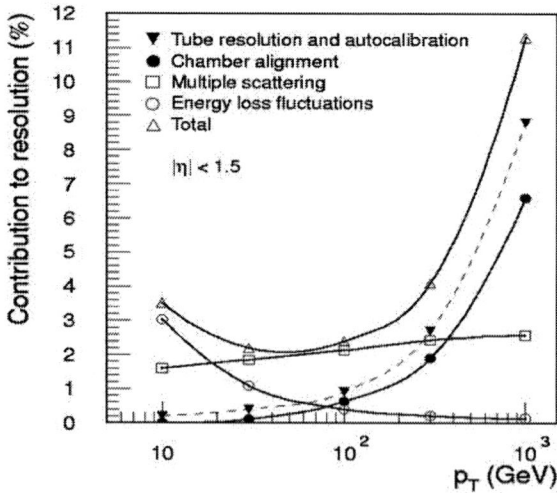

FIGURE 2. Muon Spectrometer standalone p_t resolution.

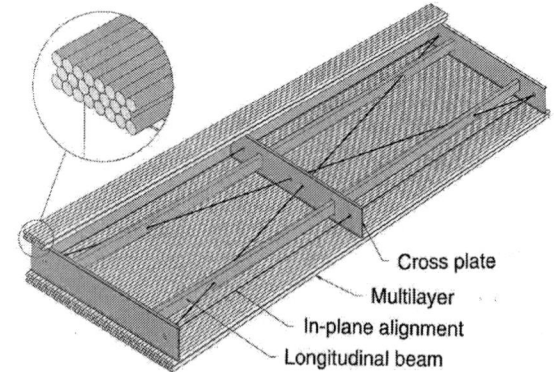

FIGURE 3. An MDT chamber scheme.

tion point. The trigger system covers the pseudorapidity range $|\eta| < 2.5$. In the barrel the trigger is provided by Resistive Plate Chambers (RPC) while in the end-cap is provided by Thin Gap Chambers (TGC). Both these technologies reach a time resolution of few nanoseconds and are also used to measure the coordinate along the MDT wire direction. In fig. 1 a section on the transverse plane of the Atlas detector is shown. In fig. 2 the relative resolution on the transverse muon momentum ($\Delta p_t/p_t$) is shown. For low momenta the contribute due to statistical fluctuation of the energy loss dominates while for p_t from few tens of GeV up to 300 GeV the multiple scattering contribute dominates. For higher p_t the effects of the detector spatial resolution on the muon momentum measurement become more important.

The Muon Spectrometer comprises 1194 MDT chambers produced in 13 production sites in 7 different countries. The MDT chamber production is now completed. The installation in the Atlas cavern is expected to be finished in the spring of 2006.

The Monitored Drift-Tube (MDT) chambers

The muon trajectory is precisely measured by MDT chambers [3]. The MDT chambers consist of six (eight in the inner stations) layers of pressurized tubes. The layers are separeted in two multilayers mounted on a light aluminum space frame as is shown in fig. 3. The tube, also in aluminum, have a diameter of 30 mm and a wall thickness of 400 μm. Each tube is filled with a gas mixture of Ar:CO_2 (93:7) at a pressure of 3 bar. The W-Re wire has a 50 μm diameter and is put at 3080 V. The tube has a gas gain of 2×10^4.

In each tube the radial distance between a charged particle track and the wire is determined by measuring the arrival time of the ionization electrons. A typical tube time spectrum is shown in fig.4. The non uniformity of the drift velocity in the gas mixture is evident from the time distribution. The start of the physical time window, t_0, corresponds to a particle that crosses the wire. The t_0 depends on the delays of the signal cables and front-end electronics and has to be computed for each tube by fitting the time spectrum with an appropriate function. The accuracy in determining the t_0 is about 0.5 ns. The end of the physical time window, t_1, corresponds to a particle that crosses the tube wall. The $t_{max} = t_1 - t_0$ depends on the tube drift properties and is the same for all the tubes. To performe the track fit it is necessary to convert the measured time in a position by use of an appropriate r-t relation. This is calculated using an iterative procedure of autocalibration, applied multilayer by multilayer.

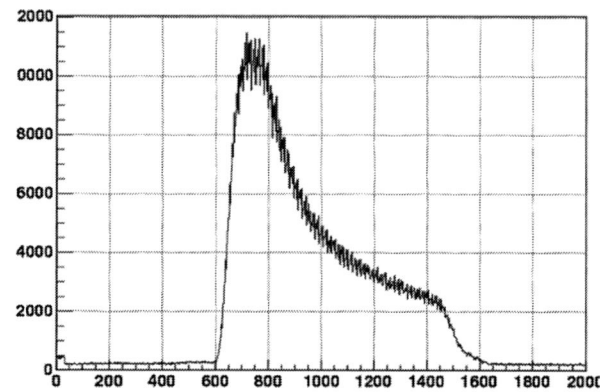

FIGURE 4. Single tube time spectrum in TDC counts.

The single tube space resolution ranges from 200μm very close to the wire to 50μm near the tube wall. The overall chamber position resolution is 40 μm. To reach this high precision the wire position in the chamber has to be known with a precision of 20μm.

The deformation of the MDT chambers and their relative position to each other in Atlas will be continuously monitored by using optical systems, RASNIK [4], incorporated into the chamber structure. This system provides a very good control on the effective chamber positions. The effects of chamber misalignment on the measurement of the muon sagitta has to be less than 40μm.

SYSTEM TEST AT H8

The Atlas Muon Spectrometer has been tested in the last years at the H8 beam line at the CERN SPS in order to test and validate the performances of the system. The setup [5] for the beam test of the summer 2004 includes 12 monitored drift tube chambers organized in one barrel sector and one end-cap octant. The barrel sector reproduces an Atlas tower positioned at an angle of 15 degrees with respect to the direction ortogonal to the H8 beam. A barrel tower consists of two Barrel Inner Large chambers (BIL), two Barrel Middle Large chambers (BML) and two Barrel Outer Large chambers (BOL). The BML and BOL chambers are equipped with the Resistive Plate Chambers (RPC) that provide trigger for muons and measure the coordinate of a track along the MDT tubes. The chambers are fully instrumented with Front End electronics (FE) read-out and with the Muon Readout Driver (mROD). Both barrel and end-cap chambers are equipped with the alignment system. Two additional barrel stations are present in the H8 area. One of these is a BIL put on a rotating support. The chamber can rotate up to 10 degrees around the vertical axis. A magnet is put between the rotating BIL and the barrel sector and bends the tracks in the orizontal plane.

The beam at H8 covers a muon momentum range from 20 GeV up to 350 GeV. This setup allowed to perform a lot of studies on the system performance [6]: integration between the different technologies of the Muon Spectrometer, test of the detector control system, noise and efficiency studies, calibration, test of the DAQ, test of the alignment system and so on.

In particular in this paper the measurement of the muon momentum and of the sagitta resolution as a function of the momentum is described in more detail. The setup used to perform the sagitta and momentum measurements in H8 is schematically shown in fig. 5, where the coordinate system convention is shown: the muon beam is along the x axis while the MDT wires are along the y axis.

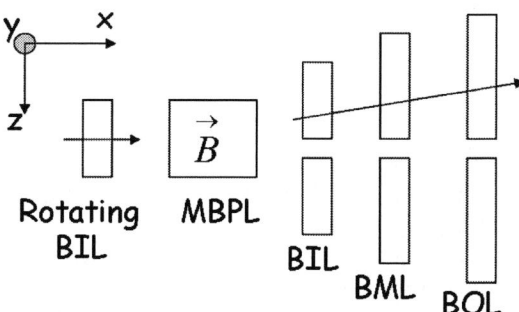

FIGURE 5. Schematic top view of the muon H8 test area at Cern.

Sagitta resolution as a function of the muon momentum

The Atlas Muon Spectrometer measures the sagitta of muon tracks from the measurements of the precision chambers in the inner, middle and outer muon stations. The track sagitta is defined as the distance between the segment reconstructed in the middle station and the line connecting the segments reconstructed in the inner and in the outer stations. The sagitta is related to the muon curvature and thus to the momentum. A muon with $p_t = 1$ TeV has a sagitta of $\simeq 500\mu$m in the average magnetic field of 0.5T of the Atlas spectrometer, thus to reach a momentum resolution of 10% the sagitta has to be measured with an accuracy of 50μm.

The resolution on the sagitta measurement takes two contributes. One is the effect of the multiple scattering due to the material that the muon has to cross in the spectrometer, that dominates at low momenta. The second contribute is due to the intrinsic resolution of the tracking chambers and is independent from the beam energy. In H8 the curvature of the tracks on the orizontal plane is provided by a magnet put between the rotating BIL and the barrel sector (fig. 5). The muon momentum is given by:

$$p(GeV) = \frac{0.3BL(Tm)}{\Delta\theta_B(rad)} \quad (1)$$

where BL is the bending power of the magnetic field, and $\Delta\theta_B = \theta_{BARREL} - \theta_{rotBIL}$ is the difference between the angle of the track reconstructed by the barrel and the angle of the segment reconstructed in the rotating BIL. In order to take into account a possible misalignment of the rotating BIL with respect to the barrel chambers due to the rotation around the vertical axis, the formula has to be corrected by subtracting the $\Delta\theta$ computed for a run with the magnetic field turned off, $\Delta\theta_0$. If there is a misalignment of the barrel chambers due to rotation of the chambers around the beam direction, the sagitta depends on the track vertical position. To take into account the effects of such misalignment, only tracks that cross the system at the same vertical position -measured by RPCs- are used.

In fig.6 the momentum distribution for a beam of nominal momentum of 120 GeV is shown. The distribution is fitted with a gaussian excluding the tail at low momenta. A measured mean value of 108 GeV is obtained with $\sigma_P =$ 4 GeV. The discrepancy between the nominal and measured momentum is mainly due to the energy loss by the beam in the material upstream the muon system.

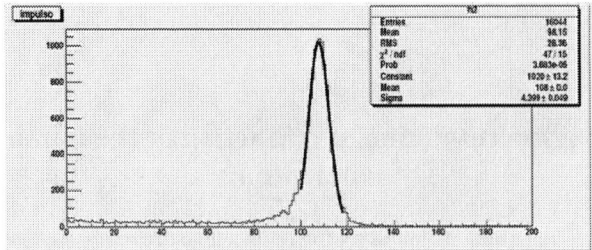

FIGURE 6. Momentum distribution for a muon beam of 120 GeV nominal momentum.

The sagitta and momentum measurements were repeated for several runs with different values of the beam energy. For each data set the sagitta resolution is determined after two cuts: only tracks in the same horizontal plane, corresponding to a particular strip of the RPCs, and with a momentum larger than $P_{mean} - 2\sigma_P$ are used. In fig.7 the sagitta resolution as a function of the beam momentum is shown.

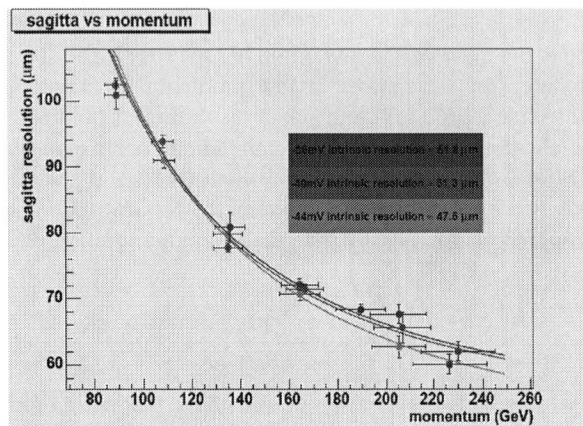

FIGURE 7. Sagitta resolution as a function of the muon momentum.

The curves in fig.7 are fitted with the function:

$$\sigma_{sag} = \sqrt{K_1^2 + \frac{K_2^2}{P}} \qquad (2)$$

where K_1 is the constant term related to the intrinsic resolution, K_2 is the term related to the multiple scattering. The K_2 term is related with the quantity $<x/X_0>$, that represents the distance crossed by the muon in radiation lengh units. The results found are reported in table 1 for the three MDT thresholds.

The results show an intrinsic resolution contribute to the

TABLE 1. Results of the fit on the sagitta resoltion versus momentum curves with the function of eq.(2).

	-36 mV	-40 mV	-44 mV
K_1 (μm)	51.8	51.3	47.5
$<x/X_0>$ (%)	31.5	31.2	34.1

sagitta resolution of about 50μm in good agreement with the expectations. The multiple scattering term is also in agreement with the expected value. No relevant dependence on the MDT thresholds is observed.

CONCLUSIONS

The Atlas Muon Spectrometer provides standalone measurement of the transverse muon momentum with a very high accuracy ($\Delta p_t / p_t =$ 10% at $p_t =$ 1 TeV) by the use of the Monitored Drift-Tube chambers as tracking detectors. Estensive test of the muon system in the last years have allowed to reach a good control on the Atlas Muon Spectrometer performances. In particular it has been shown with experimental test beam results that the intrinsic resolution of the tracking detectors provides a resolution of 50μm on the muon sagitta determination at high momenta where the multiple scattering is negligible.

REFERENCES

1. The Atlas Muon Spectrometer Technical Design Report, The Atlas Muon Collaboration,CERN/LHCC/97-22, 31 May 1997.
2. A study of Higgs boson decay H→4μ based on full simulation of the Atlas detector, By I.R. Boiko, K.V. Nikolaev, G.A. Shelkov (Dubna, JINR),. 2005. 14pp. Published in Phys.Part.Nucl.Lett.2:67-76,2005, Pisma Fiz.Elem.Chast.Atom.Yadra 2:7-20,2005.
3. MDT
4. The RASNIK/CCD 3D alignment system, ATL-MUON-94-063; ATL-M-PN-63, Geneva CERN, 15 Dec 1994.
5. System Test of the ATLAS Muon Spectrometer in the H8 Beam at the CERN SPS, ATL-MUON-2004-026; ATL-COM-MUON-2004-020; CERN-ATL-COM-MUON-2004-020, Geneva CERN, 2004.
6. The H8 muon community, Proposed measurement program for H8 2004 muon system test, ATL-COM-MUON 2004-006, 4 May 2004.

The CMS Electromagnetic Calorimeter

Riccardo Paramatti

CERN and INFN Rome1

Abstract. The electromagnetic calorimeter of the CMS experiment at LHC will consist of about 76000 Lead Tungstate crystals. Its main purpose is the very precise energy measurement of electrons and photons produced at 14 TeV centre-of-mass energy. A review of its performances and its construction status is given. Then the calibration strategy is described in details.

Keywords: calorimeter, scintillation, lead tungstate, calibration, electron
PACS: 29.40.Mc, 29.40.Vj

INTRODUCTION

The Large Hadron Collider will allow the study of pp interactions at a center of mass energy of 14 TeV. The main physics goal of the CMS experiment [1], one of the four detectors at LHC, is the quest for Higgs and SUSY particles. The maximum design luminosity foreseen at this machine is 10^{34} cm^{-2} s^{-1} and the crossing rate will be 40 MHz. The production of around 25 events per bunch crossing will result in about 1000 tracks. The radiation environment is expected to be particularly severe. In ten years of running (corresponding to a total integrated luminosity of around 10^6 pb^{-1}) detector parts will be exposed to a maximum neutron fluence of 10^{17} n/cm^2 and γ doses of 10^6 Gy. These extreme conditions imposed a long R&D phase in order to obtain high granularity, radiation resistant, fast and selective detectors.

The electromagnetic calorimeter of the CMS experiment [2] is composed of about 76000 Lead Tungstate (PWO) crystals organized in a barrel covering the central rapidity region $|\eta|<1.48$ and two endcaps which extend the coverage up to $|\eta|=3$. A preshower detector placed in front of the endcaps improves the γ - π^0 separation in this region. To increase hermeticity the barrel has a nearly pointing geometry both in ϕ and η (crystals have a 3° tilt in both directions).

LEAD TUNGSTATE (PWO)

Lead tungstate (PWO) is a very dense ($\rho = 8.28$ g/cm^3) scintillating material. The most appealing properties of PWO crystals are the fast scintillation light decay time (80% of light is emitted within 25 ns), the basic radiation resistance, the small radiation length ($X_0 = 0.89$ cm) and the small Molière radius ($R_M = 2.2$ cm) which allow the construction of a compact calorimeter with a granularity suitable for the multiplicity expected at LHC.

Nevertheless this material presents major drawbacks: the low level of light yield (100 photons/MeV, $\sim$ 0.2% with respect to NaI:Tl) imposes a read out with a photo detector with gain, the strong dependence of the light yield on temperature (-2%/ ^{o}C around 18 ^{o}C) and the high refractive index (2.29 at peak emission wavelength $\lambda = 420$ nm) which makes the extraction of light from the crystal very difficult.

A huge effort was done during R&D (1994-1998) to improve the light yield of these crystals without spoiling their fast response and to guarantee their radiation hardness at a level sufficient to preserve the requested energy resolution.

The PWO crystals show a damage induced only by electromagnetic radiation through the creation of color centers that reduce crystal transparency. The amount of the damage depends on the dose rate and reaches a stable level after a small administered dose. A partial recovery of crystal transparency happens in few hours. The scintillation mechanism remains untouched and the observed loss of light is tolerable and can be followed with a monitor system by injecting light through the crystals.

Photodetectors

Avalanche photo diodes (APD) were developed for the barrel part of the calorimeter. These silicon devices are insensitive to the 4 T magnetic field of the experiment, have an internal gain (M = 50 foreseen for CMS) essential for PWO. The quantum efficiency around the wavelength of the PWO emission peak is about 75%-80%. The amplification is obtained in a small region ($d_{eff} \sim 6$ mm) of high electric field. Two APDs of 25 mm^2 area are coupled to each crystal. In this configuration 4000 photo electrons are produced per GeV of deposited energy.

In the endcap regions radiation levels are too high to

use APDs; the longitudinal magnetic field allows there the use of vacuum photo triodes (VPT). These are single stage photo-multiplier tubes with a fine metal grid anode. Their active area is 280 mm^2, quantum efficiency is about 20% at 420 nm and a gain of 10 is expected in the 4 T magnetic field.

CONSTRUCTION

The construction of the calorimeter is a distributed process. The barrel construction started in 2002. Parts as crystals, capsules, mechanical elements for the support structure etc. are produced under the responsibility of different Institutions taking part to the project and sent after in two Regional Centers located in CERN and Rome.

The calorimeter has a modular structure. The basic sub-unit of the barrel composed by a crystal (whose dimensions vary with η but are approximately of 2cm x 2cm x 23cm) and a capsule (hosting two APDs read in parallel) glued together is inserted in a glass fiber alveolar structure. Ten sub-units fit in this structure that is closed by an aluminum tablet and constitute a "sub-module". Sub-modules are of 17 different types (as well as crystals) depending on η position. A "module" is made by 40 or 50 sub-modules mounted on a 3 cm thick aluminum grid. Modules are of 4 η types. A "super-module" is a set of 4 modules and the barrel consists of two identical halves made by 18 super-modules.

Endcaps will have a simpler structure, 138 "super-crystals" made of 25 crystals are a so-called "dee" and two of them make an endcap. Endcap crystals are larger with respect to the barrel; the transversal dimensions are $\sim$ 3cm x 3cm.

In the Regional Centers all the elements are assembled and checked step by step, following a well defined quality control protocol, the quality and the functionality of parts resulting from mounting operations. In particular all the crystals are subject to the systematic test of their geometrical shape, longitudinal transmission (related to radiation hardness), transversal transmission, light yield and uniformity of light collection. Both Centers are equipped with similar automatic machines to insure crystal quality. A detailed description of the performances of the machine installed in the INFN/ENEA Regional Center of Rome can be found in [3]. The strict control of quality at each step of the construction will insure a fully operational calorimeter at best of its potentiality and reduce the risk of the presence of dead or malfunctioning channels. The "bare" supermodule assembly and its subsequent dressing are done at CERN. Around 60% of the barrel is currently (june 2005) assembled.

ENERGY RESOLUTION

The Higgs boson hunt in the "low mass" region is the basic ground for the CMS choice to go for a crystal calorimeter. In the mass interval between 100 GeV and 150 GeV the Higgs can be favorably detected by its decay in two photons. The H$\to \gamma\gamma$ reaction is not the most abundant but the cleanest one for detection in spite of the large irreducible background. The natural width of a Higgs with a mass around 100 GeV is in the MeV region making the resolution on the invariant mass of the two photons completely determined by the energy and angular resolution of the calorimeter. Thus the discovery potential is set by the experimental resolution and H$\to \gamma\gamma$ is a benchmark reaction for the electromagnetic calorimeter of CMS. The energy resolution of a calorimeter can be parameterized by:

$$\frac{\sigma(E)}{E} = \frac{a}{\sqrt{E}} \oplus \frac{b}{E} \oplus c \qquad (1)$$

where a is the stochastic term related to any kind of Poisson-like fluctuation, b is the electronic noise term and c the so-called constant term and the symbol $\oplus$ indicates the sum in quadrature of the terms. The target values for coefficients in (1) are set to a = 0.025 GeV$^{1/2}$, b $\leq$ 0.2 GeV and c $\simeq$ 0.005 with an angular resolution $\sigma(\theta) \sim \frac{50 mrad}{\sqrt{E(GeV)}}$. With these values a Higgs of 100 GeV mass can be observed by the CMS experiment with a significance S = 5 with 30 fb^{-1} taken at luminosity 10^{33} cm^{-2}s^{-1}.

Crystal Uniformity

Key issues to keep the constant term c $\sim$ 0.5% are:

- Light collection uniformity
- Inter-calibration by monitor and physics signals at 0.5% including the effect of radiation damage

The tapered shape of crystals needed to obtain an hermetic pointing geometry induces a focusing effect in the light collection. The same amount of energy deposited far or near the photo-detector will produce a higher or lower signal respectively. This effect, named uniformity of the light collection, can be controlled by de-polishing with a given roughness one lateral face of the crystal. The ideal situation is to have a uniform light collection in the region of the maximum of the electromagnetic shower energy deposition. Due to the fact that the position of the shower maximum fluctuates, deviations from uniformity in this region will produce additional contribution to the constant term c. To maintain this contribution below 0.3% the maximum allowed deviation from uniformity

(FNUF) in the crystal region 3 X_0 to 13 X_0 was set to:

$$\frac{dLY}{X_0} = \pm 0.35\%/X_0 \quad (2)$$

The FNUF parameter is measured in laboratory using a radioactive source of Cobalt 60.

CALIBRATION

Raw intercalibration from laboratory measurements of crystal Light Yield performed during assembly together with CERN-SPS test beam results of supermodules will represent the precalibration of ECAL. Following the present construction schedule, only few supermodules will be precalibrated in the test beam. The raw intercalibration expected for channels that have not been precalibrated in the test beam is ~5%. This result is shown in Fig. 1 where the crystal intercalibration measured during crystal quality controls is compared with intercalibration obtained from test beam data.

The optimal functioning of the complete CMS detector required to reach the design calibration, will take some time. However it will be possible to improve the calibration precision just at the start-up by using the ϕ symmetry of deposited transverse energy; this method leads to a reduction of the number of intercalibration constants. Within a few days, it should be possible to determine the remaining constants using $Z \to e^+e^-$ decays.

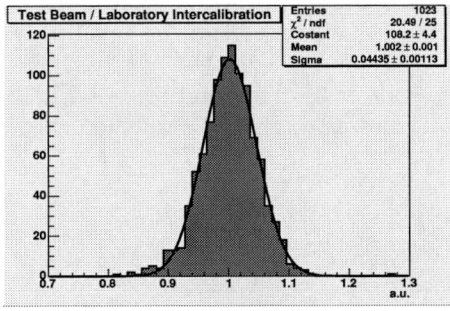

FIGURE 1. Comparison of crystal intercalibration obtained from laboratory measurements and from electron beam data.

Tracker Material and Bremsstrahlung

The big amount of material between the interaction point and ECAL represents the main difficulty in performing in situ calibration. In terms of radiation lenghts, the tracker material thickness rises from about 30% X_0 at $\eta=0$ to a peak of more than 1 X_0 at an η corresponding to the ECAL endcap region [4].

Such quantity of material gives rise to electron bremsstrahlung and photon conversion, degrading the energy resolution of electromagnetic particles; physical distributions relevant in the calibration process, such as the electron pair invariant mass peak in the Z channel or E/p resolution in the W channel, are strongly bremsstrahlung dependent. Functions of variables such as the energy sum in the 9 crystals matrix around the most energetic deposit ($E_{3\times 3}$), the 25 energy sum ($E_{5\times 5}$) or the energy of the electromagnetic cluster, arranged by a variable number of crystals [4], allow to select good electrons, in term of bremsstrahlung, using only ECAL information.

ϕ Symmetry

The proposed method has been tested with a sample of 18 million fully simulated minimum-bias events [5] corresponding to few hours at the start-up luminosity $2\cdot 10^{33} cm^{-2} s^{-1}$ assuming 1 kHz Level 1 Trigger bandwidth dedicated to the calibration.

The method is based on the following assumption: the total (transverse) energy deposited from a large number of events should be the same for all crystals in a ring at fixed η.

After an artificial miscalibration of 6%, for each crystal i a calibration constant c_i is extracted using this ratio:

$$c_i = \frac{<\sum_{all\ events} E_T>_{ring}}{(\sum_{all\ events} E_T)_{crystal\ i}} \quad (3)$$

This procedure can be iterated but already after the first iteration the spread of residual miscalibration is compatible with the expected calibration precision due to the variation of $\sum E_T$ seen in perfectly calibrated crystals. This spread is due to the fact that the ϕ symmetry of the detector is not exact; the main source of asymmetry is the inhomogeneity of tracker material. The limit on the precision achievable, without using explicit knowledge of tracker inhomogeneity, is represented as a function of η in the barrel range in Fig. 2 by square points. In the same figure the round points represent the precision which can be obtained using 18 million of minimum-bias events.

Intercalibration with Z events

At the LHC center of mass energy, $\sqrt{s} = 14$ TeV, the $Z \to e^+e^-$ cross section is ~2 nb; the clear signature of these decays will assure enough data to set the ECAL absolute energy scale and to intercalibrate different regions of the calorimeter.

The calorimeter rings can rapidly be intercalibrated using Z events by the electron pair invariant mass reconstruction; it will be possible, together with the ϕ symmetry method, to calibrate in this way the whole ECAL at

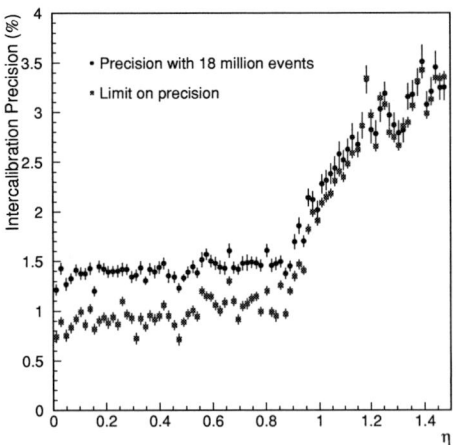

FIGURE 2. Intercalibration precision which can be obtained with 18 million minimum-bias events and systematic limit of the precision due to the ϕ asymmetry of the detector.

~2% just at the start-up and without using tracker momentum measurements. The Z channel gives energetic correlated electrons in different regions of the calorimeter, hence a large fraction of events allows to intercalibrate the endcaps with respect to the barrel. Typically the transverse momentum of the Z is low, giving essentially back-to-back electrons in the azimuthal plane. Furthermore the $Z \rightarrow e^+e^-$ rate versus η of the electrons is flat over most of the ECAL range.

Intercalibration with W events

$W \rightarrow e\nu_e$ is a very high statistics channel particularly important during the start-up low luminosity phase, when the rate will be ~10 Hz. The use of tracker measurements of isolated electron momentum will lead to the determination of the calibration coefficients for individual crystals. As in the Z channel, the electron shower involves many crystals, therefore the unscrambling of the calibration constants is required.

Electron bremsstrahlung gives a long tail in the E/p distribution, therefore some cuts should be applied to avoid using electrons which belong to the tails. As in the Z channel, a compromise must be found between hard cuts and the necessity to select an adequate number of events per crystal; the strategy foresees to use hard cuts to intercalibrate groups of crystals and then, by loose cuts application to get high statistics, to intercalibrate single crystals of the same group. In a couple of months at $L = 2 \cdot 10^{33} cm^{-2} s^{-1}$, exploiting the full tracker information, the high statistics $W \rightarrow e\nu_e$ decays will allow to reach 0.5% resolution.

Laser Monitoring

In between calibrations with physics events the evolution of the system due to the loss of transparency induced by radiation damage will be followed by a light injection system. The monitoring system foresees a laser light injection at two different wavelengths in beam gaps at a frequency of 80 Hz. A system of multiple fan-outs guarantee the uniformity of the distributed light. Radiation hard reference PN diodes can be compared to APDs signal to track changes. The full calorimeter can be checked in 20 minutes.

CONCLUSIONS

The electromagnetic calorimeter of CMS is a challenging project; an intense and rewording R&D effort was performed. The calorimeter is actually in the construction phase in CERN and Rome Regional Centers. In the meanwhile studies are going in order to understand details of the system behavior having as final aim to produce outstanding physics results.

The electron and photon reconstruction have not been described in this report; a detailed description can be found in [4].

The goal of in-situ calibration is to achieve a level of 0.5% for the energy resolution constant term. As seen the amount of material in front of ECAL complicates the calibration procedure. At the start-up it should be possible to intercalibrate at the 2% level in few days using the ϕ symmetry method together with $Z \rightarrow e^+e^-$ events. As soon as the tracker will be aligned, $W \rightarrow e\nu_e$ decays will allow to reach 0.5% resolution.

REFERENCES

1. CMS Collaboration, "The CMS Technical Proposal", CERN/LHCC 94-38, 1994.
2. CMS Collaboration, "The CMS Electromagnetic Calorimeter Technical Design Report", CERN/LHCC 97-33, 1997.
3. S.Baccaro, B.Borgia, M.Castellani, A.Cecilia, I.Dafinei, M.Diemoz, S.Guerra, E.Longo, M.Montecchi, G.Organtini, F. Pellegrino, "An automatic device for the quality control of large-scale crystal's production", Nucl. Instr. & Methods A459, pp. 278-284, 2001.
4. CMS Collaboration, "Data Acquisition & High-Level Trigger Technical Design Report", CERN/LHCC 2002-26, 2002.
5. D. Futyan, C. Seez, "Intercalibration of ECAL crystals in ϕ Using Symmetry of Energy Deposition", CMS Note 2002/031

The ATLAS Liquid Argon Electromagnetic Calorimeter

L. Carminati

INFN e dipartimento di Fisica dell' Università di Milano
Via G. Celoria 16, 20122 Milano
Email: `leonardo.carminati@mi.infn.it`

Abstract.
The construction of the ATLAS Liquid Argon Electromagnetic calorimeter has been completed and commissioning is in progress. After a brief description of the detector layout, readout electronics and calibration, a review of the present status of the integration and the detector qualification is reported. Finally a selection of performance results obtained during several test beams will be presented with particular attention to linearity, uniformity, position reconstruction and γ/π^0 separation.

Keywords: Calorimetry, Liquid Argon, ATLAS
PACS: <29.40.Vj>

INTRODUCTION

The ATLAS experiment, presently under construction, is a multi-purpose detector designed to study the physics at the CERN's future proton-proton collider called *Large Hadron Collider*. The LHC will provide collisions at a centre-of-mass energy of 14 TeV with a design luminosity from 10^{33} (*low luminosity phase*) up to 10^{34} cm^{-2}s^{-1} (*high luminosity phase*) and a bunch crossing frequency of 40 MHz. In high luminosity conditions about 20 minimum bias events per bunch crossing are expected with a total number of approximately 700 charged tracks. In order to deal with this operation environment, all sub-detectors in the experiment must be radiation-tolerant and have a fast signal readout and fine granularity to minimize the pileup effect.

The physics program is quite ambitious, going from precision measurements of W$^\pm$ and top quark masses to searches for new physics such as Higgs boson and SUSY particles. The electromagnetic calorimeter will play a crucial role in the detection of electrons and photons, jets and missing energy measurements like in the final states of Higgs boson decays H$\to \gamma\gamma$, H$\to$4e and H$\to$WW$^{(*)}$. The electromagnetic calorimeter has been primarily designed to optimally detect Higgs boson decay in photons or electrons/positrons pairs: the consequent requirements, determined from simulation ([1] and [2]), are the following:

1. Good hermeticity: largest possible acceptance with best granularity up to $\eta = 2.5.$ is required in order to observe rare decay processes
2. Good resolution of the energy measurements of photons and electrons: typically $10\%/\sqrt{E} \oplus 0.7\%$. This requirement involves some constraints on the mechanical design and electronic calibration.
3. Linearity of the response better than a few per mille which requires a good electronics calibration and the possibility to correct for upstream material energy loss.
4. Particle identification (e$^\pm$/jets or γ/π^0) and measurement of the direction of photons with a resolution of 50 mrad/$\sqrt{E}$. These requirements lead to the design of a fine lateral and longitudinal segmented detector.
5. A large dynamic range (from 20 MeV to 2 TeV per channel) which sets severe requirements on the performance of the read out electronics.

To fulfill these criteria, and having to operate in a high radiation level environment, the ATLAS collaboration has chosen a lead/liquid argon sampling calorimeter with an innovative accordion geometry. The liquid Argon tecnique ensures a good intrinsic linearity, long term stability and radiation tolerance. The accordion geometry provides a very good hermeticity while minimizes inductances in the signal path allowing the fast signal shaping required by the 25 ns buch crossing interval.

After a short description of the calorimeter geometry and signal reconstruction , the status of the detector construction and commissioning at April 2005 is reviewed. Finally a selection of the main testbeam results on production modules are reported showing that they meet LHC physics requirements.

DETECTOR DESCRIPTION

Detector layout

The ATLAS electromagnetic calorimeter is a lead/liquid Argon sampling calorimeter with accor-

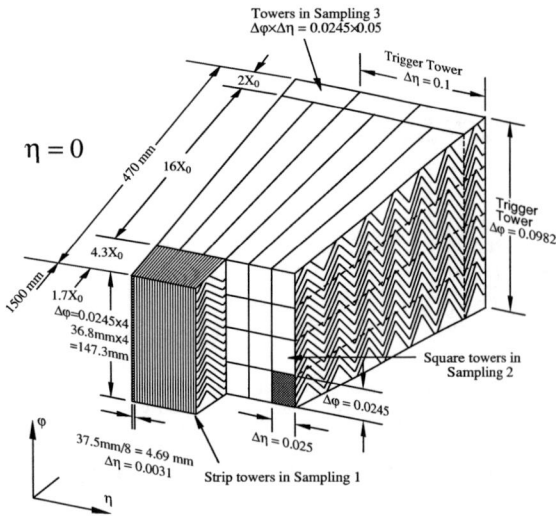

FIGURE 1. Segmentation of the EM barrel calorimeter

dion shaped electrodes and absorbers interleaved as sketched in figure 1. The absorbers are made of lead glued between thin stainless steel sheets with a total thickness of 2.2 mm. In the barrel, in order to mantain a uniform sampling term over the entire pseudorapidity range, the lead thickness decreases from 1.5 to 1.1 mm at $\eta = 0.8$. The electrodes are made of 3 copper layers insulated by 2 kapton layers ([3]). The two outer layers are connected to HV sources and polarize the LAr gap to the absorbers. The inner layer collects the signal through capacitive coupling.

The calorimeter is divided in two half-barrel wheels covering the pseudorapidity range $|\eta| \leq 1.475$ housed in a single cryostat (togheter with the superconducting coil placed in front of the calorimeter wich provides the magnetic field for the inner tracker) and two endcap detectors (each subdivided into an *outer wheel* for the $1.375 \leq |\eta| \leq 2.5$ range and an *inner wheel* from $2.5 \leq |\eta| \leq 3.2$) in two separate endcap cryostats. From a mechanical point of view each half barrel(endcap) is made of 16(8) identical modules.

In the barrel the LAr gap can be kept at constant 2.1 mm (half-gap) thickness from front to back by opening the folds of the accordion: in this way the HV value can be set to a 2 kV constant value for all gaps. In the endcap the gap increases with the radius due to geometrical reasons from 0.9 to 3.1 mm. This requires varying HV to keep the detector response indipendent of η. In a practical way the HV has 9 steps from 0.9 to 2.5 kV.

The calorimeter is segmented in ϕ, η and depth. The readout granularity in ϕ direction is obtained by summing signals of a certain number of gaps. The depth separation in compartments and the η granularity are made by divisions etched on the copper surface of the electrodes. Except for the edge zones such as the transition region at $|\eta| \simeq 1.5$ between barrel and endcaps and the inner wheel of the endcaps, the segmentation of the calorimeter presents 3 compartments in depth while both η and ϕ granularity depends on the compartment:

1. The first sampling (*strips*) ends at about $5\,X_0$ including $\simeq 1.5\,X_0$ (depending on η) of dead upstream material in the ATLAS configuration. It has a fine η - segmentation (0.003 equivalent to 5 mm at $\eta = 0$) while rather coarse in ϕ (0.1). It is optimized for γ/π^0 separation and allows a good photons direction reconstruction.

2. The second sampling (*middle*) contains most of the shower energy and ends after $22\,X_0$; its granularity is $\Delta \eta \times \Delta \phi = 0.025 \times 0.025$.

3. The third sampling (*back*) complements the calorimeter in radius and has twice granularity in η of the middle layer while the same granularity in ϕ.

In the $|\eta| \leq 1.8$ range a thin presampler detector is used to correct for energy losses in the dead material in front of the calorimeter. Further details about the mechanical description of these detector can be found in [4].

Signal reconstruction and energy calibration

When charged particles cross the gap, the ionization signal collected onto the electrodes by the drift field gives rise to a triangular current with a typical duration of 400 ns and whose initial value I_0 is proportional to the incoming particle energy. The signal is collected in the inner layer of the electrodes via capacitive coupling and sent through read-out lines to the front-end electronics where it is amplified, shaped [1], sampled at the bunch crossing frequency and stored into analog buffers. Upon a decision from the Level-1 trigger, the samples are digitized and sent to the back-end electronics via optical fibers. The signal amplitude A and the phase τ are determined from the signal samples S_i (typically 5) using a *digital filtering* technique ([5]) as

$$A = \sum_{i=1}^{5} a_i S_i \qquad A\tau = \sum_{i=1}^{5} b_i S_i \qquad (1)$$

The coefficients a_i and b_i can be analytically evaluated once the signal shape and the noise autocorrelation function are known.

---

[1] A CR-RC^2 bipolar shaper has been choosen with a three gain structure. The peaking time after shaping for a typical middle cell is $\simeq 40$ns

Each read-out channel can be calibrated electronically by sending precise exponential pulses allowing to measure the gain and linearity of each channel. However the calibration and ionization (physics) signals differ in shape and in normalization because the ionization signale is generated inside the LAr gap as a triangular current pulse while the calibration signal is exponential and it is injected at one end of the detector. The shape of the physics signal which is required for the digital filtering coefficients evaluation can be predicted from the calibration signal with different methods ([6] and [7]) by properly taking into account the differences between the two signals. After digital filtering and electronic calibration the best estimation of the signal amplitude is given in term of ionization current produced in the LAr gap.

The measure of the energy deposited by an electromagnetic showering particle requires a detailed knowledge of the intrinsic response of the calorimeter (the sampling fraction) and the inactive or dead materials distribution.

The calorimeter response f is a function of the shower depth δ ([8]) and η and can be estimated from accurate MonteCarlo simulations. The shower depth can be evaluated event by event exploiting the longitudinal segmentation of the calorimeter. The energy lost in dead materials in front of the calorimeter can be recovered by properly weighting the signal collected in the presampler. The energy deposited in the dead materials between presampler and strips can be evaluated by properly combining the measurement from the two compartments. Finally the longitudinal leakage has been parametrized as a function of η and the shower depth ([8]). The precise knowledge of the electromagnetic scale requires a detailed and reliable tuning of the detector simulation which can be achieved from the comparison with real data obtained on beam tests.

STATUS OF THE DETECTOR CONSTRUCTION

The contruction of the calorimeters is finished and both barrel and endcaps have been inserted into their cryostats. The installation of the solenoid is also finished. At the present time the whole barrel and one endcap (EMEC-C) have undergone several tests both at room temperature and cold, in order to comply with the required performance. The channel integrity has been tested by pulsing a step signal into each calibration line and measuring the output on the readout cables: only 0.03 % of faulty channels has been observed. All calibration injection resistors have been measured: the r.m.s spread is less the 0.1% well within specification and only a few resistances ($\simeq 0.15\%$) are broken or damaged. A

FIGURE 2. A view of the electromagnetic calorimeter cryostat in the ATLAS pit surrounded by hadronic tiles clorimeter.

complete test of the HV system at cold has been performed at the nominal voltage: a few sectors ($\simeq 1.5\%$ of the total acceptance) with one half-gap faulty have been identified. The absence of high voltage on one side of a sector can be recovered to first order by a factor two on the signal at the expense of a resolution deterioration by a factor of $\sqrt{2}$. No sectors with failures on both sides of the electrodes has been found. After the qualification tests the barrel calorimeter has been moved in the cavern in october 2004 (see figure 2) while the transportation of EMEC-C is scheduled by the end of 2005. The second endcap wheel (EMEC-A) is now in the cooling down phase and the qualification tests are ongoing.

PERFORMANCE AT BEAM TESTS

A total number of four barrel and three end-cap calorimeter modules (in addition to the so called "module 0"s) were exposed to electrons, muons and pions beams with energies up to 245 GeV on the H8 and H6 beamlines of CERN's SPS. A selection of the main results is reported in the following.

1. *Energy resolution and uniformity:* the energy resolution is measured from energy scans performed on several calorimeter cells. After unfolding the noise and the beam energy uncertainty from the energy spectrum, the energy resolution can be fit with the function $\sigma_E/E = a/\sqrt{E} \oplus c$: the sampling term a measured for the barrel (endcap) modules is less than 9.5 % (12.5 %) and the local constant term c less the 0.3 % (0.5%). The uniformity of the energy response of the calorimeter has also been systematically measured with beam scans in each cells along η and ϕ. For both barrel and endcap module the

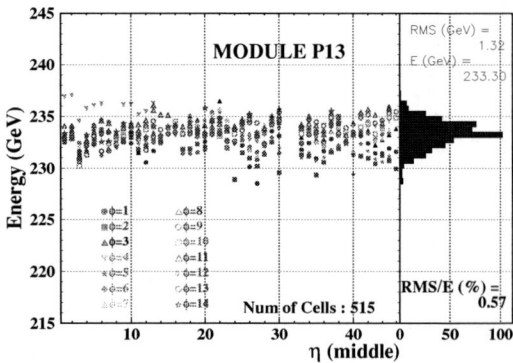

FIGURE 3. Uniformity for barrel module P13

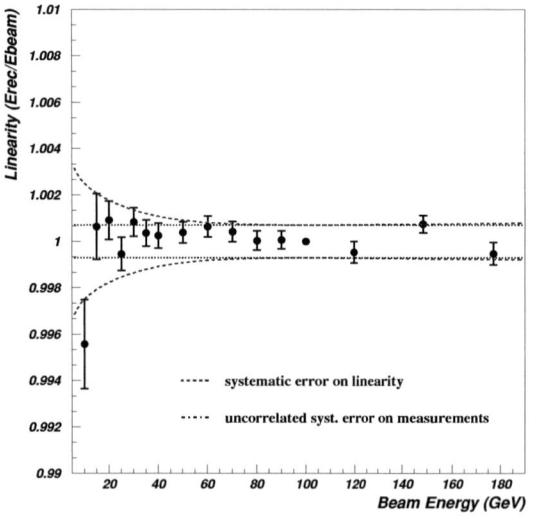

FIGURE 4. Linearity achieved in electromagnetic energy reconstruction in the barrel calorimeter

3. *Position resolution:* the electromgnetic calorimeter can provide a standalone measurement of the shower direction exploiting its longitudinal segmentation and the fine granularity of the first layer. An overall resolution on the θ angle reconstruction of $\simeq 55$ mrad$/\sqrt{E[GeV]}$ has been obtained. Such a performance is relevant for the vertex reconstruction in the H$\to \gamma\gamma$ decay since the longitudinal coordinate of the primary vertex can be reconstructed by the calorimeter with an uncertainty of $\simeq 1.6$ cm, better than the 5.6 cm dispersion expected at LHC.

4. γ/π^0 *separation:* A large potential background to the H$\to \gamma\gamma$ is due to jet-jet and γ-jet events where a jet is misidentified as a photon, especially when the jet fragments into a leading π^0. Discriminating variables based on the shower shapes in the first layer provide further rejection against these events. Using a dedicated setup to produce photons in the beam line, events faking a 50 GeV transverse momentum π^0 have been emulated by superimposing offline photon shower shapes. The measured γ/π^0 rejection is 3.54 ± 0.12 for a 90 % efficiency for isolated photons, in good agreement with the MC expectation of 3.66 ± 0.12.

CONCLUSION

The ATLAS electromagnetic calorimeter has concluded the contruction and integration period. Cold tests have been performed on the full barrel and one endcap. The tests on the the second endcap are forseen to start in the summer 2005. The quality of the calorimeter in terms of performance at testbeams and number of defective channels allow the LArg collaboration to have good expectations for a successful operation wihin ATLAS.

non uniformity is found to be below 0.6 % for almost an entire module as shown in figure 3 for a barrel module. For a local region $\Delta\eta \times \Delta\phi = 0.2 \times 0.4$ is less than 0.5 %, within specifications. Taking into account the reported modules performance the goal of a 0.7 % global constant term is expected to be achieved intercalibrating different sectors with $Z \to e^+e^-$ events.

2. *Linearity:* a special effort was made to precisely measure the linearity of the calorimeter response in the 10 - 180 GeV beam energy range ([8]). Calibration coefficients have been determined from an accurate comparison between data and GEANT4 simulations leading to a linearity < 0.1 % for energies above 40 GeV (< 0.25 % for energies > 10 GeV). as shown in figure 4.

REFERENCES

1. The ATLAS collaboration, *Detector and Physics performances Technical Design Report,* Vol. I , Tech. Rep. CERN/LHCC/99-14, CERN (1999).
2. The ATLAS collaboration, *Detector and Physics performances Technical Design Report,* Vol. II , Tech. Rep. CERN/LHCC/99-15, CERN (1999).
3. B. Aubert et al., NIM A539 (2005) 558 (2005).
4. The ATLAS collaboration, *Liquid Argon Calorimeter Technical Design Report,* Tech. Rep. CERN/LHCC/96-41, CERN (1996).
5. W.E. Cleland and E.G. Stern, NIM A338 (1994) 467 (1994).
6. D. Prieur, ATLAS note ATL-LARG-PUB-2005-001 (2005).
7. D. Banfi, M. Delmastro, M. Fanti, ATLAS note ATL-LARG-2004-004 (2004).
8. G. Graziani, ATLAS note ATL-LARG-2004-001 (2004).

The LHCb Muon System

W.Baldini, on behalf of the LHCb muon group

Istituto Nazionale di Fisica Nucleare, sezione di Ferrara, via Saragat 1 44100 Ferrara

Abstract. In this paper is described the design, the construction and the performances of several Multi Wire Proportional Chamber prototypes built for the LHCb Muon system. In particular we report results for detection efficiency, time resolution, high rate performances and ageing effect measured at the CERN T11 test beam area and at the high irradiation ENEA Casaccia Calliope Facility.

Keywords: LHCb, Muon, MWPC
PACS: 07.05.Fb, 29.40.Cs

INTRODUCTION

The LHCb muon system [1] is made of five Stations (M1-M5) of rectangular shape, covering an acceptance of 300 mrad (horizontally) and 200 mrad (vertically). It's main task is to provide fast (L0) triggering and offline muon identification. The first station, M1, is placed in front of the SPD/PS. M2-M5 follow the Hadron Calorimeter (HCAL) and are separated by iron filters. The stations cover a total area of about 435 m^2.

Each station is divided into four regions: R1-R4, with increasing distance from the beam axis and with approximately the same angular acceptance. The granularity of the readout is higher in the horizontal plane, in order to give an accurate measurement ($\delta p/p \simeq 20\%$) of P_t.

Since the muon detector is part of the L0 trigger, a detection efficiency of at least 99% per station, in a 20ns time window, is required. The system must then be fast and have a good time resolution ($\simeq 5\%$).

Given the large active area to be covered, the detector should be relatively cheap but, at the same time, robust, in order to guarantee stable working conditions for 10 years. The chosen Multi Wire Proportional Chambers (MWPC) technology satisfy all the above requirements.

MWPC DESCRIPTION AND REQUIREMENTS

The muon detector is built with 1380 4-gaps MWPCs (apart from M1, which is built with 2-gaps MWPC) and 12 triple GEM detectors, in the innermost region of station M1 (which will not be described in this paper). A wire-cathode distance of only 5 mm ensure a fast charge collection (about 10 $nsec$) while a wire spacing of 2 mm has been chosen to have a time resolution of 5ns (or less) and a good mechanical stability [2]

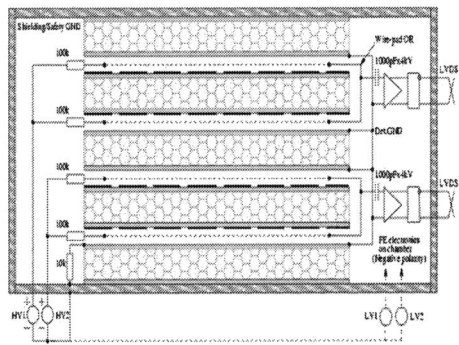

FIGURE 1. Schematic side view of a 4-gaps MWPC of the LHCb Muon detector.

MWPC description

Figure 1 shows a schematic view of a MWPC. The five panels needed to build a chamber are made by injecting a polyurethanic foam (ESADUR 120) between two $FR4$ printed circuits boards. The chambers are operated with a non flammable mixture of $A_r/CO_2/CF_4$ in the following proportions: (40 % / 40% / 20%). With this gas mixture a 6 GeV muon (this is the muon minimum energy needed to cross all the 5 stations) will generate, in average, about 50 primary electrons per gap. At the nominal voltage the gas gain is $G \simeq 5 \times 10^4$, that corresponds to a collected charge of about 0.4 pC per mip.

The four gaps are wire-ORed, in pairs, to form two independent 2-gaps before the front-end electronics, and have four independent HV lines. This guarantees redundancy and robustness since each 2-gap can be operated independently.

Depending on the granularity and on the particle rate, different regions of the detector are readout through cathode pads, wire pads or through both of them (combined readout). The granularity varies from $(0.6 \times 3.1 \ cm^2)$ in the M2R1 region, to $(24.8 \times 30.9 \ cm^2)$ in M5R4.

MWPC requirements

The main requirement imposed by the L0 trigger for each muon station (i.e. per each 4-gap chamber) is to have a detection efficiency of 99% in a $20\,ns$ time window. Given this condition we requires a 2-gap efficiency higher than 95% in a 20 ns time window, which means a 4-gaps efficiency of 99.8%. A good rate capability (up to $460\,kHz/cm^2$) and 10 years of stable operations (i.e. up to an integrated charge of $1C/cm$ in the hottest region) is also required.

THE FRONT END ELECTRONICS

The readout of the MWPC is performed using the CARIOCA chip [3], a custom designed current amplifier which also includes shaper and discriminator circuits. The CARIOCA chip exploits a $0.25\,\mu m$ CMOS radiation hard technology and allows the radout of both, cathode and wire pads signals. A short peaking time ($t_p < 15ns$) ensures a good time resolution with a reasonably low gas gain, while the shaper circuit, which is designed to have an average pulse width of $50\,ns$, minimize the inefficiency due to signal pile up.

EFFICIENCY AND TIME RESOLUTION MEASUREMENTS

Several MWPC prototypes have ben tested at the CERN T11 test beam area (3.6 GeV charged pions) in the last three years [4] in order to measure the efficiency and the time resolution.

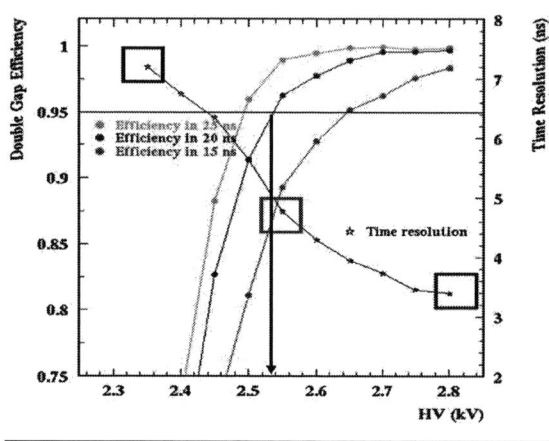

FIGURE 2. Detection efficiency and time resolution, as a funcion of the operating voltage, measured on a MWPC prototype at the CERN T11 test beam area.

Figure 2 shows the behaviour of these parameters as a function of the operating voltage, for a 2-gaps prototype with cathode readout. The three efficiency curves correspond to a time window of 25, 20, 15 ns respectively. As shown, the required efficiency of 95% in a 20 ns time window is reached already at an operating voltage of less that $2.55\,kV$. This corresponds to a time resolution of less that $5\,ns$.

Let's now make some important comments about these results. First, since from our measurements we have seen that the gas gain doubles every 100 V, for the wire pad readout, where the signal is two times bigger than cathode readout, the above requirements are reached at an operating voltage of about 2.45 kV (100 V lower with respect to the cathode readout).

Second, as shown in fig. 2, in case of a problem with one of the two 2-gaps of a MWPC, the other 2-gap can reach the same efficiency and time resolution by simply increasing the operational voltage to $2.65V$ (2.6 kV plus some 50 V safety factor). This ensure the redundancy and robustness required.

Efficiency and time resolution uniformity have also been measured, as a function of the position on the chamber surface, for a full size M3R3 prototype. From our measurements (see fig. 3, 4) we see that about 97% of the points satisfy the requirements for an operating voltage of 2.6 kV.

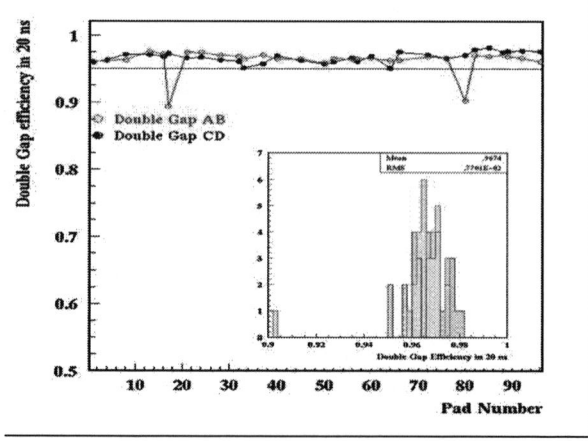

FIGURE 3. Detection efficiency measured on about 100 points on the MWPC surface.

HIGH RATE TEST

The efficiency and time resolution of a 2-gap prototype has been measured at the CERN Gamma Irradiation Facility (GIF) under high irradiation conditions. The chamber was exposed to a rate of about $1MHz$ per wire pad. We have observed a time resolution stable, and so no

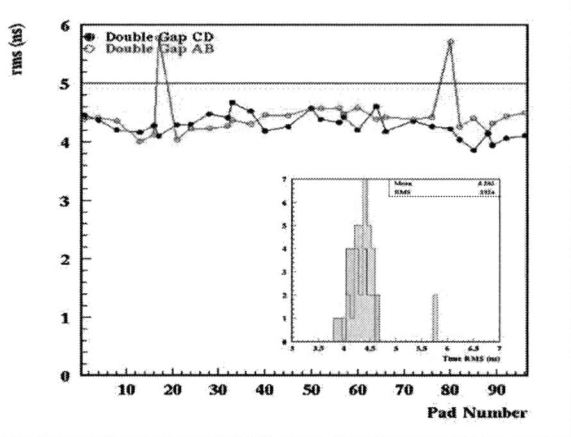

FIGURE 4. Time resolution measured on about 100 points on the MWPC surface.

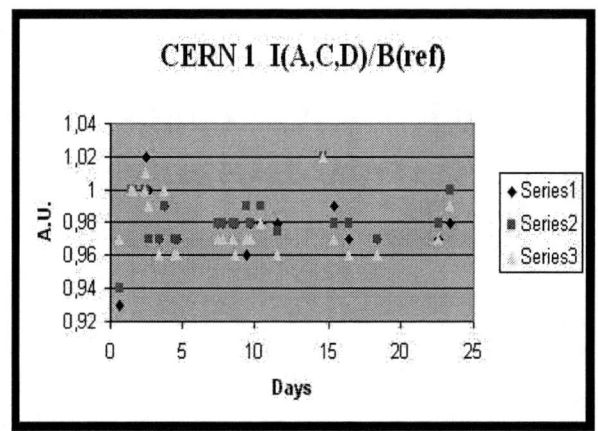

FIGURE 5. Currents of three gaps normalized to the reference gap as a function of time during the ageing test.

space charge effects are present, and an efficiency drop of about $0.4\%/100 kHz$.

AGEING TESTS

An ageing test has been performed on three prototypes at the ENEA Casaccia "Calliope" facility, with a $10^{15} Bq$, ^{60}Co source (emitting 1.25 MeV photons). The prototypes were irradiated with a dose rate of up to $0.3 Gy/hr$ for about one month, which corresponds, for different regions, to 5-10 years of LHCb run. The three chambers were operated at a gas gain $G \simeq 1 \times 10^5$, double with respect to the nominal gain. For each chamber, three gas gaps were kept always on, while the last gap was used as a reference, turning it on only for short periods. The ratio of the currents $(I_1, I_2, I_3)/I_4$ was then checked as a function of time (and so of the integrated charge). In this way we have taken into account changes in enviromental conditions and accidental variations of gas mixtures. Figure 5 shows the results for one prototype. There was no evidence of a worsening of the detector performancees in any of the three prototypes. Since the spread in the data (which is not a negative trend anyway) is about 10%, we can assume this value as the upper limit for the aging effect of the chambers.

After the ageing tests, a visual inspection of the prototypes has shown some etching effects in the FR4 frame due to the presence of ionised fluorine in the gas (the effect increases going from the gas inlet to the gas outlet). We should note, anyway, that while during the ageing test we can increase the radiation flux (and so also the concentration of fluorine ions) so that one month corresponds to 5-10 years of data taking, we cannot increase the gas flow of the same factor. This means that in the real data taking the concentration of fluorine will never be high enough to cause such an etching effect. To stay on the safer side we are considering, anyway, the possibility of reducing the fraction of CF_4 from 20% to 5%.

REFERENCES

1. The LHCb Collaboration, *LHCb Muon System Technical Design Report*, **CERN-LHCC 2001-10** (2001)
2. W.Riegler, *Detector Physics and Performances simulation of the MWPC for the LHCb Muon System*, **LHCB 2000-60** (2000)
3. W.Bonivento et al., *Nucl. Instr. and Meth.*, **A491**, 233–243 (2002).
4. M.Anelli et al., *Test of a double gap MWPC for the region 3 of the LHCb Muon System*, **LHCb-MUON 2001-120** (2001).

The CERN RD50 Collaboration: Development of Radiation-Hard Semiconductor Detectors for Super-LHC

Anna Macchiolo *on behalf of the RD50 Collaboration*

University of Firenze and INFN Firenze
Via G. Sansone, 1 I-50019 Sesto Fiorentino

Abstract. The proposed luminosity upgrade of the Large Hadron Collider (S-LHC) at CERN represents a technological challenge for the vertex detectors of the SLHC experiments since the innermost layers will receive fast hadron fluences up to 10^{16} cm^{-2}. The CERN RD50 project has been established to explore detector materials and designs that will allow to operate devices up to this limit. Among the different research lines followed by RD50 we report on the development of sensors produced with substrates like Czochralski and epitaxial silicon and on the investigation of the radiation hardness of p-type silicon detectors. Moreover innovative designs like thin, 3D and 3D-STC sensors are under evaluation in the RD50 Collaboration.

Keywords: Silicon detectors, Radiation damage, microstrip detectors, defects
PACS: 29.40.Gx, 29.40.Wk, 61.80.-x

INTRODUCTION

The Large Hadron Collider (LHC) at CERN has been designed to reach a goal luminosity of 10^{34}cm^{-2}s^{-1} after it will become operational in 2007. An upgrade of the accelerator is already envisaged (SuperLHC or SLHC) [1] where the luminosity will be increased by a factor of ten. The inner layers of the tracking detectors of the CMS and the ATLAS experiment will receive a fluence of 10^{16} cm^{-2}, after five years of SLHC operations (2500 fb^{-1}). The state of the art radiation-hard technology is represented by n$^+$-n pixels produced on Diffusion Oxygenated Float Zone silicon (DOFZ) and it survives up to fast hadron fluences of 10^{15} cm^{-2}. The CERN RD50 Collaboration [2] has been created to develop semi-conductor detectors that can be operated in the harsh environment of SLHC and it follows two main research lines: Material Engineering to suppress microscopic defects with a detrimental effect on the macroscopic detector parameters, and Device Engineering, to produce more radiation-tolerant detector geometries. In these proceedings only selected topics among the different RD50 activities are discussed.

MATERIAL ENGINEERING

The RD48 Collaboration [3] demonstrated the increased radiation-hardness of the DOFZ silicon against charged hadrons and γ-ray irradiation thanks to its higher oxygen content ($\approx 10^{17}$ cm^{-3} instead of 10^{16} cm^{-3} in the standard Fz silicon). Recently Czochralski (CZ) and Magnetic Czochralski (MCz) silicon have been studied as detector-grade material since high resistivity wafers ($\rho \approx 1$ KΩ cm) were made available. Cz and MCz silicon is grown in quartz crucibles that melt during the process and release SiO$_2$ into the fused silica. The resulting oxygen concentration is in the range 4-10x10^{17} cm^{-3}.

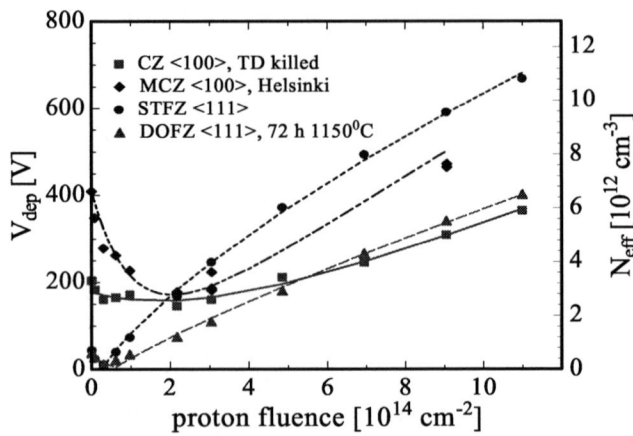

FIGURE 1. Comparison of the performances after irradiation of standard (Fz) and oxygenated (DOFZ) Float Zone with Cz and MCz silicon. The devices have been irradiated with 24 GeV/c protons.

The n-type detectors produced with Float Zone silicon (Fz) (ρ=1-6kΩcm) show a decrease of the depletion voltage V_{fd} as a function of the received irradiation fluence until they reach a minimum around 10^{13} p cm^{-2} (see Fig. 1). For larger fluences they show an inversion of the space charge sign ("type inversion") and a strong increase of the V_{fd} value. This effect is due to the creation of defects of acceptor type with a generation rate depending both on the time and the temperature at which they have been kept after irradiation. The detectors produced with n-type MCz silicon show a similar behaviour regarding V_{fd} but with a minimum at fluences around 1.5×10^{14} n cm^{-2} [4]. Annealing studies with MCz pad structures irradiated with 24 GeV/c protons at fluences up to 10^{15} p cm^{-2} have demonstrated that type inversion did not occur. Further investigations are under way at higher fluences and different proton energies, also by means of the TCT technique, to determine if eventually type inversion takes places in Cz and MCz n-type materials [5]. Since the Cz and MCz type material is not inverting in the range of fluences <10^{15} p cm^{-2} the increase of the effective doping concentration with fluence can be interpreted with the formation of donors, which overcompensate the introduction of deep acceptors [6].

Moreover the annealing behavior of Cz and MCz material is different from that observed in Fz and DOFZ silicon in the sense that V_{fd} varies less as a function of the time spent at 80 °C [7]. This indicates the Cz and MCz silicon is more stable with respect to V_{fd} during prolonged periods at room temperature and it represents an advantage for beam-off periods in SLHC condition.

The SMART Collaboration has produced micro-strip mini-sensors on Cz, MCz and Fz material of n- and p-type using the same process masks [8,9]. A picture of the wafer is shown in Fig. 2.

FIGURE 2. The SMART wafer: ten mini-sensors with different strip geometries and a strip pitch of 50 and 100 μm surrounded by test-structures like diodes, MOS, Gate-Controlled-Diodes.

The detectors have been irradiated with 24 GeV/c and 26 MeV/c protons at fluences up to 10^{16} cm^{-2}. The MCz sensors show a good performance after irradiation with respect to breakdown voltages and inter-strip capacitance that is an important parameter driving the noise of the front-end electronics. The V_{fd} values vary as a function of the received fluence with the same dependence observed in the diodes hosted on the same wafers (see Fig.3). A problem linked to the MCz substrate regards non-uniformities of the effective doping concentration before irradiation, due to the Thermal Donor Activation during the processing steps. An improvement has been achieved by changing the production method in order to avoid process temperatures around 450 °C [8]. A second difficulty concerns the p-type detectors suffering from low breakdown voltages before irradiation even if their performances are much improved after irradiation. The effect has been attributed to the p-spray technique used to maintain the inter-strip isolation and variations of this method will be applied and tested in the next production runs.

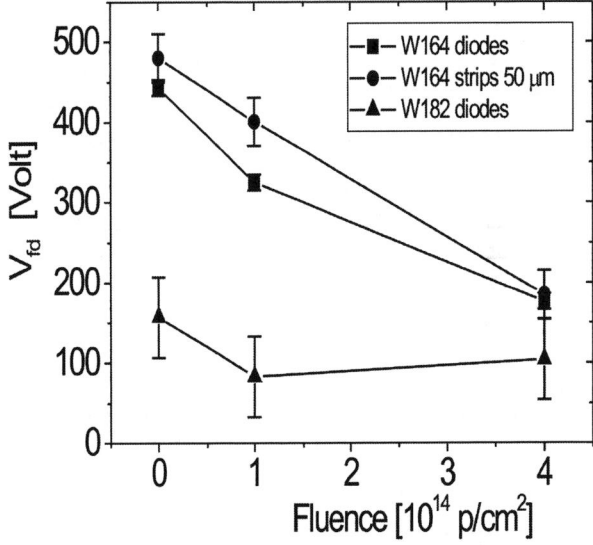

FIGURE 3. Depletion voltages for diodes and micro-strip sensors manufactured on n-and p-type material after irradiation with 24 GeV/c protons. W164 is a MCz n-type wafer and W182 is a MCz p-type wafer.

DEVICE ENGINEERING

Thin Detectors

A reduction of the pixel area is needed to maintain

the occupation rate at a manageable rate in the inner layers of the vertex detectors of SLHC. This allows to use thinner sensors without an excessive increase of the input capacitance seen by the front-end electronics. Thin detectors are intrinsically more radiation-hard, thanks to the lower leakage current ($I_L \propto V$) and depletion voltage ($V_{fd} \propto W^2$). Moreover, after irradiation with fast hadrons up to fluences of 10^{16} cm^{-2} the effective drift length of charged carriers is strongly reduced by trapping effects ($\lambda_e = 20$ μm, $\lambda_h = 10$ μm) [10]. The collected signal does no longer depend linearly on the device thickness d that can be decreased without affecting the charge collection efficiency (CCE). Two approaches have been followed within the RD50 Collaboration, the thinning of processed Fz silicon wafers with a chemical etching [11] and the growth of an epi-layer on a low resistivity (<0.02 Ωcm) Cz substrate [12, 13]. Fig.4 shows a comparison of the behavior of the two types of detectors after irradiation..

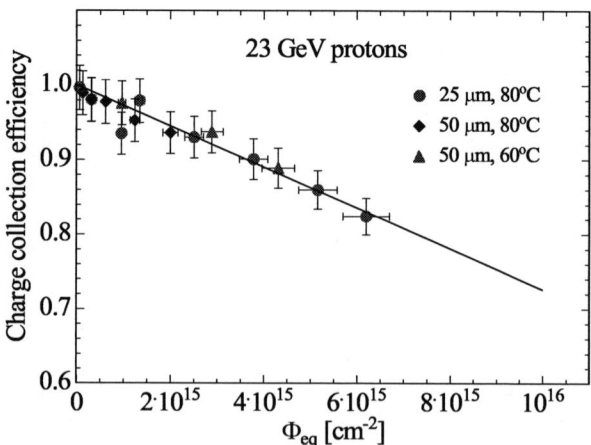

FIGURE 5. Measurements of the Charge Collection Efficiency for epi-Si diodes as a function of the received fluence. A linear fit is performed to extract the β_e parameter of 2.7×10^{-17} cm^2.

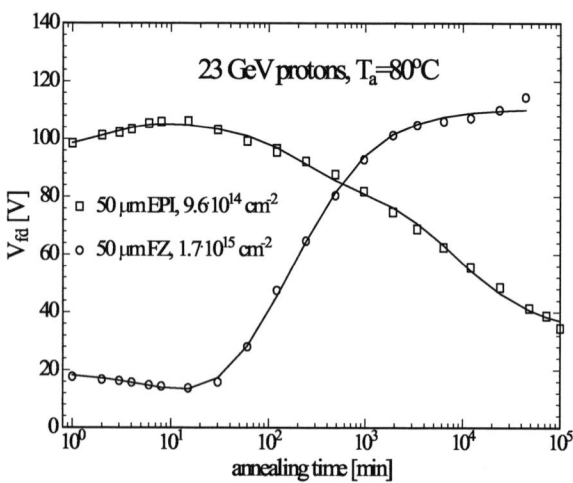

FIGURE 4. Annealing studies comparing the behavior of Fz and epi-Si diodes after irradiation. The annealing curves show that the Fz material is type inverted and the epi-Si is not.

The V_{fd} before any thermal treatment has a low value for both of them. The main difference is the behavior with respect to the annealing time suggesting that the Fz material is type inverted whereas the epitaxial material is not. Only if the epitaxial material is heavily irradiated the evolution of the radiation induced damage can cause the type inversion of the epitaxial detectors in the long-term operation, even if this effect is not observed just after irradiation.

Also in the case of the epitaxial material the superior radiation tolerance can be explained by the high oxygen concentration ([O]≈10^{17} cm^{-3}) thanks to the O diffusion from the Cz substrate during the high temperature growth process. Measurements of the charge collection efficiency with detectors processed on the epi-Si layers were performed with α-particles from a ^{244}Cm source on different samples irradiated with protons and neutrons up to fluences of 10^{16} cm^{-2} [13].

In Fig.5 results are shown for the proton irradiation where a linear dependence of the CCE degradation as a function of the fluence is visible.

p-type Detectors

A good performance in terms of the CCE after heavy irradiation is obtained with micro-strip detectors of p-type substrate by reading out the segmented n-side. Since this material does not suffer from type-inversion the signal is formed mainly by electron carriers being collected on the high electric field side. The reduced charge collection time results in less trapping and larger signals. The n^+-p-p^+ sensors, as opposite to the n^+-n detectors, do not require backplane processing, that can turns out in an excessive cost for the coverage of large areas.

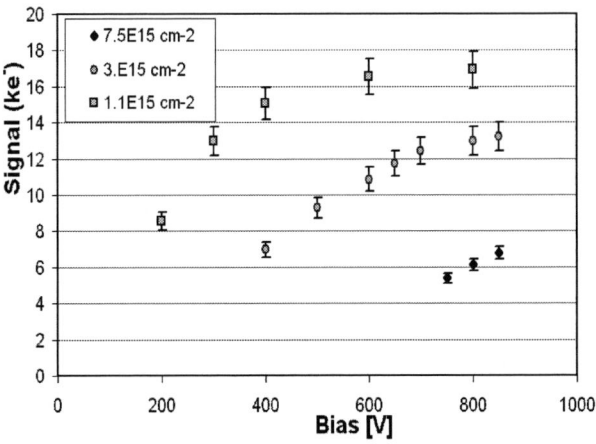

FIGURE 6. Charge collection efficiency versus applied bias voltage, after 1.1, 3 and 7.5×10^{15} cm^{-2}. The device irradiated at 3×10^{15} cm^{-2} is standard p-type substrate while the others device are oxygen-enriched.

Fig. 6 shows the results of the CCE measurements on detectors irradiated up to fluences of 7.5×10^{15} p cm^{-2} with 24 GeV/c protons [14]. The devices have been measured using a 1060 nm laser and fast electrons from a ^{106}Ru source with a 40MHz readout. Even at the highest dose the charge collected at the maximum applied bias (900 V) is > 6500 electrons corresponding to a signal over noise value of about 7.5. Recent annealing studies [15] have revealed that the CCE of heavily irradiated structures remains stable after a time equivalent to several years at Room Temperature while the V_{fd} is expected to change by a factor of four. This indicates that a complete measurement of the electric field in the detector after irradiation would be beneficial in order to extrapolate the relevant detector properties.

New Detectors

3D detectors present a lower depletion voltage with respect to planar detectors of the same thickness since the active area develops laterally. The standard 3D design [16] presents columnar electrodes of both doping types arranged in adjacent cells. The path of the electric field begins at one electrode type and ends at the closest electrode of the opposite type. The distance between the electrodes is usually small, 30-100 μm, allowing for the lateral depletion with very low depletion voltage. The long process needed for their fabrication is a drawback of 3D detectors that can be in part overcome by a new simplified architecture [17]. The proposed device (3D-STC, Single Type Column) features electrodes of the same type, n$^+$ columns on a p-type substrate. The column etching and doping is performed only once resulting in a considerable process simplification.

CONCLUSIONS

Recent developments of the RD50 Collaboration concerning defect-engineered silicon and new detector concepts have been presented. Oxygen-enriched material as Cz and epi-Si are particularly promising for their radiation-hardness. Type-inversion in p$^+$-on-n devices is absent or retarded in comparison to Fz and DOFZ. Cz silicon could be a cost-effective solution for sensors in the outer radii of the vertex detectors of the SLHC experiments at fluences up to $\approx 10^{15}$ p cm^{-2} while epi-Si material is an excellent candidate for the pixel layer in the innermost region. An alternative is offered by p-type substrates that have shown very high CCE after fluences up to 7.5×10^{15} p cm^{-2}. Innovative detector designs like the 3D or the 3D-STC are being presently developed and tested within the RD50 Collaboration.

REFERENCES

1. F. Gianotti et al., "Physics potential and experimental challenges of the LHC luminosity upgrade", hep-ph/0204087, April 2002.
2. R&D Proposal, LHCC 2002-003/P6, 15.2.2002, CERN
3. The RD48 (ROSE) Collaboration, http://cern.ch/rd48
4. M. Lozano, G. Pellegrini, C. Fleta, J. M. Rafí, M. Ullán, F. Campabadal, et al., "Comparison of radiation hardness of p-in-n, n-in-n and n-in-p silicon pad detectors on FZ, DOFZ and MCZ Si," presented at the 5th RD50 Workshop (Firenze, Italy), 14-16 October 2004. On line available: http://rd50.web.cern.ch/rd50/.
5. The SMART Collaboration, "Characterization of n and p-type diodes processed on Fz and MCz silicon after irradiation with 24 GeV/c and 26 MeV protons and with reactor neutrons", presented at the 6th RD50 Workshop (Helsinki, Finland), 2-4 June 2005. On line available: http://rd50.web.cern.ch/rd50/.
6. D. Menichelli, M. Bruzzi, and M. Scaringella, "Shallow donors in MCz-Si n- and p-type detectors at different process temperature, irradiation and thermal treatments," presented at the 6th RD50 Workshop (Helsinki, Finland), 2-4 June 2005. On line available: http://rd50.web.cern.ch/rd50/.
7. G. Pellegrini et al., paper presented at the 5th Int.Conf. on Radiation Effects on Semiconductor Materials Detectors and Devices RESMDD04, Florence, 10.-13. Oct. 2004, to be published in NIMA.
8. M. Bruzzi et al., " Process and first characterization of detectors made with radiation hard Si materials", RESMDD04, Florence, 10.-13. Oct. 2004, to be published in NIMA.
9. A. Messineo et al. "Irradiation of Fz and MCz mini-sensors of n- and p-type with 24 GeV/c and 26 MeV/c protons up to fluences of 3 10^{15} 1 MeV n cm^{-2}", presented at the 6th RD50 Workshop (Helsinki, Finland), 2-4 June 2005. On line available: http://rd50.web.cern.ch/rd50/.

10. G. Kramberger, V. Cindro, I. Mandić, M. Mikůz and M. Zavrtanik, "Effective trapping time of electrons and holes in different silicon materials irradiated with neutrons, protons and pions," *Nucl. Instrum. Methods*, vol. A481, pp. 297-305, 2002.
11. S. Ronchin, M. Boscardin, G. F. Dalla Betta, P. Gregori, V. Guarnieri, C. Piemonte, et al., "Fabrication of PIN diode detectors on thinned silicon wafers," *Nucl. Instrum. Methods*, vol. A530, pp. 134-138, 2004.
12. G. Kramberger, D. Contarato, E. Fretwurst, F. Hönninger, G. Lindström, I. Pintilie, et al., "Superior radiation tolerance of thin epitaxial silicon detectors," *Nucl. Instrum. Methods*, vol. A515, pp. 665-670, 2003.
13. G. Lindström, E. Fretwurst, F. Hönniger, G. Kramberger, M. Moll, E. Nossarzewska, et al., "Epitaxial silicon detectors for particle tracking: overview on radiation tolerance at extreme hadron fluence," presented at the 10th European Symposium on Semiconductor Detectors (Wildbad Kreuth, Germany), 12-16 June 2005, submitted for publication in *Nucl. Instrum. Methods*, vol. A.
14. G. Casse, P. P. Allport, S. Marti i Garcia, M. Lozano, and P. R. Turner, "Performances of miniature microstrip detectrors made on oxygen enriched p-type substrates after very high proton irradiation," *Nucl. Instrum. Methods*, vol. A535, pp. 362-365, 2004.
15. G. Casse, "Update of annealing measurements on heavily irradiated p-type Si sensors", presented at the 6th RD50 Workshop (Helsinki, Finland), 2-4 June 2005. On line available: http://rd50.web.cern.ch/rd50/.
16. S. I. Parker, C. J. Kenney and J. Segal, "3D - A proposed new architecture for solid-state radiation detectors," *Nucl. Instrum. Methods*, vol. A395, pp. 328-343, 1997.
17. C. Piemonte, M. Boscardin, G. F. Dalla Betta, S. Ronchin and N. Zorzi, "Development of 3D detectors featuring columnar electrodes of the same doping type," *Nucl. Instrum. Methods*, vol. A541, pp. 441-448, 2005.

The Virgo Detector

F. Acernese*, P. Amico†, M. Al-Shourbagy**, S. Aoudia‡, S. Avino*, D. Babusci§,
G. Ballardin¶, R. Barillé¶, F. Barone*, L. Barsotti**, M. Barsuglia‖, F. Beauville††,
M.A. Bizouard‖, C. Boccara‡‡, F. Bondu‡, L. Bosi†, C. Bradaschia**, S. Braccini**,
A. Brillet‡, V. Brisson‖, L. Brocco§§, D. Buskulic††, E. Calloni*, E. Campagna¶¶,
F. Cavalier‖, R. Cavalieri¶, G. Cella**, E. Chassande-Mottin‡, C. Corda**,
A.-C. Clapson‖, F. Cleva‡, J.-P. Coulon‡, E. Cuoco¶, V. Dattilo¶, M. Davier‖,
R. De Rosa*, L. Di Fiore*, A. Di Virgilio**, B. Dujardin‡, A. Eleuteri*, D. Enard¶,
I. Ferrante**, F. Fidecaro**, I. Fiori**, R. Flaminio††,¶, J.-D. Fournier‡, S. Frasca§§,
F. Frasconi**,¶, A. Freise¶, L. Gammaitoni†, A. Gennai**, A. Giazotto**,
G. Giordano§, L. Giordano*, R. Gouaty††, D. Grosjean††, G. Guidi¶¶, S. Hebri¶,
H. Heitmann‡, P. Hello‖, L. Holloway¶, S. Kreckelbergh‖, P. La Penna¶,
V. Loriette‡‡, M. Loupias¶, G. Losurdo¶¶, J.-M. Mackowski***, E. Majorana§§,
C. N. Man‡, M. Mantovani**, F. Marchesoni†, F. Marion††, J. Marque¶, F. Martelli¶¶,
A. Masserot††, M. Mazzoni¶¶, L. Milano*, C. Moins¶, J. Moreau‡‡, N. Morgado***,
B. Mours††, A. Pai§§, C. Palomba§§, F. Paoletti**,¶, S. Pardi*, A. Pasqualetti¶,
R. Passaquieti**, D. Passuello**, B. Perniola¶¶, F. Piergiovanni¶¶, L. Pinard***,
R. Poggiani**, **M. Punturo**†, P. Puppo§§, K. Qipiani*, P. Rapagnani§§, V. Reita‡‡,
A. Remillieux***, F. Ricci§§, I. Ricciardi*, P. Ruggi ¶, G. Russo*, S. Solimeno*,
A. Spallicci‡, R. Stanga¶¶, R. Taddei¶, D. Tombolato††, M. Tonelli**, A. Toncelli**,
E. Tournefier††, F. Travasso†, G. Vajente **, D. Verkindt††, F. Vetrano¶¶, A. Viceré¶¶,
J.-Y. Vinet‡, H. Vocca†, M. Yvert†† and Z.Zhang¶

*INFN - Sezione di Napoli and/or Università di Napoli "Federico II" Complesso Universitario di Monte S. Angelo Via Cintia, I-80126 Napoli, Italia and/or Università di Salerno Via Ponte Don Melillo, I-84084 Fisciano (Salerno), Italia
†INFN Sezione di Perugia and/or Università di Perugia, Via A. Pascoli, I-06123 Perugia - Italia
**INFN - Sezione di Pisa and/or Università di Pisa, Via Filippo Buonarroti, 2 I-56127 PISA - Italia
‡Department Artemis - Observatoire de la Côte d'Azur, BP 42209, 06304 Nice Cedex 4, France
§INFN, Laboratori Nazionali di Frascati Via E. Fermi, 40, I-00044 Frascati (Roma) - Italia
¶European Gravitational Observatory (EGO), Via E. Amaldi, I-56021 Cascina (PI) Italia
‖Laboratoire de l'Accélérateur Linéaire (LAL),IN2P3/CNRS-Université de Paris-Sud, B.P. 34, 91898 Orsay Cedex - France
††Laboratoire d'Annecy-le-Vieux de physique des particules Chemin de Bellevue - BP 110, 74941 Annecy-le-Vieux Cedex - France
‡‡ESPCI - 10, rue Vauquelin, 75005 Paris - France
§§INFN, Sezione di Roma and/or Università "La Sapienza", P.le A. Moro 2, I-00185, Roma
¶¶INFN - Sezione Firenze/Urbino Via G.Sansone 1, I-50019 Sesto Fiorentino; and/or Università di Firenze, Largo E.Fermi 2, I - 50125 Firenze and/or Università di Urbino, Via S.Chiara, 27 I-61029 Urbino, Italia
***LMA, 22, Boulevard Niels Bohr 69622 - Villeurbanne- Lyon Cedex France

Abstract. The Virgo Experiment is a gravitational wave interferometric detector. It consists in a Michelson interferometer with two 3 km long Fabry–Perot cavities as orthogonal arms. The installation of the detector has been completed in September 2003 and presently the apparatus is under commissioning. In this article an overview of the detector status is presented.

Keywords: Gravitational wave detectors
PACS: 04.80.Nm, 95.55.Ym

INTRODUCTION

One of the most interesting predictions of Einstein's theory of general relativity is the existence of gravitational waves (GW). The wave equation for *space-time* was proposed by Einstein in 1916, but no direct evidence has been observed yet. Recently, a new family of detectors has been developed: the interferometric GW detectors. In this case, a Michelson interferometer is set up with arms a few kilometers long that typically are optically multiplied in length by using Fabry-Perot cavities. Optical components (mirrors) are suspended to realize a set of *free falling* reference masses that – ideally – are only perturbed by their interaction with gravitational waves. Because of the tidal nature of GWs, the length of the interferometer arms are modulated by the passage of a gravitational wave which produces a signal at the output of the interferometer. Several detectors of this kind are operative or under an advanced commissionig phase: LIGO in the USA, Geo600 in Germany, Tama in Japan and Virgo, funded by the CNRS (France) and INFN (Italy) in Italy.

DETECTOR DESCRIPTION

The Virgo detector consists in a Michelson interferometer (ITF) with 3 km long Fabry–Perot cavities in the arms. The optical scheme of the ITF is reported in figure 1. The laser is a Nd:YAG laser with 22 Watt of power and $1.064 \mu m$ of wavelength. The frequency of the laser is stabilized using a reference cavity (RFC) made in ULE (Ultra Low Expansion) glass. The TEM_{00} mode of the laser is selected using a 144 m long triangular Fabry–Perot cavity (the Input Mode Cleaner (IMC)). The laser beam coming out this *injection system* (IS) is sent, through the power recycling mirror (PR, whose role is to increase the power circulating in the ITF, recycling the light wasted in the ITF input port), in the beam splitter (BS) mirror, where the beam is directed to the two arms. Here, two Fabry-Perot cavity are realized with the two input (WI, NI) and the two end mirrors (WE, NE). These mirrors are realized in Suprasil and Herasil fused silica respectively; they have a diameter of 350 mm and a thickness of about 100 mm. The two beams are recombined at the level of the BS and sent, through an output mode cleaner (OMC), to the output photodiode bench (EB).

All the optical components and all the light path are in high vacuum; the Virgo vacuum volume is about $7000 m^3$. The ITF main mirrors and suspension system are located inside steel towers, $9m$ high; in each tower, the mirror is kept at about 10^{-9} mbar of hydrogen partial pressure, while the remaining suspension mechanics is kept at 10^{-7} mbar. The beam path, along the 3 km arms, is inside a $1.2m$ diameter tube, at 10^{-7} mbar of pressure. This guarantees a low contamination of the mirrors, a low disturbance of the mirror rest position and a low fluctuation of the light optical path.

The suspension system

The amplitude of a GW impinging on the Earth surface from an astrophysical source, in term of *space–time* strain, is expected to lie typically in the range of $h < 10^{-21}$. Traslating this value in terms of mirror displacement ΔL, we obtain $\Delta L \simeq L_{arm} \cdot h < 10^{-18} m$. It is evident that several are the noise sources that can mime the gravitational signal. At low frequency, the dominant noise source is the seismic vibration of the mirrors. To reduce the effect of this noise source on the detector output, several chains of seismic filters have been used to suspend the Virgo optics. In figure 2 a scheme of the attenuation system (super–attenuator – SA), used to suspend the main Virgo mirrors, is reported. It consists in an Inverted Pendulum (IP), that filters the very low frequency, in several seismic filters and in a Last Stage (LS) or payload; a detailed description of this system can be found in [1] and [2]. The horizontal seismic vibrations are attenuated thanks to the multistage pendulum behavior of this system: in a over-simplified way, if N is the number of stages and ω_0 is the (common) resonant frequency, the attenuation, for $\omega >> \omega_0$ is $x_{Mirror}/x_{ground} \simeq (\omega_0/\omega)^{2N}$. The vertical seismic vibration is attenuated thanks to the seismic filters (SF). A SF consists in a rigid steel cylinder (70 cm diameter, 18 cm high) suspended as close as possible to its center of mass. In the bottom part of this cylinder a system of triangular cantilever springs is clamped. The lower SF is suspended to the previus one through a steel wire, with a complex system of magnetic antisprings used to lower the vertical resonance frequency. The payload consists in a steering element called, "marionetta", to wich every mirror of the ITF is suspended together with a recoil mass. A detailed description of the LS can be found in [3]. The mirror suspension design has been optimized [4] to reduce another source of noise: the thermal noise. The mirror is suspended through a two wires loop, made of C85 steel, selected because of its high breaking strength and low dissipation angle values [5].

The Super–Attenuator control

The SA is a complex device that filters the seismic noise at frequency higher than its internal resonance frequencies; above 4 Hz the residual seismic noise, at the mirror level, is below the expected suspension thermal noise. But, between 0.04 and 2.5 Hz, the SA resonances can amplify the seismic noise, resulting in a residual mo-

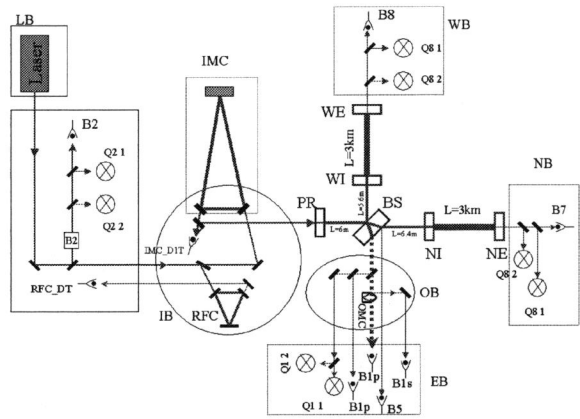

FIGURE 1. Optical scheme of the Virgo interferometer

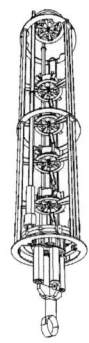

FIGURE 2. Virgo Super–Attenuator scheme

tion of the mirror of tens of microns. To suppress this kind of motion, a multidimensional active control (the so–called *inertial damping* [6]) has been implemented in the DC-5 Hz range. Three linear variable differential transformer (LVDT) position sensors and three accelerometers, placed at the top of the IP, monitor the relative motion of the IP respect to an external frame. The LVDT signal is used to correct the very low frequency drifts (DC–30 mHz) and the accelerometer signal allow a wideband reduction of the noise in the region of the SA resonances (30 mHz–5 Hz). The actuation is performed through a system of three coils and permanent magnets located on the top of the IP. The mirror residual motion, after the *inertial damping* application is about $0.1\mu m$ in 10 s[6]. The *inertial damping* cannot control some internal angular modes of the suspension involving the mirror. Therefore, a dedicated last stage suspension local control (LC) system has been implemented [7]. The LC consists in a ground–based CCD-camera readout system and in an internal–force action applied within the mirror suspension. The LC system works in two different modes: in a *coarse* mode, the CCD-camera reads the longitudinal and angular position of the mirror thanks to the light, generated by an halogen lamp, reflected by four diffusive markers attached to the mirror. They consist in four fused silica disks, 20 mm of diameter (10 mm for the BS and PR mirrors), 4 mm of thickness, attached to the mirror surface through a *silicate bonding*[8] technique. On the top face of the marker there is a metallic coating and 7 macor inserts diffuse the light of the lamp. In the *fine* mode, operative when the mirror oscillations have been enough damped with the *coarse* one, the CCD-camera reads the spots of a 633 nm wavelength laser–diode system, reflected by the mirror. The LC is able to reduce the angular residual motion down to a level lower than $1\mu rad$, enough to permit the ITF locking and to start the linear alignment system.

Hierarchical control implementation

The stringent noise requirements require the adoption of a hierarchical strategy in the detector control after the locking acquisition. In fact, external disturbances, like the tidal fluctuation of the mirror distance ($\delta L \simeq 10^{-4}$ m over 3 km) or the resonant oscillation of the mirrors driven by the residual seimic noise ($\delta L \simeq 10^{-4}$ m) are well beyond the locking requirements ($\delta L < 10^{-11}$ m) at low frequency. Hence, in order to keep the ITF locked, a feedback control must be applied from DC to few tens of Hz to the suspended optics; this control cannot be fully applied directly to the suspended mirror because the actuation noise would affect the detector sensitivity. Hence, the actuation on the suspended optics is divided in three frequency ranges: at very low frequency (DC–0.01 Hz) the tidal breath is compensated acting on the top of the inverted pendulum (IP). Between 0.01 Hz and 8 Hz the mirror fluctuation are controlled acting on the marionetta and then, up to 50 Hz the feedback acts directly on the mirror magnetic actuators.

INTERFEROMETER COMMISSIONING

The assembling and integration phase of the detector has been concluded in September 2003. Since that date the ITF is in a commissioning phase that should bring the detector to a sensitivity close to the nominal one through a complex process of continuous noise reduction and stability improvement of all the control systems that are governing the machine. The Virgo commissioning activity has been organized in steps of increasing complexity: the separate commissioning of the North and West Fabry-Perot cavities, followed by the commissioning of the recombined Michelson Fabry-Perot ITF, and finally the commissioning of the recycled Michelson Fabry-Perot ITF. Short periods of continuous data-taking (the so–called commissioning runs) have taken place every two to three months since November 2003, in order to check the evolution of the detector and the consequent progress in the level of sensitivity. Five commissioning runs have been performed so far; in the final part the last commissioning run (C5) the ITF has been operated in a fully recycled configuration, since the full locking of the ITF has been acquired the 26th of October 2004. The locking acquisition procedure in Virgo is different from the techniques adopted in the other experiments and it is really innovative: the ITF is locked outside the working point for the dark fringe. In this way a good fraction of light escapes through the output port and the power build-up in the recycling cavity is low. Then the ITF is adiabatically brought on to the dark fringe, that is the operating point where the sensitivity to the signal is maximized and the effect of the common noise sources (like laser power fluctuations) is minimized. This locking technique is referred to as *variable finesse* [9], because the finesse of the recycling cavity changes during the lock acquisition path.

In figure 3 the sensitivity progresses are reported for the different commissioning runs (C1–C5); the improvement of the detector sensitivity in these steps was mainly due to the introduction in the detector configuration of components, present in the project design, but still not activated. In C5 the detector was running in an almost complete configuration, with the recycling cavity operating, but not automatically aligned. The progresses between the C5 curve and the latest curve reported (1st of July 2005) are mainly due to the reduction of the noise contributions due to the controls acting on the SA and to some electronics (phase noise in the demodulation electronics). The attained sensitivity of the detector is still far from the expected one, but the progress rate is similar or even better than the improvement speed shown by competitor detectors that are now in a more advanced condition. The reduction of the current noise level will be attained thanks to the effects of several actions: the automatic alignment of the ITF in the recycled configuration will be finally introduced (it is now under commissioning); this procedure should act in all the frequency range, but, mainly, it will reduce the low frequency noise. The high frequency noise floor will be lowered through the reduction of the residual phase noise due to the electronics and through the injection in the ITF of the design laser light power (about 10 W respect to the 0.7 W currently entering in the ITF) thanks to an upgrade of the current injection system.

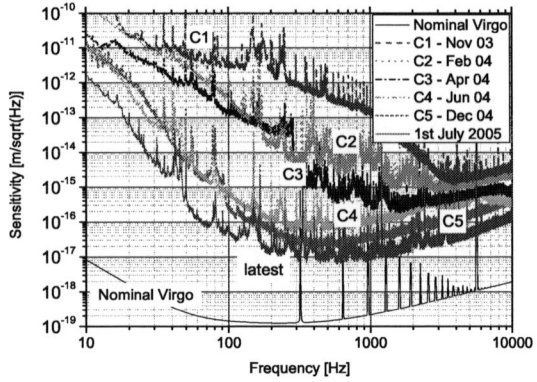

FIGURE 3. Sensitivity (in $\left[m/\sqrt{Hz}\right]$ units) of the Virgo detector during the commissioning runs. The latest sensitivity reached by the detector is reported too.

CONCLUSIONS

The Virgo detector construction has been completed in September 2003 and it is now under an advanced commissioning phase. The detector sensitivity is improving toward the expected design sensitivity, that should be reached before the end of 2006. Since that date, the Virgo detector will operate as one of the most sensitive nodes of the global network of gravitational wave detectors in the Earth.

REFERENCES

1. S.Braccini *et al.*, *Class. Quant. Grav.* **19** (2002) 1623
2. G.Ballardin *et al.*, *Rev. Sci. Inst.* **72** (2001) 3643
3. F. Acernese *et al.*, *Class. Quant. Grav.* **21**, (2004) 425
4. G. Cagnoli *et al.*, *Rev. Sci. Instr.* **71** (2000) 2206
5. G. Cagnoli *et al.*, *Phys. Lett.* **A 255** (1999) 230
6. G. Losurdo *et al.*, *Class. Quant. Grav.* **19** (2002) 1631
7. F. Acernese *et al.*, *Astroparticle Phys.* **20**, (2004) 617
8. P. Amico *et al.*, *Rev. Sci. Inst.* **73** (2002) 3318
9. L. Barsotti et al, Status of Virgo, Class. and Quant. Grav., proc.GWDAW 2004, to be published.

Triple-GEM Detectors for the Innermost Region of the LHCb Muon Apparatus

M. Poli Lener

On behalf of the LHCb-GEM group [1]

Abstract. We present in this paper the mechanical construction procedures, the tools and the relative quality check of a triple-GEM detector. This kind of detector is the result of R&D activity performed for the study of detectors for the hard radiation environment of the innermost region, around the beam pipe, of the first muon station of the LHCb experiment. We also present the performances of the chamber final design, operated with $Ar/CO_2/CF_4$ (45/15/40) gas mixture, obtained at PS beam facility at CERN.

Keywords: GEM, tracking, aging, discharge
PACS: 29.40.Cs, 29.40.Gx

INTRODUCTION

The GEM (*Gas Electron Multiplier*) [1] used in these detectors is a thin (50μm) kapton foil, copper clad on each side, chemically perforated with an high density bi-conical holes, with an external (internal) diameter of 70μm (50μm) and a pitch of 140μm. The GEM holes act as multiplicative channels for gaseous detectors by applying a voltage difference between the two copper surfaces. The cross section of a triple-GEM detectors, together with the labelling used in this paper, is shown in Fig. 1.

This detector has been developed [2] to equip the region around beam pipe (R1) of the first muon station (M1) of the LHCb experiment [3]. The M1R1 will be equipped with 12 stations (two independent detectors per station, logically OR-ed) covering an area of about of 0.6 m^2. This small area corresponds to $\sim$ 20% of the triggered muons.
Because the experiment requires for a very fast detector response, the triple-GEM detectors must be operated with high drift velocity and high ionization yield gas mixtures. Among the tested gas mixtures the $Ar/CO_2/CF_4$ (45/15/40) [4] largely fulfills the LHCb requirements.

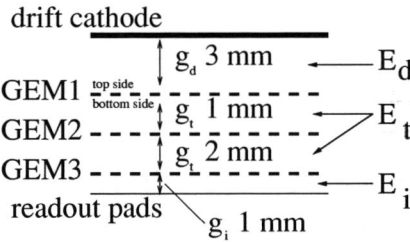

FIGURE 1. Cross-section of the triple-GEM detector. E_d, E_t and E_i are the drift, transfer and induction fields respectively; g_d, g_t, and g_i are the drift, transfer and induction gaps respectively.

The requirements for the detectors in M1R1 are:

- a rate capability up to $\sim$ 500 kHz/cm^2;
- each station, constituted by two detectors logically OR-ed [2], must have an efficiency $\geq$ 96% in a 20 ns time window;
- the pad cluster size, i.e. the number of adjacent detector pads fired when a track crosses the detector, should not be larger than 1.2, for a 10×25 mm^2 pad size.

In addition the detector must tolerate, without damages or performance losses, an integrated charge of $\sim$ 0.88 C/cm^2 in 10 years of operation at a gain of $\sim 6\times 10^3$ and an average particle flux of 184 kHz/cm^2 for an average machine luminosity of 2×10^{32} $cm^{-2}s^{-1}$.
Due to the large amount of CF_4 (40%) present in the gas mixture, many testes have been performed [6] in order to check the compatibility between the construction

---

[1] M. Alfonsi, G. Bencivenni, P. de Simone, F. Murtas - Laboratori Nazionali di Frascati, INFN, Frascati;
W. Bonivento, A. Cardini, D. Raspino, B. Saitta - Sezione INFN di Cagliari, Cagliari;
D. Pinci - Sezione INFN di Roma1, Roma;
Coresponding author: Marco Poli Lener,
E-mail: marco.polilener@lnf.infn.it

[2] This improves time resolution and provides some redundancy.

FIGURE 2. View of the class 1000 cleanrom.

FIGURE 3. The gas tight plexiglass box used for the HV test of GEM folis

materials (both for detector and gas system) and the gas mixture.

In the first two sections we describe the tools, the construnction procedure and quality checks on the final chambers, while in the last one, we show the station performance obtained at PS beam facility at CERN.

CHAMBER CONSTRUCTION AND TOOLS

All the construction operations are performed inside a class 1000 clean room equipped for the construction of the detectors, Fig. 2. The whole assembly procedure has been defined in each single step.

The GEM foils are manufactured by the CERN-EST-DEM workshop following our global geometrical design. The foil has an active area of 20×24 cm^2. In order to reduce the energy stored on the GEM and the discharge propagation, one side of the foil has been divided in six sectors ($\sim 66 \times 240$ mm^2) while the other side is not segmented. The separation between sectors is 200μm.
The GEM foils are optically inspected with a microscope in order to check the right perforation of the kapton and/or the copper surfaces. Then each sector of the GEM foil is tested with high voltage in order to check their quality. Such a test is performed in a gas tight box, Fig. 3, flushed with nitrogen in order to keep the relative humidity at $\sim 25\%$ level. The voltage is applied to each sector through a 100 MΩ limiting resistor, to avoid damages to the GEM foils in case of discharges. The acceptance requirement is a maximum leakage current less than 3 nA (on each sector) at 600 Volts.

The FR4 frames are visual inspected and the broken fibers are carefully removed. Then the frames are cleaned in a ultrasonic bath and successively dried in a oven at 80^0C for 12 hours.

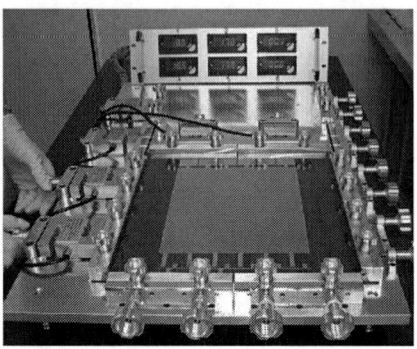

FIGURE 4. The GEM foil under stretching. The connection to the six sectors are visible

The GEM foils that pass the HV test, are stretched with a specific tool (Fig. 4). The GEM foil is clamped with jaws equipped with a plastic O-ring. The mechanical tension (18 kg/jaw, two jaws per side) applied at the edge of the foil is monitored with gauge-meters. Then the frame is glued on the stretched GEM foil using the Araldite 2012 which has a good electrical behaviour and suitable handling properties (work life: 5min, curing time: 2 h). The aging properties of this glue has been study with a global irradiation test [6]. Each GEM foil is framed following the above procedure. The frames thickness are 1, 2 and 3 mm.

The three framed GEMs are glued on the top of the cathode PCB with the sequence: 3 mm frame, then 1 mm and 2 mm (Fig. 5). The chamber is finally closed gluing the last 1 mm (bare) frame and the pad PCB This assembly operation is performed on a machined Al-alloy reference plane equipped with four reference pins. On the top of the whole sandwich a load of 80 kg is uniformly applied for 24h, as required for epoxy polimerization (Araldite AY103 + HD991 hardener).

FIGURE 5. The framed GEMs are assembled on the cathode PCB with the right sequence (3mm, 1 mm, 2 mm), facilitating with the reference pins.

The chamber is then completed with the final soldering of the GEM HV connections to the cathode, gas connectors and the external faraday cage.

QUALITY CHECKS

Various tests are performed in order to check both the quality of the chamber mechanics and the detector performances.

The gas leak rate measurement is generally referred to that one of a reference chamber (same volume and a gas leak < 5 mbar/day) in order to take into account atmospheric pressure and temperature variations. Both the test and the reference chambers are inflated in parallel, up to an overpressure of $\sim$10 mbar. A gas leak of the order of a few mbar/day has been achieved, corresponding with a humidity level in the gas mixture (supplying the chamber in open mode, with a flow rate of 80 cc/min) of the order of 50 ppm per volume [3].

The gain uniformity of the chamber is tested with an X-ray gun facility, Fig. 6. The current signal induced on each pad (192 pads per chamber), is read-out with a 1 nA sensitivity current-meter and corrected for the temperature and the pressure variations. The water content and the temperature of the gas mixture are monitored with a probe mounted on the gas line outlet. The atmospheric pressure is monitored outside the gas line with another probe. The chamber is mounted on an X-Y plane moved with computer controlled step-motors. The gain measurement is done automatically and are about 12 hours long. The measured gain uniformity is

---

[3] This value ensures no water contamination of the gas mixtures.

FIGURE 6. Picture of the X-ray gun facility used to measure the gain uniformity of the chamber.

better than 10% (6% excluding edge effects due to the finite diameter of the X-ray collimator, $\sim$5mm).

Finally, the station is realized by coupling, through the reference pin holes and with the cathodes faced one to each other, two of such 20×24 cm^2 chambers. The front-end electronics, based on the CARIOCA chip [7], are mounted on the four side of the station. Taking into account the pad size of 10×25 mm^2, 384 electronic channels are needed to readout the whole station.
The performaces of a station are then tested with a cosmic ray setup, as shown in Fig. 7. The trigger is the coincidence of two scintillator layers which cover all the active area of the station. In order to track the muon position, two layers of drift tubes are also used (r.m.s.$\sim$150 μm).

EXPERIMENTAL MEASUREMENTS

The results shown in this section have been obtained at the PS beam facility at CERN with the final station module, built in agreement with the previous described procedure.
The module has been operated with the Ar/CO$_2$/CF$_4$ (45/15/40) gas mixture, which is characterized by large electron drift velocity ($\sim$ 10 cm/μs) and with a configuration gap fields $E_d/E_{t1}/E_{t2}/E_i$ = 3.5/3.5/3.5/5 kV/cm, thus optimizing either time performances or electron transparency of the detector [5].
The best time distribution of the station (two chamber logically OR-ed) is shown in Fig. 8 and gave a time resolution better than 3 ns (r.m.s.). The corresponding time

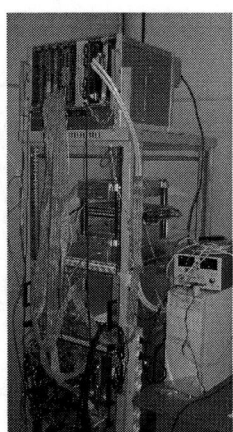

FIGURE 7. Picture of the cosmic ray setup used to measure the final station performances.

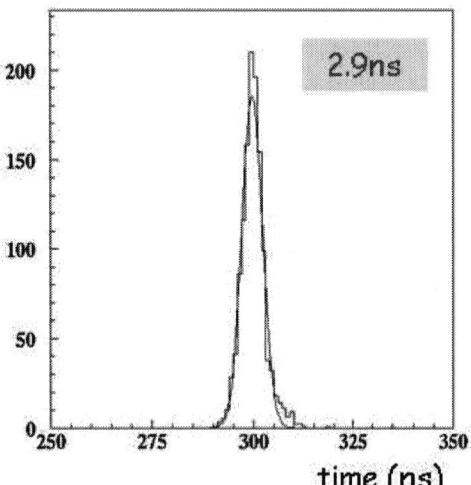

FIGURE 8. The best time distribution of the station at V_{gem}=1370 Volts.

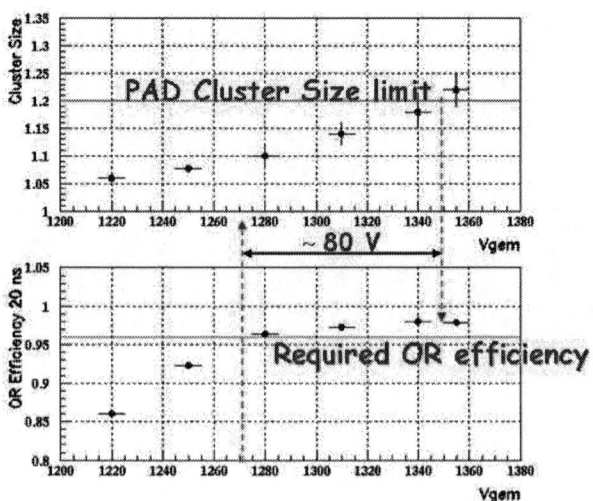

FIGURE 9. Results obtained at the PS beam facility at CERN: top) cluster size as function of the sum of the voltages applied to the three GEM foils (V_{gem}); bottom) efficiency in 20 ns time window as function of V_{gem}.

CONCLUSIONS

The triple-GEM detector operated with the fast Ar/CO_2/CF_4 (45/15/40) gas mixtures are suitable to operate in the harsh environment around the beam pipe and largely fulfills the requirements of the LHCb experiment. The construction chamber procedures is narrowed and doesn't show critical points in terms of robustness, assembly steps and gas tightness; the tools and quality checks performed are well defined; the results obtained at the PS beam facility at CERN with the final station module, built in agreement with the assembly procedure and check requirements, confirm those obtained with the prototypes.

resolution of the single chamber is better than 5 ns, that must be compared with the standard Ar/CO_2 (70/30) gas mixture, commonly used by other authors in GEM detectors, which exhibits an r.m.s. of ∼10 ns [4].

Fig. 9 shows the cluster size and the station efficiency in 20 ns time window as function of the sum of the voltages applied to the three GEM foils. These results are in good agreement with those previously obtained with the prototypes [4]. A large and safe working region (defined as the HV range between the onset of the efficiency plateau and the voltage at which the pad cluster size is under 1.2) of about 80 Volts, corrisponding to a factor 3 in gain, is found at the electronic threshold corrisponding to 2-3 fC.

REFERENCES

1. F. Sauli, Nucl. Instrum. Meth. **A386** (1997) 531.
2. LHCb Collaboration, "LHCb: Addendum to muon system technical design report", CERN LHCC MUON 2003-002, LHCb TDR 4 Addendum 1, 15 January 2003.
3. LHCb Muon System Technical Design Report, CERN LHCC 2001-010, LHCb TDR 4, (2001).
4. M. Alfonsi et al., Nucl. Instrum. Meth. **A518** (2004) 106.
5. G. Bencivenni et al., Nucl. Instrum. Meth. **A488** (2002) 493.
6. M. Alfonsi et al., Studies of etching effects on triple-GEM detectors operated with CF_4-based gas mixtures, presented at Rome 2004 IEEE conference (Oct 16-22) and submitted to Transaction on Nuclear Science.
7. W. Bonivento et al., Nucl. Instrum. Meth. **A491** (2002) 233.

Liquid Xenon Calorimetry: Status and Perspectives

Donato NICOLÒ

Dipartimento di Fisica, Università di Pisa & I.N.F.N., sez. di Pisa
Largo B.Pontecorvo,3 56127 PISA

Abstract. Liquid Xenon calorimetry is considered as one of the most promising and versatile techniques in High Energy Physics, to be used either in accelerator experiments or in searches for Physics beyond the Standard Model ($\beta\beta_{0\nu}$, Dark Matter). The main features are reviewed in relation to the different Physics items involved, with particular emphasis on Lepton Flavour Violation search.

PACS: 13.35.Bv; 23.40.-s; 29.40.Mc; 29.40.Vj

SCINTILLATION IN LIQUID XENON

Liquid Xenon (LXe from now on), as well as other liquid rare gases, is considered as an excellent medium for radiation detectors because of its scintillating properties. Luminescence emission by rare gases has been extensively studied for gas, liquid and solid phases for many years[1]. The mechanism involves the production of the excimer state Xe_2^* induced by ionizing radiation through the two following schemes:

$$Xe^* + Xe \rightarrow Xe_2^* \rightarrow 2Xe + h\nu \quad (1)$$

or

$$\begin{aligned}
Xe^+ + Xe &\rightarrow Xe_2^+, \\
Xe_2^+ + e^- &\rightarrow Xe + Xe^{**}, \\
Xe^{**} &\rightarrow Xe^* + \text{heat} \\
Xe^* + Xe &\rightarrow Xe_2^* \rightarrow 2Xe + h\nu
\end{aligned} \quad (2)$$

where $h\nu$ represent an UV photon.

A high light yield (comparable to that of NaI(Tl)), combined with a fast decay component (about 10 times as fast as NaI) is peculiar of this process. Moreover, the excimeric emission makes noble liquids transparent to their own scintillation light, as the bound Xe^2 molecule is absent in the ground state. We will see below that the energy resolution of LXe detectors is strongly related to the transparency of the scintillatiin medium.

With respect to Ar and Kr, Xe is the optimal choice for calorimetry because of its high atomic number and density (which implies a shorter radiation length). Moreover, it has a higher boiling temperature, which requires a lower cryogenic power. The main features of LXe are listed in Tab.(1).

TABLE 1. Main properties of LXe.

Density	2.95 g/cm^3
Boiling and melting points	165 K, 161 K
Light yield	42000 photons/MeV [2]
Radiation length	2.77 cm
Decay-time	4.2 ns,* 22 ns, 45 ns [3]
Peak emission wavelength	175 nm
Scintillation absorption length	> 100 cm
Scattering length (Rayleigh)	~ 40 cm
Refractive index	1.56 [4]

* Due to the different time behavior of excitation and recombination.

Optical properties

While self-absorption should be negligible, light attenuation may occur as a result of diffusion (Rayleigh scattering) and absorption by impurities. The latter is a critical issue in the case of large detectors, as it heavily affects both photo-statistics and uniformity of the energy response function. The dominant absorber in the range of interest is water vapor, whose absorption spectrum is shown in Fig.(1) in comparison to the emission spectrum of LXe. It follows that even a 1 ppm concentration of water in LXe is sufficient to shorten the absorption length down to a few cm.

This was recently pointed out by people working on MEG [5] (see next section for details), which measured the amount of scintillation light collected by an array of photomultipliers located at different distances from a set of α-sources. The transparency (see Fig.(2)) was shown to improve by removing the residual water by means of a purification system. An exponential fit to the final distribution provides a lower limit to the absorption length $\lambda > 95$ cm at 90% C.L., with a sensitivity limited by the

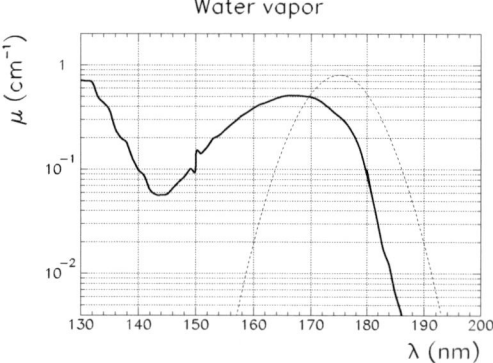

FIGURE 1. Absorption coefficient of 1 ppm water vapor dissolved in LXe as a function of wavelength, superimposed to the LXe emission spectrum.

dimensions (50 cm) of the sample[1].

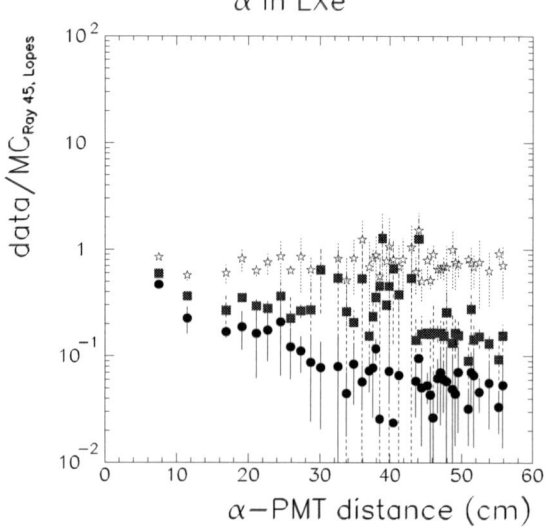

FIGURE 2. Absorption profile of LXe scintillation light in the Large Prototype of the LXe calorimeter of the MEG experiment at three different purification stages: at the beginning (circle), after two weeks (squares) and after one month (stars).

THE PRESENT: THE MEG EXPERIMENT

The MEG experiment will start operating next year at PSI, where the most intense DC muon beam in the world is currently available, to search for Lepton Flavor Viola-

tion through the decay $\mu^+ \to e^+ \gamma$. This search is considered as one of the most sensitive probe of Supersymmetry, with a predicted sensitivity (10^{-13}) to the branching ratio improved by two orders of magnitude about with respect to the current limit[6]. For details on the detector see [7].

This sensitivity can be achieved thanks to unprecedented detector performances at these energies (e^+ and γ move back-to-back and both with energy equal to 52.8 MeV, *i.e.* half the mass of muon). In particular, the resolution on photon energy and direction plays a key role in background suppression and required a lot of effort in the last years on research and development on LXe calorimetry, which was never carried out with such a large volume ($V = 800$ l) before.

R&D work on the photon detector has been accomplished by using a 100 l prototype, deep enough ($\sim 18X_0$) to fully contain the γ e.m. shower. It was first used for PMT calibration and to study the main optical properties of LXe. Both light and α-sources were used to this purpose. In addition, events associated with environmental radioactivity were collected to provide a calibration of the energy scale in the MeV region. Fig.3 shows the measured spectrum, where γ-lines from ^{40}K and ^{208}Tl emerge on an exponentially decreasing background.

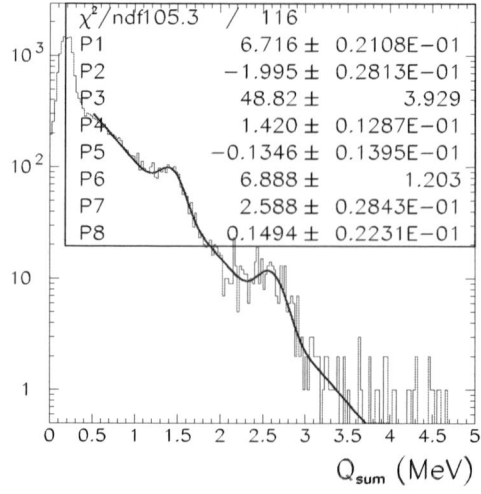

FIGURE 3. Energy spectrum of radioactive background events; γ-lines from ^{40}K and ^{208}Tl are identified.

Energy and timing resolution

More recently (fall 2003 and 2004) this "Large Proptotype" was exposed to 55 MeV γs from decays of π^0 from charge exchange reaction ($\pi^- p \to \pi^0 n$) on a liquid Hydrogen target, to test the detector behaviour under

---

[1] The purity of different samples used in the past might explain the contradictory results in the literature concerning LXe emission yield.

conditions similar to $\mu \to e\gamma$ decay. The photon tagging was performed by using a NaI detector on the opposite side with respect to the π^- target.

The results obtained for energy and timing resolution are shown in Figg.4 and 5. The energy distribution was obtained by applying simple topological cuts (distance from the photon spot centre < 1.5 cm and depth > 3 cm) to exclude photons interacting respectively with the sidewalls of a Lead collimator and with the front wall of the LXe prototype. The resolution turned out to 4.8% FWHM, dominated by escape effects on the low-energy tail (the right edge, which does not depend on these effects, is 1.2% wide), which is close to the experimental goal. The timing resolution was studied as a function of the energy deposit in the calorimeter and found to improve with photostatistics, as expected. The FWHM obtained for 55 MeV photons is 90 ps. By combining this with the uncertainty on the γ interaction vertex inside the calorimeter, one obtains an overall resolution of 110 ps on the γ time-of-flight, in agreement with expectations.

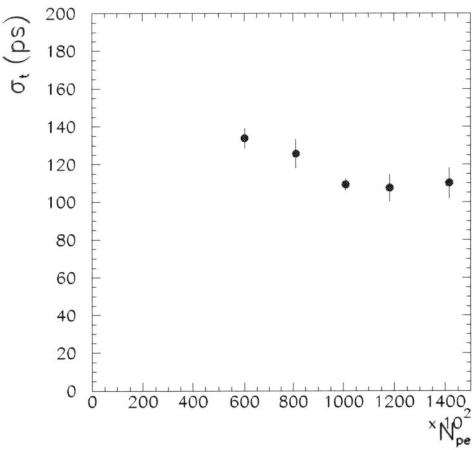

FIGURE 5. Timing resolution as a function of photoelectron statistics. A 55 MeV energy deposit corresponds to 100000 photoelectrons about.

In both cases, a simultaneous measurement of ionization and scintillation further enhances the energy resolution. Moreover, the different amplitude/time response provides a powerful tool for an efficient separation of nuclear recoils (which is associated with WIMPs or neutrons) from electron recoils (related to gammas or electrons), thus enhancing the background suppression.

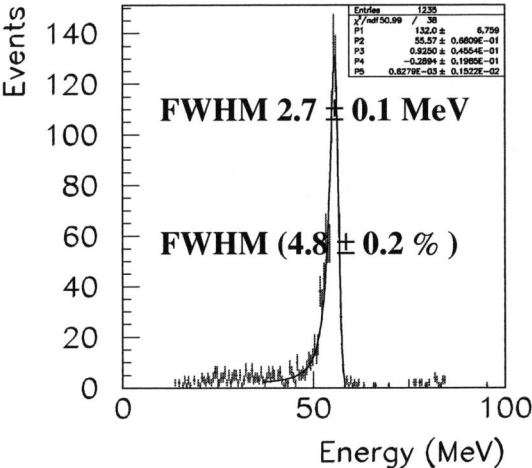

FIGURE 4. Energy spectrum for 55 MeV events induced by γs from π^0 decay.

FUTURE PROJECTS

Another advantage of using LXe instead of other liquid noble gases consists in a negligible content in long-lived radioactive isotopes (which is instead a drawback of Kr and Ar, for instance). This feature, combined with the excellent energy resolution, makes LXe suited also for low energy spectroscopy. In particular, it is worth citing two projects:

- XENON, a search for Dark Matter WIMPs[8];
- EXO, a search for neutrinoless double-beta ($\beta\beta 0\nu$) decays.

The EXO project

As an instance, let us consider the case of EXO, a project aiming to detect $\beta\beta 0\nu$ decays by using large amounts of Xe isotopically enriched in the isotope 136, in order to probe the Majorana nature of neutrinos. The goal sensitivity is a half-life $T^{1/2}_{\beta\beta 0\nu} = 8.3 \cdot 10^{26}$ yr, more than three orders of magnitude lower than the current limit. The signature for these decays relies on a combined identification of the daughter ion, $^{136}Ba^{++}$, which is accomplished by using a laser tagging technique, and the $2e^-$ final state, with the electrons sharing the total energy available $Q = 2.479$ MeV. Also in this case, rejection of the background induced by "ordinary" double beta decays ($\beta\beta 2\nu$), depends on the capability to reconstruct energies close to the end-point. The energy resolution needed for such a sensitive search is of the order of a few percents at energies around 1 MeV.

People involved in EXO recently showed that a combination of ionization and scintillation signals allows a substantial improvement of calorimetric energy resolution with respect to that obtained with ionization only[9].

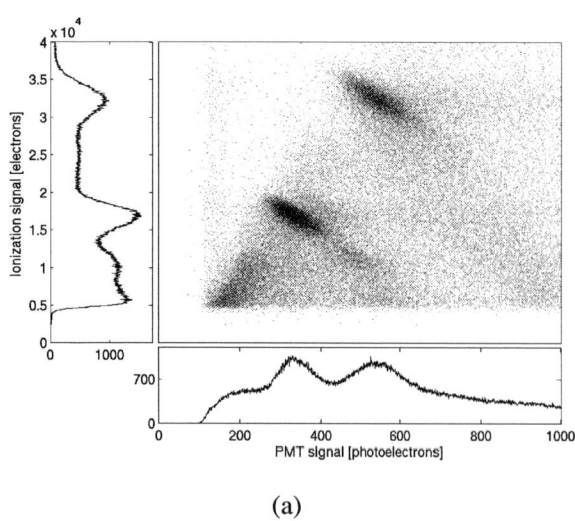 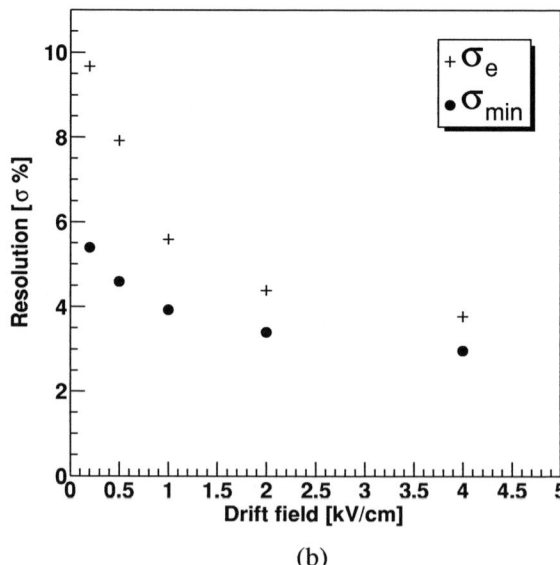

FIGURE 6. (a) Correlation between luminescence and ionization in ^{136}Xe for the two γ-lines from a ^{207}Bi source. (b) Energy resolution obtained with ionization only (σ_e) and with combined ionization+scintillation σ_{min} for 570 keV γs as a function of the drift field. The improvement looks like more significant at lower fields.

The underlying idea is that an additional contribution to ionization fluctuations (apart from Poisson statistics) comes from the electron-ion recombination mechanism (see scheme 2) responsible for luminescence in liquid rare gases. This is shown in Fig.6a, where ionization electrons are plotted vs. the number of photoelectrons for events associated with 570 and 1064 keV γ-lines from a ^{207}Bi source. The scatter-plot exhibits a strong anti-correlation between the two, with a linear coefficient $-0.8 < \rho < -0.6$ slightly dependent on the drift electrical field. So scintillation can be used to suppress this source of fluctuations; as evident in Fig.6b, the energy resolution can be improved up to 3% at 570 keV.

CONCLUSIONS

LXe calorimetry is one of the most promising technique to be used in High Energy Physics. Detection of scintillation light was shown to play a decisive role in improving both energy and timing resolution. Although scintillation mechanism has been extensively studied in liquid rare gases for many years, only recent measurements of their optical properties proved that LXe is transparent enough to be efficiently used for calorimetry also in large detectors ($V \sim 1m^3$). This opens the possibility to search for rare events predicted in Physics beyond the Standard Model (such as $\mu \to e\gamma$ decay, WIMPs or $\beta\beta 0\nu$) with sensitivity improved by about two orders of magnitudes with respect to current upper limits.

REFERENCES

1. T. Doke and K. Masuda, Nucl. Instr. and Meth. **A420** (1999) 62;
   E. Morikawa *et al.*, J. Chem. Phys. **91** (1989) 1469;
   J. Jortner *et al.*, J. Chem. Phys. **42** (1965) 4250.
2. T. Doke, Nucl. Instr. and Meth. **A327** (1993) 113.
3. A. Hitachi *et al.*, Phys. Rev. **B27** (1983) 5279.
4. L. M. Barkov *et al.*, Nucl. Instr. and Meth. **A379** (1996) 482.
5. A. Baldini *et al.*, Nucl. Instr. and Meth. **A545** (2005) 753.
6. M. L. Brooks *et al.*, Phys. Rev. Lett. **83** (1999) 1521.
7. A. Baldini *et al.*, "The MEG experiment: search for the $\mu \to e\gamma$ decay at PSI" (2002) available at http://meg.psi.ch.
8. E. Aprile *et al.*, arXiv:astro-ph/0207670 (2002).
9. EXO collaboration, available at http://www-project.slac.stanford.edu/exo/default.htm;
   E. Conti *et al.*, Phys. Rev. **B68** (2003) 054201.

IFAE 2005 – List of participants

Giovanni	Abbiendi	INFN Bologna
Oscar	Adriani	Università e INFN Firenze
Sebastiano	Albergo	Università e INFN Catania
Matteo	Alfonsi	INFN LNF
Roberto	Aloisio	INFN LNGS
Attilio	Andreazza	Università e INFN Milano
Carla	Aramo	INFN Napoli
Patrizia	Azzi	INFN Padova
Wander	Baldini	INFN Ferrara
Marcello	Barisonzi	NIKHEF
Guillaume	Batigne	INFN Torino
Matteo	Beccaria	Università e INFN Lecce
Franco	Bedeschi	INFN Pisa
Riccardo	Bellan	Università di Torino
Fabio	Bellini	Università Roma "La Sapienza" & INFN Roma
Giovanni	Bencivenni	INFN LNF
Tommaso	Boccali	SNS Pisa
Andrea	Bocci	Università di Firenze
Mirko	Boezio	INFN Trieste
Marcella	Bona	Università e INFN Torino
Laura	Borrello	Università e INFN Pisa
Graziano	Bruni	INFN Bologna
Francesca	Bucci	Università di Pisa
Michele	Caffo	INFN Bologna
Marta	Calvi	Università di Milano Bicocca e INFN
Francesca	Capozzi	Università di Milano Bicocca e INFN
Angelo	Carbone	INFN Bologna
Leonardo	Carminati	Università di Milano
Massimo	Casarsa	INFN Trieste
Gianluca	Cerminara	Università e INFN Torino
Giorgio	Chiarelli	INFN Pisa
Massimiliano	Chiorboli	Università e INFN Catania
Marco	Ciuchini	INFN Roma III
Vitaliano	Ciulli	Università di Firenze
Carlo	Civinini	INFN Firenze
Gianluca	Comune	Università di Berna
Rosa	Coniglione	INFN LNS
Gennaro	Corcella	CERN
Giorgio	Cortiana	Università di Padova

Fabio	Cossutti	INFN Trieste
Salvatore	Costa	Università e INFN Catania
Silvia	Dalla Torre	INFN Trieste
Nicola	D'Ambrosio	INFN Napoli
Sandro	De Cecco	INFN Roma
Guglielmo	De Nardo	Università di Napoli "Federico II"
Giuseppe	Della Ricca	Università e INFN Trieste
Pasquale	Di Bari	Max Planck fur Phisik, Munich
Carla	Distefano	INFN LNS
Fabrizio	Fabbri	INFN Bologna
Riccardo	Faccini	Università di Roma "La Sapienza" e INFN Roma
Rossella	Ferrandes	Università e INFN Bari
Massimo	Ferrario	INFN LNF
Fernando	Ferroni	Università di Roma "La Sapienza"
Lorenzo	Foa'	SNS e INFN Pisa
Nicolao	Fornengo	Università e INFN Torino
Mario	Galanti	Università e INFN Catania
Erika	Garutti	DESY
Chiara	Genta	Università e INFN Firenze
Alessio	Ghezzi	Università Milano Bicocca e INFN Milano
Andrea	Giammanco	SNS e INFN Pisa
Mario Paolo	Giordani	Università di Udine e INFN Trieste-Udine
Domenico	Giordano	Università e INFN Bari
Gianfrancesco	Giudice	CERN
Pietro	Govoni	INFN Milano
Alessia	Gruzza	Università di Parma
Giuseppe	Iacobucci	INFN Bologna
Aldo	Ianni	INFN LNGS
Antonio	Insolia	Università di Catania
Gianluca	Lamanna	Università e INFN Pisa
Tommaso	Lari	INFN Milano
Alfio	Lazzaro	INFN Milano
Sandra	Leone	INFN Pisa
Luigi	Li Gioi	Università di Roma "La Sapienza" e INFN Roma
Anna	Macchiolo	Università e INFN Firenze
Mario	Macrì	INFN Genova
Marcello	Maggi	INFN Bari
Nicolo'	Magini	Università di Firenze
Michelangelo	Mangano	CERN, PH-TH
Cristiano	Marchettini	INFN Firenze
Antonio	Marrone	Università e INFN Bari
Matteo	Martini	INFN LNF
Antonio	Masiero	Università di Padova
Davide	Meloni	INFN Roma 1
Federico	Mescia	Università Roma Tre e INFN LNF
Andrea	Messina	INFN Roma 1
Ernesto	Migliore	Università e INFN Torino
Emilio	Migneco	Università di Catania e INFN LNS
Stefano	Miscetti	LNF INFN
Daniele	Montanino	Università di Lecce

Stefano	Moretti	University of Southampton
Aldo	Morselli	INFN Roma 2
Donato	Nicolo'	Università e INFN Pisa
Francesco	Noto	INFN Catania
Marco	Paganoni	Università di Milano Bicocca e INFN
Armando	Palmeri	INFN Catania
Riccardo	Paramatti	CERN e INFN Roma1
Fabrizio	Parodi	Università e INFN Genova
Massimo	Passera	Università e INFN Padova
Francesca	Pastore	INFN Roma
Marisa	Pedretti	Università dell'Insubria e INFN Milano
Sergio	Petrera	Università e INFN L'Aquila
Paolo	Piattelli	INFN LNS
Davide	Piccolo	INFN Napoli
Maurizio	Pierini	LAL Orsay
Ofelia	Pisanti	Università di Napoli "Federico II"
Marco	Poli Lener	INFN LNF
Alessandro	Polini	INFN Bologna
Antonello	Polosa	Università di Bari
Alexis	Pompili	Università e INFN Bari
Renato	Potenza	Università e INFN Catania
Antonino	Pullia	Università Milano-Bicocca
Michele	Punturo	INFN Perugia
Shahram	Rahatlou	Università di Roma "La Sapienza"
Riccardo	Ranieri	Università e INFN Firenze
Marco	Rescigno	INFN Roma 1
Giulia	Ricciardi	Università degli Studi di Napoli
Giorgio	Riccobene	INFN LNS
Enrico	Robutti	INFN Genova
Andrea	Romanino	CERN
Francesco	Romano	Politecnico e INFN Bari
Alessandra	Romero	Università di Torino
Marcello	Rotondo	Università e INFN Padova
Matteo	Sani	Università e INFN Firenze
Piera	Sapienza	INFN LNS
Walter	Scandale	CERN
Marco	Serone	SISSA
Lucia	Silvestris	INFN Bari
Francesca	Soramel	Università di Udine
Antonio	Stamerra	Università di Siena e INFN Pisa
Luca	Stanco	INFN Padova
Concetta	Sutera	INFN Catania
Roberto	Tenchini	INFN Pisa
Francesco	Terranova	INFN LNF
Guido	Tonelli	Università di Pisa
Michele	Treccani	Università di Pavia
Luca	Trentadue	Università e INFN Parma
Alessia	Tricomi	Università e INFN Catania
Cristina	Tuvè	Università e INFN Catania
Vincenzo	Vagnoni	INFN Bologna

Stefano	Veneziano	INFN Roma
Silvia	Ventura	INFN LNF
Monica	Verducci	CERN
Francesco	Vissani	INFN LNGS
Pasquale	Zema	CERN e Università della Calabria

Author Index

A

Adriani, O., 263
Aloisio, R., 236
Aramo, C., 240
Azzi, P., 66

B

Baldini, W., 299
Barisonzi, M., 85
Batigne, G., 279
Beccaria, M., 127
Bedeschi, F., 24
Berezinsky, V. S., 236
Boccali, T., 271
Bucci, F., 177

C

Calvi, M., 36
Carbone, A., 189
Carloni Calame, C. M., 58
Carminati, L., 295
Casarsa, M., 201, 275
Cerminara, G., 97
Comune, G., 123
Corcella, G., 78
Cortiana, G., 111
Costantini, M. L., 219

D

D'Ambrosio, N., 267
De Cecco, S., 197
Di Bari, P., 224
Diglio, S., 93
Distefano, C., 248

F

Faccini, R., 30
Ferroni, F., 3
Fornengo, N., 107

G

Galanti, M., 135
Garutti, E., 259

Ghezzi, A., 146
Giammanco, A., 81
Giordani, M. P., 115
Giordano, D., 119
Gruzza, A., 143
Guidice, G. F., 150

H

Hsu, S.-C., 275

I

Ianni, A., 215
Illana, J. I., 252

L

Lamanna, G., 165
Lari, T., 131
Leone, S., 62
Li Gioi, L., 185
Lipeles, E., 275

M

Macchiolo, A., 302
Magini, N., 89
Mangano, M. L., 8
Marrone, A., 207
Martini, M., 161
Masip, M., 252
Meloni, D., 252
Mescia, F., 157
Messina, A., 70
Migliore, E., 283
Montagna, G., 58

N

Neubauer, M., 275
Nicolò, D., 315
Nicrosini, O., 58

P

Paramatti, R., 291
Parodi, F., 20

Passera, M., 74
Pedretti, M., 228
Petrera, S., 48
Pierini, M., 193
Pisanti, O., 232
Poli Lener, M., 311
Polini, A., 101
Punturo, M., 307

R

Ricciardi, G., 173
Robutti, E., 169
Rotondo, M., 181

S

Sarkar, S., 275
Scandale, W., 15
Serone, M., 139
Serpico, P. D., 232

Sfiligoi, I., 275
Stamerra, A., 244

T

Tenchini, R., 42
Terranova, F., 211
Tonazzo, A., 93
Treccani, M., 58
Trentadue, L., 55

V

Ventura, S., 287
Verducci, M., 93
Vissani, F., 219

W

Würthwein, F., 275